Signals and Communication Technology

More information about this series at http://www.springer.com/series/4748

Markus Rupp · Stefan Schwarz
Martin Taranetz

The Vienna LTE-Advanced Simulators

Up and Downlink, Link and System Level Simulation

Springer

Markus Rupp
Institute of Telecommunications
TU Wien
Vienna
Austria

Martin Taranetz
Institute of Telecommunications
TU Wien
Vienna
Austria

Stefan Schwarz
Institute of Telecommunications
TU Wien
Vienna
Austria

ISSN 1860-4862 ISSN 1860-4870 (electronic)
Signals and Communication Technology
ISBN 978-981-10-0616-6 ISBN 978-981-10-0617-3 (eBook)
DOI 10.1007/978-981-10-0617-3

Library of Congress Control Number: 2016931607

Printed on acid-free paper

This Springer imprint is published by SpringerNature
The registered company is Springer Science+Business Media Singapore Pte Ltd.

Preface

A simulation is something nobody believes, except the person who made it.
An experiment is something everybody believes, except the person who made it.

Freely modified after Albert Einstein

The Vienna LTE Simulators started as a project in 2009. Our first concept included two simulators, a link and a system level simulator for Downlink (DL). With one postdoc, five Ph.D., and some five Master level students we started this endeavor and soon realized that it would take longer than expected. After one year, we had basic functionality working and were continuously puzzled by inconsistent results. The more we tested, the more we found inconsistencies. Once we were satisfied with overcoming most of those hurdles, we decided to make the first release freely available on the Internet. Although we included from the very beginning an acknowledgment that the simulator is only to use for academic research, not all of the downloading parties took their promise seriously. At this point we did not care much about it but were mostly happy about the feedback we obtained. We started a web forum on which everybody could freely post troubles that were encountered and to our utmost surprise even people outside of our group showed up as experts and helped us in solving our 'mysteries'. After some two years we counted ten thousands of downloads, by far much more than serious users and it was time to become more professional about it. Both simulators had evolved into a product and after spending several hundred thousand Euros to pay all the students' salaries, we needed to see something in return. Also, the continuation of the project was in serious financial jeopardy at some time periods. Our download handling thus changed into a business part that allowed only downloads for obvious academic usage, confirmed by deans or department heads of existing universities, while all others had to pay a fee. The fees have been kept moderately low in comparison to commercial products on the market. We basically traded responsibility against price. Some companies decided not to use our 'cheap' tools as we were not willing to guarantee compensatory payment in case of faults, some companies required 24 h switching boards to have someone answering their questions; all of such desires we could not possibly satisfy. Nevertheless, word had passed around that our simulators were quite useful and that we behaved as trustworthy partners, taking requests seriously and repairing reported bugs in the next revisions. Finally, many 3rd Generation Partnership Project (3GPP) companies had decided to use the tools as platform for exchanging ideas. Our simulators serve as reference to compare against their own ideas.

While there were sporadic requests for Uplink (UL) simulators as well, for a long time we could not respond positively as we were not having sufficient personnel as well as money to pay extra students. Finally, in 2012 we decided to devote also effort into a link level UL simulator. All three simulators together now have experienced more than 50,000 downloads and have become reliable tools for companies as well as academia worldwide. As we still continue working out our own research tasks, the simulators are permanently improved and enhanced by more and more capabilities. This book is intended to provide an overview of our own research activities related to these three simulators. We are aware of hundreds of serious efforts of other research groups that are covering fields we could not and will not. Listing all of those would certainly blast this book. From the very beginning, quality issues were important for us and even if it hampered down the publication rate, we devoted ourselves to some requirements. These are as follows:

Reproducibility has become an increasingly important issue in the past years. As systems become more and more complex and thus complicated, it becomes more and more difficult to repeat results of others and even reproducing our own results is often difficult after some time has passed. To facilitate reproducibility we have therefore launched open access 'Vienna LTE Simulators' to provide a common platform for researchers and engineers. Furthermore, most of our research papers are associated with code that can be downloaded and easily repeated on other researchers' personal computers.

Our Web Forum represents a focal point for information exchange and provides us with permanent requests by third-party users. Through this measure many bugs are found and, as we take these seriously, we permanently repair them and provide new improved releases. By this means we have been tested by several hundreds of experts. Not many software providers can say that about their products.

Quality insurance is important to us. How do you know your Monte Carlo (MC) runs are sufficient? We always include confidence intervals with every averaging point to measure how reliable our simulation is. This often leads to significant speed improvements as there is no need to average more data once the confidence intervals have become very small. Standard bootstrapping techniques are being applied to compute such intervals.

Speed of simulation is a relevant factor. As multi-core machines are becoming cheaper and cheaper, it makes sense to parallelize the software and speedup simulation significantly. Some time-consuming decoders have also been ported to the graphical boards featuring hundreds of cores, which are now available in every personal computer as standard equipment; this allows to boost performance even more.

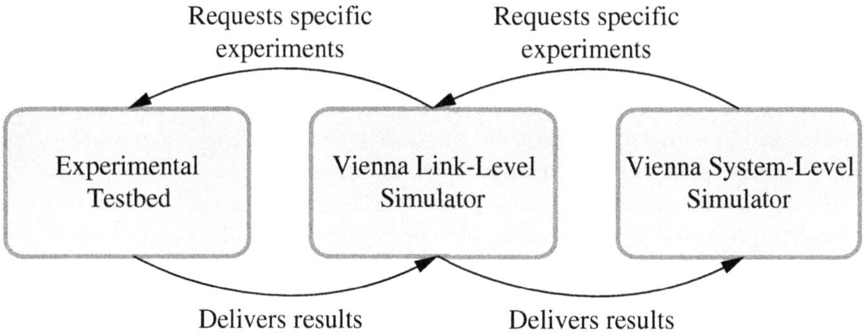

Fig. 1 Applied evaluation chain, cross-validating amongst experimental results from testbed measurements and simulation outcomes from the Vienna Link Level and System Level Simulators

Evaluation is particularly important when modeling is very abstract. We thus conduct experimental research for which the link level simulators in up- and downlink provide key functionality. In the same way the link level simulator provides the reference for the system level simulator, Fig. 1 describes the scenarios. While many more ensemble values can be generated by the link level simulator, only a few are selected to be evaluated by the testbed. Similarly, the system level simulator selects small setups with a few base stations and/or a few users that can still be computed at link level precision in reasonable time.

How to Read This Book

Part I of this book is about long-term evolution (LTE) and long-term evolution-advanced (LTEA) link level techniques. As the downlink direction has experienced high demand and interest from our partners and collaborators, a lot of research work has been conducted by the DL simulator and most chapters thus focus on the downlink. This first part of the book reports some highlights including single-user (SU) and multi-user (MU) as well as single-input single-output (SISO) and multiple-input multiple-output (MIMO) transmissions, and also reports on optimal pilot pattern for high-speed communications as well as synchronization issues. One chapter is devoted to experiments showing how the link level simulator can provide input to the measurement testbed. Also basic results on orthogonal frequency division multiplexing (OFDM) transmissions, not only valid for LTE, are reported and validated by measurements. Furthermore, one chapter deals with our newest tool, the UL link level simulator, and explains some interesting novel results obtained by it.

The second part of this book is on system level simulations. From early on system level simulations have been in high demand in industry and academia, since people wanted to find answers for situations when many base stations and hundreds of users were involved. How the mathematical abstraction of such large cellular networks can be accomplished to speedup simulations by a factor of several hundred without sacrificing precision is explained in this part of the book. We explain some new theories of how to abstract large cities equipped with small cells, that is, heterogeneous networks, and demonstrate the capabilities of our system level simulator on applications such as train and car transmissions.

Both parts are in general a rather independent read; if the reader is interested in comparing link and system level results, he is recommended to study both.

Vienna
December 2015

Markus Rupp
Stefan Schwarz
Martin Taranetz

Acknowledgments

This book is the outcome of many years of research and teaching in the field of signal processing and wireless communications.

We thank many people from A1 Telekom for their long-term support; specifically, we express our appreciation to W. Müllner, W. Wiedermann, T. Ergoth, W. Weiler, M. Kogelbauer, and T. Baumgartner.

Our friends and colleagues at Kathrein Werke KG have no small impact on our work; in particular we thank A. Kathrein, G. Schell, R. Gabriel, G. Saala, and J. Rumold for their support and input.

Recently, we started to collaborate more intensively with General Motors who introduced us to interesting communication problems in vehicular scenarios. We thank T. Philosof and K. Scheim for this.

Very special thanks go to Christoph F. Mecklenbräuker who supported many of our efforts within his Christian Doppler Laboratory "Wireless Technologies for Sustainable Mobility."

This book would not have been possible without the support of many helpful people from Springer; our thanks especially go to C. Baumann and I. Schildmann.

Contents

Part I Link Level Simulations and Experiments

Part II System Level Simulations

Authors and Contributors

About the Authors

Markus Rupp was born in 1963 in Völklingen, Germany. He received his Dipl.-Ing. degree in 1988 at the University of Saarbrücken, Germany and his Dr.-Ing. degree in 1993 at the Technische Universität Darmstadt, Germany, where he worked with Eberhardt Hänsler on designing new algorithms for acoustical and electrical echo compensation. From November 1993 until July 1995, he had a postdoctoral position at the University of Santa Barbara, California, with Sanjit Mitra where he worked with Ali H. Sayed on a robustness description of adaptive filters with impact on neural networks and active noise control. From October 1995 until August 2001 he was a member of Technical Staff in the Wireless Technology Research Department of Bell-Labs at Crawford Hill, NJ, where he worked on various topics related to adaptive equalization and rapid implementation for IS-136, 802.11 and UMTS. Since October 2001 he is full professor for Digital Signal Processing in Mobile Communications at the TU Wien where he served as Dean from 2005–2007 and is appointed from 2016–2019. He was associate editor of IEEE Transactions on Signal Processing from 2002–2005, is currently associate editor of JASP EURASIP Journal of Advances in Signal Processing, and JES EURASIP Journal on Embedded Systems. He is a Fellow of the IEEE and was elected AdCom member of EURASIP from 2004 to 2012, serving as president of EURASIP from 2009–2010. He has authored and co-authored more than 500 scientific papers including 15 patents on adaptive filtering, and wireless communications. E-mail: mrupp@nt.tuwien.ac.at

Stefan Schwarz was born in Neunkirchen, Austria, in 1984. He received his BSc in Electrical Engineering and his Dipl.-Ing. degree (MSc equivalent) in Telecommunications Engineering with highest distinctions in 2007 and 2009, respectively, both at TU Wien. He also received his Dr. techn. degree (Ph.D. equivalent) in Telecommunications Engineering with highest distinctions in 2013 at TU Wien. In 2010 he received the honorary prize of the Austrian Minister of Science and Research, for excellent graduate of scientific and artistic universities and in 2014 he received the INiTS award in the category Information and Communication Technologies for innovative scientific works with prospect for economic applicability. From 2008 to 2014 he was working as a project assistant at the Mobile Communications group of Prof. Markus Rupp at the Institute of Telecommunications of TU Wien, focusing on link and system level simulation of LTE/LTE-A networks and acting as one of the lead developers of the Vienna LTE Simulators. In 2012 he visited The University of Texas at Austin as research scholar, collaborating with Prof. Robert W. Heath, Jr. on limited feedback algorithms for distributed antenna systems. Since 2015 he is employed as postdoctoral researcher (University Assistant) at the Institute of Telecommunications, leading the research collaboration with General Motors. He also serves as the Head of the Christian Doppler laboratory on Dependable Wireless Connectivity for the Society in Motion. His research interests include the broad fields of wireless communications and signal processing. E-mail: sschwarz@nt.tuwien.ac.at

Martin Taranetz was born in 1986 in Amstetten, Austria. He received the B.Sc. degree in Electrical Engineering and Dipl.-Ing degree (M.Sc. equivalent) in Telecommunications with highest distinctions from the TU Wien, Vienna, Austria, in 2008 and 2011, respectively. He also received his Dr.-techn. degree (Ph.D. equivalent) in Telecommunications Engineering with highest honors from the TU Wien in 2015. In his dissertation, he focused on system level modeling and evaluation of heterogeneous cellular networks. Since 2015, he is employed as project assistant in the Mobile Communications group of the Institute of Telecommunications, TU Wien. From January 2014 to April 2014 he was a visiting researcher with the Wireless Networking and Communications Group at The University of Texas at Austin. His research interests include the broad fields of wireless communications and signal processing. He is a reviewer for IEEE Transactions on Wireless Communications and IEEE Transactions on Signal Processing. E-mail: mtaranet@nt.tuwien.ac.at

Contributors

Fjolla Ademaj was born in 1989 in Istog, Kosovo. She received the B.Sc. degree in Electrical Engineering and the M.Sc. degree in Telecommunications from Faculty of Electrical and Computer Engineering, University of Prishtina, Kosovo, in 2011 and 2014, respectively. Since 2014 she is working as project assistant at the Institute of Telecommunications, TU Wien, and is currently pursuing her Ph.D. degree in Telecommunications Engineering. Her research interests are in the field of wireless communications, in modeling of 3D wireless channels by system-level tools, full dimension MIMO, and 3D beamforming techniques. E-mail: fademaj @nt.tuwien.ac.at

Martin Lerch was born in Vienna, Austria in 1978. After attending a technical high school, he studied electrical and communication engineering at TU Wien where he received his master's degree in 2008. During this time and later, Martin developed several database, Web, and desktop applications before he returned to TU Wien in 2011 to work on the Vienna MIMO Testbed at the Institute of Telecommunications. Martin's work focuses on the development of new measurement methodologies for static and high mobility mobile communication scenarios. E-mail: mlerch@nt.tuwien.ac.at

Martin Müller was born in Fürstenfeldbruck, Germany. He received his B.Eng. degree in Electrical Engineering and Telecommunications from the DHBW Ravensburg, Germany, in 2009 and his Dipl.-Ing degree in Telecommunications with highest distinctions from the TU Wien, Vienna, Austria, in 2013. Since 2011 he is working as project assistant at the Institute of Telecommunications, TU Wien, and is currently pursuing the Ph.D. degree in Telecommunications Engineering. His research interests include supplying mobile access in train and highway environments and wireless communications in indoor scenarios. E-mail: mmueller@nt.tuwien.ac.at

Ronald Nissel was born 1985 in Vienna, Austria. He finished his Bachelor degree in Electrical Engineering in 2010 and his Dipl.-Ing. degree (MSc equivalent) in Telecommunications in 2013, both with highest distinctions at the TU Wien. Additionally, he received a Bachelor degree in Economics from the Vienna University of Economics and Business in 2010 and a Mag.rer.soc.oec. degree (MSc equivalent) in Economics from the University of Vienna in 2015 (with highest distinction). Currently he is working towards his PhD degree at the TU Wien. From 2011 to 2013 he was working as a project member at the Mobile Communications group of Prof. Markus Rupp at the Institute of Telecommunications of TU Wien, focusing on testbed measurements. Since 2013 he is employed as pre-Doc researcher (Project Assistant) at the Institute of Telecommunications. His research focuses on high-velocity transmission as well as testbed measurements and in particular the comparison of theoretical expectations with measurement results. E-mail: rnissel@nt.tuwien.ac.at

Stefan Pratschner was born in 1990, in Vienna, Austria. He started his studies at the TU Wien, Austria, in 2010, where he received his BSc in 2014. Since 2013 Stefan Pratschner is working at the Institute of Telecommunications, TU Wien, as a project assistant in the mobile communication group. His research interests include the LTE-A uplink and its simulation at link level. Currently he is working toward his master degree at the TU Wien. E-mail: spratsch@nt.tuwien.ac.at

Illia Safiulin was born in 1991 in Donetsk, Ukraine. In 2008 he started studying Applied Physics in Donetsk National University, Ukraine, where he received his BSc and master degrees with highest distinctions in 2012 and 2013, respectively. In 2013 he started his Master's degree in Telecommunications at the TU Wien. Additionally from 2014 he has been employed as project assistant at the Institute of Telecommunications in the Mobile Communications group. His research interests are vehicular communication and scheduling optimization. E-mail: illia.safiulin@nt.tuwien.ac.at

Michal Šimko was born in Bratislava, Slovakia, in October 1985. In June 2009, he received his Dipl.-Ing. degree (with highest honors) and in May 2013 his Ph.D. degree (with highest honors), both from TU Wien, Vienna, Austria. In June 2011, he received also a Bachelor degree in Industrial/Organizational and Social Psychology from the Comenius University, Bratislava, Slovakia. Since August 2009 he has been working as project assistant at the Institute of Telecommunications, TU Wien. During August and September 2010 he was visiting researcher at the Department of Electrical Engineering of Linköping University, Sweden. From January 2012 to June 2012, he was visiting researcher at the Signal Processing Laboratory of Federal University of Rio de Janeiro, Brazil. His research interests include channel estimation techniques, pilot pattern optimization, and communications under time-variant channels. He has been awarded the Best Student Paper Award by IEEE VTS Society at IEEE VTC 2011 Spring in Budapest, Hungary. E-mail: michal@simko.biz

Qi Wang was born in Beijing, China, in November 1982. In 2001, she received the Bachelor degree in Telecommunication Engineering from Beijing University of Posts and Telecommunications. From August 2005 to October 2007, she studied in Linköping University in Linköping, Sweden, where she obtained the Master's degree in Computer Science and Engineering. From November 2007 to December 2012, she worked as university assistant at the Institute of Communications and Radio-Frequency Engineering of the TU Wien. Her research interests are in the field of wireless communications, in particular, synchronization aspects of OFDM systems. E-mail: wangqiasia@gmail.com

Erich Zöchmann was born in 1989 in Horn, Austria. In 2015, he finished his master's degree in Electrical Engineering (Telecommunications) from TU Wien. His master's thesis is entitled "Generalized LASSO in Array Signal Processing." He is currently employed as project assistant at the Institute of Telecommunications in the Mobile Communications group, headed by Markus Rupp. His research interests include sparse modeling, optimization, array processing, and wireless communications for 4G and beyond. E-mail: erich.zoechmann@nt.tuwien.ac.at

Notation

Table 1 Mathematical notation

Symbol	Annotation
$f_X(\cdot)$	Probability density function of X
$F_X(\cdot)$	Cumulative distribution function of X
$\mathbb{E}[\mathbb{X}]$	Expected value of X
Var[X]	Variance of X
$\mathcal{N}_{\mathbb{C}}(\mathbf{a}, \sigma^2\mathbf{I})$	Complex-valued Gaussian distribution with mean **a** and variance σ^2
$\Gamma[k, \theta]$	Gamma distribution with shape k and scale θ
$\mathcal{W}_N^{\mathbb{C}}(M, \Gamma)$	Central complex-valued Wishart distribution of dimension N,having M degrees of freedom and scale matrix Γ
$\mathcal{W}_N^{-\mathbb{C}}(M, \Gamma)$	Inverse complex-valued Wishart distribution of dimension N, having M degrees of freedom and scale matrix Γ.
$\mathbf{v} \in \mathbb{R}^{d \times 1}$	Real-valued column vector of length d
$[\mathbf{v}]_l$	l-th element of vector **v**
$[\mathbf{V}]_{l,k}$	l, k-th element of matrix **V**
$1_{a \times b}$	All one matrix of dimension $a \times b$
$\mathbf{e}_k$	Canonical base vector with a single one entry at position k
Φ	Point process on $\mathbb{R}^d$
$\mathcal{B}(x, r)$	Ball with center $x \in \mathbb{R}^2$ and radius $r > 0$
diag (**V**)	Vector **v** formed out of diagonal elements of matrix **V**
Diag (**v**)	Diagonal matrix **V** formed out of elements of vector **v**
$tr\,(\mathbf{V})$	Scalar representing the sum of diagonal elements of matrix **V**
$\otimes$	Kronecker product
$\circ$	Hadamard point-wise product

Table 2 Important parameters and variables

Variable	Dimension	Description
$\mathbf{A}_{\mathrm{LS}}$	$\mathbb{C}^{N_{\mathrm{D}}\times N_{\mathrm{P}}}$	LS estimator matrix
$\mathbf{A}_{\mathrm{LMMSE}}$	$\mathbb{C}^{N_{\mathrm{D}}\times N_{\mathrm{P}}}$	LMMSE estimator matrix
$c(t,\tau)$	$\mathbb{C}$	Time-variant channel impulse response
$C(t,f)$	$\mathbb{C}$	Time-variant transfer function
$C^{(S)}(\nu,\tau)$	$\mathbb{C}$	Spreading function
$C^{(D)}(\nu,f)$	$\mathbb{C}$	Doppler variant transfer function
D_{t}	$\mathbb{N}$	Pilot spacing in time
D_{f}	$\mathbb{N}$	Pilot spacing in frequency
$\mathbf{D}$	$\mathbb{C}^{N_{\mathrm{c}}\times N_{\mathrm{c}}}$	Equivalent diagonal channel matrix for time-invariant channel
$\mathbf{D}_n$	$\mathbb{C}^{N_{\mathrm{c}}\times N_{\mathrm{c}}}$	Equivalent channel matrix for time-variant channel
Δt	$\mathbb{R}_+$	Sampling interval
Δf	$\mathbb{R}_+$	Subcarrier spacing
f	$\mathbb{R}$	Continuous frequency
$\mathbf{F}_k$	$\mathbb{C}^{N_{\mathrm{T}}\times N_{\mathrm{L}}}$	Precoding matrix at subcarrier k
$\mathbf{F}_{n,k}$	$\mathbb{C}^{N_{\mathrm{T}}\times N_{\mathrm{L}}}$	Precoding matrix at time instant n and subcarrier k
$\mathbf{F}_{n,k}[u,i]$	$\mathbb{C}^{N_{\mathrm{T}}\times N_{\mathrm{L}}}$	Precoding matrix at time instant n and subcarrier k of user u and cell i
$\mathbf{G}_k$	$\mathbb{C}^{N_{\mathrm{L}}\times N_{\mathrm{R}}}$	Channel equalizer at subcarrier k
$\mathbf{G}_{n,k}$	$\mathbb{C}^{N_{\mathrm{L}}\times N_{\mathrm{R}}}$	Channel equalizer at time instant n and subcarrier k
$\mathbf{G}_{n,k}[u,i]$	$\mathbb{C}^{N_{\mathrm{L}}\times N_{\mathrm{R}}}$	Channel equalizer at time instant n and subcarrier k of user u and cell i
$\mathbf{G}_{n,k}^{(\mathrm{ct})}[u,i]$	$\mathbb{C}^{N_{\mathrm{R}}\times N_{\mathrm{L}}}$	Conjugate transpose of receive filter $\mathbf{G}_{n,k}[u,i]$
$\mathbf{H}_k$	$\mathbb{C}^{N_{\mathrm{R}}\times N_{\mathrm{T}}}$	Channel matrix at subcarrier k
$\mathbf{H}^{(\mathrm{eff})}$	$\mathbb{C}^{N_{\mathrm{R}}N_{\mathrm{c}}\times N_{\mathrm{L}}N_{\mathrm{c}}}$	Channel matrix (precoding included)
$\mathbf{H}_k^{(\mathrm{eff})}$	$\mathbb{C}^{N_{\mathrm{R}}\times N_{\mathrm{L}}}$	Channel matrix (precoding included) at subcarrier k
$\mathbf{H}^{(\mathrm{diag})}$	$\mathbb{C}^{N_{\mathrm{R}}N_{\mathrm{c}}\times N_{\mathrm{T}}N_{\mathrm{c}}}$	Diagonalized channel matrix
$\mathbf{H}_{n,k}$	$\mathbb{C}^{N_{\mathrm{R}}\times N_{\mathrm{T}}}$	Channel matrix at time instant n and subcarrier k
$\mathbf{H}_{n,k}^{(\mathrm{eff})}$	$\mathbb{C}^{N_{\mathrm{R}}\times N_{\mathrm{L}}}$	Channel matrix (precoding included) at time instant n and subcarrier k
$\mathbf{H}_{n,k}[u,i]$	$\mathbb{C}^{N_{\mathrm{R}}\times N_{\mathrm{T}}}$	Channel matrix at time instant n and subcarrier k of user u and cell i
$\mathbf{H}_{n,k}^{(\mathrm{ct})}[u,i]$	$\mathbb{C}^{N_{\mathrm{T}}\times N_{\mathrm{R}}}$	Conjugate transpose of channel matrix $\mathbf{H}_{n,k}[u,i]$
$\mathbf{H}_{n,k}^{(\mathrm{eff})}[u,i]$	$\mathbb{C}^{N_{\mathrm{L}}\times N_{\mathrm{R}}}$	Channel matrix (receive filter included) at time instant n and subcarrier k of user u and cell i
$\mathbf{H}_{n,k}^{(\mathrm{ect})}[u,i]$	$\mathbb{C}^{N_{\mathrm{R}}\times N_{\mathrm{L}}}$	Conjugate transpose of effective channel matrix $\mathbf{H}_{n,k}^{(\mathrm{eff})}[u,i]$
$\hat{\mathbf{h}}_{\mathscr{P}}^{\mathrm{LS}}$	$\mathbb{C}^{N_{\mathrm{P}}\times 1}$	LS estimate of the channel at pilot positions
$\hat{\mathbf{H}}_{\mathscr{P}}^{\mathrm{LS}}$	$\mathbb{C}^{N_{\mathrm{P}}\times N_{\mathrm{P}}}$	LS estimate of the channel matrix at pilot positions
$\hat{\mathbf{h}}_{\mathscr{P}}^{\mathrm{LMMSE}}$	$\mathbb{C}^{N_{\mathrm{P}}\times 1}$	LMMSE estimate of the channel at pilot positions
$\tilde{h}_i$	$\mathbb{C}^{N_{\mathrm{P}}\times 1}$	Estimation error of the channel at position i
$\mathbf{h}_{\mathscr{D}}$	$\mathbb{C}^{N_{\mathrm{D}}\times 1}$	Channel at data positions

(continued)

Table 2 (continued)

Variable	Dimension	Description
$\hat{\mathbf{h}}_{\mathscr{D}}$	$\mathbb{C}^{N_D \times 1}$	Channel estimate at data positions
$\hat{\mathbf{h}}_{\mathscr{D}}^{LS}$	$\mathbb{C}^{N_D \times 1}$	LS estimate of the channel at data positions
$\hat{\mathbf{h}}_{\mathscr{D}}^{LMMSE}$	$\mathbb{C}^{N_D \times 1}$	LMMSE estimate of the channel at data positions
$\mathbf{h}_{\mathscr{P}}$	$\mathbb{C}^{N_P \times 1}$	Channel at pilot positions
$\hat{\mathbf{h}}_{\mathscr{P}}$	$\mathbb{C}^{N_P \times 1}$	Channel estimate at pilot positions
k	$\mathbb{N}$	Subcarrier index
l	$\mathbb{N}$	Transmission layer index
ℓ_k	$\mathbb{N}$	Number of spatial streams over time
$\ell_k[u, i]$	$\mathbb{N}$	Number of spatial streams at subcarrier k for user u in cell i
$\ell_{n,k}[u, i]$	$\mathbb{N}$	Number of spatial streams at time n and subcarrier k for user u in cell i
n	$\mathbb{Z}$	OFDM symbol index
N_c	$\mathbb{N}$	Number of active subcarriers
N_f	$\mathbb{N}$	Number OFDM symbols/subframe
N_{FFT}	$\mathbb{N}$	Max. number of subcarriers = FFT size
N_{cp}	$\mathbb{N}$	Number of samples per CP
N_O	$\mathbb{N}$	Oversampling factor
N_h	$\mathbb{N}$	Number of channel taps
N_s	$\mathbb{N}$	Number of samples per OFDM symbols
N_{sl}	$\mathbb{N}$	Number of OFDM symbols/slot
N_D	$\mathbb{N}$	No. of OFDM data symbols/slot
N_P	$\mathbb{N}$	Number of OFDM pilot symbols/slot
N_L	$\mathbb{N}$	No. of transmission layers
N_T	$\mathbb{N}$	No. of transmit antennas
N_R	$\mathbb{N}$	No. of receive antennas
N_{TTI}	$\mathbb{N}$	No. of transmission time intervals
P_D	$\mathbb{R}_+$	Data symbol power
$P_{n,k}^D$	$\mathbb{R}_+$	Data symbol power at subcarrier k and OFDM symbol n
P_{ICI}	$\mathbb{R}_+$	ICI power
$P_{n,k}^{ICI}$	$\mathbb{R}_+$	ICI power at subcarrier k and OFDM symbol n
$P_{n,k}[i]$	$\mathbb{R}_+$	Transmit power at time instant n and subcarrier k of cell i
P_P	$\mathbb{R}_+$	Pilot power
P_S	$\mathbb{R}_+$	Transmit power
P_V	$\mathbb{R}_+$	Noise power
$r(t)$	$\mathbb{C}$	Continuous received signal
$r_{n,k}$	$\mathbb{C}$	Received data symbol at subcarrier k and OFDM symbol n
$\mathbf{r}_{n,k}$	$\mathbb{C}^{N_R \times 1}$	Received data symbol vector at subcarrier k and OFDM symbol n
$\mathbf{r}_{n,k}[u, i]$	$\mathbb{C}^{N_R \times 1}$	Received data symbol vector at subcarrier k and OFDM symbol n for user u and cell i
$\mathbf{r}_n$	$\mathbb{C}^{N_c \times 1}$	Received data symbol vector at OFDM symbol n

(continued)

Table 2 (continued)

Variable	Dimension	Description
$\mathbf{r}_{\mathscr{P}}$	$\mathbb{C}^{N_P \times 1}$	Received symbol vector at pilot positions
$\mathbf{r}_{\mathscr{P},0}$	$\mathbb{C}^{N_P \times 1}$	Received symbol vector at pilot positions in slot 0
$\mathbf{r}_{\mathscr{P},1}$	$\mathbb{C}^{N_P \times 1}$	Received symbol vector at pilot positions in slot 1
$\mathbf{R}_{\mathbf{h}_{\mathscr{D}},\mathbf{h}_{\mathscr{P}}}$	$\mathbb{C}^{N_D \times N_P}$	CCF matrix of data and pilot channel
$\mathbf{R}_{\mathbf{h}_{\mathscr{P}},\mathbf{h}_{\mathscr{D}}}$	$\mathbb{C}^{N_P \times N_D}$	CCF matrix of pilot and data channel
$\mathbf{R}_{\mathbf{h}_{\mathscr{P}},\mathbf{h}_{\mathscr{P}}}$	$\mathbb{C}^{N_P \times N_P}$	ACF matrix of pilot channel
$\mathbf{R}_{\mathbf{h}_{\mathscr{D}},\mathbf{h}_{\mathscr{D}}}$	$\mathbb{C}^{N_D \times N_D}$	ACF matrix of data channel
t	$\mathbb{R}$	Continuous time
T	$\mathbb{R}_+$	OFDM symbol duration without cyclic prefix
T_{cp}	$\mathbb{R}_+$	Cyclic prefix duration
T_s	$\mathbb{R}_+$	OFDM symbol duration in the time domain
$\mathbf{W}_N$	$\mathbb{C}^{N \times N}$	DFT matrix of dimension N
$x(t)$	$\mathbb{C}$	Continuous transmit signal
$x_{n,k}$	$\mathbb{C}$	Transmitted data symbol at subcarrier k and OFDM symbol n
$\mathbf{x}_{n,k}$	$\mathbb{C}^{N_T \times 1}$	Transmitted data symbol vector at subcarrier k and OFDM symbol n
$\mathbf{x}_{n,k}[u, i]$	$\mathbb{C}^{N_T \times 1}$	Transmitted data symbol vector at time instant n and subcarrier k of user u and cell i
$\mathbf{x}_n$	$\mathbb{C}^{N_c \times 1}$	Transmitted data symbol vector at subcarrier k and OFDM symbol n
$\mathbf{x}_{\mathscr{D}}$	$\mathbb{C}^{N_D \times 1}$	Vector with stacked data symbols
$\mathbf{x}_{\mathscr{P}}$	$\mathbb{C}^{N_P \times 1}$	Vector with stacked pilot symbols
$\mathbf{x}_{\mathscr{P}}^{(l)}$	$\mathbb{C}^{N_c \times 1}$	Vector with stacked pilot symbols in uplink layer l
$\mathbf{X}_{\mathscr{P}}^{(l)}$	$\mathbb{C}^{N_c \times N_c}$	Matrix obtained from Diag $(\mathbf{x}_{\mathscr{P}}^{(l)})$
$\mathbf{x}_{\mathscr{L}}$	$\mathbb{C}^{N_c \times 1}$	Vector with Zadoff–Chu uplink pilot sequence
$\mathbf{x}_{\mathscr{P},0}$	$\mathbb{C}^{N_P \times 1}$	Vector with pilot symbols in slot 0
$\mathbf{x}_{\mathscr{P},1}$	$\mathbb{C}^{N_P \times 1}$	Vector with pilot symbols in slot 1
$y_{n,k}$	$\mathbb{C}$	Received symbol after equalizer at subcarrier k and OFDM symbol n
$\mathbf{y}_{n,k}$	$\mathbb{C}^{N_R \times 1}$	Received symbol vector after equalizer at subcarrier k and OFDM symbol n
$\mathbf{y}_{n,k}[u, i]$	$\mathbb{C}^{N_R \times 1}$	Received symbol vector after equalizer at time instant n and subcarrier k of user u and cell i
$\mathbf{y}_{n,k}^{ICI}$	$\mathbb{C}$	ICI vector at subcarrier k and OFDM symbol n
$v_{n,k}$	$\mathbb{C}$	Noise symbol at subcarrier k and OFDM symbol n
v_k	$\mathbb{C}$	Noise symbol at subcarrier k
$\mathbf{v}_{n,k}$	$\mathbb{C}^{N_R \times 1}$	Noise vector at subcarrier k and OFDM symbol n
$\mathbf{v}_n$	$\mathbb{C}^{N_c \times 1}$	Noise vector at OFDM symbol n

Acronyms

2D	Two-dimensional
3D	Three-dimensional
3GPP	3rd Generation Partnership Project
ACK	ACKnowledged
ALMMSE	Approximate Linear Minimum Mean Square Error
AMC	Adaptive Modulation and Coding
AP	Access Point
ARQ	Automatic Repeat reQuest
ASE	Area Spectral Efficiency
AWGN	Additive White Gaussian Noise
BCa	Bias-Corrected and accelerated
BCCH	Broadcast Control Channel
BCH	Broadcast Channel
BD	Block Diagonalization
BEP	Bit Error Probability
BER	Bit Error Ratio
BF	Beam Forming
BICM	Bit Interleaved Coded Modulation
BICMLR	Bit Interleaved Coded Modulation with Linear Receiver
BLEP	Block Error Probability
BLER	Block Error Ratio
BM	Boolean Model
BPP	Binomial Point Process
BPSK	Binary Phase Shift Keying
BS	Base Station
CA	Content Aware
CAM	Cooperative Awareness Message
CAPEX	Capital Expenditure
CAS	Centralized Antenna System
CB	Code Block
CC	Chase Combining

CCCH	Common Control Channel
CCDF	Complementary Cumulative Distribution Function
CCI	Co-Channel Interference
CDD	Cyclic Delay Diversity
CDF	Cumulative Distribution Function
CDI	Channel Direction Indicator
CDMA	Code-Division Multiple Access
CE	Channel Estimation
CF	Characteristic Function
CFO	Carrier Frequency Offset
CIR	Carrier to Interference Ratio
CL	Closed Loop
CLMI	Closed Loop Mutual Information
CLMILR	Closed Loop Mutual Information with Linear Receiver
CLSM	Closed Loop Spatial Multiplexing
CLT	Central Limit Theorem
CM	Coded Modulation
CoMP	Coordinated Multi-Point
COST	European COoperation in the field of Scientific and Technical research
CP	Cyclic Prefix
CPE	Common Phase Error
CPICH	Common Pilot CHannel
CPP	Control Plane
CPU	Central Processing Unit
CQI	Channel Quality Indicator
CRC	Cyclic Redundancy Check
CRLB	Cramér-Rao Lower Bound
CSG	Closed Subscriber Group
CSI	Channel State Information
CSI-RS	Channel State Information Reference Signal
CSIT	Channel State Information at the Transmitter
CTC	Convolutional Turbo Code
CU	Channel Use
CVQ	Channel Vector Quantization
CW	Codeword
DAS	Distributed Antennas System
DCCH	Dedicated Control Channel
DCH	Dedicated Channel
DFT	Discrete Fourier Transform
DHCP	Dynamic Host Configuration Protocol
DiffServ	Differentiate Service
DL	Downlink
DLSCH	Downlink Shared Channel
DMRS	Demodulation Reference Signals

DoF	Degrees of Freedom
DPC	Dirty Paper Coding
DPCCH	Dedicated Physical Control CHannel
DPDCH	Dedicated Physical Data CHannel
DS	Double-Stream
DSA	Dynamic Subcarrier Allocation
DSTTDSGRC	Double Space-Time Transmit Diversity with Sub-Group Rate Control
DTCH	Dedicated Traffic Channel
DTxAA	Double Transmit Antenna Array
ECDF	Empirical Cumulative Distribution Function
ECR	Effective Code Rate
EDGE	Enhanced Data Rates for Global system for mobile communications Evolution
EESM	Exponential Effective Signal to Interference and Noise Ratio Mapping
EGC	Equal Gain Combining
eICIC	Enhanced Intercell Interference Coordination
eMBMS	Evolved Multimedia Broadcast Multicast Service
eNodeB	Evolved Base Station
EPC	Evolved Packet Core
ESM	Effective Signal to Interference and Noise Ratio Mapping
ETSI	European Telecommunications Standard Institute
EUTRA	Evolved Universal Terrestrial Radio Access
EUTRAN	Evolved Universal Terrestrial Radio Access Network
EVD	Extreme Value Distribution
EVehA	Extended Vehicular A
FAP	Femtocell Access Point
FBI	FeedBack Information
FD	Full Dimensional
FD-CDM	Frequency Domain Code Division Multiplexing
FDD	Frequency Division Duplex
FDMA	Frequency Division Multiple Access
FEC	Forward Error Correction
FFO	Fractional Frequency Offset
FFT	Fast Fourier Transform
FIFO	First In, First Out
FIR	Finite Impulse Response
FPGA	Field-Programmable Gate Array
FTP	File Transfer Protocol
GGSN	Gateway-General packet radio service Support Node
GIG	Generalized Integer Gamma
GOF	Goodness Of Fit
GOP	Group Of Pictures
GPRS	General Packet Radio Service

GPS	Global Positioning System
GSCM	Geometry-based Stochastic Channel Model
GSM	Global System for Mobile communications
GSMR	Global System for Mobile communications—Rail(way)
HARQ	Hybrid Automatic Repeat reQuest
HCN	Heterogeneous Cellular Network
HeNB	Home evolved base station
HI	High Isolation
HPP	Homogeneous Point Process
HSDPA	High-Speed Downlink Packet Access
HSDPCCH	High-Speed Dedicated Physical Control CHannel
HSDSCH	High-Speed Downlink Shared CHannel
HSPA	High-Speed Packet Access
HSPDSCH	High-Speed Physical Downlink Shared CHannel
HSS	Home Subscriber Server
HSSCCH	High-Speed Shared Control CHannel
HST	High Speed Train
HSUPA	High-Speed Uplink Packet Access
IA	Interference Alignment
IC	Interference Cancelation
ICE	Inter-City Express
ICI	Inter-Carrier Interference
ICIC	Intercell Interference Coordination
IDE	Integrated Development Environment
IDFT	Inverse Discrete Fourier Transform
IEEE	Institute of Electrical and Electronics Engineers
IFFT	Inverse Fast Fourier Transform
IFO	Integer Frequency Offset
iid	independent and identically distributed
IMS	IP Multimedia Subsystem
IMT	International Mobile Telecommunications
IMT-A	International Mobile Telecommunications-Advanced
IP	Internet Protocol
IR	Incremental Redundancy
ISI	Inter-Symbol Interference
ISM	Industrial, Scientific and Medical
ISO	International Standard Organization
ITU	International Telecommunication Union
ITU-R	International Telecommunication Union—Radiocommunication Sector
JFI	Jain's Fairness Index
KS	Kolmogorov–Smirnov
LAN	Local Area Network
LDPC	Low Density Parity Check
LEP	Link Error Prediction

LLR	Log-Likelihood Ratio
LMMSE	Linear Minimum Mean Square Error
LMMSE-MAP	Linear Minimum Mean Square Error Maximum A Posteriori
LOS	Line-Of-Sight
LS	Least Squares
LT	Laplace Transformation
LTE	Long-Term Evolution
LTEA	Long-Term Evolution-Advanced
LTER	LTE for Railway
MAC	Multiple Access Channel
MACd	Medium Access Control dedicated
MAChs	Medium Access Control for High-Speed Downlink Packet Access
MAP	Maximum A-Posteriori
maxCI	Maximum Carrier-to-Interference ratio
MB	MacroBlock
MBMS	Multimedia Broadcast Multicast Service
MBSFN	Multimedia Broadcast/Multicast Single Frequency Network
MBS	Macrocell Base Station
MC	Monte Carlo
MCCH	Multicast Control Channel
MCH	Multicast Channel
MCS	Modulation and Coding Scheme
MET	Maximum Eigenmode Transmission
MEX	Matlab EXecutable
MGF	Moment Generating Function
MI	Mutual Information
MIESM	Mutual Information Effective Signal to Interference and Noise Ratio Mapping
MIMO	Multiple-Input Multiple-Output
MISO	Multiple-Input Single-Output
ML	Maximum Likelihood
MLE	Maximum Likelihood Estimation
MME	Mobility Management Entity
MMSE	Minimum Mean Square Error
mm-wave	millimeter-wave
MoRSE	Mobile Radio Simulation Environment
MPEG	Moving Picture Expert Group
MRC	Maximum Ratio Combining
MRT	Maximum Ratio Transmission
MSE	Mean Square Error
MTCH	Multicast Control Channel
MTU	Maximum Transfer Unit
MU	Multi-User
MUMIMO	Multi-User Multiple-Input Multiple-Output

MVU	Minimum Variance Unbiased
NACK	Non-ACKnowledged
NAL	Network Abstract Layer
NAS	Non-Access Stratum
NBAP	NodeB Application Part
NDI	New Data Indicator
NI	No Isolation
NLOS	Non-Line of Sight
NodeB	Base station
NSAW	N Stop And Wait
NSN	Nokia Siemens Networks
OA	Open Acess
OFDM	Orthogonal Frequency Division Multiplexing
OFDMA	Orthogonal Frequency Division Multiple Access
OLMI	Open Loop Mutual Information
OLSM	Open Loop Spatial Multiplexing
OOP	Object-Oriented Programming
OPEX	Operational Expenditure
OSG	Open Subscriber Group
OSI	Open Systems Interconnection
O-to-I	Outdoor-to-Indoor
PAPR	Peak-to-Average Power Ratio
PARC	Per-Antenna Rate Control
PBCH	Physical Broadcast Channel
PC	Personal Computer
PCCC	Parallel Concatenated Convolutional Code
PCCH	Paging Control Channel
PCCPCH	Primary Common Control Physical Channel
PCH	Paging Channel
PCI	Precoding Control Indicator
PCIsock	Peripheral Component Interconnect
PCP	Poisson Cluster Process
PCPs	Poisson Cluster Processes
PCRF	Policy and Charging Rules Function
PDCCH	Physical Downlink Control Channel
PDCP	Packet Data Convergence Protocol
pdf	Probability density function
PDN	Packet Data Network
PDP	Power Delay Profile
PDSCH	Physical Downlink Shared Channel
PDU	Packet Data Unit
PedA	Pedestrian A
PedB	Pedestrian B
PER	Packet Error Ratio
PF	Proportional Fair

PGFL	Probability Generating Functional
PGW	Packet Data Network Gateway
PHY	Physical
PLR	Packet Loss Ratio
PMCH	Physical Multicast Channel
PMI	Precoding Matrix Indicator
PN	Pseudo Noise
PP	Point Process
PPs	Poisson Processes
PPP	Poisson Point Process
PPPs	Poisson Point Processes
PPS	Pulse Per Second
PRACH	Physical Random Access Channel
PSCH	Primary Synchronization Signal
PSD	Power Spectral Density
PSK	Phase-Shift Keying
PSNR	Peak Signal-to-Noise Ratio
PSS	Primary Synchronization Signal
PUCCH	Physical Uplink Control Channel
PUSCH	Physical Uplink Shared Channel
QAM	Quadrature Amplitude Modulation
QBC	Quantization-Based Combining
QBICM	QuantizedBit Interleaved Coded Modulation
QCIF	Quarter Common Intermediate Format
QoE	Quality of Experience
QoS	Quality of Service
QPP	Quadratic Permutation Polynomial
QPSK	Quadrature Phase Shift Keying
QS	Quadratic Smoothing
QSBICM	Quantized and Shifted Bit Interleaved Coded Modulation
RAC	Random Closed Set
RACH	Random Access Channel
RAID	Redundant Array of Independent Disks
RAN	Radio Access Network
RAS	Receive Antenna Selection
RB	Resource Block
RE	Resource Element
RF	Radio Frequency
RFO	Residual Frequency Offset
RI	Rank Indicator
RLC	Radio Link Control
RLS	Recursive Least Squares
RMS	Root Mean Square
RN	Relay Node
RNC	Radio Network Controller

ROI	Region Of Interest
ROP	Random Object Process
RR	Round Robin
RRC	Radio Resource Control
RRCfilt	Root Raised Cosine filter
RRE	Remote Radio Equipment
RRH	Remote Radio Head
RRM	Radio Resource Management
RRU	Remote Radio Unit
RS	Reference Signal
RS232	Recommended Standard 232
RS-CC	Reed–Solomon Convolutional Code
RTP	Real-time Transport Protocol
RV	Random Variable
Rvidx	Redundancy version index
RVQ	Random Vector Quantization
RX	Receiver
SAE	System Architecture Evolution
SBICM	Shifted Bit Interleaved Coded Modulation
SC-FDM	Single-Carrier Frequency Division Multiplexing
SC-FDMA	Single-Carrier Frequency Division Multiple Access
SCH	Synchronization Channel
SCM	Spatial Channel Model
SCME	Spatial Channel Model Extension
SDMA	Spatial Division Multiple Access
SDU	Service Data Unit
SER	Symbol Error Ratio
SGSN	Serving-General packet radio service Support Node
SGW	Serving Gateway
SIC	Successive Interference Cancelation
SID	Size Index Identifier
SIMO	Single-Input Multiple-Output
SINR	Signal to Interference and Noise Ratio
SIR	Signal-to-Interference Ratio
SISO	Single-Input Single-Output
SM	Spatial Multiplexing
SNR	Signal-to-Noise Ratio
SQBC	Subspace Quantization-Based Combining
SQP	Sequential Quadratic Programming
SRS	Sounding Reference Signals
SS	Single-Stream
SSCH	Secondary Synchronization Signal
SSD	Soft Sphere Decoder
SSS	Secondary Synchronization Signal
STBC	Space-Time Block Code

STMMSE	Space-Time Minimum Mean Squared Error
STTD	Space-Time Transmit Diversity
SU	Single-User
SUMIMO	Single-User Multiple-Input Multiple-Output
SUS	Semi-Orthogonal User Selection
SVD	Singular Value Decomposition
TAS	Transmit Antenna Selection
TB	Transport Block
TBS	Transport Block Size
TDD	Time Division Duplex
TFC	Transport Format Combination
TFCI	Transport Format Combination Indicator
TFT	Traffic Flow Template
TM	Transmission Mode
TOS	Type Of Service
TPC	Transmit Power-Control
TR	Transmission Rate
TrCH	Transport CHannel
TSN	Transmission Sequence Number
TTI	Transmission Time Interval
TU	Typical Urban
TX	Transmitter
TxAA	Transmit Antenna Array
TxD	Transmit Diversity
UDP	User Datagram Protocol
UE	User Equipment
UL	Uplink
ULA	Uniform Linear Array
ULSCH	Uplink Shared Channel
UMa	Urban Macro cell
UMa-H	UMa-High rise
UMi	Urban Micro cell
UMTS	Universal Mobile Telecommunications System
UP	User Plane
USB	Universal Serial Bus
UTRA	Universal mobile telecommunications system Terrestrial Radio Access
UTRAN	Universal mobile telecommunications system Terrestrial Radio Access Network
VBLAST	Vertical Bell Laboratories Layered Space-Time
VCEG	Video Coding Expert Group
VCL	Video Coding Layer
VehA	Vehicular A
VHSIC	Very High Speed Integrated Circuits
VoIP	Voice over IP

WB-CLMI	Wideband Closed Loop Mutual Information
WB-CLMI-LR	Wideband Closed Loop Mutual Information with Linear Receiver
WCDMA	Wideband Code-Division Multiple Access
WiMAX	Worldwide Inter-Operability for Microwave Access
WINNER	Wireless World Initiative New Radio
WLAN	Wireless Local Area Network
WSSUS	Wide-Sense Stationary Uncorrelated Scattering
XPR	Cross Polarization Power Ratio
ZC	Zadoff-Chu
ZF	Zero Forcing
ZFBF	Zero Forcing Beam Forming

Part I
Link Level Simulations and Experiments

Up and Downlink–From Simulations to Experiments

In this first part of the book the Vienna Link Level (LL) Simulators are described. The first chapter provides basics of LL simulations, introduces the most common variables and parameters as well as the transceiver structures that are applied in Long-Term Evolution (LTE) and Long-Term Evolution-Advanced (LTEA). We focus here mostly on the Downlink (DL) of LTE as most results reported in later chapters are related to DL transmissions.

Chapter 2 focuses on the DL synchronization, in particular how to optimally estimate frequency offsets under various channel conditions. The interested reader can find more details on synchronization issues and frequency offset impact from Qi Wang's Ph.D. thesis available at http://theses.eurasip.org/theses/470/performance-evaluation-of-practical-ofdm-systems/.

In Chap. 3 Ronald Nissel reports on his findings in the context of very high mobility. As we are able to reproducibly measure at velocities of 400km/h, we were interested in the performance of optimal decoding and Orthogonal Frequency DivisionMultiplexing (OFDM) architectures under such extreme conditions.

Chapter 4 provides insight into optimal pilot pattern design. Here a method is described that allows to match the pilot pattern optimally to a given Root MeanSquare (RMS) delay spread and Doppler spread. Also the transmit power for the pilots is selected optimally to achieve maximal mutual information. More details are available in Michal Šimko's Ph.D. thesis http://theses.eurasip.org/theses/488/pilot-pattern-optimization-for-doubly-selective/.

Chapter 5 reviews the basic principles of link adaptation and linear precoding for single-user Multiple-Input Multiple-Output (MIMO) transmission and evaluates LTE's DL Single-User (SU) MIMO transmission modes. The focus of this chapter is on designing optimal limited feedback algorithms to enable efficient operation with quantized channel state information at the base station.

Chapters 6 and 7 put the scope on multi-user MIMO transmission. In Chap. 6 channel state information quantization algorithms for block-diagonalization precoding are developed, exploiting the subspace structure of the precoder design to enable efficient quantization on the Grassmann manifold. Memoryless and predictive Grassmannian quantization algorithms are presented, with the latter utilizing temporal channel correlation to improve quantization accuracy. In Chap. 7 these quantizers are extended to OFDM transmission over frequency-selective channels. Furthermore, systems with excess antennas at the receivers are considered and receive antenna combiners that minimize the impact of channel state information quantization errors are developed. The content of Chaps. 5–7 are taken from the Ph.D. thesis of Stefan Schwarz; more detailed treatments of such techniques are available in http://theses.eurasip.org/theses/514/limited-feedback-transceiver-design-for-downlink/.

The Uplink (UL) simulator is the youngest of the Vienna LTE simulators. We were first hesitant to start this project as our main resources were bound to DL simulations, but the requests for a UL LL simulator became stronger and stronger from the community. We first tried to spin the effort off by colleagues from TU Brno [1] but they were also suffering a lack of students. In 2012 we finally decided to set up some resources and, based on a few students' work, we eventually got started.

Our initial opinion was that the UL is a straightforward copy of the DL with just some particularities such as Single-Carrier Frequency Division Multiple Access (SC-FDMA), but soon we learned that there are significant differences in both worlds and concepts that proved to work well in the DL did not work equally in the UL. In Chap. 8 Erich Zöchmann and Stefan Pratschner report on their findings for channel estimation and interpolation in UL scenarios.

Early on there has always been the demand for experimental work to validate link level simulations. In Chap. 9 Martin Lerch reports on LTE transmission scenarios and displays the truly obtained throughput.

Reference

[1] J. Blumenstein, J. Ikuno, J. Prokopec, and M. Rupp, Simulating the long term evolution uplink physical layer, in Proceedings ELMAR, pp. 141–144, Zadar, (2011)

Chapter 1
Link Level Simulation Basics

Markus Rupp, Stefan Schwarz and Martin Taranetz

This chapter serves to provide the basic background on Orthogonal Frequency Division Multiplexing (OFDM) transmission and Multiple-Input Multiple-Output (MIMO) transmissions for Multi-User (MU). We will introduce the most important parameters used later on in the following chapters. The experts in Long Term Evolution (LTE) can simply skip this chapter.

1.1 Some Basics on Wireless Channels and Notations

We model wireless channels by a two-dimensional time-variant channel impulse response $c(t, \tau)$, reflecting that for each time instant t, we obtain a different impulse response $c(\tau)$. Given such channel, we can perform Fourier transformations on both parameters t and τ either individually or jointly. Figure 1.1 displays the four different representations. The most important transform is the time-variant transfer function $C(t, f)$ obtained by Fourier transformation of the lag parameter τ. This function is very much suitable for the OFDM transmissions of LTE and will be used extensively. Applying the Fourier transform on the first parameter t, we compute the so-called spreading function $C^{(S)}(\nu, \tau)$. Finally, applying the Fourier transform on both parameters results in the so-called Doppler-variant transfer function $C^{(D)}(\nu, f)$.

As we are using sampled values in digital form, it makes sense to introduce the corresponding sampled versions. Sampling with $(\Delta t, \Delta\tau, \Delta f, \Delta\nu)$, we obtain

M. Rupp (✉) · S. Schwarz · M. Taranetz
Institute of Telecommunications, TU Wien, Vienna, Austria
e-mail: mrupp@nt.tuwien.ac.at

S. Schwarz
e-mail: sschwarz@nt.tuwien.ac.at

M. Taranetz
e-mail: martin.taranetz@tuwien.ac.at

M. Rupp et al., *The Vienna LTE-Advanced Simulators*,
Signals and Communication Technology, DOI 10.1007/978-981-10-0617-3_1

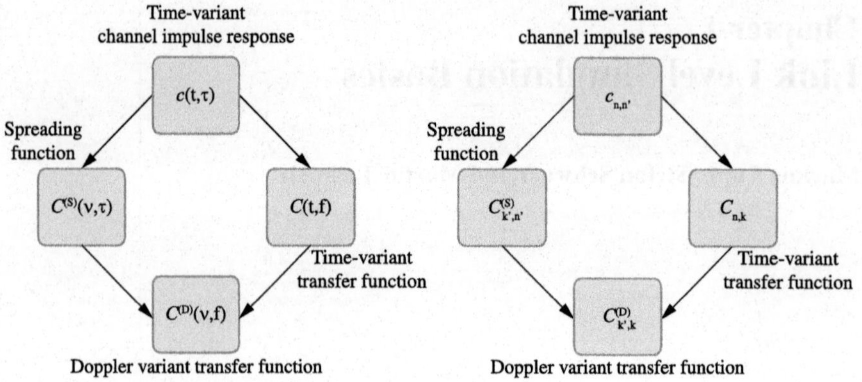

Fig. 1.1 The Fourier transformations of the time-variant channel impulse response $c(t, \tau)$. *Left* time-continuous description; *right* time-discrete version

$$c_{m,m'} = c(m\Delta t, m'\Delta\tau), \tag{1.1}$$
$$C_{m,k} = C(m\Delta t, k\Delta f), \tag{1.2}$$
$$C^{(S)}_{k',m'} = C^{(S)}(k'\Delta\nu, m'\Delta\tau), \tag{1.3}$$
$$C^{(D)}_{k',k} = C^{(D)}(k'\Delta\nu, k\Delta f), \tag{1.4}$$

respectively. In this form the four terms describe two-dimensional sequences of infinite dimensions. More insight into these four functions can be found in the textbook [1]. While the time index m typically remains unbounded, we will only employ a limited bandwidth and thus the frequency index k will be limited by the number of subcarriers that we design the OFDM transmission system for.

1.2 OFDM Basics

Consider data symbols $x_{n,k}$; $k = 0, 1, \ldots, N_c - 1$ to be transmitted over a duration T at time instant n in a Single-Input Single-Output (SISO) point-to-point transmission:

$$s(t) = \sum_{n=-\infty}^{\infty} \sum_{k=0}^{N_c-1} x_{n,k} g_{T,k}(t - nT), \tag{1.5}$$

thus being modulated with help of a pulse shaping filter

$$g_{T,k}(t) = \begin{cases} \frac{1}{\sqrt{T}} e^{j2\pi k \frac{t}{T}} & 0 \le t < T \\ 0 & \text{else.} \end{cases} \tag{1.6}$$

In LTE $T = 66.7\,\mu$s. For this specific pulse shaping filter, the modulation resembles a so-called OFDM modulation that can be implemented very efficiently by a Fast Fourier Transform (FFT) algorithm. Here, the number of subcarriers N_c defines how many individual streams of data $x_{n,k}$; $k = 0, 1, \ldots, N_c - 1$ are being transmitted. Note that $1/T = \Delta f$ in (1.6) defines the subcarrier spacing. As not all subcarriers are modulated with data symbols $x_{n,k}$, the true number is typically larger and is equivalent to the order $N_{\text{FFT}} > N_c$ of the FFT operation. The physical subcarrier spacing is simply calculated by dividing the entire transmit bandwidth B_c by the number of active subcarriers N_c, thus $\Delta f = \frac{B_c}{N_c} = \frac{B}{N_{\text{FFT}}}$. Note, however, that the bandwidth $B > B_c$ required for the transmission is larger as some subcarriers may not be modulated but are required for different reasons, e.g., to facilitate the provisioning of analogue filters that keep a security distance to carriers of other service providers. Note that the given bandwidth B is what is offered to a service provider to use but OFDM modulation theoretically requires an infinite bandwidth due to the unlimited time-frequency product of such signals. The demarcation to the next service provider's frequency range in practice is achieved by additional analogue filters. To understand why the bandwidth of OFDM signals is unlimited, simply recall that

$$g_{T,k}(t) = g_{T,k}(t) \cdot \text{rect}_T(t). \tag{1.7}$$

The symbols are thus multiplied by a rectangular pulse in time (here $\text{rect}_T(t) = 1$ for the symbol duration T and zero outside) which corresponds to the convolution with a sinc function in frequency. The spectrum is defined by the squared magnitude of such sinc function and thus the occupied bandwidth is infinite.

Once the generated signal $s(t)$ is transmitted over a time-variant channel with impulse response $c(t, \tau)$, we obtain at the receiver antenna the distorted signal

$$r(t) = \int_0^{\tau_{\max}} c(t, \tau)s(t - \tau)d\tau + v(t), \tag{1.8}$$

corrupted by the channel as well as additive noise $v(t)$. To accommodate a channel with maximal duration $\tau_{\max}$, we introduce a so-called Cyclic Prefix (CP), that is simply a fractional repetition of the transmit signal, placed in front of the signal. In principle, it could also be a cyclic postfix, repeated after the symbol but in LTE a CP was selected. In the Downlink (DL) of LTE two possible values of a CP are allowed, a short one of duration $T_{\text{cp}} = 4.7\,\mu$s and a long one of $T_{\text{cp}} = 16.7\,\mu$s. To be precise, when applying the short CP, the first symbol has a CP of 5.2 μs, the following six are 4.7 μs long. The reason for the different CP length of the first symbol is to make the overall slot length in terms of time units divisible by 15 360. Seven OFDM symbols together with their CPs make up one slot ($N_f = 7$). Then the pattern starts again. The long CP is to be employed in high Root Mean Square (RMS) delay spread scenarios. If this is used, only six symbols make up for one slot ($N_f = 6$). With a typical CP length of 4.7 μs and an FFT length of $N_{\text{FFT}} = 1024$, we find $N_{\text{cp}} = 72$. In this example the CP length has a duration of 72 samples.

The pulse shaping with CP can be best modeled by modifying the pulse shaping filter to:

$$g_{T_s,k}(t) = \begin{cases} \frac{1}{\sqrt{T}} e^{j2\pi k \frac{t}{T}} & -T_{cp} \le t < T \\ 0 & \text{else} \end{cases}. \tag{1.9}$$

With such a pulse shape, the actual signal duration becomes a bit longer: $T_s = T + T_{cp}$. Applying this pulse shape and assuming that the channel impulse response is not exceeding T_{cp}, the convolution of (1.8) now reads during the period T of the OFDM symbol at time instant n:

$$r_T(t) = \int_0^{T_{cp}} c(t,\tau) \sum_{k=0}^{N_c-1} x_{n,k} g_{T_s,k}(t-\tau-nT_s) d\tau + v(t) \tag{1.10}$$

$$= \frac{1}{\sqrt{T}} \int_0^{T_{cp}} c(t,\tau) \sum_{k=0}^{N_c-1} x_{n,k} e^{j2\pi k \frac{t-\tau-nT_s}{T}} d\tau + v(t) \tag{1.11}$$

$$= \sum_{k=0}^{N_c-1} \frac{1}{\sqrt{T}} \int_0^{T_{cp}} c(t,\tau) e^{-j2\pi k \frac{\tau}{T}} d\tau x_{n,k} e^{j2\pi k \frac{t-nT_s}{T}} + v(t) \tag{1.12}$$

$$= \frac{1}{\sqrt{T}} \sum_{k=0}^{N_c-1} C(t, k\Delta f) x_{n,k} e^{j2\pi k \frac{t-nT_s}{T}} + v(t), \tag{1.13}$$

with $C(t, f)$ denoting the time-variant transfer function from Eq. (1.2). Due to the CP extension, no Inter-Symbol Interference (ISI) occurs from the previous symbols and as a further consequence, the time-variant transfer function $C(t, k\Delta t)$ fully describes the transmission system.

At the receiver end we sample the observed signal, say with N_{FFT} samples per period T. Starting at the beginning of the symbol the sampling period per symbol runs from $t_m = (n-1)T_s + T_{cp} + m\Delta t$, $m = 0, 1, \ldots, N_{FFT} - 1$, with $\Delta t = \frac{T}{N_{FFT}} = \frac{T_s}{N_s}$. The receiver thus performs at every time instant n a Fourier transform of the sampled symbol:

$$r_{n,k} = \frac{\sqrt{T}}{N_{FFT}} \sum_{m=0}^{N_{FFT}-1} r_T(t_m = m\Delta t + (n-1)T_s + T_{cp}) e^{-j2\pi \frac{mk}{N_{FFT}}} + v_{n,k} \tag{1.14}$$

$$= \frac{1}{N} \sum_{m=0}^{N-1} \sum_{k'=0}^{N_c-1} C(m\Delta t + (n-1)T_s + T_{cp}, k'\Delta f)$$

$$\cdot x_{n,k'} e^{j2\pi k' \frac{m\Delta t + (n-1)T_s + T_{cp} - nT_s}{T}} e^{-j2\pi \frac{mk}{N_{FFT}}} + v_{n,k}, \tag{1.15}$$

$$= \frac{1}{N_{FFT}} \sum_{m=0}^{N_{FFT}-1} \sum_{k'=0}^{N_c-1} C(m\Delta t + (n-1)T_s + T_{cp}, k'\Delta f) x_{n,k'} e^{j2\pi \frac{(k'-k)m}{N_{FFT}}} + v_{n,k}. \tag{1.16}$$

We now reformulate the term during $m\Delta t + (n-1)T_s + T_{cp} = \Delta t(m - N_{FFT} + nN_s)$ which allows us to introduce a time-variant transfer function that only depends on discrete parameters: $C_{m-N_{FFT}+nN_s,k'} = C(m\Delta t + (n-1)T_s + T_{cp}, k'\Delta f)$. This allows now a much more compact formulation

$$r_{n,k} = \frac{1}{N_{FFT}} \sum_{m=0}^{N_{FFT}-1} \sum_{k'=0}^{N_c-1} C_{m-N_{FFT}+nN_s,k'} x_{n,k'} e^{j2\pi \frac{(k'-k)m}{N_{FFT}}} + v_{n,k}. \tag{1.17}$$

Figure 1.2 depicts the scenario showing a time-frequency resource grid with symbols and their CP associated with the magnitude of a potential time-variant transfer function $C_{m,k}$. The figure also contains the placement of the pilot positions as they are specified in the LTE standards for DL transmissions.

Understanding Eq. (1.17) now is crucial for the understanding of OFDM transmissions over time-variant channels. We would like to obtain $r_{n,k} = x_{n,k}$ only hampered by the additive noise term $v_{n,k}$. If we have a time-invariant transfer function $C_{k'}$, the dependence of the channel on m and n disappears. Only for $k = k'$, one element remains and we obtain $r_{n,k} = C_k x_{n,k} + v_{n,k}$, or in form of a diagonal channel matrix $\mathbf{D}$ containing all elements of C_k on its diagonal. This allows us to formulate a simple transmission model from transmit symbols $x_{n,k}$ to observed noisy symbols $r_{n,k}$ after the FFT operation at the receiver in vector notation:

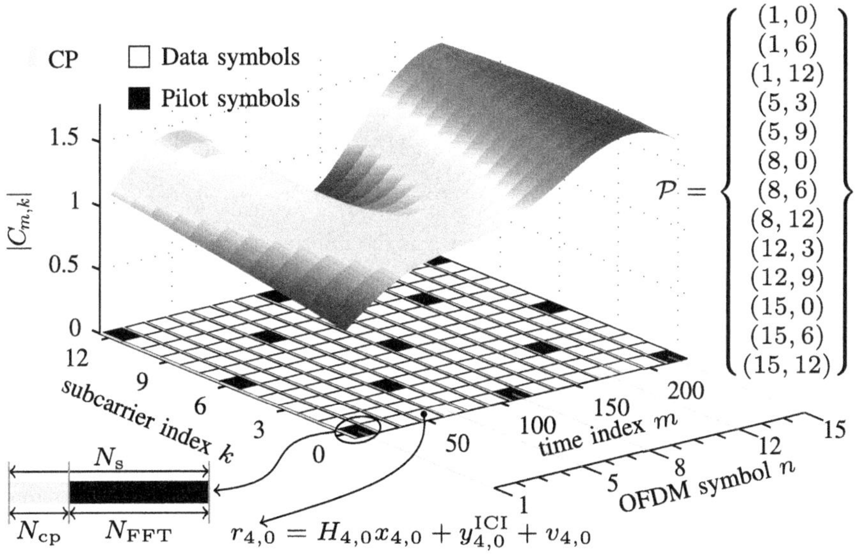

Fig. 1.2 A possible realization of the sampled time-variant transfer function $C_{m,k}$ for $N_c = 13$, $N_f = 15$, $N = 13$ and $N_{cp} = 2$. Data symbols $x_{n,k}$ are transmitted over a $N_c \times N_f$ frequency-time grid, whereas $H_{n,k}$ denotes the piecewise time-average of $C_{m,k}$. The pilot symbol positions are chosen according to the LTE standard for DL transmissions

$$\mathbf{r}_n = \mathbf{D}\mathbf{x}_n + \mathbf{v}_n, \tag{1.18}$$

where we concatenated all data symbols $x_{n,k}$ at time instant n over all subcarriers in vector $\mathbf{x}_n \in \mathbb{C}^{N_c \times 1}$ and all sampled received values $r_{n,k}$ after FFT in vector $\mathbf{r}_n \in \mathbb{C}^{N_c \times 1}$. The additive noise is considered $\mathbf{v}_n \sim \mathcal{N}_{\mathbb{C}}\left(0, P_V \mathbf{I}_{N_c}\right)$ as sampled white Gaussian noise after a uniform FFT transformation preserves its statistical properties. Note that the noise statistics are preserved by the Fourier transform and simply reuse the same notation $v_{n,k}$. Once the channel taps, that is the diagonal values of $\mathbf{D}$ are known, the corresponding receiver can then simply apply its inverse and so obtain good and low complex estimates of $x_{n,k}$:

$$\hat{\mathbf{x}}_n = \mathbf{D}^{-1}\mathbf{r}_n = \mathbf{r}_n + \mathbf{D}^{-1}\mathbf{v}_n, \tag{1.19}$$

well-known as the Zero Forcing (ZF) receiver (one-tap receiver for SISO transmissions).

As soon as the channel is time-variant, however, the situation becomes somewhat more complicated. Equation (1.17) tells us now that all data symbols $x_{n,k}$ at time instant n of all N_c subcarriers have an influence on the decoded value. Even in the absence of noise, there exists an Inter-Carrier Interference (ICI) that corrupts the data and prevents simple detection schemes. We can still describe the mapping from $\mathbf{x}_n$ to $\mathbf{r}_n$ by a matrix but such matrix is no longer diagonal and also dependent on time-instant n:

$$\mathbf{r}_n = \mathbf{D}_n\mathbf{x}_n + \mathbf{v}_n, \tag{1.20}$$

where we identify all elements of matrix $\mathbf{D}$ from (1.17)

$$[\mathbf{D}_n]_{k,k'} = \frac{1}{N_{\mathrm{FFT}}} \sum_{m=0}^{N_{\mathrm{FFT}}-1} C_{m-N_{\mathrm{FFT}}+nN_s,k'} e^{j2\pi \frac{(k'-k)m}{N_{\mathrm{FFT}}}}, \quad k, k' = 0, 1, \ldots, N_c - 1. \tag{1.21}$$

We can now interpret the elements of $\mathbf{D}_n$ as the Fourier transformed values of the time-variant transfer function. In particular the elements on the main diagonal $k = k'$, are the piecewise time-averaged values $H_{n,k}$ of the channel transfer function $C_{m,k}$ at time instant n. With $\mathbf{h}_n = \operatorname{diag}(\mathbf{D}_n)$, we can thus equivalently formulate Eq. (1.20) in a very compact form in which now $\mathbf{x}_n$, $\mathbf{r}_n$ and $\mathbf{v}_n \in \mathbb{C}^{N_c \times 1}$:

$$\mathbf{r}_n = \operatorname{Diag}(\mathbf{h}_n)\,\mathbf{x}_n + \mathbf{y}_n^{\mathrm{ICI}} + \mathbf{v}_n, \tag{1.22}$$

where we split the data symbol part in a diagonal part associated with $\mathbf{h}_n$ and a non-diagonal part $\mathbf{y}_n^{\mathrm{ICI}}$ that depends on channel and data. For time-invariant channels, we find that $\mathbf{y}_n^{\mathrm{ICI}} = \mathbf{0}$.

At this point several important comments are required:

- The presented concept here requires a perfect synchronization to start the sampling exactly at the beginning of the OFDM symbol. However, due to the CP the sampling can also start a bit earlier as the symbol is preserved as long as the sampling points are not earlier than $T_{cp} - \tau_{max}$ before the symbol starts.

- The presentation here is for a simple SISO system, the extension to MIMO transmissions with N_T transmit and N_R receive antennas is straightforward; the time-variant matrix $\mathbf{D}_n$ becomes multidimensional, its main diagonal elements $H_{n,k}$ then become $N_R \times N_T$ dimensional matrices $\mathbf{H}_{n,k}$. The m, lth entry of this matrix is referred to as $[\mathbf{H}_{n,k}]_{m,l}$; $m = 1, 2, \ldots, N_R$; $l = 1, 2, \ldots, N_T$.
- We did not mention transmit power so far. If it is included, Eq. (1.22) is modified to

$$\mathbf{r}_n = \text{Diag}(\mathbf{h}_n)\sqrt{P_S}\mathbf{x}_n + \mathbf{y}_n^{\text{ICI}} + \mathbf{v}_n, \tag{1.23}$$

which also scales the ICI term $\mathbf{y}_n^{\text{ICI}}$. In a MIMO transmission also precoding matrices $\mathbf{F}_{n,k}$ can be included for each $\mathbf{x}_{n,k}$ at every entry of the time-frequency grid (n, k) and we obtain

$$\mathbf{r}_{n,k} = \mathbf{H}_{n,k}\mathbf{F}_{n,k}\mathbf{x}_{n,k} + \mathbf{y}_{n,k}^{\text{ICI}} + \mathbf{v}_{n,k}, \tag{1.24}$$

where the precoding matrix includes the transmit power term. Once an equalizer $\mathbf{G}_{n,k}$ is applied, the corresponding signal is

$$\mathbf{y}_{n,k} = \mathbf{G}_{n,k}\left(\mathbf{H}_{n,k}\mathbf{F}_{n,k}\mathbf{x}_{n,k} + \mathbf{y}_{n,k}^{\text{ICI}} + \mathbf{v}_{n,k}\right). \tag{1.25}$$

- Similarly to the presented concept here also Single-Carrier Frequency Division Multiplexing (SC-FDM) can be described, in which an FFT operation followed by an inverse FFT causes a mixing of the symbols including pilots over the subcarriers. The FFT operation is typically employed twice after the reception to include an equalizer stage. Multiple FFT and inverse FFT schemes keep the transmission properties more or less equivalent due to the linearity of the channel. However, non-linear power amplifiers as well the different treatment of the additive noise cause some differences in the behavior when compared to OFDM transmissions. Due to smaller Peak-to-Average Power Ratio (PAPR) values, the SC-FDM transmission mode is employed in the Uplink (UL) of LTE and Long Term Evolution-Advanced (LTE-A) while OFDM is used in the DL. A more detailed description of SC-FDM can be found in Chap. 8.

At this point we like to refer the interested reader to some classical textbooks, see, e.g., [2–4], for more details on OFDM transmissions.

1.3 Advanced Features: ICI Modeling

As shown in Eq. (1.22) the ICI component is relatively complex as it contains data as well as channel components. Although often argued that due to a sum of multiple independent sources, the Central Limit Theorem would hold and the corresponding signal is Gaussian distributed, the nature of this signal is not easily described as such. It certainly can be described by a Gaussian mixture process [5]. Nevertheless, once multiple transmit antennas come into play, ICI can become more Gaussian and with additional coding the obtained ICI impacts throughput of OFDM transmission systems very much like Gaussian distortions with proper signal power. This section describes how we model ICI in system level simulations and we validate our results for high speed channels, as it may occur in high speed trains or even transmissions to airplanes.

1.3.1 Analytical Model

The impulse response of the channel is described by $c_{m,m'}$ as described in Eq. (1.1), the indexes denoting the tap gain of the m'th tap at time instant m. We can express $c_{m,m'}$ as: $c_{m,m'} = L \cdot \widetilde{c}_{m,m'}$, where L denotes macroscopic pathloss and shadow fading that remains a constant for a longer time period and $\widetilde{c}_{m,m'}$ represents microscopic fading. We consider for this a so-called Wide-Sense Stationary Uncorrelated Scattering (WSSUS) channel model [1, 6], where the channel can be characterized by its statistics, such as the Power Delay Profile (PDP) and its scattering function [1]. According to these assumptions, and further assuming a two-dimensional propagation scenario, the autocorrelation function of the channel at time indexes m_1, m_2 and channel taps m_1', m_2' is

$$\mathbb{E}\{c_{m_1,m_1'} c^*_{m_2,m_2'}\} = c|L|^2 J_0\Big(2\pi f_d \Delta t(m_1 - m_2)\Big)\delta(m_1' - m_2'), \tag{1.26}$$

where c is a normalization constant, $J_0(\cdot)$ denotes the zeroth-order Bessel function of the first kind, f_d is the Doppler frequency and Δt is the sampling time. The inverse Fourier transform of the Doppler spectrum is its autocorrelation function which is in two-dimensional propagation scenarios well described by the zeroth-order Bessel function $J_0(\cdot)$ of the first kind [7].

We can write Eq. (1.22) as

$$r_{n,k} = H_{n,k}\sqrt{P_S} x_{n,k} + y^{\mathrm{ICI}}_{n,k} + v_{n,k}, \quad k = 0, 1, \ldots, N_c. \tag{1.27}$$

As derived in [8], the power of the ICI term can be calculated as

$$\begin{aligned}\mathbb{E}\{P^{\mathrm{ICI}}_{n,k}\} = \frac{|L|^2 P_S}{N^2_{\mathrm{FFT}}} \sum_{m=0,m\neq k}^{N_c-1} \sum_{m_1=0}^{N_{\mathrm{FFT}}-1} \sum_{m_2=0}^{N_{\mathrm{FFT}}-1} J_0\Big(2\pi f_d \Delta t(m_1 - m_2)\Big) \\ \cdot \exp\left[j\frac{2\pi(m_1 - m_2)(m - k)}{N_{\mathrm{FFT}}}\right].\end{aligned} \tag{1.28}$$

We recognize that the ICI power is a function of the subcarrier index k. Further we will indicate the overall ICI power at subcarrier k as $P_k^{\mathrm{ICI}} = \mathbb{E}\{|y_{n,k}^{\mathrm{ICI}}|^2\}$. Obviously here, the averaging has been performed over all possible data and thus the dependency on symbol n is lost.

1.3.2 Extension to MIMO Transmissions

In this subsection we extend our transmission system model to MIMO transmissions. Let $\mathbf{F}_k$ be an $N_T \times N_L$ dimensional semi-unitary precoding matrix. Assuming equal power allocation at transmit antennas (according to the LTE standard) and making an assumption that the channels between the transmit antennas at each base station are statistically independent, we can extend our model as

$$\mathbf{r}_{n,k} = \mathbf{H}_{n,k}\mathbf{F}_{n,k}\mathbf{x}_{n,k} + \mathbf{y}_{n,k}^{\mathrm{ICI}} + \mathbf{v}_{n,k} \tag{1.29}$$

where $\mathbf{r}_{n,k}$ is the received vector of size $N_R \times 1$, $\mathbf{H}_{n,k}$ is the averaged channel matrix of size $N_R \times N_T$, $\mathbf{v}_{n,k}$ is the noise vector and vector $\mathbf{y}_{n,k}^{\mathrm{ICI}}$ introduces the total ICI power at each time instant n.

Under an isotropic precoder assumption, the autocorrelation of the transmit signals can be expressed as

$$\mathbb{E}\{(\mathbf{F}_{n,k}\mathbf{x}_{n,k})(\mathbf{F}_{n,k}\mathbf{x}_{n,k})^H\} = \frac{P_{\mathrm{S}}}{N_{\mathrm{T}}}\mathbf{I} \tag{1.30}$$

where $\mathbb{E}\{\|\mathbf{x}_{n,k}\|_2^2\} = N_{\mathrm{L}}$ and $\mathbb{E}\{\mathbf{F}_{n,k}\mathbf{F}_{n,k}^H\} = \frac{P_{\mathrm{S}}}{N_{\mathrm{T}}}\mathbf{I}$. Based on this assumption, the ICI terms from different antennas are uncorrelated. The ICI on receive antenna i, i.e., element $[\mathbf{y}_{n,k}^{\mathrm{ICI}}]_i$ of vector $\mathbf{y}_{n,k}^{\mathrm{ICI}}$, is obtained by summing up the contributions from all N_{T} transmit antennas:

$$[\mathbf{y}_{n,k}^{\mathrm{ICI}}]_i = \sum_{m=1}^{N_{\mathrm{T}}} y_{n,k}^{\mathrm{ICI},m} \frac{1}{N_{\mathrm{T}}}. \tag{1.31}$$

This finally allows to compute ICI in MIMO transmissions and then add it as additional Gaussian noise in a simple model. Even though the ICI is not Gaussian distributed, it might converge to Gaussian due to Central limit theorem for sufficiently large N_{T}. This may become handy for so-called massive MIMO technologies that are being discussed for the 5th Generation in wireless cellular transmissions.

1.3.3 Performance Evaluation

In this subsection we provide the validation of our ICI analytical model. As already mentioned at the receiver the ICI noise can be treated as additional complex-valued Gaussian noise with distribution

$$\mathbf{y}_{n,k}^{\mathrm{ICI}} \sim \mathcal{N}_{\mathbb{C}}\left(0,\, P_{n,k}^{\mathrm{ICI}} \cdot \mathbf{I}\right). \tag{1.32}$$

To obtain a similar performance to fast fading simulations, however, it is necessary to additionally decrease the length of the fading blocks in case of block fading simulations of high user velocity systems, to account for the reduced coherence time of the channel. Below we demonstrate the validity of the implemented ICI abstraction model by comparing the throughput obtained with fast-fading simulations, i.e., when the channel varies during one OFDM symbol causing ICI, to the performance achieved with block-fading simulations and adding Gaussian noise to emulate the ICI effects. To demonstrate the validity of the implemented ICI abstraction, we compare results with the Vienna LTE link level simulator that allows a very detailed modeling of the various signal sources. In Figs. 1.3 and 1.4 we compare the achievable throughput and BLER. We recognize that with appropriate ICI calculation and shortened block-fading length we can almost perfectly mimic the fast-fading behaviour and, as a consequence, simulate with high accuracy systems with high mobility users. Similar performance can be observed in the case of MIMO transmissions which is depicted in Fig. 1.5.

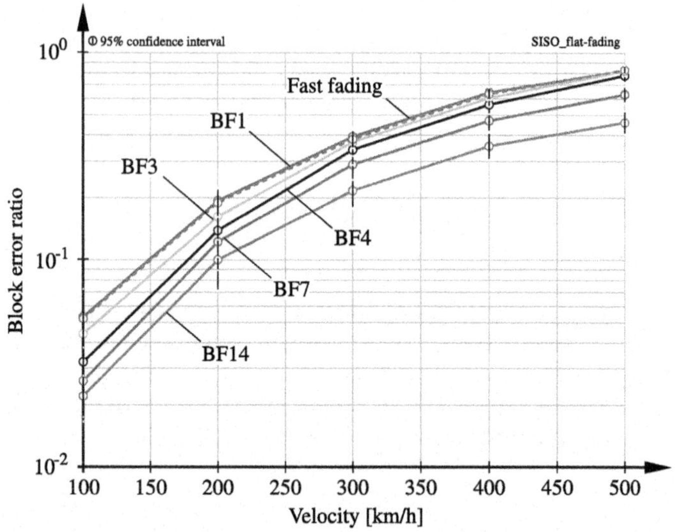

Fig. 1.3 Comparison of BLER between fast-fading and block-fading simulation with shortened block-fading length. Simulations of block-fading channels with block fading length of 1, 3, 4, 7 and 14 OFDM symbols are considered

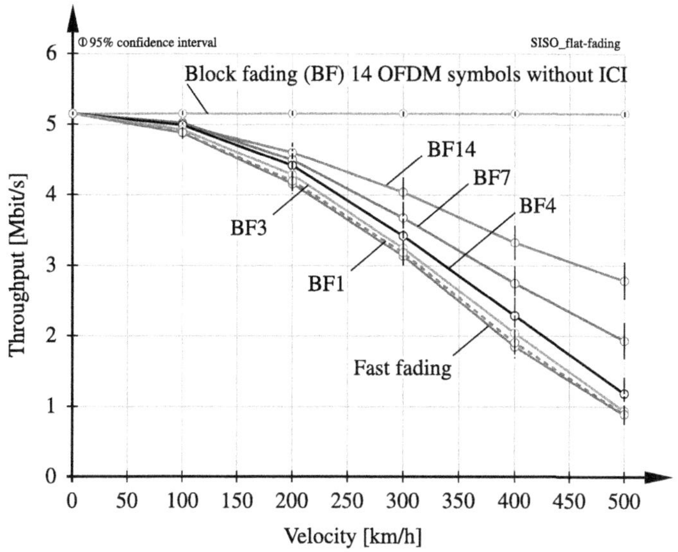

Fig. 1.4 Justification of ICI abstraction model at 2 GHz center frequency. Simulations of block-fading channels with block fading length of 1, 3, 4, 7 and 14 OFDM symbols are considered

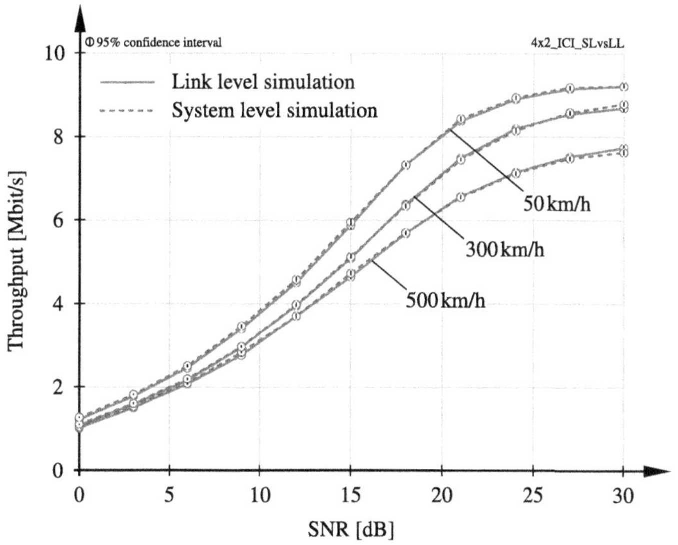

Fig. 1.5 Comparison of 4 × 2 MIMO transmissions with system and link level simulations

1.4 LTE in a Nutshell

Understanding how OFDM transmissions work, makes the essential part of LTE but a lot more details are required in order to understand what options LTE and LTE-A offer. The entire transmitter architecture for DL transmissions is depicted in Fig. 1.6 about which we have only explained the last box in the previous section. Going from left to right, the pure data bits are going through an interleaver, shuffling the bit locations to reduce burst errors and introducing coding as well as various modulation schemes. In LTE 32 so called Modulation and Coding Schemes (MCSs) are defined. They all are a pair of coding and modulation schemes. Here, four modulation schemes are in use all based on Quadrature Amplitude Modulation (QAM): Binary Phase Shift Keying (BPSK), 4-QAM, 16-QAM and 64-QAM. Note that in many 3rd Generation Partnership Project (3GPP) documents modulation schemes referred to BPSK and Quadrature Phase Shift Keying (QPSK) are actually QAM modulation schemes. As only the symbol constellation in the I-Q plane is of interest, the more modulation details do not matter here. A turbo coder, very similar to the 3GPP scheme for 3rd generation allows to vary the code rate between 1/2 in many steps up to 5/6. The selection of the MCS is made adaptively based on the Channel Quality Indicator (CQI) which is feedback information from the receivers. As the selection of the MCS is adaptive, it is called Adaptive Modulation and Coding (AMC). More details on how the CQI value is selected, are explained further ahead in Sect. 1.6. The next phase in Fig. 1.6 is the MIMO processing required if transmitted over more than a single antenna. LTE supports $N_{\mathrm{T}} = 2$ and $N_{\mathrm{T}} = 4$ transmit antennas, LTE-A even eight of them. Depending on the Rank Indicator (RI) which is also feedback information delivered from the User Equipments (UEs), the transmitter selects the number of layers (or streams) over which the data stream is transferred. In LTE a set of $N_{\mathrm{f}} = 7$ OFDM symbols at six adjacent subcarriers is combined to form a so-called Resource Element (RE). These seven symbols span the duration of 0.5 ms, called a slot. In the

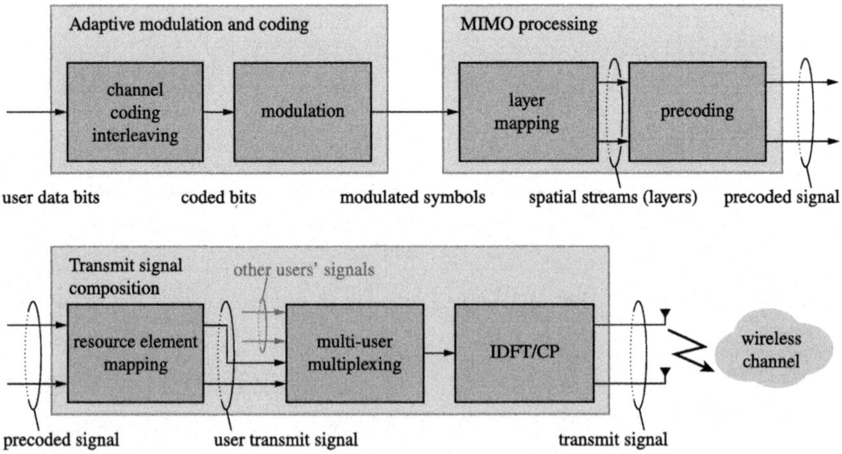

Fig. 1.6 MIMO OFDM transceiver architecture

investigations of this book we even allow the scheduler a finer granularity, i.e., each symbol in the time-frequency plane is assigned to a number $\ell_{n,k}[u, 0]$ of layers it will make use of. By this user u is assigned its resources. With each data stream comes a so-called precoding matrix $\mathbf{F}_{n,k}[u, 0]$ selected with help of a third piece of feedback information, the Precoding Matrix Indicator (PMI).

With all this information on resource allocation for each user u, the data streams are mapped into the time-frequency plane, signaling information is added as well and finally the Inverse Discrete Fourier Transform (IDFT) operation with the additive CP as explained in the previous section is performed. There is many textbooks on LTE available that we recommend the reader to for more details, see e.g., [9–11]. The UL behaves very similar and is described in more detail in Chap. 8.

The receiver architecture is more or less the transmitter in backward order. Receivers are called UE in LTE terminology and can have an arbitrary number of receiver antennas N_R. The first step after the Discrete Fourier Transform (DFT) operation and removing of the CP is to apply an equalizer $\mathbf{G}_{n,k}$ to minimize interference between layers and potentially reduce noise impact. A transmission chain can thus be abstracted to the simple form

$$\mathbf{r}_{n,k} = \mathbf{G}_{n,k}\mathbf{H}_{n,k}\mathbf{F}_{n,k}\mathbf{x}_{n,k} + \mathbf{v}_{n,k}, \tag{1.33}$$

of which we understand now that for each point (n, k) of the time-frequency plane the entire transmission including the wireless channel and the OFDM modulation is condensed in an equivalent channel matrix $\mathbf{H}_{n,k} \in \mathbb{C}^{N_\mathrm{R}\times N_\mathrm{T}}$, the N_T antennas at the transmitter are controlled by the precoding matrix $\mathbf{F}_{n,k}$ and the receiver applies an equalizer $\mathbf{G}_{n,k}$ on its N_R receive antennas. If the base station is concerned (typically in the DL), often the channel and the equalizer are combined as effective channel

$$\mathbf{H}_{n,k}^{(\mathrm{eff})} = \mathbf{G}_{n,k}\mathbf{H}_{n,k}. \tag{1.34}$$

However, if the receiver is in the focus, it may make sense to combine the precoding with the channel and define $\mathbf{H}_{n,k}^{(\mathrm{eff})} = \mathbf{H}_{n,k}\mathbf{F}_{n,k}$. However, if this happens we will refer to it in the text. In general we use the short form (1.34).

1.5 Input–Output Relationship of a Cellular User

A single point-to-point connection from a base station, in LTE called Evolved Base Station (eNodeB), to a UE does not make an LTE network yet. The modeling of a cellular network consisting of $I + 1$ cells is under consideration in the following. In Fig. 1.7 we depict a single eNodeB serving multiple UEs in three sectors, which employs so called Remote Radio Units (RRUs). Such RRUs are becoming more and more common as they offer low cost solutions with increased coverage. They typically consist of a set of antennas with power amplifiers and are connected via fiber glass to a single eNodeB that does all the processing of data.

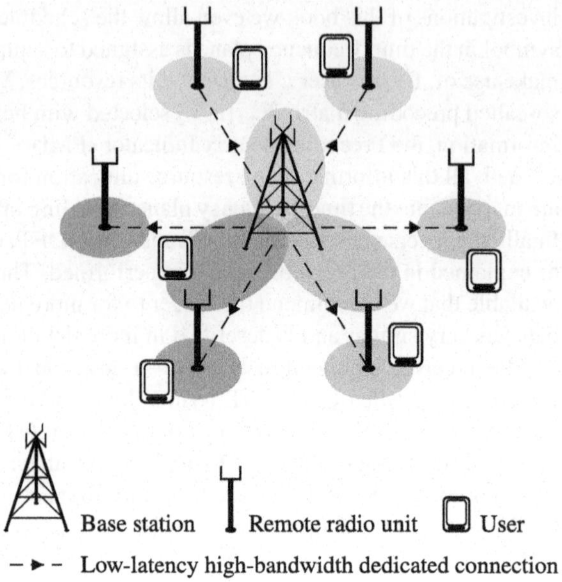

Fig. 1.7 Segment of an exemplary cellular network serving multiple users over sectorized macro base stations and remote radio units

In each cell $i \in \{0, \ldots, I\}$, the attached users are served via a central antenna array that is located at the macro base station and consists of $N_{\mathrm{T}}[0, i]$ antenna elements. Note that we denote the central eNodeB by index 0, while the surrounding base stations have higher indexes. Additionally, the cell is supplied with R_i RRUs that are distributed over the cell area, as illustrated in Fig. 1.7. RRU $r \in \{1, \ldots, R_i\}$ is equipped with an antenna array comprising $N_{\mathrm{T}}[r, i]$ antenna elements. The total number of transmit antennas available in cell i is denoted $N_{\mathrm{T}}[i] = \sum_{r=0}^{R_i} N_{\mathrm{T}}[r, i]$.

In cell i, a total number of U_i users is served. The receive antenna array of user $u \in \mathscr{U}_i$, $\mathscr{U}_i = \{1, \ldots, U_i\}$ is composed of $N_{\mathrm{R}}[u, i]$ antennas. The channel between user u in cell i and RRU r in cell j at OFDM subcarrier N_{FFT} and symbol-time k is described by an $N_{\mathrm{R}}[u, i] \times N_{\mathrm{T}}[r, j]$ dimensional complex-valued channel matrix $\mathbf{H}_{n,k}^{(r,j)}[u, i]$, employing the equivalent complex baseband representation of the OFDM wireless communication system, see, e.g., [2–4]. The channels to the central antenna array of cell j as well as all R_j RRUs are combined in the matrix

$$\mathbf{H}_{n,k}^{(j)}[u, i] = \left[\mathbf{H}_{n,k}^{(0,j)}[u, i], \ldots, \mathbf{H}_{n,k}^{(R_j,j)}[u, i]\right] \in \mathbb{C}^{N_{\mathrm{R}}[u,i] \times N_{\mathrm{T}}[j]}. \tag{1.35}$$

To simplify notation, the superscript $^{(j)}$ is omitted whenever in-cell channels, $j = i$, are considered. Also, if the description of an algorithm or method is independent of the subcarrier index N_{FFT} and/or the symbol-time index k, the respective index is dropped. Such situations are particularly indicated in the text. Following LTE notations, the subcarrier and symbol-time index pair (n, k) is denoted as a RE.

The set of users that is selected by the scheduling algorithm of cell i to be served on a given RE is denoted $\mathscr{S}_{n,k}[i] \subseteq \mathscr{U}_i$. The employed Single User Multiple-Input

Multiple-Output (SU-MIMO) and Multi User Multiple-Input Multiple-Output (MU-MIMO) schedulers are described in detail in Chaps. 5 and 6, respectively. The number of served users in cell i on RE (n, k) is given by $S_{n,k}[i] = |\mathscr{S}_{n,k}[i]|$. The transmit symbol vector intended for user $u \in \mathscr{S}_{n,k}[i]$ is written as $\mathbf{x}_{n,k}[u, i] \in \mathbb{C}^{\ell_k[u,i]\times 1}$, with $\ell_k[u, i] \leq N_{\mathrm{R}}[u, i]$ being the number of data-streams spatially multiplexed to user u. In general, the number of streams per user $\ell_k[u, i]$ could potentially change over subcarriers. With the considered LTE compliant transceiver architecture shown in Fig. 1.6 this is not possible, because the mapping of the user data onto spatial streams (layer mapping) is performed before the mapping onto REs. Hence, $\ell_k[u, i]$ can at most change from one OFDM symbol to the next, provided that each OFDM symbol is coded individually. In LTE the number of streams per user is constant for the duration of one slot consisting of seven OFDM symbols ($N_{\mathrm{f}} = 7$), provided the short CP is employed. It is assumed that such restrictions are handled by the scheduler.

The total number of streams of cell i transmitted on RE (n, k) is denoted $\ell_{n,k}[i] = \sum_{u\in\mathscr{S}_{n,k}[i]} \ell_k[u, i]$. To account for the maximum possible spatial multiplexing capabilities of the base station in cell i, the number of streams is constrained as $\ell_{n,k}[i] \leq N_{\mathrm{T}}[i]$. Although $\ell_k[u, i]$ is independent of the subcarrier index, the total number of streams depends on N_{FFT} due to the frequency selective multi-user scheduling. The transmit symbol vector is normalized as

$$\mathbb{E}\left(\mathbf{x}_{n,k}[u, i]\,\mathbf{x}_{n,k}[u, i]^{\mathrm{H}}\right) = \mathbf{I}_{\ell_k[u,i]}. \tag{1.36}$$

Prior to transmission over the wireless channel, the user symbol vector $\mathbf{x}_{n,k}[u, i]$ is precoded with a precoding matrix $\mathbf{F}_{n,k}[u, i] \in \mathbb{C}^{N_{\mathrm{T}}[i]\times\ell_k[u,i]}$, mapping the $\ell_k[u, i]$-dimensional transmit symbol vector onto the $N_{\mathrm{T}}[i]$ transmit antennas. Notice that the allocation of the available transmit power $P_{n,k}[i]$ among users and spatial streams is considered in the precoding matrices, as detailed in Chaps. 5 and 6. The precoders are obtained such that $P_{n,k}[i]$ is conserved, irrespective of the number of users and spatial streams. In the simulation results presented in the following chapters, power loading over REs is not considered: $P_{n,k}[i] = P_i,\ \forall n, k$. Better performance can be achieved by allocating the available transmit power over subcarriers following a water-filling power allocation policy [12], provided the required Channel State Information at the Transmitter (CSIT) is available.

With this notation, the $N_{\mathrm{R}}[u, i]$-dimensional received signal vector of user u in cell i at RE (n, k) is

$$\begin{aligned}\mathbf{r}_{n,k}[u, i] &= \mathbf{H}_{n,k}[u, i]\mathbf{F}_{n,k}[u, i]\mathbf{x}_{n,k}[u, i] + \mathbf{H}_{n,k}[u, i] \sum_{\substack{s\in\mathscr{S}_{n,k}[i]\\ s\neq u}} \mathbf{F}_{n,k}[s, i]\mathbf{x}_{n,k}[s, i]\\ &\quad + \underbrace{\sum_{j=0, j\neq i}^{I} \mathbf{H}_{n,k}^{(j)}[u, i] \sum_{s\in\mathscr{S}_{n,k}[j]} \mathbf{F}_{n,k}[s, j]\mathbf{x}_{n,k}[s, j] + \mathbf{z}_{n,k}[u, i]}_{\tilde{\mathbf{z}}_{n,k}[u,i]},\end{aligned} \tag{1.37}$$

where the Additive White Gaussian Noise (AWGN) added at the receiver is denoted $\mathbf{z}_{n,k}[u,i] \in \mathcal{N}_{\mathbb{C}}\left(\mathbf{0}, \sigma_z^2\, \mathbf{I}_{N_{\mathrm{R}}[u,i]}\right)$. The intended signal of user u is represented by the first summand in this equation. In-cell interference between the streams spatially multiplexed to several users in the same cell i is captured in the second summand. Out-of-cell interference from other cells $j \neq i$ of the cellular network is taken into account in the third term on the right hand side of Eq. (1.37). The sum of out-of-cell interference and receiver noise is called the effective noise vector $\tilde{\mathbf{z}}_{n,k}[u,i]$.

The users are assumed to employ linear receive filters to equalize their respective channels and to separate the spatially multiplexed data-streams from each other and from the interference caused by the transmission to other users. The $\ell_k[u,i] \times N_{\mathrm{R}}[u,i]$ dimensional receive filtering matrix applied by user u in cell i is written as $\mathbf{G}_{n,k}[u,i]$. Applying this matrix to the received signal vector, the estimated symbol vector is obtained as

$$\begin{aligned}\mathbf{y}_{n,k}[u,i] = \mathbf{G}_{n,k}[u,i]\,\mathbf{r}_{n,k}[u,i] &= \mathbf{G}_{n,k}[u,i]\mathbf{H}_{n,k}[u,i]\mathbf{F}_{n,k}[u,i]\mathbf{x}_{n,k}[u,i] \\ &+ \mathbf{G}_{n,k}[u,i]\mathbf{H}_{n,k}[u,i] \sum_{\substack{s\in\mathcal{S}_{n,k}[i] \\ s\neq u}} \mathbf{F}_{n,k}[s,i]\mathbf{x}_{n,k}[s,i] + \mathbf{G}_{n,k}[u,i]\,\tilde{\mathbf{z}}_{n,k}[u,i].\end{aligned} \tag{1.38}$$

The product of channel matrix and receive filter is referred to as effective channel matrix

$$\mathbf{H}^{(\mathrm{eff})}_{n,k}[u,i] = \mathbf{G}_{n,k}[u,i]\mathbf{H}_{n,k}[u,i]. \tag{1.39}$$

In Chaps. 6 and 7, we will employ an alternative representation of channel matrices and receive filters using their conjugate transposes:

$$\mathbf{H}^{(\mathrm{ct})}_{n,k}[u,i] = \mathbf{H}_{n,k}[u,i]^{\mathrm{H}} \in \mathbb{C}^{N_{\mathrm{T}}[i]\times N_{\mathrm{R}}[u,i]}, \tag{1.40}$$

$$\mathbf{G}^{(\mathrm{ct})}_{n,k}[u,i] = \mathbf{G}_{n,k}[u,i]^{\mathrm{H}} \in \mathbb{C}^{N_{\mathrm{R}}[u,i]\times \ell_k[u,i]}, \tag{1.41}$$

$$\begin{aligned}\mathbf{H}^{(\mathrm{ect})}_{n,k}[u,i] = \mathbf{H}^{(\mathrm{eff})}_{n,k}[u,i]^{\mathrm{H}} &= (\mathbf{G}_{n,k}[u,i]\mathbf{H}_{n,k}[u,i])^{\mathrm{H}} \\ &= \mathbf{H}^{(\mathrm{ct})}_{n,k}[u,i]\mathbf{G}^{(\mathrm{ct})}_{n,k}[u,i] \in \mathbb{C}^{N_{\mathrm{T}}[i]\times \ell_k[u,i]}.\end{aligned} \tag{1.42}$$

This representation facilitates highly intuitive subspace interpretations in the context of multi-user MIMO transmission and considerably shortens notation for precoder calculation. The corresponding input–output relationship reads

$$\begin{aligned}\mathbf{y}_{n,k}[u,i] &= \mathbf{H}^{(\mathrm{ect})}_{n,k}[u,i]^{\mathrm{H}}\mathbf{F}_{n,k}[u,i]\mathbf{x}_{n,k}[u,i] \\ &+ \mathbf{H}^{(\mathrm{ect})}_{n,k}[u,i]^{\mathrm{H}} \sum_{\substack{s\in\mathcal{S}_{n,k}[i] \\ s\neq u}} \mathbf{F}_{n,k}[s,i]\mathbf{x}_{n,k}[s,i] + \mathbf{G}^{(\mathrm{ct})}_{n,k}[u,i]^{\mathrm{H}}\,\tilde{\mathbf{z}}_{n,k}[u,i].\end{aligned} \tag{1.43}$$

Considering the transceiver architecture of Fig. 1.6, all signal processing steps starting from the spatial streams at the output of the layer mapper up to the received signal, obtained at the receiver after the inverse OFDM processing and the equalization, are incorporated in the input–output relationships (1.38) and (1.43). The non-linear mappings involved in the AMC stage are not covered by the system model, but are considered in more detail in Chap. 5.

1.6 Instantaneous Post-equalization SINR

The supported transmission rate of communication channels is frequently subject to significant fluctuations over time and frequency. Such fading effects are especially pronounced in cellular communication systems, where multipath propagation caused by reflections and refractions, shadowing of the radio signal due to obstacles, and movement of the users and/or obstacles can result in variations of the signal strength in the order of tens of decibels [13]. A common method to respond to these fading effects is the application of transmission rate adaptation, e.g., by means of AMC as employed in LTE, such as to match the current data rate to the channel conditions. As detailed throughout Chap. 5, rate adaptation in practical systems can be based on the instantaneous per-stream Signal to Interference and Noise Ratio (SINR) experienced after the receive filter. Considering the input–output relationship of Eq. (1.38), the instantaneous post-equalization SINR of stream $\nu \in \{1, \ldots, \ell_k[u, i]\}$ of user u in cell i is obtained as

$$\beta_{n,k}[\nu, u, i] = \frac{S_{n,k}[\nu, u, i]}{Z_{n,k}[\nu, u, i] + I_{n,k}^{(\text{self})}[\nu, u, i] + I_{n,k}^{(\text{in})}[\nu, u, i] + I_{n,k}^{(\text{out})}[\nu, u, i]}, \tag{1.44}$$

$$S_{n,k}[\nu, u, i] = \left|\mathbf{g}_{n,k}[\nu, u, i]^{\mathrm{H}}\mathbf{H}_{n,k}[u, i]\mathbf{f}_{n,k}[\nu, u, i]\right|^2,$$

$$Z_{n,k}[\nu, u, i] = \sigma_z^2 \left\|\mathbf{g}_{n,k}[\nu, u, i]\right\|^2,$$

$$I_{n,k}^{(\text{self})}[\nu, u, i] = \sum_{\mu=1,\mu\neq\nu}^{\ell_k[k]} \left|\mathbf{g}_{n,k}[\nu, u, i]^{\mathrm{H}}\mathbf{H}_{n,k}[u, i]\mathbf{f}_{n,k}[\mu, u, i]\right|^2,$$

$$I_{n,k}^{(\text{in})}[\nu, u, i] = \sum_{\substack{s\in\mathcal{S}_{n,k}[i]\\ s\neq u}} \left\|\mathbf{g}_{n,k}[\nu, u, i]^{\mathrm{H}}\mathbf{H}_{n,k}[u, i]\mathbf{F}_{n,k}[s, i]\right\|^2,$$

$$I_{n,k}^{(\text{out})}[\nu, u, i] = \sum_{j=0,j\neq i}^{I} \sum_{s\in\mathcal{S}_{n,k}[j]} \left\|\mathbf{g}_{n,k}[\nu, u, i]^{\mathrm{H}}\mathbf{H}_{n,k}[u, i]\mathbf{F}_{n,k}[s, j]\right\|^2,$$

where $\mathbf{g}_{n,k}[\nu, u, i]$ and $\mathbf{f}_{n,k}[\nu, u, i]$ denote the νth column of $\mathbf{G}_{n,k}[u, i]^{\mathrm{H}}$ and $\mathbf{F}_{n,k}[u, i]$, respectively. Notice that the statistical independence of the data symbols of the streams of a user according to Eq. (1.36) has been exploited to obtain Eq. (1.44), and that the symbols of different users are too assumed as statistically independent

$$\mathbb{E}\left(\mathbf{x}_{n,k}[u,i]\mathbf{x}_{s,j}[n,k]^{\mathrm{H}}\right) = \mathbf{0}_{\ell_k[u,i]\times\ell_k[s,j]}, \text{ whenever } u \neq s \vee i \neq j. \qquad (1.45)$$

In Eq. (1.44), the useful signal power of stream ν is represented by the term $S_{n,k}[\nu,u,i]$ in the numerator. The residual interference power between the streams of the user after equalization is given by $I_{n,k}^{(\mathrm{self})}[\nu,u,i]$. The in-cell interference power on stream ν from other users that are served in parallel in the same cell i is denoted $I_{n,k}^{(\mathrm{in})}[\nu,u,i]$, and the out-of-cell interference power from other cells that operate at the same frequency is captured in the term $I_{n,k}^{(\mathrm{out})}[\nu,u,i]$. Notice that depending on the considered transmission and reception strategy, some of these terms are equal to zero, e.g., for SU-MIMO there is no in-cell interference. In Chaps. 5 and 6 this general SINR expression is further specialized to account for the considered transceiver architectures.

References

1. A. Molisch, *Wireless Communications* (Wiley, New York, 2005)
2. S. Haykin, *An Introduction to Analog and Digital Communications* (Wiley, New York, 1989)
3. T. Chiueh, P. Tsai, *OFDM Baseband Receiver Design for Wireless Communications* (Wiley, New York, 2008)
4. Y. Cho, J. Kim, W. Yang, C. Kang, *MIMO-OFDM Wireless Communications with MATLAB* (Wiley, Singapore, 2010)
5. T. Wang, J. Proakis, E. Masry, J. Zeidler, Performance degradation of OFDM systems due to Doppler spreading. IEEE Trans. Wirel. Commun. **5**(6), 1422–1432 (2006). doi:10.1109/TWC.2006.1638663
6. P. Bello, Characterization of randomly time-variant linear channels. IEEE Trans. Commun. Syst. **11**(4), 360–393 (1963)
7. R. Clarke, A statistical theory of mobile-radio reception. Bell Syst. Tech. J. **47**(6), 957–1000 (1968)
8. Y.-S. Choi, P.J. Voltz, F.A. Cassara, On channel estimation and detection for multicarrier signals in fast and selective Rayleigh fading channels. IEEE Trans. Commun. **49**(8), 1375–1387 (2001)
9. E. Dahlman, S. Parvall, J. Skold, P. Beming, *3G Evolution: HSPA and LTE for Mobile Broadband* (Academic Press, Orlando, 2007)
10. S. Sesia, M. Baker, I. Toufik, *LTE, the UMTS Long Term Evolution: From Theory to Practice* (Wiley, New York, 2009)
11. S. Caban, C. Mehlführer, M. Rupp, M. Wrulich, *Evaluation of HSDPA and LTE: From Testbed Measurements to System Level Performance* (Wiley, UK, 2012)
12. J.A.C. Bingham, Multicarrier modulation for data transmission: an idea whose time has come. IEEE Commun. Mag. **28**(5), 5–14 (1990)
13. D. Tse, P. Viswanath, *Fundamentals of Wireless Communication*, Wiley series in telecommunications (Cambridge University Press, Cambridge, 2005)

Chapter 2
Downlink Synchronization

Qi Wang

In this chapter, we present a framework for link performance evaluation of a Long Term Evolution (LTE) downlink with imperfect carrier frequency synchronization. This framework interconnects three performance metrics, namely mean squared error of the carrier frequency offset estimation, post-equalization signal-to-interference-plus-noise ratio, and eventually bit-interleaved coded modulation capacity. With the presented framework, the throughput loss from a residual Carrier Frequency Offset (CFO) estimation error can be analytically determined, given standardized OFDM transmission parameters. In order to validate this mathematical model, extensive link level simulations were carried out using a standard compliant LTE link level simulator. The comparison between the calculated and the simulated results exhibits a fair agreement. This model on the one hand exposes the CFO-tolerance of a standardized OFDM transmission system; on the other hand, it may serve as a means for evaluating CFO estimation algorithms designed for the LTE downlink.[1]

2.1 Introduction

As a dominant physical layer technique in the next generation wireless communication standard, 3GPP LTE, Orthogonal Frequency Division Multiplexing (OFDM) promises significant performance gain in frequency selective channels. Nevertheless, it poses a drawback, namely, sensitive to synchronization errors, such as CFO,

[1]More details on synchronization issues can be found in the Ph.D. thesis of Qi Wang, available at http://theses.eurasip.org/theses/470/performance-evaluation-of-practical-ofdm-systems/download/.

This work has been carried out while she was at the Institute of Telecommunications, TU Wien.

Q. Wang (✉)
Huawei Technologies Düsseldorf GmbH, Düsseldorf, Germany
e-mail: wangqiasia@gmail.com

M. Rupp et al., *The Vienna LTE-Advanced Simulators*,
Signals and Communication Technology, DOI 10.1007/978-981-10-0617-3_2

sampling frequency offset and symbol timing offset. Tremendous efforts were devoted to estimating synchronization errors in the digital signal processing domain. Taking the CFO for example, the performance is evaluated in terms of the Mean Square Error (MSE). Such a metric indicates the estimation performance itself, yet fails to reflect the influence of a residual estimation error on the overall system performance. In a real world communication system, the physical layer performance is eventually expressed in terms of coded throughput. Therefore, not only the performance of an individual processing block but also their overall impact on the throughput needs to be investigated.

For an OFDM system in general, the performance degradation caused by a CFO has been investigated in [1–8]. The authors of [1, 2] evaluated the degradation in terms of the Signal to Interference and Noise Ratio (SINR) in the demodulated OFDM signal. In [3–8], Bit Error Ratio (BER) of OFDM systems with CFO was analytically derived for Additive White Gaussian Noise (AWGN) [3–5] and frequency selective fading channel [6]. In [8], a capacity analysis of impaired OFDM links is presented, taking into account a variety of receiver imperfections, e.g., channel estimation errors, CFO and I/Q imbalance. The authors calculated the average mutual information of the impaired OFDM link using the probability density function derived in [7]. From a methodology point of view, these approaches evaluate the degradation induced by the Inter-Carrier Interference (ICI) exclusively and are applicable to an arbitrary OFDM system; whereas in order to evaluate a standardized system like LTE, many practical aspects need to be taken into account, such as the specified frame structure and the overall receiver design.

Performance modeling has become of interest as nowadays communication systems grow dramatically in complexity. Since simulating a perfect replica of the real system turns to be costly in terms of run time efficiency, it is necessary to combine mathematical and empirical models in the simulation-based performance evaluation. In order to reduce the simulation complexity without losing insight into the real behavior, modeling with an acceptable degree of approximations is desired. For the LTE downlink, a link quality modeling approach has been presented in [9, 10]. The authors derived bounds on achievable throughput of LTE where the post-equalization SINR was employed as the intermediate performance metric. This approach can be utilized to abstract the major physical layer behavior, whereas synchronization errors were ignored.

We apply the methodology elaborated in [9] and develop a throughput loss prediction model for the CFO impaired LTE downlink. This model takes into account the estimation performance of a CFO estimator, a linear receiver structure as well as the Bit Interleaved Coded Modulation (BICM) architecture and analytically determines the performance loss in terms of coded throughput. Validated by extensive standard compliant simulations, this evaluation model on the one hand exposes the CFO-tolerance of such a system; on the other hand, implies how accurate the carrier frequency offset estimation is required to be.

The chapter is organized as follows. A description of the evaluation model is presented in Sect. 2.2. The MSE performance of an exemplified CFO estimator is described in Sect. 2.3. The post-equalization SINR model for a CFO-impaired OFDM transmission is derived in Sect. 2.4. The BICM capacity model is briefly introduced in Sect. 2.5. Numerical validations using the standard compliant Vienna LTE Link-Level DL simulator are provided in Sect. 2.6. Conclusion can be found in Sect. 2.7.

2.2 Evaluation Model

Consider a signal processing chain on the receiver side of the LTE downlink in Fig. 2.1, the frequency synchronization block is located at the beginning of the processing chain, compensating the CFO in the time domain. In order to model the impact of a residual CFO on the resulting coded throughput at the end of the chain, analytical representations need to be found for the function blocks in between.

The signal transmission in LTE is based on a frame structure, illustrated in Fig. 2.2. The transmission resources are segmented into 10 ms frames. Each frame is divided into ten subframes. When a Cyclic Prefix (CP) of normal length is employed, a subframe consists of 14 OFDM symbols [11]. The frame structure is designed so that the signaling information can be embedded on a certain basis with a reasonable overhead. After the Fast Fourier Transform (FFT) transform, the transmission resource can be interpreted as a time-frequency grid, where Reference Signals (RSs) are embedded among data symbols.

Since the post-equalization SINR has been widely utilized for evaluating the performance of a radio link, we choose it as an intermediate step to evaluate the entire receiver chain, i.e.,

$$\text{MSE} \rightarrow \text{post-equalization SINR} \rightarrow \text{coded throughput.}$$

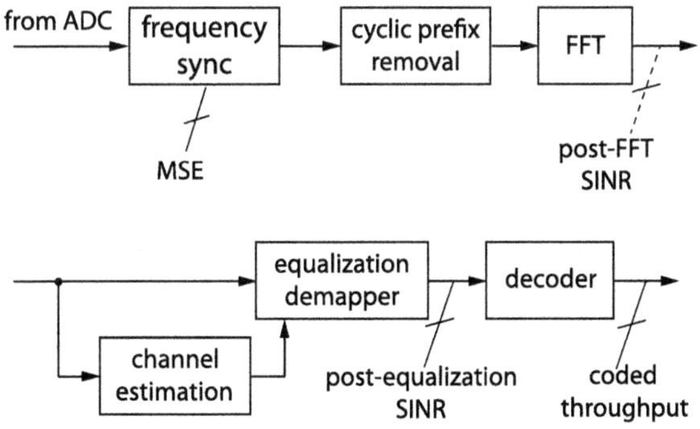

Fig. 2.1 Signal processing chain in an OFDM receiver

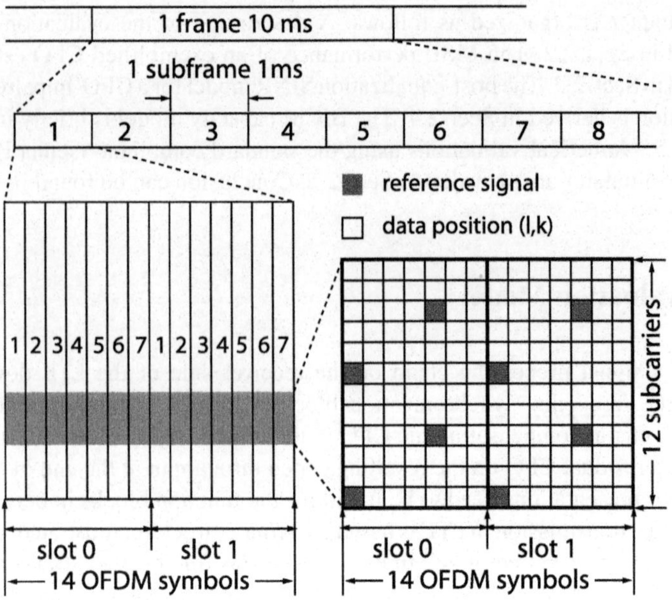

Fig. 2.2 LTE frame structure with normal length CPs, displaying a time-frequency grid in the frequency domain with RS positions

Typically, optimizations are applied based on MSE, assuming an overall optimal can be correspondingly achieved. Our analysis however, will provide an insight by modeling the joint effect of the three.

2.3 Mean Square Error of Carrier Frequency Offset Estimation

In order to compensate the CFO in a real-world OFDM transmission system, various CFO estimation schemes can be applied. Their estimation performances in terms of MSE can be mathematically determined. In [12], a generic CFO estimation scheme was investigated based on the RSs of LTE, where the normalized CFO $\varepsilon \in (-0.5, 0.5)$ is estimated firstly in the time domain then refined in the frequency domain. We focus on the frequency domain estimation in the following because it determines the overall estimation performance. Define vectors $\mathbf{r}_{\mathscr{P},0}, \mathbf{r}_{\mathscr{P},1} \in \mathbb{C}^{N_\mathrm{P}N_\mathrm{R}\times 1}$ for the received RS in slot 0 and 1, diagonal matrices $\mathbf{X}_0 = \mathrm{diag}\left(\mathbf{x}_{\mathscr{P},0}\right), \mathbf{X}_1 = \mathrm{diag}\left(\mathbf{x}_{\mathscr{P},1}\right) \in \mathbb{C}^{N_\mathrm{P}N_\mathrm{R}\times N_\mathrm{P}N_\mathrm{R}}$ containing the corresponding RSs along their diagonals. The CFO is estimated in the frequency domain by

$$\hat{\varepsilon} = -\frac{N_c}{2\pi(N_c + N_{cp})N_{sl}} \arg\left\{\mathbf{r}^H_{\mathscr{P},1}\mathbf{X}_1\mathbf{X}_0^H\mathbf{r}_{\mathscr{P},0}\right\}, \tag{2.1}$$

where $N_{sl} \in \{6, 7\}$ is the number of OFDM symbols per slot and N_P is the number of RSs per slot. We consider this estimation scheme as an example and evaluate the throughput loss of a CFO-impaired LTE downlink with this estimation scheme employed.

The MSE of the overall CFO estimation scheme is given as

$$\mathrm{MSE}(\gamma) = \mathbb{E}\left\{|\varepsilon - \hat{\varepsilon}|^2\right\} = \frac{N_c^2}{4\pi^2(N_c + N_{cp})^2 N_{sl}^2 N_R N_P \gamma}, \tag{2.2}$$

The variable γ denotes the average Signal to Noise Ratio (SNR) at the receiver side in the frequency domain.

The estimation performance of the generic difference phase estimator has been thoroughly analyzed in [13]. A derivation of Eq. (2.2) following the work in [12] is provided. Define the channel frequency response on the RS subcarriers as $\mathbf{h} \in \mathbb{C}^{N_P N_R \times 1}$; the received RS symbols in slot 0 and 1 can be expressed as

$$\mathbf{r}_{\mathscr{P},0} = \mathbf{X}_0\mathbf{h} + \mathbf{v}_0, \tag{2.3}$$

$$\mathbf{r}_{\mathscr{P},1} = \mathrm{e}^{\mathrm{i}\cdot\phi N_{sl}}\mathbf{X}_1\mathbf{h} + \mathbf{v}_1, \tag{2.4}$$

where $\mathbf{v}_0, \mathbf{v}_1$ denote the corresponding noise vectors and $\phi = \frac{2\pi\varepsilon(N_c+N_{cp})}{N_c}$. From Eq. (2.1), there is

$$\begin{aligned}\mathbf{r}^H_{\mathscr{P},1}\mathbf{X}_1\mathbf{X}_0^H\mathbf{r}_{\mathscr{P},0} = {} & \mathrm{e}^{-\mathrm{i}\cdot\phi N_{sl}}\mathbf{h}^H\mathbf{X}_1^H\mathbf{X}_1\mathbf{X}_0^H\mathbf{X}_0\mathbf{h} \\ & + \mathrm{e}^{-\mathrm{i}\cdot\phi N_{sl}}\mathbf{h}^H\mathbf{X}_1^H\mathbf{X}_1\mathbf{X}_0^H\mathbf{v}_0 + \mathbf{v}_1^H\mathbf{X}_1\mathbf{X}_0^H\mathbf{X}_0\mathbf{h} \\ & + \mathbf{v}_1^H\mathbf{X}_1\mathbf{X}_0^H\mathbf{v}_0. \end{aligned} \tag{2.5}$$

Omitting the second-order noise term $\mathbf{v}_1^H\mathbf{X}_1\mathbf{X}_0^H\mathbf{v}_0$, $\mathbf{r}^H_{\mathscr{P},1}\mathbf{X}_1\mathbf{X}_0^H\mathbf{r}_{\mathscr{P},0}$ can be approximated as a complex Gaussian random variable, expressed as

$$\mathbf{r}^H_{\mathscr{P},1}\mathbf{X}_1\mathbf{X}_0^H\mathbf{r}_{\mathscr{P},0} \sim \mathscr{N}_{\mathbb{C}}\left(\mathrm{e}^{-\mathrm{i}\phi N_{sl}} P_S^2\mathbf{h}^H\mathbf{h}, 2P_V P_S^3\mathbf{h}^H\mathbf{h}\right), \tag{2.6}$$

where P_S and P_V denote the signal and noise power, respectively. Define

$$Y, X \sim \mathscr{N}_{\mathbb{C}}\left(0, P_V P_S^3\mathbf{h}^H\mathbf{h}\right), \tag{2.7}$$

and assume $-\frac{\pi}{2} < \phi N_{\mathrm{sl}} < \frac{\pi}{2}$, Eq. (2.1) becomes

$$\begin{aligned}\hat{\phi} &= -\frac{1}{N_{\mathrm{sl}}}\arctan\left\{\frac{\Im\left\{\mathbf{r}_{\mathscr{P},1}^{\mathrm{H}}\mathbf{X}_1\mathbf{X}_0^{\mathrm{H}}\mathbf{r}_{\mathscr{P},0}\right\}}{\Re\left\{\mathbf{r}_{\mathscr{P},1}^{\mathrm{H}}\mathbf{X}_1\mathbf{X}_0^{\mathrm{H}}\mathbf{r}_{\mathscr{P},0}\right\}}\right\}\\ &= -\frac{1}{N_{\mathrm{sl}}}\arctan\left\{\frac{P_{\mathrm{S}}^2\mathbf{h}^{\mathrm{H}}\mathbf{h}\cdot\sin(-\phi N_{\mathrm{sl}})+Y}{P_{\mathrm{S}}^2\mathbf{h}^{\mathrm{H}}\mathbf{h}\cdot\cos(-\phi N_{\mathrm{sl}})+X}\right\}\\ &\approx -\frac{1}{N_{\mathrm{sl}}}\arctan\left\{\frac{P_{\mathrm{S}}^2\mathbf{h}^{\mathrm{H}}\mathbf{h}\cdot\sin(-\phi N_{\mathrm{sl}})}{P_{\mathrm{S}}^2\mathbf{h}^{\mathrm{H}}\mathbf{h}\cdot\cos(-\phi N_{\mathrm{sl}})}\right\}\\ &\quad -\frac{1}{N_{\mathrm{sl}}}\cdot\frac{Y\cos(-\phi N_{\mathrm{sl}})-X\sin(-\phi N_{\mathrm{sl}})}{P_{\mathrm{S}}^2\mathbf{h}^{\mathrm{H}}\mathbf{h}}\\ &= \phi-\frac{1}{N_{\mathrm{sl}}}\cdot\frac{Y\cos(-\phi N_{\mathrm{sl}})-X\sin(-\phi N_{\mathrm{sl}})}{P_{\mathrm{S}}^2\mathbf{h}^{\mathrm{H}}\mathbf{h}}\end{aligned}\tag{2.8}$$

by applying a first-order Taylor expansion. Plugging in Eq. (2.7), we obtain

$$\begin{aligned}\hat{\phi} &\sim \mathscr{N}_{\mathbb{C}}\left(\phi,\frac{P_{\mathrm{V}}P_{\mathrm{S}}^3\mathbf{h}^{\mathrm{H}}\mathbf{h}\cos^2(-\phi N_{\mathrm{sl}})+P_{\mathrm{V}}P_{\mathrm{S}}^3\mathbf{h}^{\mathrm{H}}\mathbf{h}\sin^2(-\phi N_{\mathrm{sl}})}{N_{\mathrm{sl}}^2P_{\mathrm{S}}^4\mathbf{h}^{\mathrm{H}}\mathbf{h}\mathbf{h}^{\mathrm{H}}\mathbf{h}}\right)\\ &\sim \mathscr{N}_{\mathbb{C}}\left(\phi,\frac{P_{\mathrm{V}}}{N_{\mathrm{sl}}^2P_{\mathrm{S}}\mathbf{h}^{\mathrm{H}}\mathbf{h}}\right).\end{aligned}\tag{2.9}$$

Therefore, the estimator is unbiased and

$$\begin{aligned}\mathrm{MSE}(\gamma) &= \mathbb{E}\left\{|\varepsilon-\hat{\varepsilon}|^2\right\} = \frac{N_{\mathrm{c}}^2}{4\pi^2(N_{\mathrm{c}}+N_{\mathrm{cp}})^2}\mathbb{E}\left\{|\phi-\hat{\phi}|^2\right\}\\ &= \frac{N_{\mathrm{c}}^2P_{\mathrm{V}}}{4\pi^2(N_{\mathrm{c}}+N_{\mathrm{cp}})^2N_{\mathrm{sl}}^2P_{\mathrm{S}}\mathbf{h}^{\mathrm{H}}\mathbf{h}} = \frac{N_{\mathrm{c}}^2}{4\pi^2(N_{\mathrm{c}}+N_{\mathrm{cp}})^2N_{\mathrm{sl}}^2N_{\mathrm{P}}N_{\mathrm{R}}\gamma},\end{aligned}\tag{2.10}$$

where the average SNR

$$\gamma = \frac{P_{\mathrm{S}}\mathbf{h}^{\mathrm{H}}\mathbf{h}}{N_{\mathrm{P}}N_{\mathrm{R}}P_{\mathrm{V}}}.\tag{2.11}$$

In classical estimation theory, the MSE of an unbiased estimator for ε is lower bounded as

$$\mathrm{MSE}_{\varepsilon} = \frac{N_{\mathrm{c}}^2}{4\pi^2(N_{\mathrm{c}}+N_{\mathrm{cp}})^2}\mathrm{MSE}_{\phi} \geq \frac{N_{\mathrm{c}}^2}{4\pi^2(N_{\mathrm{c}}+N_{\mathrm{cp}})^2}\cdot\frac{1}{J(\phi)},\tag{2.12}$$

where the Fisher information

$$J(\phi) = -\mathbb{E}\left\{\frac{\partial^2}{\partial\phi^2}\Lambda(\phi)\right\}.\tag{2.13}$$

Given the notation in Eqs. (2.3) and (2.4), we characterize the RS-based CFO estimation problem by the log-likelihood function

$$\begin{aligned}\Lambda(\phi) &= \ln f(\mathbf{r}_{\mathscr{P},0}, \mathbf{r}_{\mathscr{P},1}; \phi) \\ &= \ln \frac{1}{\pi^{2N_\mathrm{P}N_\mathrm{R}} \det(\mathbf{R})} \exp\left\{-\left[\mathbf{r}^\mathrm{H}_{\mathscr{P},0}\ \mathbf{r}^\mathrm{H}_{\mathscr{P},1}\right]\mathbf{R}^{-1}\begin{bmatrix}\mathbf{r}_{\mathscr{P},0}\\ \mathbf{r}_{\mathscr{P},1}\end{bmatrix}\right\},\end{aligned} \tag{2.14}$$

with

$$\mathbf{R} = \mathbb{E}\left\{\begin{bmatrix}\mathbf{r}_{\mathscr{P},0}\\ \mathbf{r}_{\mathscr{P},1}\end{bmatrix}\left[\mathbf{r}^\mathrm{H}_{\mathscr{P},0}\ \mathbf{r}^\mathrm{H}_{\mathscr{P},1}\right]\right\}. \tag{2.15}$$

Plugging Eqs. (2.3) and (2.4) into (2.14), after arithmetic manipulations, Eq. (2.13) becomes

$$J(\phi) = \frac{N_\mathrm{sl}^2 P_\mathrm{S} \mathbf{h}^\mathrm{H}\mathbf{h}}{P_\mathrm{V}}. \tag{2.16}$$

This leads to the Cramér-Rao Lower Bound (CRLB)

$$\begin{aligned}\mathrm{MSE}_\varepsilon &\geq \frac{N_\mathrm{c}^2 P_\mathrm{V}}{4\pi^2(N_\mathrm{c}+N_\mathrm{cp})^2 N_\mathrm{sl}^2 P_\mathrm{S}\mathbf{h}^\mathrm{H}\mathbf{h}} \\ &= \frac{N_\mathrm{c}^2}{4\pi^2(N_\mathrm{c}+N_\mathrm{cp})^2 N_\mathrm{sl}^2 N_\mathrm{P} N_\mathrm{R}\gamma}.\end{aligned} \tag{2.17}$$

Compared to Eq. (2.10), the lower bound of the estimation variance is attained.

2.4 Signal to Interference and Noise Ratio Modeling

The authors of [1, 2] investigated the impact of a CFO on OFDM systems where such impact means exclusively the degradation in terms of the post-FFT SINR, shown in Fig. 2.1. In [14], a post-equalization SINR model was presented. This measure is of importance, because it directly determines the theoretically possible throughput I via Shannon's formula:

$$I \approx \log_2(1+\mathrm{SINR}). \tag{2.18}$$

In an urban scenario with low to medium mobility, it can be shown that the channel is quasi-static within the duration of one subframe (1 ms). Therefore, time-invariant channel estimation and equalization can be applied on a subframe basis. Following the analysis in [14], we constrain the evaluation within one subframe and consider a residual CFO which is normalized to the standardized subcarrier spacing, denoted by $\varepsilon \in (-0.5, 0.5)$. Let n be the OFDM symbol index within a subframe, k the subcarrier index, N_{T} the number of transmit antennas and N_{R} the number of receive antennas. We denote the transmitted signal vector by $\mathbf{x}_{n,k} \in \mathbb{C}^{N_{\mathrm{L}}\times 1}$, the precoded channel matrix in the frequency domain by $\mathbf{H}_k^{(\mathrm{eff})} \in \mathbb{C}^{N_{\mathrm{R}}\times N_{\mathrm{L}}}$, the received signal by $\mathbf{r}_{n,k} \in \mathbb{C}^{N_{\mathrm{R}}\times 1}$ and the AWGN by $\mathbf{v}_{n,k} \in \mathbb{C}^{N_{\mathrm{R}}\times 1}$. We use here the abbreviation $\mathbf{H}_k^{(\mathrm{eff})} = \mathbf{H}_k\mathbf{F}_k$, where the channel matrix $\mathbf{H}_k \in \mathbb{C}^{N_{\mathrm{R}}\times N_{\mathrm{T}}}$ and the precoding matrix $\mathbf{F}_k \in \mathbb{C}^{N_{\mathrm{T}}\times N_{\mathrm{L}}}$. Given the block fading assumption, the channel matrix within one subframe stays constant, independent from the OFDM symbol index n. When the system is impaired by a CFO, the signal transmission can be described as

$$\begin{aligned}\mathbf{r}_{n,k} &= I(0,\varepsilon)\cdot \mathrm{e}^{\mathrm{i}\Phi(\varepsilon,n)}\cdot \mathbf{H}_k^{(\mathrm{eff})}\mathbf{x}_{n,k} \\ &\quad + \sum_{p\neq k} I(p-k,\varepsilon)\cdot \mathrm{e}^{\mathrm{i}\Phi(\varepsilon,n)}\cdot \mathbf{H}_p^{(\mathrm{eff})}\mathbf{x}_{n,p} + \mathbf{v}_{n,k},\end{aligned} \tag{2.19}$$

where

$$I(0,\varepsilon) = \frac{\sin(\pi\varepsilon)}{N_{\mathrm{c}}\sin(\pi\varepsilon/N_{\mathrm{c}})}\cdot \mathrm{e}^{\mathrm{i}\frac{\pi\varepsilon(N_{\mathrm{c}}-1)}{N_{\mathrm{c}}}}, \tag{2.20}$$

$$I(p-k,\varepsilon) = \frac{\sin[\pi(p-k+\varepsilon)]}{N_{\mathrm{c}}\sin[\pi(p-k+\varepsilon)/N_{\mathrm{c}}]}\cdot \mathrm{e}^{\mathrm{i}\frac{\pi(p-k+\varepsilon)(N_{\mathrm{c}}-1)}{N_{\mathrm{c}}}}, \tag{2.21}$$

$$\mathrm{e}^{\mathrm{i}\Phi(\varepsilon,n)} = \mathrm{e}^{\mathrm{i}\frac{2\pi\varepsilon n(N_{\mathrm{c}}+N_{\mathrm{cp}})}{N_{\mathrm{c}}}}. \tag{2.22}$$

Here, the factor $I(0,\varepsilon)\cdot \mathrm{e}^{\mathrm{i}\Phi(\varepsilon,n)}$ introduces time-variant distortion to the desired signal term besides the channel response. However, since the system is assumed on a subframe basis to be static, the receiver is designed to be time invariant on a subframe basis; in other words, a universal channel estimate which is independent of the time index n is to be obtained using all RSs shown in Fig. 2.2.

For simplicity, we assume that the perfect and static channel knowledge is available at each subframe. A Zero Forcing (ZF) equalizer at subcarrier k is then given as

$$\mathbf{G}_k = \left(\mathbf{H}_k^{(\mathrm{eff})^{\mathrm{H}}}\mathbf{H}_k^{(\mathrm{eff})}\right)^{-1}\mathbf{H}_k^{(\mathrm{eff})^{\mathrm{H}}}. \tag{2.23}$$

Thus, the estimated data symbol after equalization can be expressed as

$$\begin{aligned}\hat{\mathbf{x}}_{n,k} &= \mathbf{G}_k\cdot \mathbf{r}_{n,k} = I(0,\varepsilon)\cdot \mathrm{e}^{\mathrm{i}\Phi(\varepsilon,n)}\cdot \mathbf{x}_{n,k} \\ &\quad + \underbrace{\mathbf{G}_k\sum_{p\neq k} I(p-k,\varepsilon)\cdot \mathrm{e}^{\mathrm{i}\Phi(\varepsilon,n)}\cdot \mathbf{H}_p^{(\mathrm{eff})}\mathbf{x}_{l,p}}_{\mathbf{y}_{n,k}^{\mathrm{ICI}}} + \underbrace{\mathbf{G}_k\mathbf{v}_{n,k}}_{\tilde{\mathbf{v}}_{n,k}} \\ &= I(0,\varepsilon)\cdot \mathrm{e}^{\mathrm{i}\Phi(\varepsilon,n)}\cdot \mathbf{x}_{n,k} + \mathbf{y}_{n,k}^{\mathrm{ICI}} + \tilde{\mathbf{v}}_{n,k},\end{aligned} \tag{2.24}$$

where $\mathbf{y}_{n,k}^{\mathrm{ICI}}$ denotes the ICI and $\tilde{\mathbf{v}}_{n,k}$ is the equalized noise vector. Let N_{L} denote the number of transmission layers which is indexed by $l = 0, 1, \ldots, N_{\mathrm{L}} - 1$, the $\mathrm{SINR}_{n,k}^{(l)}$ on the lth layer can be found by

$$\mathrm{SINR}_{n,k}^{(l)}(\varepsilon, \mathbf{H}_k^{(\mathrm{eff})}) = \frac{\left[\mathbf{x}_{n,k}\mathbf{x}_{n,k}^{\mathrm{H}}\right]_{(l,l)}}{\left[(\hat{\mathbf{x}}_{n,k} - \mathbf{x}_{n,k})(\hat{\mathbf{x}}_{n,k} - \mathbf{x}_{n,k})^{\mathrm{H}}\right]_{(l,l)}}, \tag{2.25}$$

where $[\,\cdot\,]_{(i,j)}$ denotes the entry on the ith row and jth column of the given matrix. We denote the average signal power on each subcarrier and each layer by P_{S} and the corresponding noise power by P_{V}. Plugging Eq. (2.24) into (2.25), we obtain a closed form expression of the post-equalization SINR on the lth layer at Resource Element (RE) (n, k), shown in Eq. (2.27). Since the system assumes block fading on a subframe bases, the so-called Common Phase Error (CPE) in Eq. (2.22) which increases linearly with the time index n causes a signal distortion term. As suggested in [14], this is the dominant term compared to the ICI.

$$\mathrm{SINR}_{n,k}^{(l)}(\varepsilon, \mathbf{H}_k^{(\mathrm{eff})}) \tag{2.26}$$

$$= \frac{P_{\mathrm{S}}}{\left[\mathbf{y}_{n,k}^{\mathrm{ICI}}\mathbf{y}_{n,k}^{\mathrm{ICI}\,\mathrm{H}}\right]_{(l,l)} + \left[\tilde{\mathbf{v}}_{n,k}\tilde{\mathbf{v}}_{n,k}^{\mathrm{H}}\right]_{(l,l)} + |I(0,\varepsilon)\cdot \mathrm{e}^{\mathrm{i}\Phi(\varepsilon,n)} - 1|^2 \cdot \left[\mathbf{x}_{n,k}\mathbf{x}_{n,k}^{\mathrm{H}}\right]_{(l,l)}}$$

$$= \frac{P_{\mathrm{S}}}{\mathrm{ICI} + \underbrace{P_{\mathrm{V}}\left[\mathbf{G}_k^{\mathrm{H}}\mathbf{G}_k\right]_{(l,l)}}_{\text{noise}} + \underbrace{P_{\mathrm{S}}|I(0,\varepsilon)\cdot \mathrm{e}^{\mathrm{i}\Phi(\varepsilon,n)} - 1|^2}_{\text{signal distortion}}}, \tag{2.27}$$

$$\mathrm{ICI} = P_{\mathrm{S}} \sum_{p \neq k} |I(p-k,\varepsilon)|^2 \left[\mathbf{G}_k^{\mathrm{H}}\mathbf{G}_k\mathbf{H}_p^{(\mathrm{eff})}\mathbf{H}_p^{(\mathrm{eff})\mathrm{H}}\right]_{(l,l)}. \tag{2.28}$$

2.5 Bit Interleaved Coded Modulation Capacity

In general, a BICM architecture is obtained by concatenating channel coding with modulation mapping through a bit interleaver. Such a scheme allows combinations of any channel code with any arbitrary modulation alphabet [15]. Based on this architecture, LTE employs 4, 16 or 64-Quadrature Amplitude Modulation (QAM) and a rate 1/3 turbo code that is appropriately rate matched to achieve the desired code rates as defined in [16]. The capacity of BICM systems is well known, though not in closed-form [17]. In Fig. 2.3, BICM capacity of the three LTE-defined modulation alphabets (4-QAM, 16-QAM, 64-QAM) are plotted. Analogous to [10], a function $f(\mathrm{SINR})$ is introduced to describe the maximum efficiency over all available modulation alphabets.

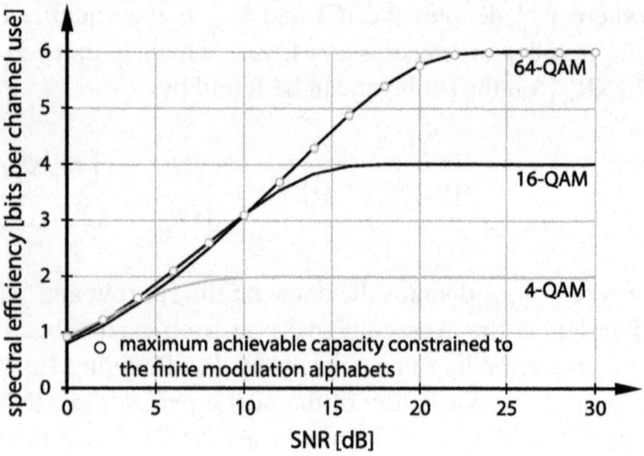

Fig. 2.3 BICM capacity of 4, 16 and 64-QAM modulation

Given the SINR model in Sect. 2.2, the spectral efficiency of an LTE downlink transmission suffering from a CFO ε can be expressed as $f(\mathrm{SINR}_{n,k}^{(l)}(\varepsilon, \mathbf{H}_k^{(\mathrm{eff})}))$, where $\mathrm{SINR}_{n,k}^{(l)}(\varepsilon, \mathbf{H}_k^{(\mathrm{eff})})$ is plugged in from Eq. (2.27). The index (n, k) denotes an RE which is devoted to data transmission; in other words, overhead such as RSs, Primary Synchronization Signal (PSS), Secondary Synchronization Signal (SSS) and guard bands are excluded. Therefore, the average spectral efficiency that can be achieved at each transmission layer is written as

$$\bar{B}(\varepsilon) = \frac{1}{N_{\mathrm{D}} N_{\mathrm{L}}} \sum_{(n,k)} \sum_{l} f(\mathrm{SINR}_{n,k}^{(l)}(\varepsilon, \mathbf{H}_k^{(\mathrm{eff})})), \tag{2.29}$$

where N_{D} is the number of available data REs.

Given the MSE analysis in Sect. 2.3, a theoretical residual estimation error can be assumed, labeled as $\bar{\varepsilon} = \sqrt{\mathrm{MSE}(\gamma)}$. Thus, a theoretically achievable BICM capacity can be expressed as

$$\begin{aligned} B(\gamma) &= \sum_{(n,k)} \sum_{l} f(\mathrm{SINR}_{n,k}^{(l)}(\bar{\varepsilon}, \mathbf{H}_k^{(\mathrm{eff})})) \\ &= \sum_{(n,k)} \sum_{l} f(\mathrm{SINR}_{n,k}^{(l)}(\sqrt{\mathrm{MSE}(\gamma)}, \mathbf{H}_k^{(\mathrm{eff})})). \end{aligned} \tag{2.30}$$

This capacity bound takes into account the finite set of Modulation and Coding Schemes (MCSs) suggested in [16], a linear receiver structure and the limitation of the CFO estimation performance, while it ignores other aspects such as a suboptimal channel coding, selection of suitable precoding matrix and number of transmit streams. Since we are only interested in the throughput difference between the zero-CFO case and the CFO-compensated case, these imperfect modeling aspects cause the same effect in both cases. The throughput loss, being the difference of them two, can be calculated as

$$\Delta B(\gamma) = \sum_{(n,k)} \sum_{l} f(\mathrm{SINR}_{n,k}^{(l)}(0, \mathbf{H}_k^{(\mathrm{eff})})) - B(\gamma). \tag{2.31}$$

2.6 Numerical Results

In this section, we validate the analytical models presented in Sects. 2.3, 2.4 and 2.5 by standard compliant simulations of LTE downlink using the *Vienna LTE Link Level Simulator* [18]. The parameter setting is shown in Table 2.1. All presented simulation examples are made available for downloading.

2.6.1 Mean Square Error

Figure 2.4 shows the calculated MSE curves and the simulated estimation performance of the estimation scheme in Sect. 2.3. Generally speaking, the overall MSE is determined by the estimation in the frequency domain. The simulated curves follow the calculation except in the lower SNR region, due to the fact that the estimation errors from the time-domain estimation exceed the estimation range of the estimator

Table 2.1 Simulation parameters for results in Sect. 2.6

Parameter	Value
Channel bandwidth	1.4 MHz
FFT size (N_c)	128
No. data subcarriers	72
Subcarrier spacing	15 kHz
Carrier frequency	2.5 GHz
CP length (N_{cp})	[10, 9] (normal [11])
Transmission setting $N_R \times N_T$	$1 \times 1, 2 \times 2$
Transmission mode	Spatial multiplexing
Precoding	Identity
Channel model	ITU Pedestrian B [19]
CFO introduced (ε)	0, 0.14159 ... subcarrier spacing
Channel knowledge	Perfect
Equalizer	Zero Forcing (ZF)

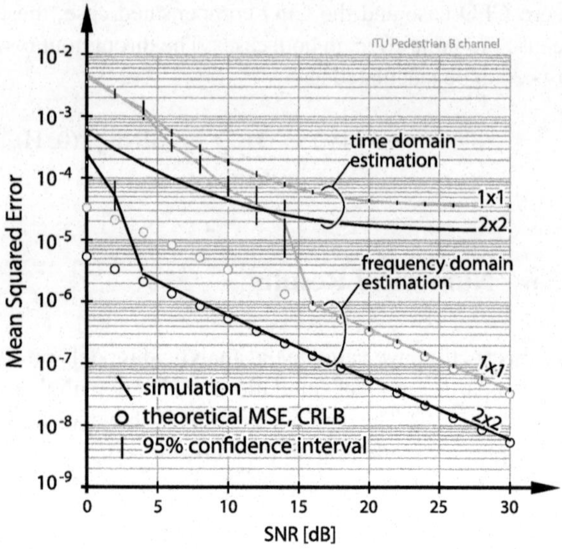

Fig. 2.4 Simulated and calculated MSE curves of the CFO estimation scheme

in the frequency domain. This effect, unfortunately, is not included in the theoretical analysis of the estimation performance.

2.6.2 Post-equalization Signal to Interference and Noise Ratio

In order to validate Eq. (2.27), we introduced 20 logarithmically spaced CFOs which are normalized to the subcarrier spacing, namely 15 kHz in LTE. Neither an estimation nor a compensation procedure was applied at this stage. For better visualization of the impact from the CFOs, the SNR is fixed at $\gamma = 30\,\text{dB}$. The resulting post-equalization SINR curves are plotted in Fig. 2.5 and compared to those obtained using Eq. (2.27).

Figure 2.5 shows that calculated results match well with those from the standard compliant simulation. This indicates that Eq. (2.27) can be used as a valid characterization of the system behavior.

2.6.3 Average Spectral Efficiency

The average spectral efficiency in Eq. (2.29) degrades as the post-equalization SINR decreases correspondingly. Given a series of deterministic CFOs, this degradation calculated using Eq. (2.29) is shown in Fig. 2.6. The results are based on 200 channel

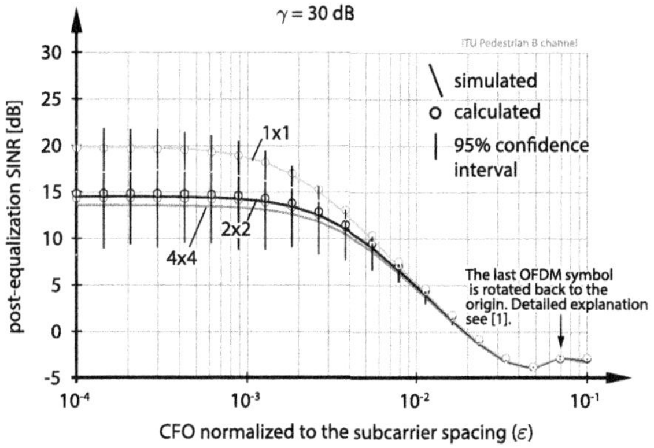

Fig. 2.5 Post-equalization SINR under increasing levels of residual CFOs in ITU Pedestrian B channel. The relatively large confidence intervals are due to the frequency selectivity over the data subcarriers

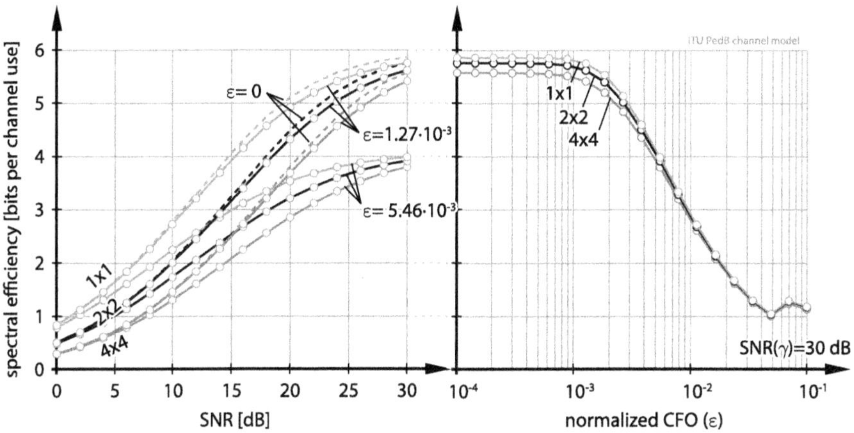

Fig. 2.6 Degradation in average spectral efficiency (per layer for the MIMO cases) due to the residual CFOs

realizations. For the MIMO cases, results are exhibited on a per transmission layer basis.

The subfigure on the left-hand side shows the theoretical degradation in average spectral efficiency subjected to a fixed CFO at different SNR levels. Two CFOs are introduced as examples where $\varepsilon = 1.27 \times 10^{-3}$ corresponds to 19.1 Hz and $\varepsilon = 5.46 \times 10^{-3}$ to 81.8 Hz given the subcarrier spacing 15 kHz. Compared to the zero-CFO case, it can be observed that the higher SNR region where higher efficiency is aimed, appears to be more sensitive to the CFO. Moreover, the impacts on Single-Input Single-Output (SISO) and MIMO systems are fairly equal on a per layer basis.

In the subfigure on the right-hand side, the SNR is fixed at $\gamma = 30$ dB in order to visualize the impact under CFOs of increasing magnitudes. The average spectral efficiency starts to decrease around $\varepsilon = 1 \times 10^{-3}$, approximately. A similar behavior can be observed for the SISO and MIMO cases, although the average spectral efficiency per layer is slightly lower for the multiple antenna scenario due to the incremental noise enhancement from a ZF equalizer.

2.6.4 *Coded Throughput Loss*

As a comparison to the calculated BICM capacity, we simulated coded throughput of the LTE downlink. The fifteen MCSs indicated by Channel Quality Indicators (CQIs) are implemented, shown in Table 2.2. In the LTE downlink, User Equipments (UEs) provide wideband feedbacks to the eNodeB so that the MCS can be adapted to the actual channel quality. In our experiment, the CQI feedback is forced to be optimal by selecting the MCS that delivers the highest throughput for each channel realization.

Figures 2.7 and 2.8 exhibit the results obtained for a SISO and a 2×2 MIMO LTE DL. In the upper figures, coded throughputs of ideally synchronized transmissions are compared to the CFO-compensated case. With the CFO estimation scheme in [12] applied, the loss between the two cases is hardly visible, especially for the MIMO case. The corresponding achievable BICM capacity curves confirm such an

Table 2.2 Modulation scheme, Effective Code Rate (ECR) and efficiency for each of the Channel Quality Indicators (CQIs) of the LTE standard

CQI Index	Modulation	ECR	Data (bit/symbol)
0	Out of range		
1	4-QAM	0.08	0.15
2	4-QAM	0.12	0.23
3	4-QAM	0.19	0.38
4	4-QAM	0.30	0.60
5	4-QAM	0.44	0.88
6	4-QAM	0.59	1.18
7	16-QAM	0.37	1.48
8	16-QAM	0.48	1.91
9	16-QAM	0.60	2.41
10	64-QAM	0.46	2.73
11	64-QAM	0.55	3.32
12	64-QAM	0.65	3.90
13	64-QAM	0.75	4.52
14	64-QAM	0.85	5.12
15	64-QAM	0.93	5.55

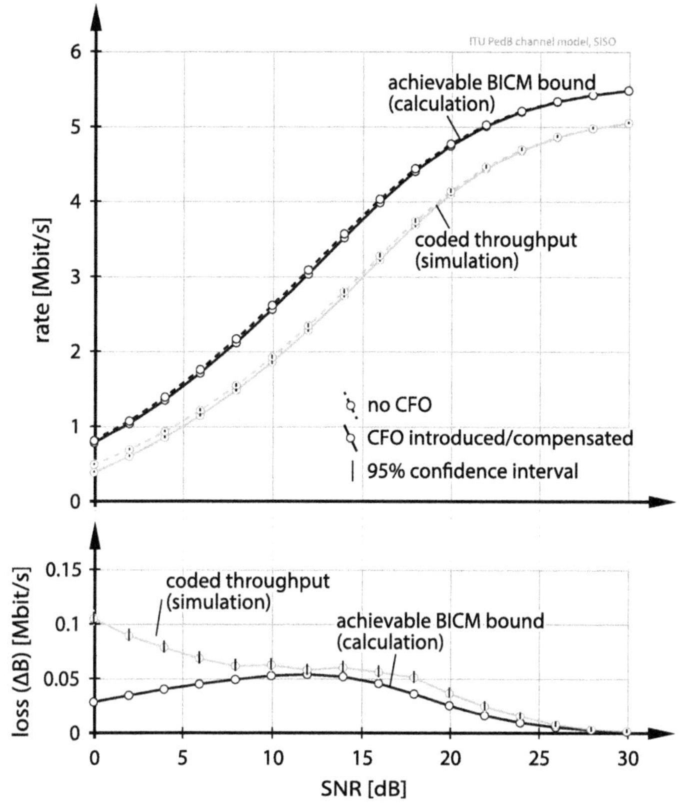

Fig. 2.7 Achievable BICM capacity and simulated coded throughput for an SISO LTE DL under CFO, 5000 subframes Monte Carlo simulation

observation. Note that there are absolute differences between calculated capacity curves and simulated coded throughput, it is due to the imperfect channel code.

In the lower subfigures of Figs. 2.7 and 2.8, the absolute coded throughput loss between the no-CFO and the CFO-compensated case are plotted. The absolute differences in the upper figure cancels out when calculating the relative loss. In the higher SNR region, it can be observed that the simulated coded throughput loss follows the trend of the analytical calculation. However, mismatches appear in the lower SNR region, which agrees with the MSE performance shown in Fig. 2.4. Since the theoretical MSE analysis fails to model the overflow in the frequency domain estimation, an increasing loss in the simulated overall throughput can be observed.

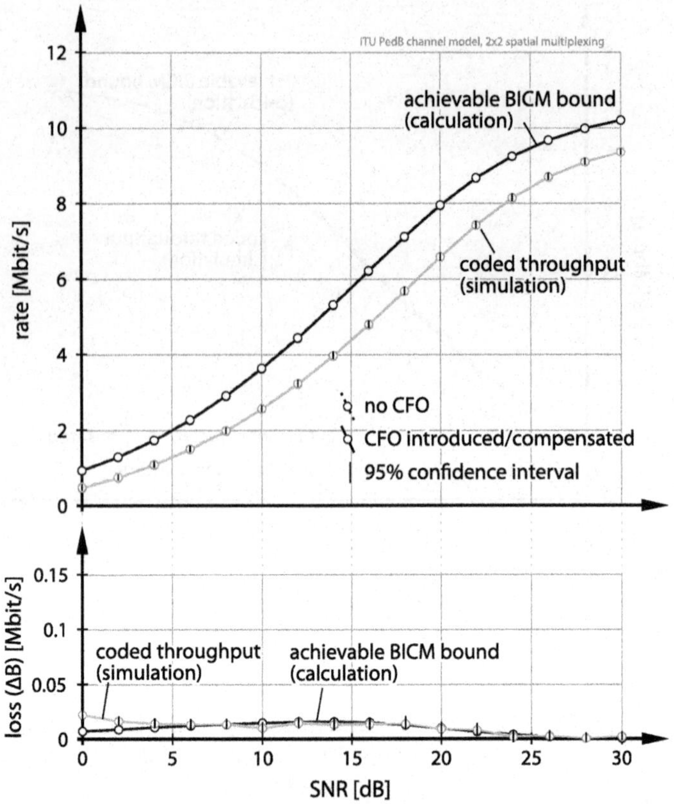

Fig. 2.8 Achievable BICM capacity and simulated coded throughput for a 2 $\times$ 2 spatial multiplexing LTE DL under CFO, 5000 subframes Monte Carlo simulation

2.7 Conclusion

In this chapter, we propose a throughput loss prediction model for a CFO-impaired LTE downlink. This model interconnects the three performance metrics, namely MSE, post-equalization SINR and the BICM capacity bound. Given the theoretical estimation performance of an arbitrary CFO estimator, the resulting performance loss in terms of coded throughput can be analytically determined with acceptable accuracy, avoiding extensive time-consuming link performance simulations. Validated by standard compliant link level simulations, a good agreement can be found. Since for a practical OFDM system, it is more important to find a sufficient estimator than the 'best', this model may serve as an evaluation tool for such purpose. Moreover, the example estimation scheme in [12] is shown to be sufficient for the LTE downlink with multiple antenna configuration in frequency selective scenarios.

References

1. B. Stantchev, G. Fettweis, Time-variant distortion in OFDM. IEEE Commun. Lett. **4**, 312–314 (2000)
2. J. Lee, H.-L. Lou, D. Toumpakaris, J. Cioffi, SNR analysis of ofdm systems in the presence of carrier frequency offset for fading channels. IEEE Trans. Wirel. Commun. **5**(12), 3360–3364 (2006)
3. K. Sathananthan, C. Tellambura, Probability of error calculation of OFDM systems with frequency offset. IEEE Trans. Commun. **49**(11), 1884–1888 (2001)
4. T. Pollet, M. Van Bladel, M. Moeneclaey, BER sensitivity of OFDM systems to carrier frequency offset and Wiener phase noise. IEEE Trans. Commun. **43**(234), 191–193 (1995)
5. P. Dharmawansa, N. Rajatheva, H. Minn, An exact error probability analysis of OFDM systems with frequency offset. IEEE Trans. Commun. **57**(1), 26–31 (2009)
6. L. Rugini, P. Banelli, BER of OFDM systems impaired by carrier frequency offset in multipath fading channels. IEEE Trans. Wirel. Commun. **4**(5), 2279–2288 (2005)
7. M. Krondorf, G. Fettweis, Bit error rate calculation for OFDM with synchronization errors in time and frequency selective fading channels, in *Proceedings of 13th European Wireless Conference (EW'07)*, Paris, France, Apr 2007
8. M. Krondorf, G. Fettweis, OFDM link performance analysis under various receiver impairments, in *EURASIP Journal on Wireless Communications and Networking*, vol. 2008, Article ID 145279, 2008
9. S. Schwarz, M. Šimko, M. Rupp, On performance bounds for MIMO OFDM based wireless communication systems, in *Proceedings of IEEE Signal Processing Advances in Wireless Communications SPAWC 2011*, June 2011
10. S. Caban, C. Mehlführer, M. Rupp, M. Wrulich, *Evaluation of HSDPA and LTE: From Testbed Measurements to System Level Performance*, 1st edn. (Wiley, New York, 2012)
11. Technical Specification Group Radio Access Network, E-UTRA; physical channels and modulation, 3GPP, Technical Report TS 36.211 Version 9.1.0, Mar 2010
12. Q. Wang, C. Mehlführer, M. Rupp, Carrier frequency synchronization in the downlink of 3GPP LTE, in *Proceeding of the 21st Annual IEEE International Symposium on Personal, Indoor and Mobile Radio Communications (PIMRC'10)* Istanbul, Turkey, 2010
13. M. Sandell, D. McNamara, S. Parker, Analysis of frequency-offset tracking in MIMO OFDM systems. IEEE Trans. Commun. **54**(8), 1481–1489 (2006). doi:10.1109/TCOMM.2006.878841
14. Q. Wang, M. Rupp, Analytical link performance evaluation of LTE downlink with carrier frequency offset, in *Conference Record of the 45th Asilomar Conference on Signals, Systems and Computers, 2011 (Asilomar-2011)*, Pacific Grove, USA, Nov. 2011
15. G. Caire, G. Taricco, E. Biglieri, Bit-interleaved coded modulation. IEEE Trans. Inf. Theory **44**(3), 927–946 (1998)
16. Technical Specification Group Radio Access Network, E-UTRA; physical layer procedures, 3GPP, Technical Report TS 36.211 Version 9.2.0, June 2010
17. G. Caire, G. Taricco, E. Biglieri, Capacity of bit-interleaved channels. Electron. Lett. **32**(12), 1060–1061 (1996)
18. C. Mehlführer, J.C. Ikuno, M. Šimko, S. Schwarz, M. Wrulich, M. Rupp, The Vienna LTE simulators—enabling reproducibility in wireless communications research. EURASIP J. Adv. Signal Process. (2011)
19. Members of ITU, Recommendation ITU-R M.1225: Guidelines for evaluation of radio transmission technologies for IMT-2000, International Telecommunication Union (ITU), Technical Report, 1997

Chapter 3
Symbol Detection in High Speed Channels

Ronald Nissel

3.1 Introduction

Mobile wireless communication channels are characterized by time-varying multi-path propagation [1], that is, due to multiple scatterers, the electromagnetic signal can propagate along several different paths which causes frequency-selectivity and time-selectivity. Orthogonal Frequency Division Multiplexing (OFDM) as being utilized in Long Term Evolution (LTE) and Long Term Evolution-Advanced (LTE-A) down-link transmissions can efficiently deal with frequency-selective channels caused by multi-path delays by inserting a so-called Cyclic Prefix (CP), see also Chap. 1 for more details. However, in high mobility scenarios, as they appear for example in car/train to infrastructure communication, the channel can become so time-selective, that it changes significantly within one OFDM symbol. In this case the subcarriers are no longer orthogonal and lead to Inter-Carrier Interference (ICI) which can only be combated by sophisticated and costly algorithms [2].

In a first step, we consider ICI as an additional Gaussian noise term which allows us to derive a closed-form expression for the Bit Error Probability (BEP). Such closed-form expressions are useful because they provide a much deeper insight into the system and allow efficient performance optimization with respect to specific parameters such as the pilot symbol power.

In order to combat ICI we need an accurate estimate of the time-variant channel that also requires more knowledge of the interfering terms. In principle, we can distinguish between Channel Estimation (CE) methods utilizing either one OFDM symbol [3] or multiple OFDM symbols [4–6]. The first method usually employs a basis expansion model to reduce the number of unknown variables and requires a clustered pilot symbol structure, which is not compliant with most standards. The second method, on the other hand, initially ignores ICI and interpolates the estimated

R. Nissel (✉)
Institute of Telecommunications, TU Wien, Vienna, Austria
e-mail: rnissel@nt.tuwien.ac.at

M. Rupp et al., *The Vienna LTE-Advanced Simulators*,
Signals and Communication Technology, DOI 10.1007/978-981-10-0617-3_3

time-averaged channel impulse responds of several OFDM symbols, e.g., linearly [4], by Least Squares (LS) polynomial fitting [5] or by Least Squares (LS) discrete prolate spheroidal fitting [6]. Once the channel is estimated, the ICI at pilot positions can be reduced in order to increase the accuracy of an iterative channel estimation. We follow such second approach, based on multiple OFDM symbols. In contrast to [4–6] we interpolate the estimated channel of several OFDM symbols according to the Minimum Mean Square Error (MMSE) criteria, thus obtaining the MMSE channel estimation of the sampled time-variant transfer function. MMSE channel estimation was investigated for example in [7] and [8] but they considered only one OFDM symbol and their estimation process includes ICI, making it rather complex. We, on the other hand, treat ICI as an additional noise term which is then stepwise reduced by our iterative ICI mitigation technique that combines channel estimation, equalization and ICI cancellation. Iterative channel estimation was, e.g., discussed in [9] where the authors model the channel variations by a polynomial basis expansion and use the estimated data symbols to estimate the ICI. We, however, use the estimated data symbols only to cancel ICI and, in the final iteration step, as pilot symbols. Many authors [10–12] proposed low-complexity equalizers which exploit the underlying structure of the ICI. However, the main focus of this chapter is channel estimation so that we employ an ordinary full block MMSE equalizer [13] to combat ICI.

By employing the Vienna Wireless Testbed (for more details, see Chap. 9), we show the applicability of our ICI mitigation technique in real world physical channels. Testbeds which support mobile receivers typically place the measurement equipment in a car [14–16]. However, such method offers only a limited velocity and lacks repeatability and controllability, which constitutes the main problem for a fair comparison of different transmission techniques. To avoid these drawbacks we developed a transmission system [17, 18] in which the receive antenna rotates around a central pivot. Compared to a linear guide we avoid the repeated acceleration and deceleration step, so that even high velocities can be achieved. Our setup represents a scenario in which the rotational speed is low relative to the transmission time, so that the antenna moves approximately linear during the transmission. Assume for example a velocity of 100 km/h (4.4 revolutions per second for a 1 m arm). Then, for a transmission time of 1 ms, the antenna moves only by $4.4\,\text{rps} \times 1\,\text{ms} \times 360° = 1.6°$ which is close enough to a linear movement. To the best of our knowledge, such high velocity measurement setup is unique in the world. The main advantages are repeatability, controllability and cost efficiency. The main disadvantage, on the other hand, is the indoor location of our receiver, so that our setup does not necessarily accurately reflect the channel characteristics of a typical high velocity application.

3.2 OFDM in High Speed Channels

In this section, we model the effect of time-variant channels [1] on OFDM transmissions. In particular, we will derive equations to specify the ICI power and show how it depends on the subcarrier index.

We start our discussion with Eq. (1.22) of Chap. 1:

$$\mathbf{r}_n = \text{Diag}\left(\mathbf{h}_n\right)\mathbf{x}_n + \mathbf{y}_n^{\text{ICI}} + \mathbf{v}_n \tag{3.1}$$

$$r_{n,k} = H_{n,k}x_{n,k} + y_{n,k}^{\text{ICI}} + v_{n,k}, \tag{3.2}$$

which shows the linear operation of transmissions including the Fast Fourier Transform (FFT) operation at the receiver end. We can specify the received signal power given by $P_{\text{S}_{\text{RX}}} = \mathbb{E}\{|H_{n,k}x_{n,k}|^2\}$, ICI power by $P_{n,k}^{\text{ICI}} = \mathbb{E}\{|y_{n,k}^{\text{ICI}}|^2\}$and noise power by $P_{\text{V}} = \mathbb{E}\{|v_{n,k}|^2\}$. In Sect. 3.3, we assume that the ICI terms $y_{n,k}^{\text{ICI}}$ are uncorrelated and Gaussian distributed, even though the real distribution is given by a weighted Gaussian mixture due to random data symbols [19].

To characterize the ICI power we need a statistical description of OFDM matrix $\mathbf{D}_n$, Eq. (1.21). However, in order to describe the correlation of all elements of the two-dimensional channel transfer function $C_{m,k}$ in Eq. (1.21), we have to perform vectorization, making the description by a large correlation matrix feasible. We structure the correlation matrix by sorting the vectorized channel elements accordingly, so that the correlation matrix can be found in a straightforward manner, see for example Eq. (3.56). Thus, we define a new vector $\mathbf{c}_n \in \mathbb{C}^{N_\text{c}N_\text{FFT}\times 1}$ in which we stack all elements of the sampled time-variant transfer function, relevant to OFDM symbol n, according to:

$$\mathbf{c}_n = \begin{bmatrix} C_{nN_\text{s}+N_\text{cp},0} \\ \vdots \\ C_{nN_\text{s}+N_\text{cp},N_\text{c}-1} \\ C_{nN_\text{s}+N_\text{cp}+1,0} \\ \vdots \\ C_{nN_\text{s}+N_\text{cp}+N_\text{FFT}-1,N_\text{c}-1} \end{bmatrix}. \tag{3.3}$$

We can then find a sparse matrix $\tilde{\mathbf{W}}_k \in \mathbb{C}^{N_\text{c}\times N_\text{c}N_\text{FFT}}$ ($100\frac{N_\text{c}-1}{N_\text{c}}$ percent of the elements are zero), so that OFDM matrix $\mathbf{D}_n$, Eq. (1.21), can be rewritten in matrix notation as:

$$\mathbf{D}_n = \begin{bmatrix} \left(\tilde{\mathbf{W}}_0\mathbf{c}_n\right)^\text{T} \\ \vdots \\ \left(\tilde{\mathbf{W}}_{N_\text{c}-1}\mathbf{c}_n\right)^\text{T} \end{bmatrix}, \tag{3.4}$$

whereas we obtain $\tilde{\mathbf{W}}_k$ $(k = 0, 1, 2 \ldots, N_\text{c} - 1)$ by comparing Eqs. (3.3) and (3.4) to (1.21), resulting in:

$$\begin{aligned}\tilde{\mathbf{W}}_k =& \frac{1}{N_\text{FFT}} \exp\left\{-\frac{\text{j}2\pi}{N_\text{FFT}}\left(\mathbf{1}_{1\times N_\text{FFT}} \otimes \begin{bmatrix} k & 0 & 0 \\ 0 & \ddots & 0 \\ 0 & 0 & k-(N_\text{c}-1) \end{bmatrix}\right)\right. \\ &\left.\circ \left(\left[0 \cdots N_\text{FFT}-1\right] \otimes \mathbf{I}_{N_\text{c}}\right)\right\} \circ \left(\mathbb{1}_{1\times N_\text{FFT}} \otimes \text{diag}\left(\mathbb{1}_{1\times N_\text{c}}\right)\right).\end{aligned} \tag{3.5}$$

Here, the $\exp\{\cdot\}$ operator is applied at each element of the matrix, the operations $\otimes$ and $\circ$ are the Kronecker product, respectively the Hadamard (point-wise) product, $\mathbf{I}_{N_c}$ denotes the identity matrix of size N_c and $\mathbb{1}_{1\times N_{\text{FFT}}}$ an all-one vector of size $1 \times N_{\text{FFT}}$. Let us further split the matrix $\tilde{\mathbf{W}}_k$ into a signal part $\tilde{\mathbf{W}}_k^{\text{S}}$ and an ICI part $\tilde{\mathbf{W}}_k^{\text{ICI}}$:

$$\tilde{\mathbf{W}}_k = \tilde{\mathbf{W}}_k^{\text{S}} + \tilde{\mathbf{W}}_k^{\text{ICI}}. \tag{3.6}$$

Matrix $\tilde{\mathbf{W}}_k^{\text{ICI}}$ consists of nearly the same elements as $\tilde{\mathbf{W}}_k$, with the difference that the kth row is set to zero. Therefore, $\tilde{\mathbf{W}}_k^{\text{ICI}}$ describes only the ICI part of our system, that is, $y_{n,k}^{\text{ICI}} = \left(\tilde{\mathbf{W}}_k^{\text{ICI}}\mathbf{c}_n\right)^{\text{T}}\mathbf{x}_n$. Similar, $\tilde{\mathbf{W}}_k^{\text{S}}$ consists only of the signal part of our transmission system, that is, only the kth row of $\tilde{\mathbf{W}}_k$, while all other elements are zero. In particular, we have $H_{n,k}x_{n,k} = \left(\tilde{\mathbf{W}}_k^{\text{S}}\mathbf{c}_n\right)^{\text{T}}\mathbf{x}_n$. The signal power $P_{n,k}^{\text{S}_{\text{RX}}}$ and the ICI power $P_{n,k}^{\text{ICI}}$ are then given by:

$$P_{n,k}^{\text{S}_{\text{RX}}} = \operatorname{tr}\left\{\tilde{\mathbf{W}}_k^{\text{S}}\mathbf{R}_{\mathbf{c}_n}\left(\tilde{\mathbf{W}}_k^{\text{S}}\right)^{\text{H}}\right\} \tag{3.7}$$

$$P_{n,k}^{\text{ICI}} = \operatorname{tr}\left\{\tilde{\mathbf{W}}_k^{\text{ICI}}\mathbf{R}_{\mathbf{c}_n}\left(\tilde{\mathbf{W}}_k^{\text{ICI}}\right)^{\text{H}}\right\}, \tag{3.8}$$

whereas $\mathbf{R}_{\mathbf{c}_n} = \mathbb{E}\{\mathbf{c}_n\mathbf{c}_n^{\text{H}}\} \in \mathbb{C}^{N_c N_{\text{FFT}} \times N_c N_{\text{FFT}}}$ denotes the correlation matrix (see Sect. 3.5 for an example). For a Wide-Sense Stationary Uncorrelated Scattering (WSSUS) channel, the signal power is independent of the subcarrier position k and the time position n, that is, $P_{n,k}^{\text{S}_{\text{RX}}} = P_{\text{S}_{\text{RX}}}$. On the other hand, the ICI power depends on k if the number of samples N_{FFT} is larger than N_c, due to aliasing. For a given time autocorrelation function and the limit case of $N_{\text{FFT}} \to \infty$, an analytical expression for the signal power can be found by [20]:

$$P_{\text{S}_{\text{RX}}}^{\text{uniform}} = \frac{\cos(2\pi\nu_{\max}T) + 2\pi\nu_{\max}T\,\text{Si}(2\pi\nu_{\max}T) - 1}{2(\pi\nu_{\max}T)^2}, \tag{3.9a}$$

$$P_{\text{S}_{\text{RX}}}^{\text{Jakes}} = {}_1F_2\left(\frac{1}{2}; \frac{3}{2}, 2; -(\pi\nu_{\max}T)^2\right). \tag{3.9b}$$

A uniform distributed Doppler spectral density is assumed for the signal power $P_{\text{S}_{\text{RX}}}^{\text{uniform}}$ and a Jakes Doppler spectrum for $P_{\text{S}_{\text{RX}}}^{\text{Jakes}}$. The functions $\text{Si}(\cdot)$ and ${}_1F_2(\cdot)$ are the sine integral function respectively the generalized hypergeometric function. Note that the signal power depends only on the normalized maximum Doppler shift $\nu_{\max}T = \frac{\nu_{\max}}{\Delta f}$. For infinitely many subcarriers, the law of conservation of energy ($P_{\text{S}_{\text{RX}}} + P_{\text{ICI}} = 1$) can be applied, so that the ICI power is finally given as:

$$P_{\text{ICI}}^{\text{uniform}} = P_{\text{S}_{\text{RX}}}^{\text{uniform}} - 1, \tag{3.10a}$$

$$P_{\text{ICI}}^{\text{Jakes}} = P_{\text{S}_{\text{RX}}}^{\text{Jakes}} - 1. \tag{3.10b}$$

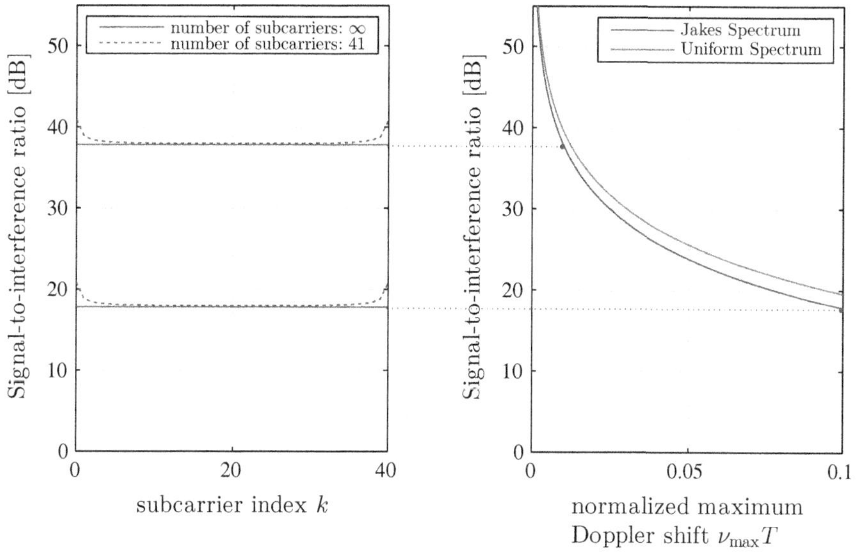

Fig. 3.1 Signal-to-Interference Ratio as a function of subcarrier index respectively Doppler shift: SIR is relatively high so that in most cases ICI can be neglected, SIR of the first and last subcarrier is approximately 3 dB higher than the center values

The right part of Fig. 3.1 shows the Signal-to-Interference Ratio (SIR) for infinitely many subcarriers (Eqs. 3.9 and 3.10). This ratio is usually very high so that in many cases the ICI can be neglected compared to the noise. Consider for example LTE (Δf =15 kHz) at 2.5 GHz. For a SIR smaller than 20 dB, the velocity has to be larger than 500 km/h ($\nu_{\max} T = 0.08$). The left figure compares the case of finite (Eqs. 3.7 and 3.8, $N_c = 41$, $N_{\text{FFT}} = 410$) with infinitely many subcarriers. Since the first and last subcarriers have interferers only at one side, the SIR is approximately 3 dB higher compared to the closed form solution. Furthermore, we observe that the SIR ratio at the middle subcarrier coincide with the closed form solution because only the few nearest neighboring subcarriers have a significant effect on the ICI.

3.3 Closed-Form BEP Expression for One-Tap Equalizer

This section is based on [18, 21] and shows how we can derive closed-form solutions for the BEP in case of one-tap equalizer and pilot-symbol-aided channel estimation. Here, the ICI is considered as an additional noise term and approximated by a Gaussian random variable while in Sect. 3.4 we will investigate more evolved data symbol estimation methods. Similar results are provided in for filter-bank multicarrier modulation [22]. Closed-form solutions are investigated in [23–25], extended forms to describe ICI in [21].

The ultimate goal of every transmission system is to recover the transmitted data symbols $x_{n,k}$. Dividing Eq. (3.2) by the estimated channel $\hat{H}_{n,k}$ leads to the zero forcing equalization and delivers an LS estimate of the transmitted data symbol $x_{n,k}$:

$$\hat{x}_{n,k} = \frac{r_{n,k}}{\hat{H}_{n,k}}. \tag{3.11}$$

For perfect channel knowledge ($\hat{H}_{n,k} = H_{n,k}$), the zero forcing equalizer corresponds to the Maximum Likelihood (ML) detection. However, for pilot-symbol-aided channel estimation such equalization is no longer optimal in the ML sense [26], but due to its simplicity still a reasonable choice. The required channel estimate in (3.11) is found by interpolation (weighted average), via the weights $\mathbf{a}_{n,k}$, of known LS estimates at pilot positions:

$$\hat{H}_{n,k} = \sum_{\{n_p,k_p\}\in\mathscr{P}} \left(\mathbf{a}^*_{n,k}\right)_{\{n_p,k_p\}} \frac{r_{n_p,k_p}}{x_{n_p,k_p}}. \tag{3.12}$$

Set $\mathscr{P}$ is a collection of the two-dimensional (2D) pilot position indexes whereas the number of pilot symbols is given by its cardinality $|\mathscr{P}| = N_\mathrm{P}$. Rewriting Eq. (3.12) in vector notation leads to:

$$\hat{H}_{n,k} = \mathbf{a}_{n,k}^H \hat{\mathbf{h}}_{\mathscr{P}}^{\mathrm{LS}}, \tag{3.13}$$

where the vector $\hat{\mathbf{h}}_{\mathscr{P}}^{\mathrm{LS}} \in \mathbb{C}^{N_\mathrm{P}\times 1}$ consists of the vectorized LS estimates at pilot positions and the vector weight function $\mathbf{a}_{n,k} \in \mathbb{C}^{N_\mathrm{P}\times 1}$ is defined by the interpolation method (for example MMSE, linear, spline). We also assume that each pilot symbol has unit power. Figure 3.2 shows a 4-Quadrature Amplitude Modulation (QAM) and

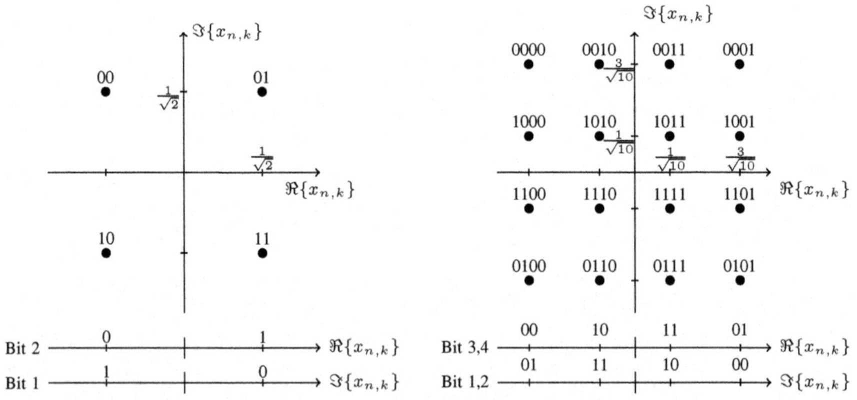

Fig. 3.2 4-QAM (*left*) and 16-QAM (*right*) applying *Gray code*

a 16-QAM signal constellation, that is, the set from which the data symbols $x_{n,k}$ are chosen. The bit mapping follows Gray-coding. The one-tap equalizer in Eq. (3.11) then leads to the following BEP for 4-QAM:

$$\begin{aligned}\mathrm{BEP}_{n,k}^{4\mathrm{QAM}}(\mathbf{a}_{n,k}) =& \frac{1}{2}\Pr\left(\Re\left\{\frac{r_{n,k}}{\hat{H}_{n,k}}\right\} < 0 \middle| x_{n,k} = \frac{1+\mathrm{j}}{\sqrt{2}}\right)\\ &+ \frac{1}{2}\Pr\left(\Re\left\{\frac{r_{n,k}}{\hat{H}_{n,k}}\right\} < 0 \middle| x_{n,k} = \frac{1-\mathrm{j}}{\sqrt{2}}\right).\end{aligned} \tag{3.14}$$

and for 16-QAM:

$$\begin{aligned}&\mathrm{BEP}_{n,k}^{16\mathrm{QAM}}(\mathbf{a}_{n,k})\\ &= \frac{1}{16}\sum_{\substack{q_r\in\\ \{1,3\}}}\sum_{\substack{q_i\in\{-3,\\ -1,1,3\}}}\Pr\left(\Re\left\{\frac{r_{n,k}}{\hat{H}_{n,k}}\right\} < 0 \middle| x_{n,k} = \frac{q_r+\mathrm{j}q_i}{\sqrt{10}}\right)\\ &+ \frac{1}{16}\sum_{\substack{q_i\in\{-3,\\ -1,1,3\}}}\left(1-\Pr\left(\frac{-2}{\sqrt{10}} < \Re\left\{\frac{r_{n,k}}{\hat{H}_{n,k}}\right\} < \frac{2}{\sqrt{10}} \middle| x_{n,k} = \frac{1+\mathrm{j}q_i}{\sqrt{10}}\right)\right)\\ &+ \frac{1}{16}\sum_{\substack{q_i\in\{-3,\\ -1,1,3\}}}\Pr\left(\frac{-2}{\sqrt{10}} < \Re\left\{\frac{r_{n,k}}{\hat{H}_{n,k}}\right\} < \frac{2}{\sqrt{10}} \middle| x_{n,k} = \frac{3+\mathrm{j}q_i}{\sqrt{10}}\right),\end{aligned} \tag{3.15}$$

In order to find closed-form expressions for the BEP, we thus have to find close-form expressions for the Cumulative Distribution Function (CDF) of the complex Gaussian ratio $\frac{r_{n,k}}{\hat{H}_{n,k}}$, which is provided by the following theorem based on:

Theorem 3.1 *Let* $H \sim \mathcal{N}_{\mathbb{C}}\left(0, P_{S_{\mathrm{RX}}}\right)$, $\hat{\mathbf{h}}_{\mathcal{P}}^{\mathrm{LS}} \sim \mathcal{N}_{\mathbb{C}}\left(\mathbf{0}, \mathbf{R}_{\hat{\mathbf{h}}_{\mathcal{P}}^{\mathrm{LS}}}\right)$ *and uncorrelated* $v \sim \mathcal{N}_{\mathbb{C}}\left(0, P_{\mathrm{V}}\right)$ *be zero mean complex-valued Gaussian random variables and vectors. Assume further that the transmission system model is given by* $r = Hx + v$, *that the channel is estimated by pilot symbols, that is,* $\hat{H} = \mathbf{a}^H\hat{\mathbf{h}}_{\mathcal{P}}^{\mathrm{LS}}$, *with* $\mathbf{a}$ *being an arbitrary complex-valued vector defined by the interpolation scheme, and that the cross-correlation vector* $\mathbb{E}\{\hat{\mathbf{h}}_{\mathcal{P}}^{\mathrm{LS}}H^*\} = \mathbf{r}_{\hat{\mathbf{h}}_{\mathcal{P}}^{\mathrm{LS}},H}$ *is given. Conditioned on* x, *the probability that the real part of the complex-valued Gaussian ratio* $r/\hat{H}$ *is smaller than a certain value* z_{R} *and at the same time the imaginary part smaller than* z_{I}, *reads:*

$$\Pr\left(\Re\left\{\frac{r}{\hat{H}}\right\} < z_\mathrm{R} \wedge \Im\left\{\frac{r}{\hat{H}}\right\} < z_\mathrm{I} \,\middle|\, x\right) = \frac{1}{4}$$

$$+\frac{\left(z_\mathrm{R}-\Re\left\{\frac{\alpha}{\beta}x\right\}\right)\left(2\tan^{-1}\left(\frac{z_\mathrm{I}-\Im\left(\frac{\alpha}{\beta}x\right)}{\sqrt{\left(z_\mathrm{R}-\Re\{\frac{\alpha}{\beta}x\}\right)^2+\gamma}}\right)+\pi\right)}{4\pi\sqrt{\left(z_\mathrm{R}-\Re\{\frac{\alpha}{\beta}x\}\right)^2+\gamma}}$$

$$+\frac{\left(z_\mathrm{I}-\Im\left\{\frac{\alpha}{\beta}x\right\}\right)\left(2\tan^{-1}\left(\frac{z_\mathrm{R}-\Re\left(\frac{\alpha}{\beta}x\right)}{\sqrt{\left(z_\mathrm{I}-\Im\{\frac{\alpha}{\beta}x\}\right)^2+\gamma}}\right)+\pi\right)}{4\pi\sqrt{\left(z_\mathrm{I}-\Im\{\frac{\alpha}{\beta}x\}\right)^2+\gamma}}, \tag{3.16}$$

with

$$\alpha = \mathbf{a}^H \mathbf{r}_{\hat{\mathbf{h}}_{\mathcal{P}}^{\mathrm{LS}},H}, \tag{3.17}$$

$$\beta = \mathbf{a}^H \mathbf{R}_{\hat{\mathbf{h}}_{\mathcal{P}}^{\mathrm{LS}}} \mathbf{a}, \tag{3.18}$$

$$\gamma = \frac{P_\mathrm{V} + P_{\mathrm{S_{RX}}} |x|^2}{\beta} - \left|\frac{\alpha}{\beta}x\right|^2. \tag{3.19}$$

Of particular interest is the case when only the projection onto one axis, say the real axis, matters. Then z_I approaches infinity, so that the $\tan^{-1}$ terms vanish and we end up with a very compact expression, see Eq. (3.26). Theorem 3.1 delivers the probability for arbitrary linear interpolation methods $\mathbf{a}$.

Let us now consider the special case of MMSE channel estimation [27]:

$$\mathbf{a}^{\mathrm{MMSE}} = \mathbf{R}_{\hat{\mathbf{h}}_{\mathcal{P}}^{\mathrm{LS}}}^{-1} \mathbf{r}_{\hat{\mathbf{h}}_{\mathcal{P}}^{\mathrm{LS}},H}, \tag{3.20}$$

which minimizes the Mean Square Error (MSE) of our channel estimation:

$$\mathbf{a}^{\mathrm{MMSE}} = \arg\min_{\mathbf{a}} \mathrm{MSE} = \arg\min_{\mathbf{a}} \mathbb{E}\{|H - \mathbf{a}^H \hat{\mathbf{h}}_{\mathcal{P}}^{\mathrm{LS}}|^2\}. \tag{3.21}$$

The variables α and β in Eqs. (3.17) and (3.18) then transform to:

$$\alpha^{\mathrm{MMSE}} = \beta^{\mathrm{MMSE}} = \mathbf{a}^H \mathbf{R}_{\hat{\mathbf{h}}_{\mathcal{P}}^{\mathrm{LS}}} \mathbf{a} = P_{\mathrm{S_{RX}}} - \mathrm{MSE}, \tag{3.22}$$

and γ in Eq. (3.19) becomes:

$$\gamma^{\mathrm{MMSE}} = \frac{P_\mathrm{V} + \mathrm{MSE}\,|x|^2}{P_{\mathrm{S_{RX}}} - \mathrm{MSE}}. \tag{3.23}$$

For perfect channel knowledge, the MSE approaches zero and therefore:

$$\alpha^{\text{perfect}} = \beta^{\text{perfect}} = P_{S_{\text{RX}}} \tag{3.24}$$

and

$$\gamma^{\text{perfect}} = \frac{P_{\text{V}}}{P_{S_{\text{RX}}}} = \frac{1}{\text{SNR}}. \tag{3.25}$$

Note that for MMSE channel estimation, the probability in Theorem 3.1 behaves the same way as for perfect channel knowledge where the signal power is lowered by the MSE and the noise power increased by MSE $|x|^2$, as shown by Eq. (3.23) and the fact that α/β becomes one, so that Theorem 3.1 only depends on γ.

The BEP equations in (3.14) and (3.15) require the projection onto the real axis, that is, $z_{\text{I}} \to \infty$, so that Theorem 3.1 transforms to:

$$\begin{aligned}\Pr\left(\Re\left\{\frac{r_{n,k}}{\hat{H}_{n,k}}\right\} < \frac{a}{\sqrt{b}}\left|\frac{q_r + \mathrm{j}q_i}{\sqrt{b}}\right.\right) = \frac{1}{2} - \frac{1}{2}&\Big[q_r\Re\{\alpha_{n,k}\} - q_i\Im\{\alpha_{n,k}\} - a\beta_{n,k}\Big] \\ &\times\bigg[\Big[(q_r^2 + q_i^2)P_{n,k}^{S_{\text{RX}}} + b(P_{n,k}^{\text{ICI}} + P_{\text{V}}) \\ &\quad - 2a\left(q_r\Re\{\alpha_{n,k}\} - q_i\Im\{\alpha_{n,k}\}\right) + a^2\beta_{n,k}\Big]\beta_{n,k} \\ &\quad - \left(q_i\Re\{\alpha_{n,k}\} + q_r\Im\{\alpha_{n,k}\}\right)^2\bigg]^{-\frac{1}{2}}\end{aligned} \tag{3.26}$$

with

$$\alpha_{n,k} = \mathbf{r}^H_{\hat{\mathbf{h}}^{\text{LS}}_{\mathcal{P}},H_{n,k}}\mathbf{a}_{n,k} \tag{3.27}$$

$$\beta_{n,k} = \mathbf{a}^H_{n,k}\mathbf{R}_{\hat{\mathbf{h}}^{\text{LS}}_{\mathcal{P}}}\mathbf{a}_{n,k}, \tag{3.28}$$

Equation (3.26) delivers immediately the required probabilities of Eqs. (3.14) and (3.15) so that we can find closed-form expressions. However, because the resulting Equations are quite lengthy we consider, without loss of generality, only the special case of of $\Im\{\mathbf{r}^H_{\hat{\mathbf{h}}^{\text{LS}}_{\mathcal{P}},H_{n,k}}\mathbf{a}_{n,k}\} = 0$ (interpolation compensates average phase shifts) and $\Re\{\mathbf{r}^H_{\hat{\mathbf{h}}^{\text{LS}}_{\mathcal{P}},H_{n,k}}\mathbf{a}_{n,k}\} > 0$ (interpolation is better than a random guess), the BEP for 4-QAM can then be written as:

$$\widetilde{\text{BEP}}^{\text{4QAM}}_{n,k}(\mathbf{a}_{n,k}) = \frac{1}{2} - \frac{1}{2\sqrt{2(P_{n,k}^{S_{\text{RX}}} + P_{n,k}^{\text{ICI}} + P_{\text{V}})\dfrac{\mathbf{a}^H_{n,k}\mathbf{R}_{\hat{\mathbf{h}}^{\text{LS}}_{\mathcal{P}}}\mathbf{a}_{n,k}}{\left(\mathbf{r}^H_{\hat{\mathbf{h}}^{\text{LS}}_{\mathcal{P}},H_{n,k}}\mathbf{a}_{n,k}\right)^2} - 1}}. \tag{3.29}$$

One advantage of having close-form expressions is the fact that they can be used for performance optimization. Let us for example minimize the BEP. In order to minimize Eq. (3.29), the generalized Rayleigh quotient inside the square root has to be minimized, or equivalently, the inverse quotient $\frac{\mathbf{a}_{n,k}^H \mathbf{r}_{\hat{\mathbf{h}}_{\mathcal{P}}^{\mathrm{LS}},H_{n,k}} \mathbf{r}_{\hat{\mathbf{h}}_{\mathcal{P}}^{\mathrm{LS}},H_{n,k}}^H \mathbf{a}_{n,k}}{\mathbf{a}_{n,k}^H \mathbf{R}_{\hat{\mathbf{h}}_{\mathcal{P}}^{\mathrm{LS}}} \mathbf{a}_{n,k}}$ maximized. Since the matrices in the nominator and denominator are Hermitian, this maximization becomes a generalized eigenvalue problem

$$\mathbf{r}_{\hat{\mathbf{h}}_{\mathcal{P}}^{\mathrm{LS}},H_{n,k}} \mathbf{r}_{\hat{\mathbf{h}}_{\mathcal{P}}^{\mathrm{LS}},H_{n,k}}^H \mathbf{a}_{n,k} = \lambda\, \mathbf{R}_{\hat{\mathbf{h}}_{\mathcal{P}}^{\mathrm{LS}}} \mathbf{a}_{n,k}. \tag{3.30}$$

The unique solution (except irrelevant scaling) for Eq. (3.30) can be found by inserting the MMSE estimation, see Eq. (3.20). The optimal interpolation for 4-QAM therefore becomes:

$$\begin{aligned} \mathbf{a}_{n,k}^{\mathrm{4QAM,minBEP}} &= \arg\min_{\mathbf{a}_{n,k}} \mathrm{BEP}_{n,k}^{\mathrm{4QAM}}(\mathbf{a}_{n,k}) \\ &= \mathbf{R}_{\hat{\mathbf{h}}_{\mathcal{P}}^{\mathrm{LS}}}^{-1} \mathbf{r}_{\hat{\mathbf{h}}_{\mathcal{P}}^{\mathrm{LS}},H_{n,k}} \\ &= \mathbf{a}_{n,k}^{\mathrm{MMSE}}. \end{aligned} \tag{3.31}$$

Following along the same lines, the optimal interpolation for 16-QAM can be found as:

$$\begin{aligned} \mathbf{a}_{n,k}^{\mathrm{16QAM,minBEP}} &= \arg\min_{\mathbf{a}_{n,k}} \mathrm{BEP}_{n,k}^{\mathrm{16QAM}}(\mathbf{a}_{n,k}) \\ &= \kappa_{n,k} \mathbf{R}_{\hat{\mathbf{h}}_{\mathcal{P}}^{\mathrm{LS}}}^{-1} \mathbf{r}_{\hat{\mathbf{h}}_{\mathcal{P}}^{\mathrm{LS}},H_{n,k}}, \end{aligned} \tag{3.32}$$

whereas for low signal-to-noise ratios the real valued scaling factor $\kappa_{n,k}$ is larger then one and for large ratios it has to be smaller. However, numerical evaluations have shown that Eq. (3.32) improves the $\mathrm{BEP}_{n,k}^{\mathrm{16QAM}}$ only by less than 1 % so that the differences between $\mathbf{a}_{n,k}^{\mathrm{16QAM,minBEP}}$ and $\mathbf{a}_{n,k}^{\mathrm{MMSE}}$ can be neglected.

As already mentioned, the BEP in case of perfect channel knowledge can also be calculated via Eq. (3.29) by selecting MMSE interpolation and setting the MSE to zero, which leads to:

$$\mathrm{BEP}_{n,k}^{\mathrm{4QAM,perfect}} = \frac{1}{2} - \frac{1}{2\sqrt{1 + 2\frac{P_{n,k}^{\mathrm{ICI}} + P_{\mathrm{V}}}{P_{n,k}^{\mathrm{S RX}}}}}. \tag{3.33}$$

3.4 Time-Variant Channel Estimation and ICI Mitigation

This section follows the idea of [28]: by assuming a continuous transmit signal, a discrete receiver structure and a sufficient long CP, the sampled time-variant transfer function straightforwardly describes the OFDM transmission, so that an accurate estimation of the channel is possibly. Compared with the usually considered impulse response, which includes frequency components not relevant for data transmission, the transfer function allows to ignore these irrelevant frequency components. By employing a full block MMSE equalizer we are then able to combat the effects of high mobility wireless channels.

Our goal is to estimate the sampled time-variant transfer function $C_{m,k}$ using known pilot symbols $\mathbf{x}_{\mathscr{P}} \in \mathbb{C}^{N_\mathrm{P} \times 1}$. The pilot symbol pattern can be arbitrary, e.g., diamond shaped, as defined in the LTE standard (see Fig. 1.2). Similar to Eq. (3.3), we stack all elements of the channel $C_{m,k}$ in a large vector $\mathbf{c} \in \mathbb{C}^{N_\mathrm{c} N_\mathrm{s} N_\mathrm{f} \times 1}$:

$$\mathbf{c} = \left[C_{0,0} \cdots C_{0,N_\mathrm{c}-1}\; C_{1,0} \cdots C_{N_\mathrm{s}N_\mathrm{f}-1,N_\mathrm{c}-1}\right]^\mathrm{T} \tag{3.34}$$

The assumption of Rayleigh fading implies that $\mathbf{c}$ is jointly complex-valued Gaussian, that is, $\mathbf{c} \sim \mathscr{N}_\mathbb{C}(0, \mathbf{R_c})$. The channel is normalized so that the diagonal elements of $\mathbf{R_c}$ become one.

According to Eqs. (3.1) and (3.2), an LS estimation of the channel (Eq. 3.37) delivers only the piecewise mean of the channel, $H_{n,k}$, corrupted by ICI and noise. The relationship between the piecewise mean channel $\mathbf{h} \in \mathbb{C}^{N_\mathrm{c} N_\mathrm{f} \times 1}$ ($\mathbf{h} = \left[\mathbf{h}_1^\mathrm{T} \cdots \mathbf{h}_{N_\mathrm{f}}^\mathrm{T}\right]^\mathrm{T}$) and the channel vector $\mathbf{c} \in \mathbb{C}^{N_\mathrm{c} N_\mathrm{s} N_\mathrm{f} \times 1}$ can be described by a sparse matrix $\mathbf{M} \in \mathbb{R}^{N_\mathrm{c} N_\mathrm{f} \times N_\mathrm{c} N_\mathrm{s} N_\mathrm{f}}$ ($100[1 - \frac{N}{N_\mathrm{c} N_\mathrm{s} N_\mathrm{f}}]$ percent of the elements are zero):

$$\mathbf{h} = \mathbf{M}\mathbf{c}. \tag{3.35}$$

with $\mathbf{M}$ given by:

$$\mathbf{M} = \left(\mathbf{I}_{N_\mathrm{f}} \otimes \left[\mathbf{0}_{1\times N_\mathrm{cp}}\; \mathbb{1}_{1\times N}\right]\frac{1}{N}\right) \otimes \mathbf{I}_{N_\mathrm{c}}. \tag{3.36}$$

The received data symbols at pilot positions $\mathscr{P}$ are stacked in a vector $\mathbf{r}_{\mathscr{P}} \in \mathbb{C}^{N_\mathrm{P} \times 1}$ and element wise divided by the pilot symbols $\mathbf{x}_{\mathscr{P}}$, so that the LS estimate of the piecewise mean channel $\hat{\mathbf{h}}_{\mathscr{P}}^\mathrm{LS} \in \mathbb{C}^{N_\mathrm{P} \times 1}$ at pilot positions becomes:

$$\hat{\mathbf{h}}_{\mathscr{P}}^\mathrm{LS} = \mathrm{Diag}\left(\mathbf{x}_{\mathscr{P}}\right)^{-1} \mathbf{r}_{\mathscr{P}}. \tag{3.37}$$

An estimation of the channel vector $\mathbf{c}$ can then be found by:

$$\hat{\mathbf{c}} = \tilde{\mathbf{A}}\hat{\mathbf{h}}_{\mathscr{P}}^\mathrm{LS}. \tag{3.38}$$

Matrix $\tilde{\mathbf{A}} \in \mathbb{C}^{N_\mathrm{c} N_\mathrm{s} N_\mathrm{f} \times N_\mathrm{P}}$ represents a general description of a linear estimation and includes, for example, basis expansion models [6] or the MMSE solution (Eq. 3.44). Note that the linear estimator in Sect. 3.3, $\mathbf{a}_{n,k}$, represents only the estimation of the piecewise mean channel and is written as vector function instead of a matrix. Figure 3.3 illustrates Eq. (3.38): estimates of the piecewise mean channel at pilot positions are interpolated/extrapolated, delivering an estimate of the time variant transfer function $C_{m,k}$.

Correlation matrix $\mathbf{R}_{\hat{\mathbf{c}}} = \mathbb{E}\{\hat{\mathbf{c}}\hat{\mathbf{c}}^\mathrm{H}\} \in \mathbb{C}^{N_\mathrm{c} N_\mathrm{s} N_\mathrm{f} \times N_\mathrm{c} N_\mathrm{s} N_\mathrm{f}}$ of our channel estimation can be written as:

$$\mathbf{R}_{\hat{\mathbf{c}}} = \tilde{\mathbf{A}}\left(\mathbf{M}_{\mathscr{P}}\mathbf{R_c}\mathbf{M}_{\mathscr{P}}^\mathrm{H} + \mathrm{Diag}\left(\mathbf{p}_{\mathscr{P}}^\mathrm{ICI}\right) + P_\mathrm{V}\,\mathbf{I}_{N_\mathrm{P}}\right)\tilde{\mathbf{A}}^\mathrm{H}, \tag{3.39}$$

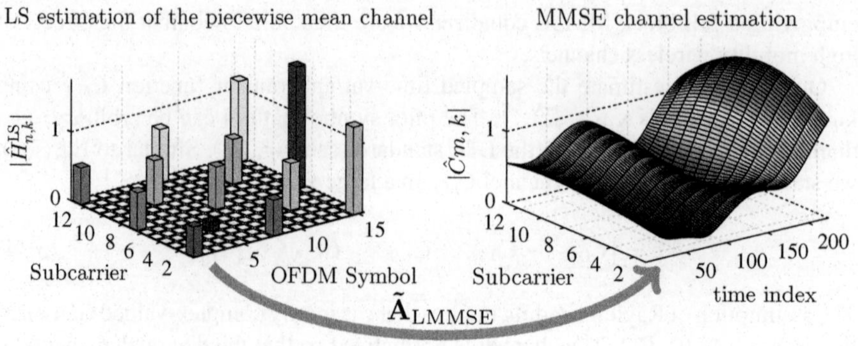

Fig. 3.3 Illustration of our time-variant channel estimation: LS estimates of the piecewise mean channel at pilot positions are interpolated respectively extrapolated (using the MMSE criterion) which delivers an estimate of the time variant transfer function $C_{m,k}$

whereas vector $\mathbf{p}^{\mathrm{ICI}}_{\mathscr{P}} \in \mathbb{R}^{N_{\mathrm{P}}\times 1}$ consists of all ICI power elements (Eq. 3.8), corresponding to the correct pilot positions. In order to determine matrix $\mathbf{M}_{\mathscr{P}} \in \mathbb{R}^{N_{\mathrm{P}}\times N_{\mathrm{c}}N_{\mathrm{s}}N_{\mathrm{f}}}$, we take N_{P} rows, at correct pilot positions, out of matrix $\mathbf{M} \in \mathbb{R}^{N_{\mathrm{c}}N_{\mathrm{f}}\times N_{\mathrm{c}}N_{\mathrm{s}}N_{\mathrm{f}}}$. The MSE of our channel estimation is then given as the diagonal elements of the following correlation matrix:

$$\mathbf{R}_{\mathbf{c}-\hat{\mathbf{c}}} = \mathbf{R}_{\mathbf{c}} - \tilde{\mathbf{A}}\mathbf{M}_{\mathscr{P}}\mathbf{R}_{\mathbf{c}} - \mathbf{R}_{\mathbf{c}}\mathbf{M}^{\mathrm{H}}_{\mathscr{P}}\tilde{\mathbf{A}}^{\mathrm{H}} + \mathbf{R}_{\hat{\mathbf{c}}}. \tag{3.40}$$

For the derivation of the MMSE channel estimation, we consider the ith element of the channel estimation $\hat{\mathbf{c}}$ individually ($i = k + (n-1)N_{\mathrm{c}}$). Let us denote the ith row of matrix $\tilde{\mathbf{A}}$ as $\breve{\mathbf{a}}_i^{\mathrm{T}} \in \mathbb{R}^{1\times N_{\mathrm{P}}}$. Equation (3.38) then transforms to:

$$[\hat{\mathbf{c}}]_i = \breve{\mathbf{a}}_i^{\mathrm{T}}\hat{\mathbf{h}}^{\mathrm{LS}}_{\mathscr{P}} \tag{3.41}$$

$$= \breve{\mathbf{a}}_i^{\mathrm{T}}\left(\mathbf{M}_{\mathscr{P}}\mathbf{c} + \mathrm{Diag}\left(\mathbf{x}_{\mathscr{P}}\right)^{-1}\left(\mathbf{y}^{\mathrm{ICI}}_{\mathscr{P}} + \mathbf{v}_{\mathscr{P}}\right)\right). \tag{3.42}$$

The orthogonal projection theorem requires:

$$\mathbb{E}\left\{([\hat{\mathbf{c}}]_i - [\mathbf{c}]_i)[\hat{\mathbf{c}}]_i^*\right\} = 0, \tag{3.43}$$

so that the MMSE solution $\tilde{\mathbf{A}}_{\mathrm{MMSE}} \in \mathbb{C}^{N_{\mathrm{c}}N_{\mathrm{s}}N_{\mathrm{f}}\times N_{\mathrm{P}}}$ becomes:

$$\tilde{\mathbf{A}}_{\mathrm{LMMSE}} = \mathbf{R}_{\mathbf{c}}\mathbf{M}^{\mathrm{H}}_{\mathscr{P}}\left(\mathbf{M}_{\mathscr{P}}\mathbf{R}_{\mathbf{c}}\mathbf{M}^{\mathrm{H}}_{\mathscr{P}} + \mathrm{diag}\{\mathbf{p}^{\mathrm{ICI}}_{\mathscr{P}}\} + P_{\mathrm{V}}\,\mathbf{I}_{N_{\mathrm{P}}}\right)^{-1} \tag{3.44}$$

Equation (3.44) together with (3.38) delivers all $N_{\mathrm{c}}N_{\mathrm{s}}N_{\mathrm{f}}$ MMSE estimates of the channel $C_{m,k}$ which can be inserted in Eq. (1.21) or (3.4) to estimate the behavior of our OFDM system. Multiplying matrix $\mathbf{M}$ with Eq. (3.44) results in the MMSE filter matrix $\mathbf{A}_{\mathrm{MMSE}} \in \mathbb{C}^{N_{\mathrm{c}}N_{\mathrm{f}}\times N_{\mathrm{P}}}$, which delivers the MMSE estimate of the piecewise mean channel $\hat{\mathbf{h}}$:

$$\mathbf{A}_{\mathrm{LMMSE}} = \mathbf{M}\tilde{\mathbf{A}}_{\mathrm{LMMSE}}. \tag{3.45}$$

This is the conventional two-dimensional MMSE solution [27], see also Eq. (3.20) in Sect. 3.3, whereas usually the matrices $\mathbf{R}_{\mathbf{h}_{\mathscr{P}}} = \mathbf{M}_{\mathscr{P}}\mathbf{R}_{\mathbf{c}}\mathbf{M}_{\mathscr{P}}^{\mathrm{H}} \in \mathbb{C}^{N_{\mathrm{P}}\times N_{\mathrm{P}}}$ and $\mathbf{R}_{\mathbf{h},\mathbf{h}_{\mathscr{P}}} = \mathbf{M}\mathbf{R}_{\mathbf{c}}\mathbf{M}_{\mathscr{P}}^{\mathrm{H}} \in \mathbb{C}^{N_{\mathrm{c}}N_{\mathrm{f}}\times N_{\mathrm{P}}}$ are assumed to be known explicitly. We, on the other hand, showed in this section how these matrices ($\mathbf{R}_{\mathbf{h}_{\mathscr{P}}}$ and $\mathbf{R}_{\mathbf{h},\mathbf{h}_{\mathscr{P}}}$) can be decomposed to describe OFDM systems in doubly-selective channels.

Let us denote $\tilde{\mathbf{A}}_n \in \mathbb{C}^{N_{\mathrm{c}}N_{\mathrm{FFT}}\times N_{\mathrm{P}}}$ as the filter matrix, consisting of those row elements of $\tilde{\mathbf{A}} \in \mathbb{C}^{N_{\mathrm{c}}N_{\mathrm{s}}N_{\mathrm{f}}\times N_{\mathrm{P}}}$ which deliver the estimated channel $\hat{\mathbf{c}}_n$ (see Eq. 3.3):

$$\hat{\mathbf{c}}_n = \tilde{\mathbf{A}}_n \hat{\mathbf{h}}_{\mathscr{P}}^{\mathrm{LS}}. \tag{3.46}$$

We now denote the column i of $\tilde{\mathbf{A}}_n$ as $\tilde{\mathbf{a}}_n^{(i)} \in \mathbb{C}^{N_{\mathrm{c}}N_{\mathrm{FFT}}\times 1}$:

$$\tilde{\mathbf{A}}_n = \left[\tilde{\mathbf{a}}_n^{(1)} \cdots \tilde{\mathbf{a}}_n^{(N_{\mathrm{P}})}\right]. \tag{3.47}$$

Combining Eqs. (3.4), (3.46) and (3.47), the estimated OFDM matrix $\hat{\mathbf{D}}_n$ becomes a linear combination of the basis matrices $\mathbf{B}_n^{(i)} \in \mathbb{C}^{N_{\mathrm{c}}\times N_{\mathrm{c}}}$:

$$\hat{\mathbf{D}}_n = \sum_{i=1}^{N_{\mathrm{P}}} \frac{y_{\mathscr{P}_i}}{x_{\mathscr{P}_i}} \underbrace{\begin{bmatrix} \left(\tilde{\mathbf{W}}_1 \tilde{\mathbf{a}}_n^{(i)}\right)^{\mathrm{T}} \\ \vdots \\ \left(\tilde{\mathbf{W}}_{N_{\mathrm{c}}} \tilde{\mathbf{a}}_n^{(i)}\right)^{\mathrm{T}} \end{bmatrix}}_{\mathbf{B}_n^{(i)}}. \tag{3.48}$$

Note that the matrices $\mathbf{B}_n^{(i)}$ can be precomputed. Once we have the estimated OFDM matrix $\hat{\mathbf{D}}_n$, the equalization is performed as follows:

$$\hat{\mathbf{x}}_n = \hat{\mathbf{D}}_n^{\mathrm{H}} \left(\hat{\mathbf{D}}_n \hat{\mathbf{D}}_n^{\mathrm{H}} + P_{\mathrm{V}}\, \mathbf{I}_{N_{\mathrm{c}}}\right)^{-1} \mathbf{r}_n. \tag{3.49}$$

For perfect channel knowledge, that is, $\hat{\mathbf{D}}_n = \mathbf{D}_n$, Eq. (3.49) delivers the MMSE estimation of the transmitted data symbols $\hat{\mathbf{x}}_n$ [13]. However, we will call it MMSE equalization even if the channel estimation is not perfect. Note that the complexity of Eq. (3.49) can be reduced by exploiting the banded structure of the OFDM matrix $\mathbf{D}_n$ [11]. It is also possible to reduce the amount of ICI by means of ICI cancellation [29]:

$$\mathbf{r}_n^{\mathrm{ICICancel}} = \mathbf{D}_n \mathbf{x}_n + \mathbf{v}_n - \hat{\mathbf{D}}_n^{\mathrm{ICI}}\, Q\left(\hat{\mathbf{x}}_n\right), \tag{3.50}$$

where matrix $\hat{\mathbf{D}}_n^{\mathrm{ICI}}$ consists of nearly the same elements as $\hat{\mathbf{D}}_n$, with the difference that the diagonal elements are all zero, thus reflecting only the ICI relevant terms. Function $Q(\cdot)$ indicates the quantization of the estimated data symbols $\hat{\mathbf{x}}_n$, that is, hard decision according to the minimum distance criteria. The quantization process is nonlinear, so that the ICI power after cancellation $P_{n,k}^{\mathrm{ICICancel}}$ cannot be straightforwardly calculated.

We approximate this power by assuming that the estimated data symbols in Eq. (3.50) are perfectly known $\hat{\mathbf{x}}_n = \mathbf{x}_n$. The ICI power after cancellation can then be found as:

$$P_{n,k}^{\text{ICICancel}} = \text{tr}\left\{ \tilde{\mathbf{W}}_k^{\text{ICI}} \mathbf{R}_{\mathbf{c}_n - \hat{\mathbf{c}}_n} \left(\tilde{\mathbf{W}}_k^{\text{ICI}} \right)^{\text{H}} \right\}, \tag{3.51}$$

whereas the correlation matrix $\mathbf{R}_{\mathbf{c}_k - \hat{\mathbf{c}}_k}$ is given by Eq. (3.40).

Our ICI mitigation method combines the OFDM matrix estimation in Eq. (3.48), the MMSE equalization in Eq. (3.49) and the ICI cancellation in Eq. (3.50). The three basic steps of our ICI mitigation method are:

- **Step 1**: We estimate the OFDM matrix $\hat{\mathbf{D}}_n$ according to Eq. (3.48) in combination with (3.44) whereby the ICI power $\mathbf{p}_{\mathcal{P}}^{\text{ICI}}$ is given by Eq. (3.8). The MMSE equalization, Eq. (3.49), then delivers the estimated data symbols $\hat{\mathbf{x}}_n$ of the first step.
- **Step 2**: The estimated data symbols $\hat{\mathbf{x}}_n$ together with the estimated matrix $\hat{\mathbf{D}}_n$, both obtained from the first step, reduce the ICI at pilot position, see Eq. (3.50), so that the piecewise mean LS channel estimates at pilot positions $\hat{\mathbf{h}}_{\mathcal{P}}^{\text{LS}}$ become more accurate: $\hat{\mathbf{h}}_{\mathcal{P}}^{\text{LS}} = \text{Diag}\,(\mathbf{x}_{\mathcal{P}})^{-1}\,\mathbf{r}_{\mathcal{P}}^{\text{ICICancel}}$. This improved channel estimation is then used in Eq. (3.48) together with (3.44) to estimate the OFDM matrix $\hat{\mathbf{D}}_n$. Compared with the first step, the ICI power $\mathbf{p}_{\mathcal{P}}^{\text{ICI}}$ in Eq. (3.44) is now lower and given by Eq. (3.51). After MMSE equalization, Eq. (3.49), we obtain the estimated data symbols $\hat{\mathbf{x}}_n$ of the second step.
- **Step 3**: Here, we treat all estimated data symbols $\hat{\mathbf{x}}_n$ as if they were pilot symbols. After ICI cancellation, using $\hat{\mathbf{x}}_n$ and $\hat{\mathbf{D}}_n$ from the second step, see Eq. (3.50), we obtain the piecewise mean LS channel estimates at all $N_{\text{c}}N_{\text{f}}$ positions by: $\hat{\mathbf{h}}^{\text{LS}} = \text{diag}\{\hat{\mathbf{x}}\}^{-1}\mathbf{r}^{\text{ICICancel}}$. In this step, we assume $N_{\text{c}}N_{\text{f}}$ "pilot symbols", so that matrix $\mathbf{M}_{\mathcal{P}}$ in Eq. (3.44) becomes $\mathbf{M}$ and N_{P} in Eq. (3.48) becomes $N_{\text{c}}N_{\text{f}}$. Additionally, the ICI power $\mathbf{p}_{\mathcal{P}}^{\text{ICI}}$ in Eq. (3.44) is further decreased compared to Step 2, because the ICI cancellation uses more accurate estimations of the OFDM matrix. Note that, if there exist data symbols of nonunit magnitude, e.g., 16-QAM, the ICI power and noise power are enhanced due to the LS estimation process. Again, MMSE estimation of the OFDM matrix, Eqs. (3.44) and (3.48), in combination with MMSE equalization, Eq. (3.48), delivers the estimated data symbols $\hat{\mathbf{x}}_n$ of the third step.

All three steps are illustrated in Figs. 3.4, 3.5 and 3.6.

3.5 Simulations and Measurements

This section presents the measurement results of [28]. In this example we use 13 subcarriers, 15 OFDM symbols and an LTE downlink pilot symbol pattern (as illustrated in Fig. 1.2). Furthermore, we assume 4-QAM, a subcarrier spacing of 15 kHz and a carrier frequency of 2.5 GHz.

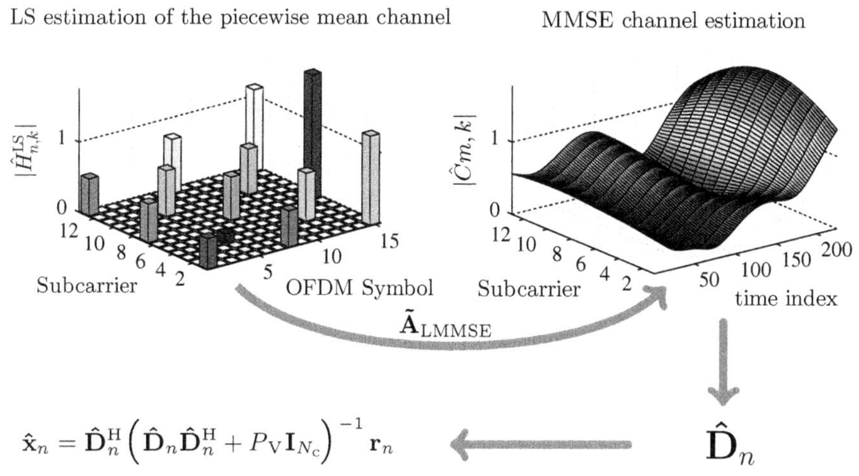

Fig. 3.4 Illustration of the first step of our ICI mitigation technique (**Step 1**)

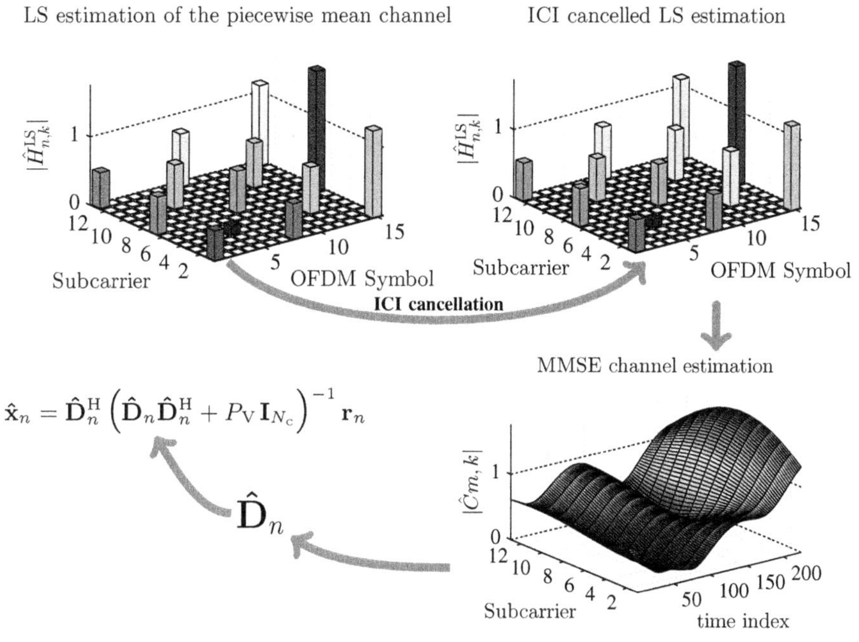

Fig. 3.5 Illustration of the second step of our ICI mitigation technique (**Step 2**)

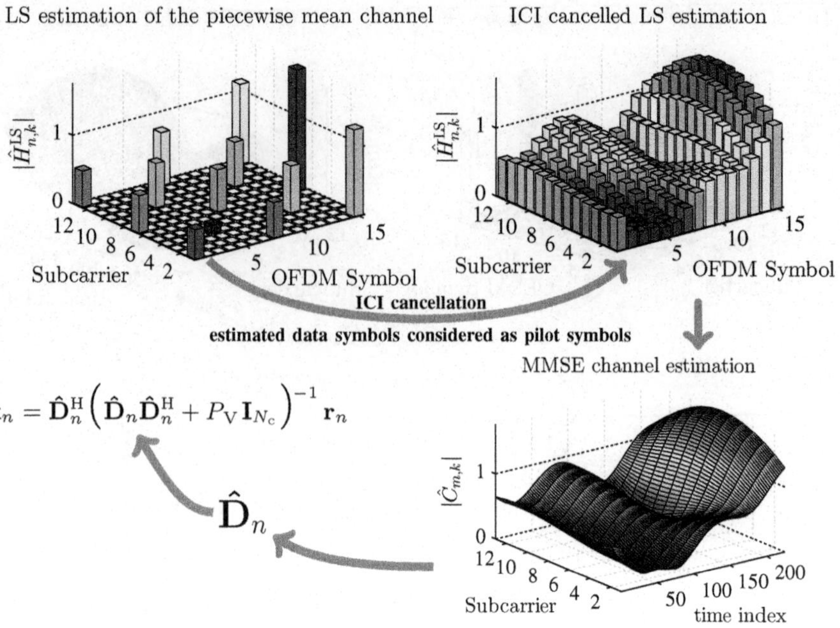

Fig. 3.6 Illustration of the third step of our ICI mitigation technique (**Step 3**)

Simulations

For our simulations we obtain the sampled time-variant transfer function $C_{m,k}$, as in Eq. (1.21), by utilizing the following WSSUS channel model [30]:

$$C_{m,k} = \frac{1}{\sqrt{I}} \sum_{i=1}^{I} e^{j\left(\theta_i + 2\pi \frac{\nu_i}{\Delta f} \frac{m}{N_{\mathrm{FFT}}} - 2\pi \frac{\tau_i}{T} k\right)}, \tag{3.52}$$

whereas the random Doppler shifts ν_i are distributed according to a Jakes Doppler spectral density, the random normalized delays $\frac{\tau_i}{T}$ are uniformly distributed between 0 and $\frac{\tau_{\max}}{T} = 0.01$ and the random phases θ_i are uniformly distributed between 0 and 2π, ensuring circularly symmetry. We assume $N_{\mathrm{FFT}} = 13$ samples, $I = 200$ channel paths and consider 50,000 realizations over which we average. Additionally, all random variables are assumed to be statistically independent, so that the frequency-time correlation function becomes:

$$\mathbb{E}\left\{C_{m_1,k_1} C^*_{m_2,k_2}\right\} = r_{C_f}[k_1 - k_2] r_{C_t}[m_1 - m_2]. \tag{3.53}$$

with (J_0 is the zero-th order Bessel function):

$$r_{C_t}[m_1 - m_2] = J_0\left(2\pi \frac{\nu_{\max}}{\Delta f} \frac{m_1 - m_2}{N_{\mathrm{FFT}}}\right) \tag{3.54}$$

$$r_{C_f}[k_1 - k_2] = \text{sinc}\left(\frac{\tau_{\max}}{T}(k_1 - k_2)\right) \mathrm{e}^{\mathrm{j}\pi \frac{\tau_{\max}}{T}(k_1 - k_2)}. \tag{3.55}$$

If the channel elements are stacked according to Eq. (3.34), correlation matrix $\mathbf{R_c} \in \mathbb{C}^{N_c N_s N_f \times N_c N_s N_f}$ becomes:

$$\mathbf{R_c} = \begin{bmatrix} r_{C_t}[0] & \cdots & r_{C_t}[(N_f N_s - 1)] \\ \vdots & \ddots & \vdots \\ r_{C_t}[-(N_f N_s - 1)] & \cdots & r_{C_t}[0] \end{bmatrix} \otimes \begin{bmatrix} r_{C_f}[0] & \cdots & r_{C_f}[(N_c - 1)] \\ \vdots & \ddots & \vdots \\ r_{C_f}[-(N_c - 1)] & \cdots & r_{C_f}[0] \end{bmatrix}. \tag{3.56}$$

Figure 3.7 shows that our theoretical BEP expressions for one-tap equalizers (Sect. 3.3) coincide with the simulated BER. The illustrated conventional methods ignore ICI and estimate the channel either by MMSE (ICI is considered as additional noise) or by linear interpolation of the three closest LS estimates at pilot position (no statistical knowledge necessary). For a high SNR, the ICI becomes the limiting factor so that the BEP of conventional methods soon saturates. This effect can be mitigated by our proposed ICI mitigation technique. The BER after the third step

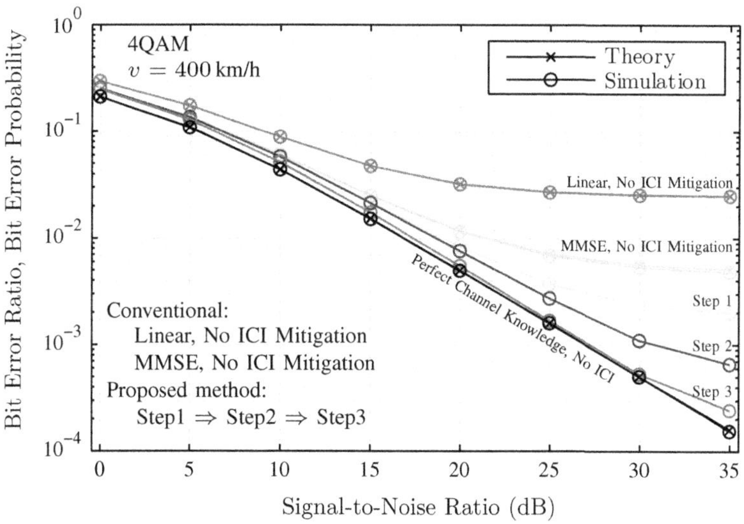

Fig. 3.7 Simulated BER and theoretical BEP as a function of SNR. Simulation confirms our analytical prediction. At high SNR, our ICI mitigation method avoids an early saturation of the BER. Each step improves the BER and Step 3 achieves a performance close to perfect channel knowledge and no ICI

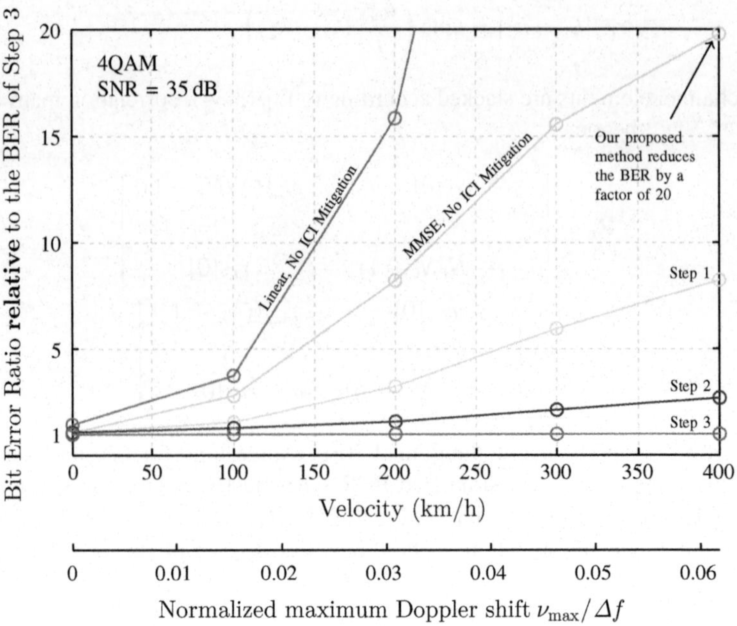

Fig. 3.8 Simulated BER *relative* to the BER of Step 3 as a function of velocity, respectively normalized maximum Doppler shift. The higher the velocity, the better our methods performs compared to conventional methods. For a velocity of zero, the BER of MMSE (no ICI mitigation), Step 1 and Step 2 coincide

comes very close to perfect channel knowledge and no ICI which corresponds to the lowest possible BER.

Figure 3.8 shows observed BER relative to the BER of Step 3, thereby delivering the improvement factor of our method. Clearly, for higher velocities, the ICI becomes more and more an impediment so that the performance of our ICI mitigation technique increases relatively to conventional methods. For time-invariant channels, i.e., zero velocity, Step 1 and 2 exhibit the same BER as the one-tap equalizer using MMSE CE (no ICI mitigation).

Measurements

Our measurements were conducted on the Vienna Wireless Testbed that has been augmented by a rotating wheel unit to generate high speed movements [18]. Transmitter and receiver are both placed indoors, in different rooms, and the number of samples after demodulation is set to 26. For a given transmit power, corresponding to a specific average SNR, we obtain different channel realizations by relocating the transmit antenna to 144 positions, equidistantly spaced over a 3 $\times$ 3 wavelength grid. The receive antenna moves at up to 400 km/h and we measure always over the same three wavelengths movement interval. To estimate the noise power at the receiver, we include two all-zero OFDM symbols, $\mathbf{x}_0 = \mathbf{0}_{N_c \times 1}$ and $\mathbf{x}_{N_f+1} = \mathbf{0}_{N_c \times 1}$. Similar, the ICI-plus-noise power is estimated by using all-zero subcarriers, $x_{n,0} = 0$ and $x_{n,N_c} = 0$. Because the second order statistics, required for the MMSE estimation, are not perfectly known, we obtain only a mismatched solution. We assume a WSSUS channel and a separable correlation function, as in Eq. (3.53). For the MMSE estimation, we cannot directly insert the estimated correlation functions because they are corrupted by noise, leading to large errors due to the matrix inversion. We thus model the delays as well as the Doppler shifts by a uniform distribution, so that the frequency correlation is given by Eq. (3.55) and the time correlation by:

$$r_{C_t}[m_1 - m_2] = \operatorname{sinc}\left(2 \frac{\nu_{\max}}{\Delta f} \frac{m_1 - m_2}{N_{\text{FFT}}}\right). \tag{3.57}$$

From the estimated frequency correlation function $\hat{r}_{C_f}$ we conclude that $\tau_{\max} = 0$. The maximum Doppler shifts $\nu_{\max}$, on the other hand, are chosen so that the theoretical ICI power and the measured ICI power coincide. At a velocity of 100 km/h, we measure the same ICI power we would expect at 98 km/h for a uniform distributed Doppler spectral density, so that the normalized maximum Doppler shift in Eq. (3.57) is set to $\frac{\nu_{\max}}{\Delta f} = 0.015$. Similar, the measurement velocity of 200 km/h corresponds to an equivalent (uniform Doppler spectral density) velocity of 187 km/h, 300–280 km/h and 400–372 km/h.

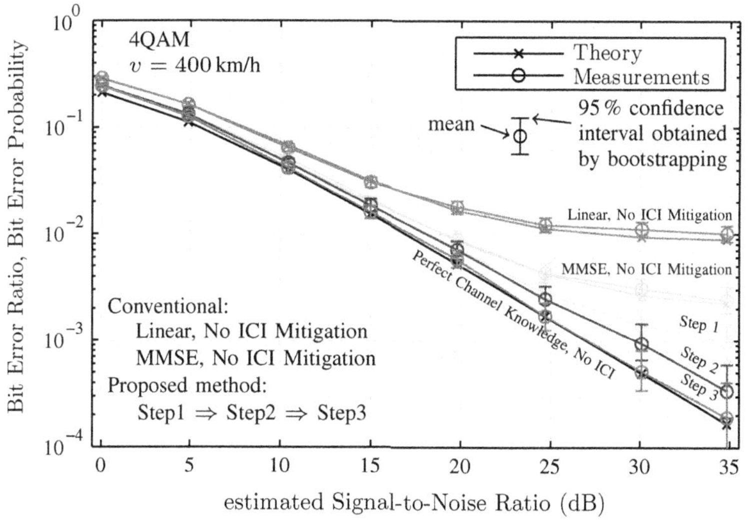

Fig. 3.9 Measured BER and theoretical BEP as a function of estimated SNR, similar to Fig. 3.7. Also in testbed measurements, our proposed ICI mitigation methods comes close to perfect channel knowledge and no ICI

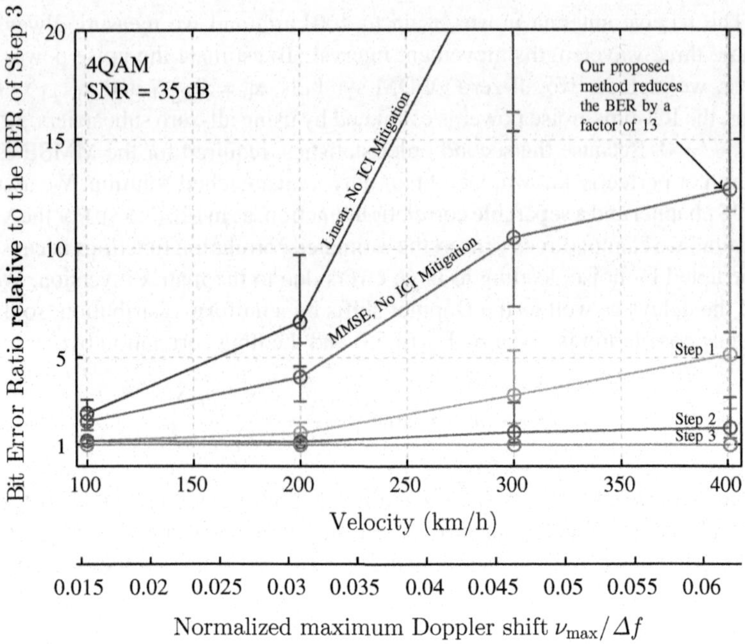

Fig. 3.10 Measured BER *relative* to the BER of Step 3 as a function of velocity. Compared to Fig. 3.8, the potential improvement of our ICI mitigation method is lower because the underlying channel parameters differ between simulation and measurements (lower delay spread due to indoor measurements, lower Doppler spread and $N_{\mathrm{FFT}} > N_{\mathrm{c}}$)

Figure 3.9 shows that our theoretical expressions for one-tap equalizers (Sect. 3.3) accurately model the true physical behavior. Each step of our proposed mitigation technique reduces the BER whereas the third step achieves a performance close to perfect channel knowledge and no ICI.

Figure 3.10 illustrates the measured BER relative to the BER of Step 3. Compared to Fig. 3.8, the one-tap equalizers perform better due to different channel parameters (lower delay spread, lower Doppler spread and $N_{\mathrm{FFT}} > N_{\mathrm{c}}$), lowering the potential improvement of our ICI mitigation method.

References

1. F. Hlawatsch, G. Matz, *Wireless Communications over Rapidly Time-Varying Channels* (Academic Press, Amsterdam, 2011)
2. F. Pena-Campos, R. Carrasco-Alvarez, O. Longoria-Gandara, R. Parra-Michel, Estimation of fast time-varying channels in OFDM systems using two-dimensional prolate. IEEE Trans. Wirel. Commun. **12**(2), 898–907 (2013). doi:10.1109/TWC.2013.010413.120624
3. Z. Tang, R.C. Cannizzaro, G. Leus, P. Banelli, Pilot-assisted time-varying channel estimation for OFDM systems. IEEE Trans. Signal Process. **55**(5), 2226–2238 (2007)

4. Y. Mostofi, D.C. Cox, ICI mitigation for pilot-aided OFDM mobile systems. IEEE Trans. Wirel. Commun. **4**(2), 765–774 (2005)
5. H. Hijazi, L. Ros, Polynomial estimation of time-varying multipath gains with intercarrier interference mitigation in OFDM systems. IEEE Trans. Veh. Technol. **58**(1), 140–151 (2009)
6. M. Simko, C. Mehlführer, T. Zemen, M. Rupp, Inter-carrier interference estimation in MIMO OFDM systems with arbitrary pilot structure, in *IEEE Vehicular Technology Conference (VTC Spring)* (2011)
7. C. Sgraja and J. Lindner, Estimation of rapid time-variant channels for OFDM using Wiener filtering, in IEEE International Conference on Communications (ICC), vol. 4, pp. 2390–2395 vol. 4, 2003, doi:10.1109/ICC.2003.1204362
8. C. Shin, J. Andrews, E. Powers, An efficient design of doubly selective channel estimation for OFDM systems. IEEE Trans. Wireless Commun. **6**(10), 3790–3802 (2007). doi:10.1109/TWC.2007.060134. Oct
9. S. Tomasin, A. Gorokhov, H. Yang, J.-P. Linnartz, Iterative interference cancellation and channel estimation for mobile OFDM. IEEE Trans. Wireless Commun. **4**(1), 238–245 (2005)
10. P. Schniter, Low-complexity equalization of OFDM in doubly selective channels. IEEE Trans. Signal Process. **52**(4), 1002–1011 (2004)
11. L. Rugini, P. Banelli, G. Leus, Simple equalization of time-varying channels for OFDM. IEEE Commun. Lett. **9**(7), 619–621 (2005). doi:10.1109/LCOMM.2005.1461683. July
12. K. Fang, L. Rugini, G. Leus, Low-complexity block turbo equalization for OFDM systems in time-varying channels. IEEE Trans. Signal Process. **56**(11), 5555–5566 (2008)
13. Y.-S. Choi, P. Voltz, F. Cassara, On channel estimation and detection for multicarrier signals in fast and selective Rayleigh fading channels. IEEE Trans. Commun. **49**(8), 1375–1387 (2001). doi:10.1109/26.939860
14. N. Miyazaki, S. Nanba, and S. Konishi, MIMO-OFDM throughput performances on MIMO antenna configurations using LTE-based testbed with 100 MHz bandwidth, in *2010 IEEE 72nd Vehicular Technology Conference Fall (VTC 2010-Fall)*, pp. 1–5 (2010). doi:10.1109/VETECF.2010.5594527
15. P. Mahasukhon, H. Sharif, M. Hempel, T. Zhou, T. Ma, Distance and throughput measurements in mobile WiMAX test bed, in *IEEE Military Communications Conference (MILCOM'2010)*, pp. 154–159 (2010)
16. K. Jang, M. Han, S. Cho, H.-K. Ryu, J. Lee, Y. Lee, S.B. Moon, 3G and 3.5 G wireless network performance measured from moving cars and high-speed trains, in *Proceedings of the 1st ACM Workshop on Mobile Internet Through Cellular Networks*, pp. 19–24. ACM (2009)
17. S. Caban, J. Rodas, J.A. García Naya, A methodology for repeatable, off-line, closed-loop wireless communication system measurements at very high velocities of up to 560 km/h, in *Proceedings of the 2011 IEEE International Instrumentation and Measurement Technology Conference (I2MTC2011)* (2011). doi:10.1109/IMTC.2011.5944019
18. R. Nissel, M. Lerch, M. Rupp, Experimental validation of the OFDM bit error probability for a moving RX antenna, in *Proceedings of IEEE VTC*, Vancouver, Canada (2014)
19. T. Wang, J. Proakis, E. Masry, J. Zeidler, Performance degradation of OFDM systems due to Doppler spreading. IEEE Trans. Wirel. Commun. **5**(6), 1422–1432 (2006). doi:10.1109/TWC.2006.1638663
20. P. Robertson and S. Kaiser, The effects of Doppler spreads in OFDM(A) mobile radio systems, in *IEEE 50th Vehicular Technology Conference (VTC Fall)*, pp. 329–333, vol. 1 (1999). doi:10.1109/VETECF.1999.797150
21. R. Nissel, S. Caban, M. Rupp, Closed-form capacity expression for low complexity BICM with uniform inputs, in *Proceedings of IEEE 26th Annual International Symposium on Personal, Indoor and Mobile Radio Communications (PIMRC)*, Hong Kong (2015)
22. R. Nissel, M. Rupp, Bit error probability for pilot-symbol aided channel estimation in FBMC-OQAM, in *Proceedings of the IEEE International Conference on Communications (ICC)*, Kuala Lumpur, Malaysia, (2016)
23. M.-X. Chang, Y.T. Su, Performance analysis of equalized OFDM systems in Rayleigh fading. IEEE Trans. Wirel. Commun. **1**(4), 721–732 (2002)

24. P. Tan, N.C. Beaulieu, Effect of channel estimation error on bit error probability in OFDM systems over Rayleigh and Ricean fading channels. IEEE Trans. Commun. **56**(4), 675–685 (2008)
25. F.J. López-Martínez, E. Martos-Naya, J.F. Paris, U. Fernández-Plazaola, Generalized BER analysis of QAM and its application to MRC under imperfect CSI and interference in Ricean fading channels. IEEE Trans. Veh. Technol. **59**(5), 2598–2604 (2010)
26. G. Taricco, E. Biglieri, Space-time decoding with imperfect channel estimation, vol. 4(4), pp. 1874–1888 (2005)
27. P. Hoeher, S. Kaiser, P. Robertson, Two-dimensional pilot-symbol-aided channel estimation by Wiener filtering, *IEEE International Conference on Acoustics, Speech, and Signal Processing*, vol. 3, pp. 1845–1848 (1997)
28. R. Nissel, M. Rupp, Doubly-selective MMSE channel estimation and ICI mitigation for OFDM systems, in *Proceedings of the IEEE International Conference on Communications (ICC'15)*, London (2015)
29. A. Molisch, M. Toeltsch, S. Vermani, Iterative methods for cancellation of intercarrier interference in OFDM systems. IEEE Trans. Veh. Technol. **56**(4), 2158–2167 (2007). doi:10.1109/TVT.2007.897628
30. P. Hoeher, A statistical discrete-time model for the WSSUS multipath channel. IEEE Trans. Veh. Technol. **41**(4), 461–468 (1992). doi:10.1109/25.182598

Chapter 4
Optimal Pilot Pattern for Downlink Transmissions

Michal Šimko

Long Term Evolution (LTE) standards define a fixed diamond shaped pilot pattern over the time frequency grid in the downlink for Multiple-Input Multiple-Output (MIMO) Orthogonal Frequency Division Multiplexing (OFDM) transmission systems. While the choice of the pattern is a compromise to cover a wide range of channels with different Root Mean Square (RMS) delays and Doppler spreads, it is also selected sparsely not to hamper capacity. In this chapter we investigate patterns, that allow a variable pilot density as well as a variable power distribution between pilots and data. Finally, both degrees of freedom are optimized and due to a certain range of trade-off we propose a small set of pilot patterns that offers a substantially higher throughput than today's.[1]

4.1 Introduction

In [1] it is shown that for a given amount of pilot-symbol overhead, in terms of Mean Square Error (MSE), it is optimal to position the pilot-symbols in a diamond-shaped constellation for a time-frequency grid in an OFDM transmission scheme. It is, however, not shown how to choose shape parameters of the diamond-shaped pilot-symbol pattern and consequently how much overhead is required for a proper channel estimation. Figure 4.1 shows an example of such a diamond-shaped pilot-symbol pattern. Diamond-shaped pilot-symbol patterns can be decomposed into two patterns with pilot-symbols equi-spaced in time and in frequency directions with distances D_t and D_f, respectively. These two patterns with equi-spaced pilot-symbols

[1]The material presented here can be accessed in more detail in the Ph.D. thesis of Michal Šimko available for download at http://theses.eurasip.org/theses/488/pilot-pattern-optimization-for-doubly-selective/.

M. Šimko (✉)
TU Wien, Vienna, Austria
e-mail: michal@simko.biz

M. Rupp et al., *The Vienna LTE-Advanced Simulators*,
Signals and Communication Technology, DOI 10.1007/978-981-10-0617-3_4

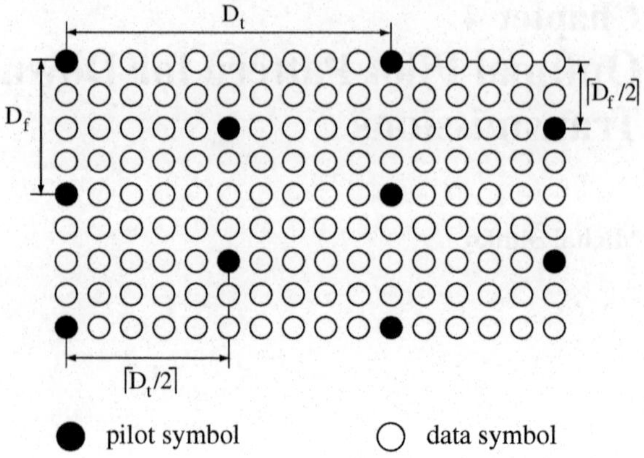

Fig. 4.1 Example of a diamond-shaped pilot-symbol pattern: the pilot-symbol pattern is fully described by two variables D_f and D_t. These parameters determine the pilot-symbol density in the time and the frequency dimension

are separated from each other by $\lceil \frac{D_\mathrm{t}}{2} \rceil$ in the time direction and by $\lceil \frac{D_\mathrm{f}}{2} \rceil$ in the frequency direction. Therefore, a diamond-shaped pattern is fully described by two variables D_f and D_t. Figure 4.1 shows an example of a diamond-shaped pilot-symbol pattern with $D_\mathrm{f} = 4$ and $D_\mathrm{t} = 10$.

Note that also LTE utilizes such a diamond shaped pilot-symbol pattern with fixed values $D_\mathrm{f} = 6$ and $D_\mathrm{t} = 7$.

When the power is evenly distributed between the data and pilot-symbols, stacked each in a vector $\mathbf{x}_\mathcal{D}$ and $\mathbf{x}_\mathcal{P}$, respectively, there is:

$$P_\mathrm{S} = \frac{1}{N_\mathrm{D}} \mathbb{E}\{\|\mathbf{x}_\mathcal{D}\|_2^2\} = 1, \tag{4.1}$$

$$P_\mathrm{P} = \frac{1}{N_\mathrm{P}} \mathbb{E}\{\|\mathbf{x}_\mathcal{P}\|_2^2\} = 1, \tag{4.2}$$

where N_D is the number of data-symbols, N_P the number of pilot-symbols and $N_\mathrm{f} = N_\mathrm{D} + N_\mathrm{P}$ the entire number of transmited symbols.

4.2 Least Squares Channel Estimation

In order to investigate the optimal choice of the pilot patterns, knowledge of the channel estimator performance with a given pilot pattern is required. The Least Squares (LS) channel estimate at the pilot-symbol positions is obtained by solving the following minimization problem

$$\hat{\mathbf{H}}_\mathcal{P}^\mathrm{LS} = \arg\min_{\mathbf{H}} \|\mathbf{r}_\mathcal{P} - \mathbf{H}\mathbf{x}_\mathcal{P}\|_2^2, \tag{4.3}$$

where the matrix $\hat{\mathbf{H}}^{\mathrm{LS}}_{\mathcal{P}} \in \mathbb{C}^{N_{\mathrm{P}} \times N_{\mathrm{P}}}$ is assumed to be diagonal, representing the channel only at the pilot-symbol positions. Vector $\mathbf{x}_{\mathcal{P}}$ contains the known pilot symbols and $\mathbf{r}_{\mathcal{P}}$ the observations. Note that, $\mathrm{Diag}\left(\hat{\mathbf{H}}^{\mathrm{LS}}_{\mathcal{P}}\right) = \hat{\mathbf{h}}^{\mathrm{LS}}_{\mathcal{P}}$, identical to Chap. 3. Obviously, such an estimator ignores the non-diagonal elements of the channel matrix and treats the potential Inter-Carrier Interference (ICI) as noise. More details on this topic can be found in [2]. The resulting MSE at the pilot-symbol positions is given as

$$\sigma^2_{e,\mathcal{P}} = \frac{P_{\mathrm{V}} + P_{\mathrm{ICI}}}{P_{\mathrm{P}}}, \tag{4.4}$$

where P_{V} denotes the noise power and P_{ICI} the ICI contribution. By increasing the power radiated at the pilot-symbols P_{P}, the channel estimates at the pilot-symbols become more reliable. The channel estimates at the data positions have to be obtained by applying a two-dimensional interpolation. In the following, we analyze the performance of a general linear channel estimator in terms of MSE. A channel estimate at an arbitrary data position using a linear interpolation is given by a weighted sum of the S nearest channel estimates (in the Euclidean sense of the time-frequency grid) at pilot positions. Let us denote the channel estimate at the jth data position by $\hat{h}_{\mathcal{D},j}$ and the channel estimate at the ith pilot position by $\hat{h}_{\mathcal{P},i}$. The lower indices i and j are used as general description of the location in the time-frequency grid. In contrast to Chap. 3, we no longer denote each time-frequency position explicitly in order to better distinguish between data symbols and pilot symbols, simplifying the optimality decision. A channel estimate at a data position j can be expressed as a weighted sum of S closest pilot-symbols

$$\hat{h}_{\mathcal{D},j} = \sum_{i \in \mathcal{P}_j} a_{j,i} \hat{h}_{\mathcal{P},i}, \tag{4.5}$$

where $\mathcal{P}_j$ denotes a set of the S nearest pilot-symbol positions to the data position j. The weight $a_{j,i}$ is a real number that indicates how much the channel estimate at the jth data position is influenced by the channel estimate at the ith pilot position.

We evaluate the performance of such a linear channel estimator by analytically deriving its theoretical MSE at the data positions. Given the definition of MSE,

$$\begin{aligned}\sigma^2_{e,j} &= \mathbb{E}\left\{\left\| h_{\mathcal{D},j} - \hat{h}_{\mathcal{D},j} \right\|^2\right\} \\ &= \mathbb{E}\left\{\left\| h_{\mathcal{D},j} \right\|^2\right\} - 2\Re\left\{\mathbb{E}\left\{h^*_{\mathcal{D},j}\hat{h}_{\mathcal{D},j}\right\}\right\} + \mathbb{E}\left\{\left\| \hat{h}_{\mathcal{D},j} \right\|^2\right\}.\end{aligned} \tag{4.6}$$

For the analytical derivation, we made the following assumptions:

- The channel power is normalized to one $\left(\mathbb{E}\left\{\left\| h_{\mathcal{D},j} \right\|^2\right\} = 1\right)$.
- The channel $h_{\mathcal{D},j}$ at a data position j and the estimation error $\tilde{h}_i = h_{\mathcal{P},i} - \hat{h}_{\mathcal{P},i}$ at a pilot position i are uncorrelated $\left(\mathbb{E}\left\{h^*_{\mathcal{D},j}\tilde{h}_i\right\} = 0\right)$.

- The channel $h_{\mathcal{P},i}$ at a pilot position i and the estimation error $\tilde{h}_{i'}$ at a pilot position i' are uncorrelated $\left(\mathbb{E}\left\{h_{\mathcal{P},i}\tilde{h}_{i'}\right\}=\right)0$.

Let us analyze the three terms in Eq. (4.6) individually. The first term is equal to one due to the system model $\mathbb{E}\left\{\left\|h_{\mathcal{D},j}\right\|^2\right\}=1$. In the second term, $\hat{h}_{\mathcal{D},j}$ can be replaced by Eq. (4.5)

$$\begin{aligned}\Re\left\{\mathbb{E}\left\{h^*_{\mathcal{D},j}\hat{h}_{\mathcal{D},j}\right\}\right\} &= \Re\left\{\mathbb{E}\left\{h^*_{\mathcal{D},j}\sum_{i\in\mathcal{P}_j}a_{j,i}\hat{h}_{\mathcal{P},i}\right\}\right\}\\ &= \Re\left\{\sum_{i\in\mathcal{P}_j}a_{j,i}\mathbb{E}\left\{h^*_{\mathcal{D},j}\left(h_{\mathcal{P},i}-\tilde{h}_i\right)\right\}\right\}\\ &= \sum_{i\in\mathcal{P}_j}a_{j,i}\Re\left\{R_{j,i}\right\},\end{aligned} \tag{4.7}$$

where $\tilde{h}_i$ represents the estimation noise at the ith pilot position. The last step in the above equation can be justified by the assumption that the channel at the jth data position and the estimation error at the ith pilot position are uncorrelated. The coefficient $R_{j,i}=\mathbb{E}\left\{h^*_{\mathcal{D},j}h_{\mathcal{P},i}\right\}$ denotes the correlation between the channels at the jth data-symbol and the ith pilot-symbol positions.

Let us proceed with the last term of Eq. (4.6), in which Eq. (4.5) is inserted and consequently the equality $\hat{h}_{\mathcal{P},i}=h_{\mathcal{P},i}-\tilde{h}_i$, which states that the channel estimate at the pilot-symbol position is given as the true channel superimposed by an estimation error. Note that due to Eq. (4.2), it can be shown that the MSE of the LS channel estimator at the pilot-symbol positions is identical to the noise power divided by the power radiated at the pilot-symbols $\frac{P_V}{P_P}$ for time-invariant channels [3]. In the case of time-variant channels, also the ICI power has to be considered. Therefore the power of $\tilde{h}_i$ is equal to $\frac{P_V+P_{ICI}}{P_P}$. We obtain:

$$\begin{aligned}\mathbb{E}\left\{\left\|\hat{h}_{\mathcal{D},j}\right\|^2\right\} &= \mathbb{E}\left\{\left\|\sum_{i\in\mathcal{P}_j}a_{j,i}\left(h_{\mathcal{P},i}-\tilde{h}_i\right)\right\|^2\right\}\\ &= \mathbb{E}\left\{\sum_{i\in\mathcal{P}_j}\sum_{i'\in\mathcal{P}_j}a_{j,i}a_{j,i'}\left(h_{\mathcal{P},i}-\tilde{h}_i\right)^*\left(h_{\mathcal{P},i'}-\tilde{h}_{i'}\right)\right\}\\ &= \sum_{i\in\mathcal{P}_j}\sum_{i'\in\mathcal{P}_j}a_{j,i}a_{j,i'}R_{i,i'}+\sum_{i\in\mathcal{P}_j}a^2_{j,i}\frac{P_V+P_{ICI}}{P_P}.\end{aligned} \tag{4.8}$$

In the last step, we applied the assumption that the estimation error and channel are uncorrelated. To summarize, Eq. (4.6) is simplified to

$$\sigma_{e,j}^2 = \mathbb{E}\left\{ \left\| h_{\mathscr{D},j} - \hat{h}_{\mathscr{D},j} \right\|^2 \right\}$$
$$= 1 - 2 \sum_{i \in \mathscr{P}_j} a_{j,i} \Re\left\{R_{j,i}\right\} + \sum_{i \in \mathscr{P}_j} \sum_{i' \in \mathscr{P}_j} a_{j,i} a_{j,i'} R_{i,i'} + \frac{P_\mathrm{V} + P_\mathrm{ICI}}{P_\mathrm{P}} \sum_{i \in \mathscr{P}_j} a_{j,i}^2. \tag{4.9}$$

A part of Eq. (4.9) is independent of the term $\frac{P_\mathrm{V}+P_\mathrm{ICI}}{P_\mathrm{P}}$, as it only depends on the weights $a_{j,i}$ and on the correlation matrix. The dependency of the correlation matrix can also be regarded as dependency on the maximum Doppler spread (or equivalently the user velocity) and RMS delay spread. The second part of the MSE depends on $\frac{P_\mathrm{V}+P_\mathrm{ICI}}{P_\mathrm{P}}$. This term is on the other hand independent of the correlation matrix. In order to obtain the overall, MSE additional averaging over all data-symbols has to be applied

$$\sigma_e^2 = \frac{1}{N_\mathrm{D}} \sum_{j=1}^{N_\mathrm{D}} \sigma_{e,j}^2. \tag{4.10}$$

The MSE error can be decomposed into a noise dependent part and a noise independent part

$$\sigma_e^2 = c_e \frac{P_\mathrm{V} + P_\mathrm{ICI}}{P_\mathrm{P}} + d, \tag{4.11}$$

where c_e is a scalar depending on the weights $a_{j,i}$, given as

$$c_e = \frac{1}{N_\mathrm{D}} \sum_{j=1}^{N_\mathrm{D}} \sum_{i \in \mathscr{P}_j} a_{j,i}^2 \geq 0. \tag{4.12}$$

The value of the variable c_e is obtained as arithmetic average over all data-symbol positions. Its depends only on the interpolation weights $a_{j,i}$ of the estimator. The constant d is a scalar depending on the weights $a_{j,i}$ as well as the correlation matrix. In the following text, it is referred to as interpolation error (variance) d, given as

$$d = \frac{1}{N_\mathrm{D}} \sum_{j=1}^{N_\mathrm{D}} \left(1 - 2 \sum_{i \in \mathscr{P}_j} a_{j,i} \Re\left\{R_{j,i}\right\} + \sum_{i \in \mathscr{P}_j} \sum_{i' \in \mathscr{P}_j} a_{j,i} a_{j,i'} R_{i,i'} \right) \geq 0. \tag{4.13}$$

The interpolation error variance d is always greater than or equal to zero. If perfect channel knowledge at the pilot-symbol positions is assumed, the resulting MSE equals he interpolation error. Therefore, the interpolation error can be interpreted as the estimation error caused by the interpolation process.

The choice of the interpolation weights $a_{j,i}$ depends on the interpolation strategy. The simplest way to obtain the interpolation weights for doubly-selective channels is a two-dimensional linear interpolation. In this case, the set of S closest pilot-symbols shrinks to those three closest pilot-symbols that define a plane. Additionally, because of geometry reasons, the sum of the interpolation weights has to be equal to one, namely $\sum_{i\in\mathscr{P}_j} a_{j,i} = 1$. An example of a two-dimensional interpolation is

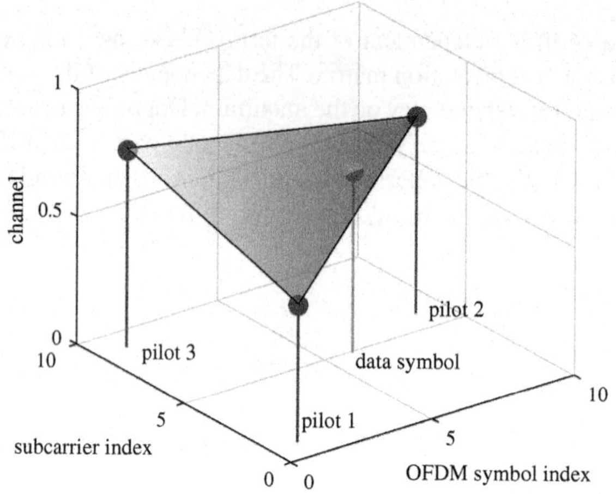

Fig. 4.2 Two-dimensional linear interpolation: channel estimate at a data position (*green line*) is obtained by spanning a two-dimensional plane that is defined by the three closest pilot-symbols (*red lines*)

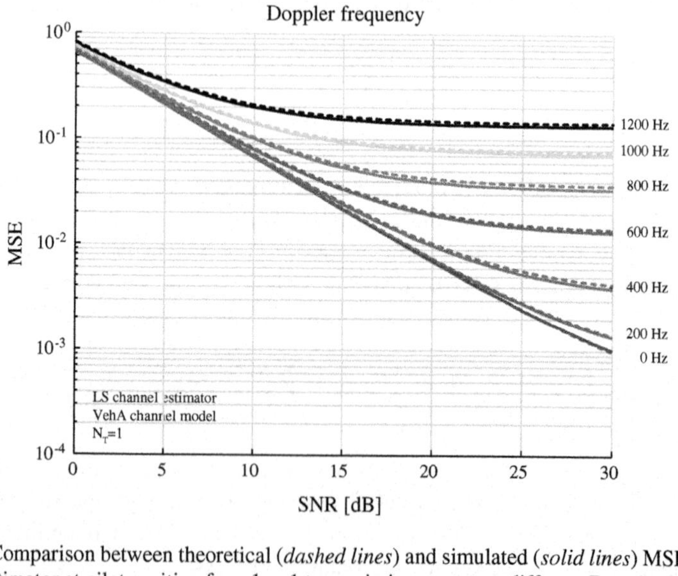

Fig. 4.3 Comparison between theoretical (*dashed lines*) and simulated (*solid lines*) MSE of an LS channel estimator at pilot position for a 1×1 transmission system at different Doppler frequencies and linear interpolation at the data positions

provided in Fig. 4.2. First, three pilot-symbols (red lines), that are closest to a data-symbol (green) line, are located. Second, a plane, that is spanned by these three pilot-symbols, is sampled at the data-symbol position. Note that due to the linear interpolation/extrapolation by a plane, some weights can become negative. More details on how to obtain the interpolation weights for the two-dimensional linear interpolation can be found in [4]. In this case, the interpolation weights depend solely on the location of the pilot-symbols and are independent of the noise variance and channel correlation.

In the case of a two-dimensional linear interpolation, the coefficient c_e depends only on the pilot pattern. In the case of a two-dimensional linear interpolation with an LTE Single-Input Single-Output (SISO) pilot-symbol pattern using an LS channel estimator at pilot positions, delivers $c_e = 0.6623$. Figure 4.3 shows the simulated (solid lines) and theoretical (dashed lines) MSE of an LS channel estimator, for which an excellent agreement can be found.

4.3 Linear Minimum Mean Square Error Channel Estimation

A better choice than the LS estimator is the Linear Minimum Mean Square Error (LMMSE) estimator. Note, however, that the LMMSE channel estimator requires knowledge of the second order statistics of the channel and the noise. It can be shown that the LMMSE channel estimate is obtained by multiplying the LS estimate at the pilot-symbol positions with a filtering matrix $\mathbf{A}_{\text{LMMSE}} \in \mathbb{C}^{N_{\text{D}} \times N_{\text{P}}}$ [5]

$$\hat{\mathbf{h}}_{\mathcal{D}}^{\text{LMMSE}} = \mathbf{A}_{\text{LMMSE}} \hat{\mathbf{h}}_{\mathcal{P}}^{\text{LS}}. \tag{4.14}$$

In order to find the LMMSE filtering matrix, the MSE

$$\sigma_e^2 = \mathbb{E}\left\{ \left\| \mathbf{h}_{\mathcal{D}} - \mathbf{A}_{\text{LMMSE}} \hat{\mathbf{h}}_{\mathcal{P}}^{\text{LS}} \right\|_2^2 \right\}, \tag{4.15}$$

has to be minimized, leading to

$$\mathbf{A}_{\text{LMMSE}} = \mathbf{R}_{\mathbf{h}_{\mathcal{D}},\mathbf{h}_{\mathcal{P}}} \left(\mathbf{R}_{\mathbf{h}_{\mathcal{P}},\mathbf{h}_{\mathcal{P}}} + \frac{P_{\text{V}} + P_{\text{ICI}}}{P_{\text{P}}} \mathbf{I} \right)^{-1}, \tag{4.16}$$

where the matrix $\mathbf{R}_{\mathbf{h}_{\mathcal{P}},\mathbf{h}_{\mathcal{P}}} = \mathbb{E}\{\mathbf{h}_{\mathcal{P}} \mathbf{h}_{\mathcal{P}}^{\text{H}}\} \in \mathbb{C}^{N_{\text{P}} \times N_{\text{P}}}$ denotes the channel correlation matrix at the pilot-symbols, and the matrix $\mathbf{R}_{\mathbf{h}_{\mathcal{D}},\mathbf{h}_{\mathcal{P}}} = \mathbb{E}\{\mathbf{h}_{\mathcal{D}} \mathbf{h}_{\mathcal{P}}^{\text{H}}\} \in \mathbb{C}^{N_{\text{D}} \times N_{\text{P}}}$ is the channel crosscorrelation matrix.

To derive the theoretical MSE, we insert Eq. (4.16) into Eq. (4.15):

$$\sigma_e^2 = \mathbb{E}\left\{\left(\mathbf{h}_{\mathscr{D}} - \left(\mathbf{R}_{\mathbf{h}_{\mathscr{D}},\mathbf{h}_{\mathscr{P}}}\left(\mathbf{R}_{\mathbf{h}_{\mathscr{P}},\mathbf{h}_{\mathscr{P}}} - \frac{P_{\mathrm{V}}+P_{\mathrm{ICI}}}{P_{\mathrm{P}}}\mathbf{I}\right)^{-1}\hat{\mathbf{h}}_{\mathscr{P}}^{\mathrm{LS}}\right)\right)\right. \\ \left.\left(\mathbf{h}_{\mathscr{D}} - \left(\mathbf{R}_{\mathbf{h}_{\mathscr{D}},\mathbf{h}_{\mathscr{P}}}\left(\mathbf{R}_{\mathbf{h}_{\mathscr{P}},\mathbf{h}_{\mathscr{P}}} - \frac{P_{\mathrm{V}}+P_{\mathrm{ICI}}}{P_{\mathrm{P}}}\mathbf{I}\right)^{-1}\hat{\mathbf{h}}_{\mathscr{P}}^{\mathrm{LS}}\right)\right)^{\mathrm{H}}\right\}. \tag{4.17}$$

After a straightforward manipulation, the average MSE at the data subcarriers is expressed as

$$\sigma_e^2 = \frac{1}{N_{\mathrm{D}}}\mathrm{tr}\left\{\mathbf{R}_{\mathbf{h}_{\mathscr{D}},\mathbf{h}_{\mathscr{D}}} - \mathbf{R}_{\mathbf{h}_{\mathscr{D}},\mathbf{h}_{\mathscr{P}}}\left(\mathbf{R}_{\mathbf{h}_{\mathscr{P}},\mathbf{h}_{\mathscr{P}}} + \frac{P_{\mathrm{V}}+P_{\mathrm{ICI}}}{P_{\mathrm{P}}}\mathbf{I}\right)^{-1}\mathbf{R}_{\mathbf{h}_{\mathscr{P}},\mathbf{h}_{\mathscr{D}}}\right\}, \tag{4.18}$$

where N_{D} is the number of data-symbols.

In order to derive interpolation error d for an LMMSE channel estimator, we neglect the expression $\frac{P_{\mathrm{V}}+P_{\mathrm{ICI}}}{P_{\mathrm{P}}}$. Such an assumption corresponds to perfect knowledge of the channel estimates at the pilot-symbol positions. In this way, interpolation error d can be directly obtained from Eq. (4.18) as

$$d = \frac{1}{N_{\mathrm{D}}}\mathrm{tr}\left\{\mathbf{R}_{\mathbf{h}_{\mathscr{D}},\mathbf{h}_{\mathscr{D}}} - \mathbf{R}_{\mathbf{h}_{\mathscr{D}},\mathbf{h}_{\mathscr{P}}}\mathbf{R}^{-1}_{\mathbf{h}_{\mathscr{P}},\mathbf{h}_{\mathscr{P}}}\mathbf{R}_{\mathbf{h}_{\mathscr{P}},\mathbf{h}_{\mathscr{D}}}\right\}. \tag{4.19}$$

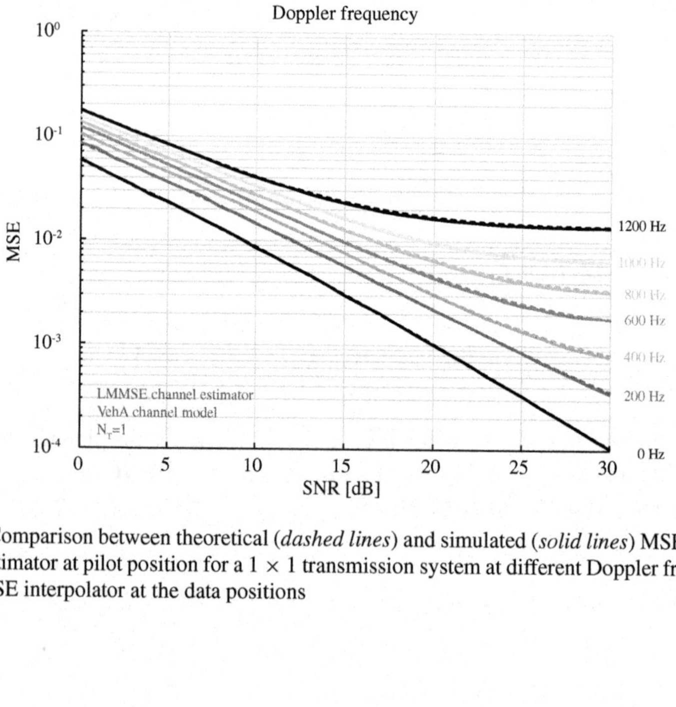

Fig. 4.4 Comparison between theoretical (*dashed lines*) and simulated (*solid lines*) MSE of an LS channel estimator at pilot position for a 1 × 1 transmission system at different Doppler frequencies and LMMSE interpolator at the data positions

An alternative approach to find the LMMSE channel estimate can be obtained using Eq. (4.5), where the set of pilot-symbols $\mathscr{P}_j$ that impact the channel estimate at a specific data position is extended from the three nearest pilot-symbols to all available pilot-symbols. However, in contrast to the previously explained two-dimensional linear interpolator, the weights $a_{j,i}$ depend on the channel correlation matrix and Signal to Noise Ratio (SNR). Therefore, for the LMMSE channel estimator, SNR dependent scalars $c_e(\mathrm{SNR})$ and $d(\mathrm{SNR})$ can be found. In [4], it is shown how to find an SNR independent approximation of the variables c_e and d.

Figure 4.4 depicts the simulated (solid line) and the analytical (dashed line) MSE of an LMMSE channel estimator for different Doppler frequencies. The simulated MSE and the analytically derived MSE show nearly perfect identity. Further generalizations in the context of multi-cell transmissions can be found in [6].

4.4 Optimal Power Allocation

We now describe how to distribute the available power among pilot- and data-symbols in an optimal manner. As the cost function, the post-equalization Signal to Interference and Noise Ratio (SINR) of a Zero Forcing (ZF) equalizer under imperfect channel knowledge is used. The general case of doubly-selective channels is considered. The resulting average post-equalization SINR for a ZF equalizer on the lth layer and kth subcarrier is derived as

$$\bar{\gamma}_{n,k} = \frac{P_{\mathrm{S}}}{\left(P_{\mathrm{V}} + P_{\mathrm{ICI}} + \sigma_e^2 P_{\mathrm{D}}\right) \mathbf{e}_n^{\mathrm{H}} \left(\mathbf{H}_k^{(\mathrm{eff})\mathrm{H}} \mathbf{H}_k^{(\mathrm{eff})}\right)^{-1} \mathbf{e}_n}, \tag{4.20}$$

with matrix $\mathbf{H}_k^{(\mathrm{eff})} = \mathbf{H}_k \mathbf{F}_k$, that is including a potential precoding matrix $\mathbf{F}_k$. In Eq. (4.20), it is assumed that the available data power P_{D} is evenly distributed over individual layers, i.e., $P_{\mathrm{S}} = \frac{P_{\mathrm{D}}}{N_{\mathrm{f}}}$. Furthermore, a-priori knowledge about the channel estimation performance in incldued in the SINR expression of a ZF equalizer by inserting Eq. (4.11) in Eq. (4.20) and decomposing the expression into two parts

$$\bar{\gamma}_{n,k} = f_h\left(\mathbf{H}_k^{(\mathrm{eff})}\right) f_{\mathrm{pow}}\left(c_e, d, P_{\mathrm{S}}, P_{\mathrm{P}}, P_{\mathrm{V}}\right), \tag{4.21}$$

with the so-called power allocation function $f_{\mathrm{pow}}\left(c_e, d, P_{\mathrm{S}}, P_{\mathrm{P}}, P_{\mathrm{V}}\right)$ being

$$f_{\mathrm{pow}}\left(c_e, d, P_{\mathrm{S}}, P_{\mathrm{P}}, P_{\mathrm{V}}\right) = \frac{P_{\mathrm{D}}}{\left(P_{\mathrm{V}} + P_{\mathrm{ICI}} + \left(c_e \frac{P_{\mathrm{V}} + P_{\mathrm{ICI}}}{P_{\mathrm{P}}} + d\right) P_{\mathrm{D}}\right)} \tag{4.22}$$

and the equalizer allocation function

$$f_h\left(\mathbf{H}_k^{(\mathrm{eff})}\right) = \frac{1}{N_\mathrm{f}\mathbf{e}_n^\mathrm{H}\left(\mathbf{H}_k^{(\mathrm{eff})\mathrm{H}}\mathbf{H}_k^{(\mathrm{eff})}\right)^{-1}\mathbf{e}_n}. \tag{4.23}$$

It was shown in [7] that the post-equalization SINR of a ZF equalizer is a random variable following a Gamma distribution for a stationary channel. Therefore, when assuming a stationary channel, the mean value of the equalizer allocation function can be obtained analytically

$$\sigma_{\mathrm{ZF},\mathbf{H}_k^{(\mathrm{eff})}} = \mathbb{E}\left\{f_h\left(\mathbf{H}_k^{(\mathrm{eff})}\right)\right\}. \tag{4.24}$$

The value of $\sigma_{\mathrm{ZF},\mathbf{H}_k^{(\mathrm{eff})}}$ is equal to $N_\mathrm{R} - N_\mathrm{T} + 1$ if neglecting antenna correlation [7, 8]. Inserting Eq. (4.24) in Eq. (4.21), the average post-equalization SINR for a ZF equalizer under imperfect channel knowledge is obtained as

$$\begin{aligned}\bar{\bar{\gamma}} &= \mathbb{E}\{\bar{\gamma}_{n,k}\} \\ &= \frac{P_\mathrm{D}}{\left(\sigma_e^2 P_\mathrm{D} + P_\mathrm{V} + P_\mathrm{ICI}\right)}\sigma_{\mathrm{ZF},\mathbf{H}_k^{(\mathrm{eff})}}.\end{aligned} \tag{4.25}$$

In order to keep the overall transmit power constant, the powers radiated at the pilot-symbols P_P and at the data-symbols P_D have to fulfill the following equation

$$P_\mathrm{P}N_\mathrm{P} + P_\mathrm{S}N_\mathrm{D} = N_\mathrm{P} + N_\mathrm{D}, \tag{4.26}$$

where the variables N_P and N_D represent the number of pilot-symbols and data-symbols in a subframe, respectively. In order to describe the interconnection between the pilot-power P_P and the data-power P_D, we introduce a variable p_off which is the power offset between the power of the pilot-symbols and the data-symbols, denoted by

$$P_\mathrm{P} = p_\mathrm{off} P_\mathrm{D}. \tag{4.27}$$

Therefore, P_P and P_D can be expressed in terms of the variables N_P, N_P, and p_off:

$$P_\mathrm{D} = \frac{N_\mathrm{P} + N_\mathrm{D}}{N_\mathrm{D} + N_\mathrm{P} p_\mathrm{off}}, \tag{4.28}$$

$$P_\mathrm{P} = \frac{N_\mathrm{P} + N_\mathrm{D}}{\frac{N_\mathrm{D}}{p_\mathrm{off}} + N_\mathrm{P}} = p_\mathrm{off} P_\mathrm{D}. \tag{4.29}$$

Inserting Eqs. (4.28) and (4.29) into the power allocation function and simplifying the expression, we obtain

$$f_{\text{pow}}\left(c_e, d, \frac{P_{\text{P}}}{p_{\text{off}}}, P_{\text{P}}, P_{\text{V}}\right) = \frac{1}{(P_{\text{V}} + P_{\text{ICI}})\left(\frac{1}{P_{\text{P}}}(p_{\text{off}} + c_e) + \frac{d}{P_{\text{V}}+P_{\text{ICI}}}\right)}$$
$$= \frac{1}{(P_{\text{V}} + P_{\text{ICI}})\left(\frac{\frac{N_{\text{D}}}{p_{\text{off}}}+N_{\text{P}}}{N_{\text{P}}+N_{\text{D}}}(p_{\text{off}} + c_e) + \tilde{d}\right)}, \tag{4.30}$$

with the variable $\tilde{d}$ being proportional to the interpolation error d (see Eq. (4.13)) and given as

$$\tilde{d} = \frac{d}{P_{\text{V}} + P_{\text{ICI}}}. \tag{4.31}$$

Note that Eq. (4.30) is independent of the channel realization.

Let us proceed to the definition of the optimization problem. The target is to find an optimal value of p_{off} that maximizes the post-equalization SINR in Eq. (4.20) while keeping the overall transmit power constant. The optimization problem can be formulated mathematically as

$$\begin{aligned} \underset{p_{\text{off}}}{\text{maximize}} \quad & \bar{\gamma}_{n,k} \\ \text{subject to} \quad & N_{\text{D}} P_{\text{D}} + N_{\text{P}} P_{\text{P}} = N_{\text{D}} + N_{\text{P}}. \end{aligned} \tag{4.32}$$

In order to maximize the post-equalization SINR with respect to the variable p_{off}, the power allocation function in Eq. (4.30) has to be maximized with respect to the variable p_{off}. The terms $P_{\text{V}} + P_{\text{ICI}}$ and $\tilde{d}$ are larger than or equal to zero and independent of the variable p_{off}. Thus, in order to maximize the power allocation function, the following expression has to be minimized

$$f(P_{\text{P}}, P_{\text{D}}) = \frac{1}{P_{\text{D}}} + \frac{c_e}{P_{\text{P}}}, \tag{4.33}$$

which we refer to as simplified power allocation function. The simplified power allocation function can be further simplified when inserting Eqs. (4.28) and (4.29) leading to

$$\bar{f}(p_{\text{off}}) = \frac{\frac{N_{\text{D}}}{p_{\text{off}}} + N_{\text{P}}}{N_{\text{P}} + N_{\text{D}}}(p_{\text{off}} + c_e). \tag{4.34}$$

The above expression depends solely on the power offset among the pilot- and data-powers not on the actual powers. The minimum of the simplified power allocation

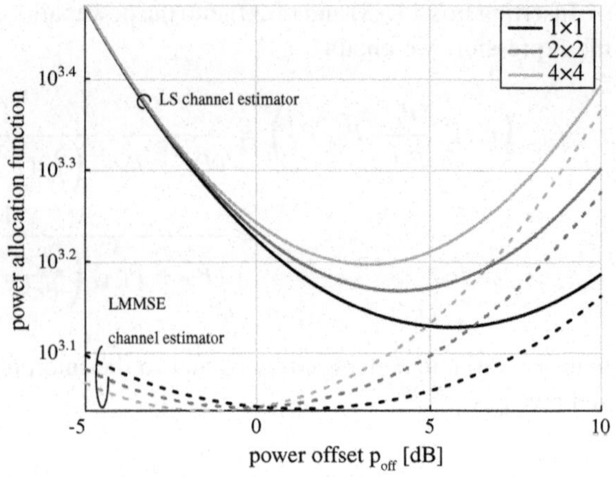

Fig. 4.5 The simplified power allocation function $f(p_{\text{off}})$ for different antenna configurations: the simplified power allocation function $f(p_{\text{off}})$ versus power offset for two state-of-the-art linear channel estimators. The functions have distinct minimum points that correspond to the optimal choice of the power offset

function in Eq. (4.34) can simply be found by differentiation, resulting in the optimal value of the variable p_{off}

$$p_{\text{off,opt}} = \sqrt{\frac{N_{\text{D}} c_e}{N_{\text{P}}}}. \tag{4.35}$$

Figure 4.5 shows an example of the simplified power allocation function from Eq. (4.34) for LS and LMMSE channel estimators for various numbers of transmit antennas utilizing LTE pilot patterns. All of these functions have distinct minimum points. These minimum points correspond to the maximization of the post-equalization SINR. Note that although N_{D} and N_{P} depend on the utilized bandwidth, the minimum of $f(p_{\text{off}})$ is independent of it, since N_{D} and N_{P} scale with the same constant with increasing bandwidth and actually only their ratio is what matters.

4.4.1 Simulation Results

In the following part, we present simulation results and discuss the performance of LTE transmission systems with different pilot-symbol powers under doubly-selective channels. Table 4.1 presents the most important simulator settings. The upper part of the table list simulator settings that remain identical for all simulations, the lower part will change during the following investigations. The performance of the system is demonstrated at an SNR value of 10 dB. Note that as derived earlier, the optimal value of the variable p_{off} is independent of the value of SNR.

Simulation results showing throughput performance for 1×1, 2×2, and 4×4 antenna configurations are shown in Fig. 4.6 for LS and LMMSE channel estimators. In this example, the Doppler frequency is set to 230 Hz and SNR = 10 dB. Little

Table 4.1 Simulator settings

Parameter	Value
Bandwidth	1.4 MHz
Number of data subcarriers	72
FFT size	128
CP duration	$\approx 4.76\,\mu s$
Number of transmit antennas	1, 2, 4
Number of receive antennas	1, 2, 4
Receiver type	ZF
Transmission mode	Open-loop spatial multiplexing
Channel type	ITU VehA [9]
MCS	Adaptive
SNR	10 dB
Doppler frequency	230 Hz

Lower part: settings for power distribution simulations

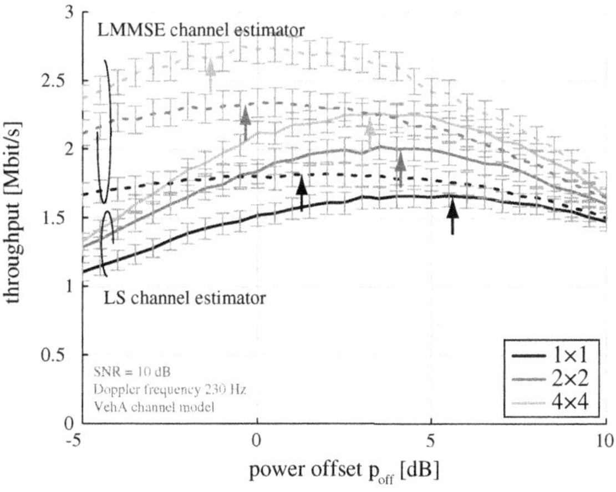

Fig. 4.6 Throughput of LTE system versus p_{off} using different channel estimators and various antenna setups. *Little arrows* always indicate the theoretically derived optimal value of the variable p_{off}, that maximizes the post-equalization SINR. Confidence intervals are 95 %

arrows always indicate the theoretically derived optimal value of the variable p_{off} that maximizes the post-equalization SINR. The simulation results show an excellent match with the analytical solution. The precision of the simulated throughput curves is indicated by 95 % confidence intervals. Using the optimal value of power offsets between pilot- and data-symbols results in throughput maximization. Moving away from this value, a throughput loss can be observed. However, this loss is usually not severe and the relatively broad maximum indicates a high robustness against inaccurate power distribution between pilot- and data-symbols. This is valid especially for a highly accurate channel estimator such as the LMMSE channel estimator.

A negative value of the variable p_{off} (in dB) corresponds to the reduction of the power radiated at the pilot-symbols and an increasing power radiated at the data-symbols. Such a negative value is optimal in the case of four transmit antennas applying an LMMSE estimator. This kind of channel estimator is of superb performance and therefore requires less power at the pilot-symbols to obtain a high quality channel estimate.

Considering a single transmit antenna with an LS channel estimator, the optimal value of $p_{\text{off,opt}} = 5.61\,\text{dB}$ may be considered rather high. However, due to the low number of pilot-symbols compared to the number of data-symbols, the overall power radiated at the pilot-symbols is increased approximately by 6 % compared to the case of the uniform power distribution.

An OFDM transmission system that does not utilize different power levels for its pilot- and data-symbols, corresponds to $p_{\text{off}} = 0\,\text{dB}$. In Fig. 4.6 can be observed, that using optimal power distribution results in throughput gains up to 10 %. For the LMMSE channel estimators, only a small gain can be observed because the values of $p_{\text{off,opt}}$ are close to 1 (0 dB). Therefore, a system with equal power radiated at the pilot- and data-symbols is already close to the optimum. Although the provided results are shown in the context of the current LTE standard, the presented concept is not limited to it and can principally be applied to any MIMO OFDM based system.

4.5 Optimal Pilot Patterns

At this point, it is possible to analytically express the performance of a linear channel estimator as a function of D_{f} and D_{t} for diamond-shaped pilot-symbol patterns. With this knowledge it is possible to maximize the post-equalization SINR in Eq. (4.21) simply by maximizing the power allocation function in Eq. (4.22) [10]. However, the optimal values of D_{f} and D_{t} and the optimal power distribution between the pilot- and the data-symbols cannot be found exclusively by maximizing the post-equalization SINR. Such an approach leads to a solution with minimum distances between adjacent pilot-symbols in the time and frequency directions, which would decrease the available bandwidth for data transmission. Therefore, another cost function is required that includes a penalty due to the bandwidth occupied by the pilot-symbols. The constrained channel capacity is thus a natural choice for the new cost function since it provides a more accurate estimate of the expected throughput than capacity [11, 12]

$$\mathscr{C} = B_s \sum_{n=1}^{N_{\text{f}}} \sum_{k} \log_2 \left(1 + \gamma_{n,k}\right), \tag{4.36}$$

where B_s is the bandwidth utilized for the data transmission of a subcarrier and $\gamma_{n,k}$ is the post-equalization SINR at a layer n and a subcarrier k. The second sum in the above equation is a sum over all subcarriers carrying data-symbols.

The constrained capacity in Eq. (4.36) is realistic as a waterfiling solution for a temporally changing channel is not feasible. Note also that for typically measured frequency selective MIMO channels the difference between the waterfilling solution (capacity) and the proposed constrained version is very small [13]. The impact of precoding matrices can be included in the channel estimation, as we consider here. Variable $\gamma_{n,k}$ represents the instantaneous post-equalization SINR. However, Eq. (4.36) cannot be directly utilized as a cost function, since it would require the knowledge of the instantaneous channel matrix and its estimation error. These are however not available and thus we utilize an ergodic capacity in terms of its expectation value. Such ergodic capacity requires the a-priori knowledge of statistics and is in general difficult to evaluate. However, its upper bound [14, 15] obtained by applying Jensen's inequality when inserting the mean post-equalization SINR Eq. (4.25) in the constrained capacity expression Eq. (4.36), results in

$$\mathscr{C} \leq \bar{\mathscr{C}}, \tag{4.37}$$

$$\bar{\mathscr{C}} = B\left(D_\mathrm{f}, D_\mathrm{t}\right)\log_2\left(1 + \bar{\bar{\gamma}}\right), \tag{4.38}$$

$$= B\log_2\left(1 + f_\mathrm{pow}\left(c_e, d, P_\mathrm{S}, P_\mathrm{P}, P_\mathrm{V}\right)\sigma_{\mathrm{ZF},\mathbf{H}_k^{(\mathrm{eff})}}\right). \tag{4.39}$$

Due to simplicity reasons in the above equations, we omit the dependency of the variables $B\left(D_\mathrm{f}, D_\mathrm{t}\right)$, $c_e\left(D_\mathrm{f}, D_\mathrm{t}\right)$, and $d\left(D_\mathrm{f}, D_\mathrm{t}, \mathbf{R}_\mathrm{h}\right)$ on the variables D_t, D_f, and $\mathbf{R}_\mathrm{h}$ in the above equations. The ultimate target from physical-layer perspective is to maximize throughput. However, throughput is very difficult if not impossible to track analytically [12], which would allow a low complexity solution. Therefore, an analytical performance measure is required that allows to predict throughput including channel estimation error. The presented upper bound of the constrained capacity fulfils these requirements.

4.5.1 *Unit Pilot Power*

In the following, we present throughput simulation results and compare the performance of two competitive wireless transmission systems in order to quantify the performance gain provided by optimal pilot patterns compared to conventional fixed pilot patterns. The first system utilizes LTE compliant pilot patterns for the purposes of channel estimation. The second system is the same in all parameters as the first system except for the pilot patterns. It utilizes optimal pilot patterns with unit power distribution among pilot- and data-symbols. We repeated the simulations according to the settings in Table 4.1 with some parameter selected differently as listed in Table 4.2. In order to generate channels with an arbitrary RMS delay spread, we utilized the model presented in [16]. For generating channels with an arbitrary Doppler spread, we employed the modified Rosa Zheng model, (see, e.g., the appendix of [17]).

Figure 4.7a illustrates throughput as a function of Doppler frequency for various MIMO setups at a fixed SNR value of 14 dB and a fixed RMS delay spread of 400 ns. The throughput of the system with LTE pilot patterns is approximately constant

Table 4.2 Simulator settings for optimal pilot patterns with unit power distribution and optimal power distribution

Parameter	Value
Channel type	RMS delay spread = 400 ns
MCS	Adaptive
SNR	14 dB, [0...30 dB]
Doppler frequency	200 Hz, [0...1200 Hz]

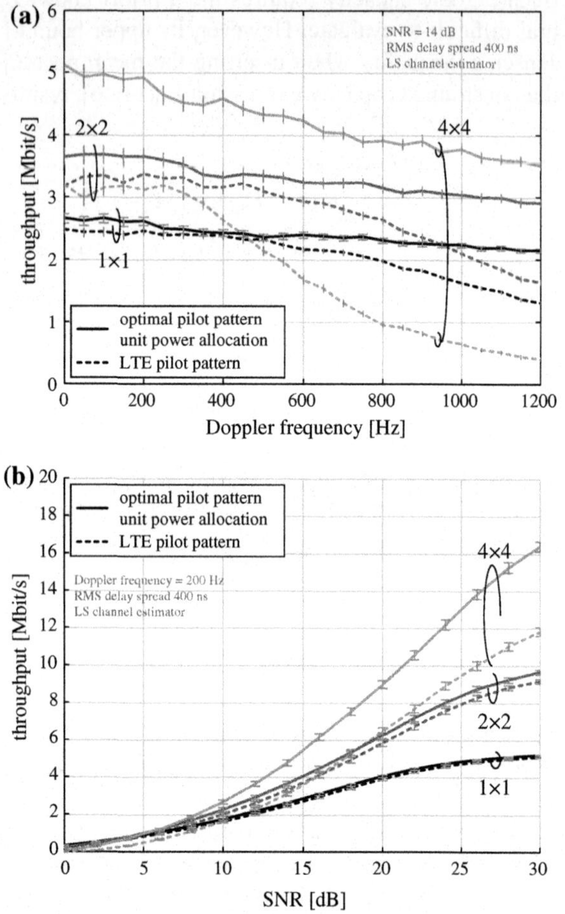

Fig. 4.7 Performance of optimal pilot patterns with unit power distribution: *Dashed lines* represent systems utilizing LTE pilot patterns. *Solid lines* represent systems utilizing optimal pilot patterns with unit power distribution. **a** Throughput versus Doppler frequency for a fixed SNR of 14 dB and a fixed RMS delay spread of 400 ns. **b** Throughput versus SNR for a Doppler frequency of 200 Hz and a fixed RMS delay spread of 400 ns

up to a certain value of Doppler frequency and then is begins to degrade. This Doppler frequency value shifts to lower values as the number of transmit antennas is increased. The performance degradation also worsens with an increasing number of transmit antennas. The performance drop with an increasing Doppler frequency of a 4×4 system with LTE pilot pattern is mainly caused due to the pilot pattern placed on the third and fourth transmit antennas that does not allow a precise estimation of time-variant channels. The systems utilizing optimal pilot patterns always outperform the corresponding LTE systems. Let us consider a 1×1 transmission system. The throughput approximately linearly decreases with an increasing Doppler frequency. At around a Doppler frequency of 350 Hz, a gap between the two competing systems is the smallest from the considered scenarios and therefore we can conclude that the LTE pilot pattern for a single transmit antenna is close to an optimal pilot pattern with unit power distribution at a Doppler frequency of 350 Hz. The performance of the remaining antenna setups with optimal pilot patterns behaves in a similar manner: approximately linear throughput loss with an increasing Doppler frequency. Note that the throughput loss is more significant as the number of transmit antennas is increased. This is caused mainly by the fact that for a higher amount of transmit antennas, there are more channel coefficients to estimate and therefore also more pilot-symbols are required. At Doppler frequencies around 950 Hz, an SISO system utilizing optimal pilot patterns outperforms a 2×2 MIMO system with the LTE pilot pattern. Thus, at high Doppler spreads optimal pilot patterns allow to save hardware cost and complexity that are inherently connected with MIMO applications.

Figure 4.7b illustrates throughput versus SNR for various MIMO setups utilizing either LTE pilot patterns or optimal pilot patterns with unit power distribution among data- and pilot-symbols. In this simulation, the RMS delay spread and the Doppler frequency are fixed to 400 ns and 200 Hz, respectively. Considering the dashed curves, which represent the systems utilizing LTE pilot patterns, a typical behavior can be observed: increasing the number of transmit and receive antennas boosts the overall throughput of the systems. However, at a low SNR value of 0 dB, it is more beneficial to transmit only with a single antenna, because the potential multiplexing benefits cannot be exploited at such a low SNR value. The solid lines represent systems utilizing optimal pilot patterns. These always outperform their corresponding counterparts. Note that for higher values of the Doppler frequency, the 4×4 MIMO system with LTE pilot pattern performs poorly, due to the LTE pilot design on the third and fourth transmit antennas. The standard compliant pilot pattern enables to estimate time-variant channels with a very poor quality [18].

4.5.2 *Optimal Pilot Power*

Naturally, the adaptively selected pilot pattern can now also be combined with optimal power allocation strategies. We thus repeat the previous experiment with optimal power allocation. Figure 4.8a presents throughput simulation results versus Doppler frequency for various MIMO setups at a fixed SNR of 14 dB and a fixed RMS delay spread of 400 ns. The dashed lines represent a transmission system utilizing pilot patterns defined by the LTE standard for various numbers of transmit antennas. The solid lines represent the competing systems utilizing optimal pilot patterns with optimal power distribution among pilot- and data-symbols. The system with optimal pilot patterns outperforms the corresponding benchmark system in the considered Doppler spread range. With increasing Doppler frequency the gap between an LTE

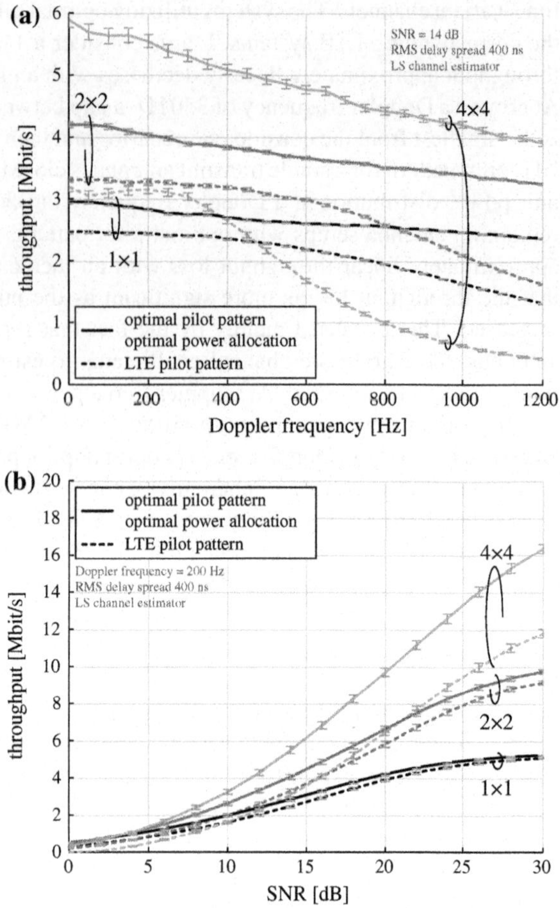

Fig. 4.8 Performance of optimal pilot patterns with optimal power distribution. **a** Throughput performance versus Doppler frequency for a fixed SNR of 14 dB and a fixed RMS delay spread of 400 ns. Throughput as a function of Doppler frequency: *Dashed lines* represent systems utilizing LTE pilot patterns. *Solid lines* represent systems utilizing optimal pilot patterns with optimal power distribution. **b** Throughput performance versus SNR for a fixed Doppler frequency of 200 Hz and a fixed RMS delay spread of 400 ns. Throughput as a function of SNR: *Dashed lines* represent systems utilizing LTE pilot patterns. *Solid lines* represent systems utilizing optimal pilot patterns with optimal power distribution

system and the competing system widens. The throughput increase grows with an increasing number of transmit antennas. Similar to the case of optimal pilot patterns with unit power distribution, the throughput loss of the optimal systems is approximately linear with an increasing Doppler frequency. The loss is more significant as the number of transmit antennas is increased. A comparison of these results with the results from Fig. 4.7a yields a significant improvement of the system utilizing also optimal power distribution and not only optimal pilot patterns.

In the following, we investigate the performance of the considered systems versus SNR for a fixed Doppler and RMS delay spreads. Figure 4.8b shows throughput for various MIMO setups for an RMS delay spread of 400 ns and a Doppler frequency of 200 Hz. The dashed lines represent systems utilizing pilot patterns provided by LTE. The solid lines represent systems utilizing optimal pilot patterns with optimal power distribution. The optimal systems always outperform the LTE benchmark systems. For the 1×1 system, the throughput gain is approximately constant up to an SNR value of 20 dB. After this value, the gain becomes smaller when further increasing SNR. For the 2×2 MIMO setup, the system utilizing optimal pilot patterns outperforms the LTE system by an approximately constant throughput offset in the whole considered SNR range. The situation differs for a 4×4 system. In this case, the pilot pattern provided by LTE fails to estimate precisely enough time-variant channels and therefore the gap between the optimal system and the LTE system widens with increasing SNR. Based on the behavior of the gap size, we can conclude that the LTE pilot patterns for a single and two transmit antennas are close to optimal for high SNR values. Note that these SNR values are far beyond a typical operation point. Thus, the LTE pilot patterns are designed for rather extreme and unrealistic situations in order to operate suboptimally but in a wide range of scenarios. Again it is worth to compare with Fig. 4.7b.

4.6 Adaptive Pilot Sets

Previously, we demonstrated how to design an optimal pilot-symbol pattern for a given SNR value and a given channel autocorrelation matrix. The channel autocorrelation matrix can be decomposed into a time correlation matrix and a frequency correlation matrix [2]. These two correlation matrices depend on the RMS delay spread and maximum Doppler spread, respectively. Therefore, an optimal pilot-symbol pattern is given by a triple of SNR, Doppler frequency, and RMS delay spread values.

In this section, we describe a concept of adaptive pilot-symbol patterns that adjust to varying channel conditions. Furthermore, we investigate the feedback requirements for adaptive pilot-symbol patterns applied in MIMO OFDM systems. The main goal is to support a wide range of Doppler spreads and RMS delay spreads while keeping the number of allowed pilot-symbol patterns at a minimum.

Let us consider a state-of-the-art LTE system for a moment. This system for wireless transmission allows to adapt coding rate, modulation alphabet, precoding, and some other important parameters of the transmission according to the quality of the channel. The main idea in LTE is the usage of the so-called Channel Quality Indicator (CQI) that is reported by the user equipment back to an Evolved Base Station (eNodeB). The CQI is not only a measure of the channel quality, in consequence it also defines two important transmission properties, the coding rate and the modulation alphabet. There are 15 different CQIs defined in LTE. The CQI corresponds to an Additive White Gaussian Noise (AWGN) equivalent SNR value of a channel realization. Therefore, for each CQI value, an optimal pilot pattern should be defined. In this way, no additional feedback overhead is created to distinguish different SNR values, since the feedback for CQI is already implemented in the standard feedback channel.

In order to allow the pilot pattern to adapt to a varying user mobility, pilot patterns for various Doppler spreads (user velocities) have to be defined. Later in this section, we investigate a number of different pilot-symbol patterns required to support a wide range of Doppler spreads. A typical LTE system shall support users moving with velocities up to 500 km/h, which corresponds to a Doppler frequency of approximately 1150 Hz at a carrier frequency of 2.5 GHz. Therefore, we divide the range of the Doppler frequencies between 0 and 1200 Hz into N_{fr} bins and for each bin, an optimal pilot-symbol pattern is defined for the center value of the corresponding bin.

Finally, in order to allow the pilot-symbol patterns to adapt to frequency selectivity of the channel, optimal pilot-symbol patterns have to be designed for different values of RMS delay spread. Typical values of RMS delay spread range between 0 and 800 ns. We divide this range of RMS delay spread into N_{t} bins and for each bin, an optimal pilot-symbol pattern is defined for the center value of the corresponding bin.

Since the same pilot-symbol pattern is used for the entire transmission bandwidth, the extra feedback requirements caused by the adaptive pilot-symbol patterns are $\log_2(N_{\mathrm{fr}})\log_2(N_{\mathrm{t}})$ bits, if coded by brute force. In case of a multi-user transmission, $\log_2(N_{\mathrm{fr}})\log_2(N_{\mathrm{t}})$bits need to be reserved for each user. Note that since a single pilot-symbol pattern is used across the entire transmission bandwidth, its feedback requirements are negligible compared to other narrowband feedback indicators such as CQI, Precoding Matrix Indicator (PMI), and Rank Indicator (RI).

4.6.1 Simulation Results

In the following, we present simulation results and compare the throughput of a system with adaptive pilot-symbol patterns with different bin granularities, against a system with a pilot-symbol pattern defined by LTE standards and unit distribution of power between data- and pilot-symbols. Figure 4.9a illustrates throughput as a function of RMS delay spread for a fixed Doppler frequency of 0 Hz at SNR values of 14 and 30 dB. The blue dashed curves represents the throughput for an SISO LTE transmission system at a given SNR value. The throughput is approximately constant versus RMS delay spread for the LTE system. The blue solid curve represents a system with optimal pilot patters with perfect granularity, i.e., an optimal pilot pattern is generated for each RMS delay value. This system outperforms the standard compliant LTE system in the whole considered RMS delay range. Let us focus on the green solid curve in the 14 dB SNR scenario. This curve represents a system utilizing optimal pilot patterns, but in contrast to the previous case, the RMS delay spread range is divided into two bins, and optimal pilot-symbol patterns generated for the center points are utilized in the corresponding bins. The system with only two different pilot patterns in the considered RMS delay spread range shows the same performance as the competitive system utilizing a much higher number of pilot-symbol patterns. Therefore, we can conclude that for the considered situation of a fixed SNR of 14 dB and a fixed Doppler frequency of 0 Hz, only one bit of extra feedback is required, i.e., $N_{\mathrm{t}} = 2$. Considering the green curve at an SNR value of 30 dB, a throughput drop occurs compared to the perfect case at an RMS delay spread of 350 ns. Therefore, we divide the RMS delay spread operation range into four bins, represented by the red curve. In this case the performance is the same as in the case of perfect granularity.

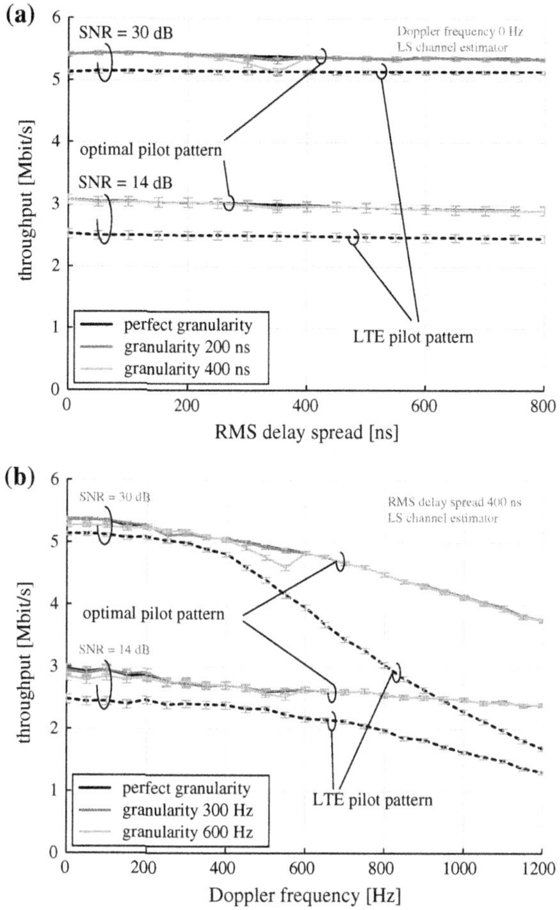

Fig. 4.9 Throughput comparison of an LS with optimal pilot patterns with various feedback granularities at SNR of 14 and 30 dB. **a** Throughput as a function of RMS delay spread for a fixed SNR of 14 and 30 dB, and a fixed Doppler frequency of 0 Hz. Two bits of extra wide-band feedback are sufficient to support a range of RMS delay spread up to 800 ns. **b** Throughput as a function of Doppler frequency for a fixed SNR of 14 dB and a fixed RMS delay spread of 400 ns. Two bits of extra wide-band feedback are sufficient to support a range of Doppler frequency up to 1200 Hz

Therefore, we can conclude that with two bits of extra wide-band feedback the RMS delay spread operation range up to 800 ns can be served.

Figure 4.9b shows throughput versus Doppler frequency for a fixed RMS delay spread of 400 ns at SNR of 14 and 30 dB. The blue dashed curves represents throughput for an SISO LTE transmission system at a given SNR value. A throughput drop occurs as the Doppler frequency increases. The blue solid curve represents a system with optimal pilot patters with perfect granularity. This system outperforms the standard compliant LTE system in the whole considered Doppler frequency range. Let us focus on the green solid curve it the 14 dB SNR scenario. This curve represents a system utilizing optimal pilot patterns, but in contrast to the previous case, the Doppler frequency range is divided into two bins, and optimal pilot-symbol patterns generated for the center points are utilized in the corresponding bins. The system utilizing only two different patterns in the considered Doppler frequency range shows poorer performance than the system with perfect granularity. The performance loss can be observed at two points: first, at low values of Doppler frequency, where too many pilots are utilized for channel estimation and therefore they cannot serve for data transmission. A second point is, at a Doppler frequency of 550 Hz, where the channel is not estimated properly. Therefore, we divide the Doppler frequency range into four bins. The system utilizing four different pilot patterns is represented by the red curve. The performance of such a system is the same as of the system with perfect granularity. Therefore, we can conclude that in order to support Doppler frequencies up to 1200 Hz at an SNR value of 14 dB, four different pilot patterns are required. In order to draw more general conclusions, let us consider the simulation results at a higher SNR value of 30 dB. In general, at higher SNR values the system is more sensitive to the utilized pilot patterns, since more precise channel estimates are required. The curves at an SNR of 30 dB show the same behavior as the 14 dB. Thus, four different pilot patterns are sufficient to support the desired Doppler frequency range. More details about adaptive pilot patterns can be found in [4, 19, 20].

References

1. J. Choi, Y. Lee, Optimum pilot pattern for channel estimation in OFDM systems. IEEE Trans. Wirel. Commun. **4**(5), 2083–2088 (2005). doi:10.1109/TWC.2005.853891
2. M. Šimko, C. Mehlführer, M. Wrulich, M. Rupp, Doubly Dispersive Channel Estimation with Scalable Complexity, in *Proceedings of the 14th International ITG Workshop on Smart Antennas (WSA 2010)*, Bremen (2010), pp. 251–256. doi:10.1109/WSA.2010.5456443
3. M. Šimko, S. Pendl, S. Schwarz, Q. Wang, J.C. Ikuno, M. Rupp, Optimal pilot symbol power allocation in LTE, in *Proceedings of the 74th IEEE Vehicular Technology Conference (VTC2011-Fall)*, San Francisco (2011). doi:10.1109/VETECF.2011.6092929
4. M. Šimko, Pilot pattern optimization for doubly-selective MIMO OFDM transmissions, Ph.D. Thesis, TU Wien, 2013
5. S. Omar, A. Ancora, D. Slock, Performance analysis of general pilot-aided linear channel estimation in LTE OFDMA systems with application to simplified MMSE schemes, in *Proceedings of the IEEE 19th International Symposium on Personal, Indoor and Mobile Radio Communications (PIMRC 2008)*, Cannes (2008), pp. 1–6. doi:10.1109/PIMRC.2008.4699619
6. M. Šimko, M. Rupp, Optimal pilot symbol power allocation in multi-cell scenarios of LTE, in *Conference Record of the Fourtyfifth Asilomar Conference on Signals, Systems and Computers*, Pacific Grove (2011)
7. D. Gore, R. Heath Jr., A. Paulraj, On performance of the zero forcing receiver in presence of transmit correlation, in *Proceedings of the IEEE International Symposium on Information Theory (ISIT 2002)*, Lausanne (2002), p. 159. doi:10.1109/ISIT.2002.1023431
8. P. Li, D. Paul, R. Narasimhan, J. Cioffi, On the distribution of SINR for the MMSE MIMO receiver and performance analysis. IEEE Trans. Inf. Theory **52**(1), 271–286 (2006). doi:10.1109/TIT.2005.860466
9. ITU, Recommendation ITU-R M.1225: Guidelines for evaluation of radio transmission technologies for IMT-, systems, International Telecommunication Union, 1998 (2000)
10. M. Šimko, Q. Wang, and M. Rupp, Optimal pilot symbol power allocation under time-variant channels, EURASIP J. Wirel. Commun. Netw. (2012). doi:10.1186/1687-1499-2012-225
11. C. Mehlführer, S. Caban, M. Rupp, Cellular system physical layer throughput: how far off are we from the Shannon bound? IEEE Wirel. Commun. **18**(6), 54–63 (2011). doi:10.1109/MWC.2011.6108334

12. S. Schwarz, M. Šimko, M. Rupp, On performance bounds for MIMO OFDM based wireless communication systems, *in Proceedings of the IEEE Signal Processing Advances in Wireless Communications*, San Francisco (2011)
13. M. Rupp, J. Garcia-Naya, C. Mehlführer, S. Caban, L. Castedo, On mutual information and capacity in frequency selective wireless channels, in *Proceedings of the IEEE International Conference on Communications (ICC 2010)*, Cape Town (2010). doi:10.1109/ICC.2010.5501942
14. A. Forenza, M. McKay, A. Pandharipande, R. Heath, I. Collings, Adaptive MIMO transmission for exploiting the capacity of spatially correlated channels. IEEE Trans. Veh. Technol. **56**(2), 619–630 (2007). doi:10.1109/TVT.2007.891427
15. S. Loyka, A. Kouki, New compound upper bound on MIMO channel capacity. IEEE Commun. Lett. **6**(3), 96–98 (2002). doi:10.1109/4234.991144
16. K. Hassan, T. Rahman, M. Kamarudin, F. Nor, The mathematical relationship between maximum access delay and the RMS delay spread, in *Proceedings of the 7th International Conference on Wireless and Mobile Communications (ICWMC 2011)*, Luxembourg (2011), pp. 18–23
17. T. Zemen, C. Mecklenbräuker, Time-variant channel estimation using discrete prolate spheroidal sequences. IEEE Trans. Signal Process. **53**(9), 3597–3607 (2005). doi:10.1109/TSP.2005.853104
18. M. Šimko, P.S.R. Diniz, Q. Wang, M. Rupp, Adaptive pilot-symbol patterns for MIMO OFDM systems. IEEE Trans. Wirel. Commun. **12**(9), 4705–4715 (2013)
19. M. Šimko, P.S.R. Diniz, M. Rupp, Design requirements of adaptive pilot-symbol patterns, in *Proceedings of the ICC Workshop: beyond LTE-A*, Budapest (2013)
20. M. Šimko, Q. Wang, M. Rupp, Optimal pilot pattern for time variant channels, in *Proceedings of the IEEE International Conference on Communications (ICC 2013)*, Budapest (2013)

Chapter 5
Single User MIMO LTE Transmission with Quantized Feedback

Markus Rupp, Stefan Schwarz and Martin Taranetz

Since the introduction of single-user spatial multiplexing in radio communications in the early nineties of the last century by A. Paulraj and T. Kailath [1], Single User Multiple-Input Multiple-Output (SU-MIMO) is heavily adopted in commercial systems to leverage the theoretically established advantages of multiple antennas at the transmitter and the receiver for improving the transmission rate (spatial multiplexing) and the reliability (diversity) of the communication channel [2–4]. The highest performance in Multiple-Input Multiple-Output (MIMO) communications is achieved if instantaneous Channel State Information (CSI) is available at both, the transmitter and the receiver. A comprehensive overview of results on the Shannon capacity of SU-MIMO with different assumptions about the availability of CSI at the transmitter and the receiver is provided in [5]. The focus of this chapter is hence on the acquisition of instantaneous CSI. It is assumed that the receivers obtain CSI autonomously using channel estimation, while the base station relies on finite rate feedback links from the users to obtain instantaneous Channel State Information at the Transmitter (CSIT).

In this chapter, the most commonly implemented variant of SU-MIMO in commercial wireless communications, that is, codebook based linear precoding [6], is considered. With codebook based precoding efficient CSI feedback is facilitated, enabling reasonably close to optimal performance with an acceptable feedback overhead [7]. The capacity of the point-to-point Additive White Gaussian Noise (AWGN) MIMO channel with perfect instantaneous CSI at the transmitter and the receiver can be achieved by Singular Value Decomposition (SVD) based unitary precoding and reception along with power loading across the eigenmodes of the channel [8].

M. Rupp (✉) · S. Schwarz · M. Taranetz
Institute of Telecommunications, TU Wien, Vienna, Austria
e-mail: mrupp@nt.tuwien.ac.at

S. Schwarz
e-mail: sschwarz@nt.tuwien.ac.at

M. Taranetz
e-mail: martin.taranetz@tuwien.ac.at

M. Rupp et al., *The Vienna LTE-Advanced Simulators*,
Signals and Communication Technology, DOI 10.1007/978-981-10-0617-3_5

In the considered wireless communication system, motivated by the approach standardized in Long Term Evolution (LTE), a coarse approximation of SVD based precoding is implemented. In this approach the unitary precoders are restricted to a pre-specified set (codebook) of matrices and the continuous power loading is replaced with an on-off switching of spatial data streams, effectively leading to a codebook of semi-unitary precoders. Notice, a $p \times q$ matrix $\mathbf{U}$ with $q \leq p$ is called semi-unitary if $\mathbf{U}^{\mathrm{H}}\mathbf{U} = \mathbf{I}_q$. Based on this codebook, the preferred precoder that optimizes a given performance criterion is determined by the users and this information is efficiently conveyed to the base station by signaling the codebook index together with the number of spatial streams that are switched on, i.e., the transmission rank. This idea of implicit CSI feedback can be straightforwardly extended to MIMO Orthogonal Frequency Division Multiplexing (OFDM), by providing feedback for each Resource Element (RE) separately. To reduce the CSI feedback overhead, the correlation of the wireless channel in the time-frequency domain can be exploited by means of clustered feedback [9] or feedback interpolation [10]. Again motivated by the approach standardized in LTE, clustered feedback is considered. Here, the optimal precoder is determined for a set of REs, denoted as Resource Block (RB), effectively scaling down the feedback overhead by the size of the set in the time domain and in the frequency domain. The feedback clustering considered in this chapter is detailed in Sect. 5.2.

In addition to CSI feedback for MIMO precoding, feedback for transmission rate adaptation is another important topic of this chapter. Transmission rate adaptation is achieved by means of Adaptive Modulation and Coding (AMC), supporting a set of pre-specified combinations of modulation alphabets and coding rates of the Forward Error Correction (FEC), denoted as Modulation and Coding Schemes (MCSs), to cover the projected operating regime of the cellular network. The binary channel coder and the modulation mapper are joined over a bit-interleaver, forming a Bit Interleaved Coded Modulation (BICM) architecture. BICM is employed by many of today's waveform communication systems, due to its high flexibility in allowing to combine virtually any binary code with any modulation format, while providing performance close to Shannon capacity [11, 12]. With the considered system architecture of Fig. 1.6, the same MCS is applied on all REs that are assigned to a user. Hence, an average channel quality measure must be defined that realistically represents the achievable transmission rate over a multitude of OFDM subcarriers. To this end, effective Signal to Interference and Noise Ratio (SINR) averaging, a technique commonly employed for link to system level abstraction [13], is used to determine the average channel quality, which is utilized as CSI feedback for transmission rate adaptation.

If multiple users are served in a cell, CSI about the achievable transmission rate can additionally be exploited during multi-user scheduling to achieve a multi-user diversity gain. It has been shown that the sum capacity over Rayleigh fading channels with opportunistic scheduling scales double logarithmically with the number of users, due to the statistically independent fading of the different users; see, e.g., [14]. Leveraging the multi-user diversity in Orthogonal Frequency Division Multiple Access (OFDMA) requires time-frequency selective feedback from the users. Similarly to the precoder feedback, clustered feedback is useful to reduce the CSI feedback overhead of the achievable rate feedback as well. The trade-off between the feedback granularity and the achieved multi-user diversity gain is investigated in a simulation-based study at the end of this chapter.

This chapter is organized as follows: In Sect. 5.1, the concepts of AMC are described in more detail and the principles of codebook based precoding and capacity achieving SVD based precoding are reviewed, considering a system architecture in accordance with the LTE standard. Taking into account the constraints mentioned above, a discrete joint-optimization problem for finding the optimal feedback indicators for the achievable transmission rate, the precoders and the transmission rank is defined in Sect. 5.2. To reduce the complexity of an exhaustive search, a suboptimal sequential solution of the optimization problem is derived. Also, an antenna subset selection algorithm is proposed that is useful for improving the performance of codebook based precoding in Distributed Antennas Systems (DASs). In Sect. 5.3, the throughput and Block Error Ratio (BLER) performance of the proposed feedback selection algorithms is investigated by means of Monte-Carlo simulations. Furthermore, the performance is compared to theoretical throughput bounds that take into account the constraints imposed by the considered technology. The details of these bounds are provided in [15, Appendix D]. The methods presented in this chapter are published in [16–20].

5.1 Principles of Link Adaptation and Linear Precoding

5.1.1 Link Adaptation

In cellular communications, the quality of the transmission link is subject to significant variations over time and frequency, due to macroscopic (pathloss, shadowing) and microscopic (multipath interference) fading and due to interference from neighboring cells [8]. A common approach to counteract these effects and to improve the reliability and data rate of the transmission is link adaptation, where the transmission parameters of the communication link are adapted to account for the current channel conditions. The preferred link adaptation method of early Code-Division Multiple Access (CDMA) systems is fast power control, where variations in the channel gain are compensated by the transmit power of the base station. More recently, e.g., in High-Speed Packet Access (HSPA) and LTE, link adaptation is implemented through AMC, improving the network capacity by exploiting the fading conditions of the wireless channel [21]. In all cases, link adaptation is based on CSIT.

Here, link adaptation by means of AMC in accordance with the LTE standard [22, 23] is considered. The corresponding BICM system architecture of LTE is illustrated in Fig. 5.1, presenting a more detailed block-diagram of the AMC part of Fig. 1.6. The user input data is processed according to the following steps:

- Mapping of payload bits onto codewords. In LTE, the number of codewords $C_{\ell_k[u,i]}$ of a user is restricted as $C_{\ell_k[u,i]} \leq 2$ and is implicitly determined by the transmission rank, i.e., the number of spatial data streams $\ell_k[u, i]$.

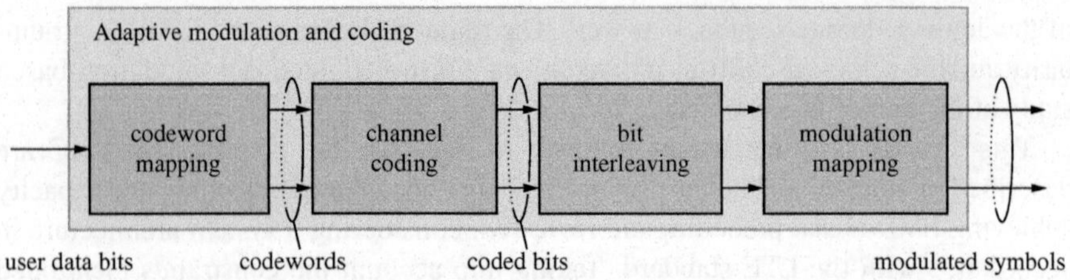

Fig. 5.1 AMC based on a BICM architecture according to the LTE specifications

- Channel coding of the codewords. The coding rate of each codeword can be chosen independently in LTE, accounting for the varying channel quality over spatial streams.
- Bit interleaving of the coded bits. Independent interleaving is applied to each codeword in LTE. Bit interleaving is employed to reduce the impact of error bursts due to channel fading [24], improving the performance of the FEC.
- Mapping of the coded bits onto modulated symbols. In AMC, a set of modulation alphabets $\mathscr{A}$ is supported to enable robust transmission in case of bad channel conditions and high spectral efficiency in case of good channel quality. In LTE, 4 Quadrature Amplitude Modulation (QAM), 16 QAM and 64 QAM are implemented.

Following the AMC stage, the modulated symbols corresponding to the codewords are mapped onto the spatial streams by the layer mapping shown in Fig. 1.6. In case that $\ell_k[u,i] > C_{\ell_k[u,i]}$, a codeword is divided onto multiple spatial streams [22].

The capacity of BICM is derived for memoryless channels in [11]. In Fig. 5.2, a comparison between the capacity achieved with the BICM architecture and the Shannon capacity is shown for a Single-Input Single-Output (SISO) AWGN channel. The three BICM curves correspond to the performance of 4/16/64 QAM. The

Fig. 5.2 Shannon capacity versus BICM capacity and the efficiency of LTE over an AWGN channel

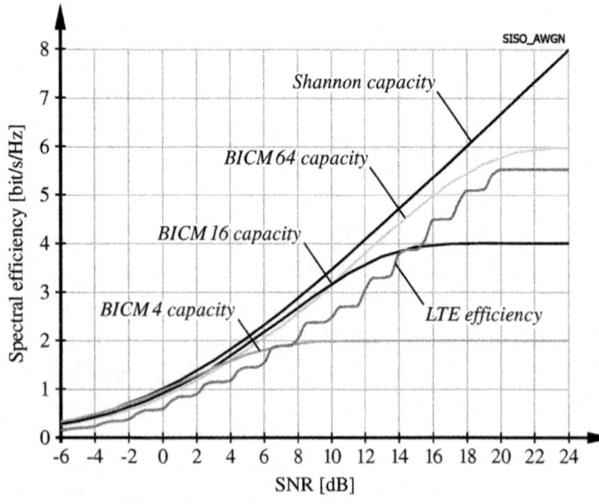

saturation of the BICM capacity at high Signal to Noise Ratio (SNR) occurs due to the finite number of symbols in the modulation alphabets. It can be seen that the theoretical spectral efficiency obtained with BICM is close to the Shannon capacity (implying Gaussian signaling) over a wide SNR range, explaining the popularity of BICM for commercial implementations. Notice though that powerful channel codes and complex detection algorithms are required in practice to achieve close to optimal performance with BICM, involving soft-information exchange between the symbol demapper and the channel decoder and possibly iterative detection [12, 25]. Also shown in the figure is the performance achieved with the MCSs defined in the LTE specifications [23] as obtained from link level simulations employing the Vienna LTE link level simulator [26]. No iterations between the soft-output symbol demapper and the soft-input channel decoder are considered in these simulations and perfect CSI at the receiver is assumed for data detection. The LTE efficiency curve is obtained as the maximum over the 15 MCSs of LTE. The efficiency is calculated only from frequency bins that are used for data transmission, i.e., the additional loss caused by the overhead for reference symbols is not taken into account. The values at the saddle points of this curve are equal to the peak spectral efficiencies of the individual MCSs. A loss of approximately 2 dB is encountered by LTE compared to the BICM capacity. Over a SISO AWGN channel the SNR range from −10 to 20 dB can effectively be covered by the link adaptation of LTE. The incurred loss of LTE compared to the BICM capacity can be attributed to the imperfect operation of the channel code.

5.1.2 Linear Precoding for Single User Muliple-Input Multiple-Output

In SU-MIMO transmission, only one user is served per cell on a given time-frequency resource. Hence, the in-cell interference term $I^{(\mathrm{in})}_{v,u,i}[n,k]$ in the SINR expression (1.44) is equal to zero. In this chapter it is assumed that the out-of-cell interference is treated as additional Gaussian noise by the transmitter and the receiver, and that the users are able to estimate the power of the effective noise defined in Eq. (1.37). Notice that Gaussianity of the out-of-cell interference may not be fulfilled in all cases, e.g., if there are only a few dominant interferers. Then better performance can be achieved with receivers that estimate the interference statistics [27]. However, Gaussianity of the out-of-cell interference can be justified by the central limit theorem, considering the increasing density of cellular networks. With this simplifying assumption, the general input-output relationship of user u in cell i according to Sect. 1.5 is reduced to a point-to-point AWGN MIMO channel, possibly with distributed transmit antennas

$$\mathbf{y}_{n,k}[u,i] = \mathbf{G}_{n,k}[u,i]\mathbf{H}_{n,k}[u,i]\mathbf{F}_{n,k}[u,i]\mathbf{x}_{n,k}[u,i] + \mathbf{G}_{n,k}[u,i]\,\tilde{\mathbf{z}}_{n,k}[u,i], \tag{5.1}$$

where $\tilde{\mathbf{z}}_{n,k}[u,i] \sim \mathcal{N}_{\mathbb{C}}\left(\mathbf{0},\, P_{\tilde{z}}\,\mathbf{I}_{N_{\mathrm{R}}[u,i]}\right)$.

SVD Based Precoding and Equalization

The capacity of the point-to-point AWGN MIMO channel with perfect CSI at the transmitter and the receiver can be achieved by SVD based precoding and equalization [8]. Although perfect CSI is unrealistic in practice, the method still provides a good benchmark for comparing the performance of limited feedback techniques and it is hence briefly reviewed.

The compact-form SVD of the channel matrix $\mathbf{H}_{n,k}[u,i]$ at RE $[n,k]$ can be written as

$$\mathbf{H}_{n,k}[u,i] = \mathbf{U}_{n,k}[u,i]\boldsymbol{\Sigma}_{n,k}[u,i]\mathbf{V}_{n,k}[u,i]^{\mathrm{H}}, \tag{5.2}$$
$$\boldsymbol{\Sigma}_{n,k}[u,i] = \operatorname{diag}\left(\sigma_{n,k}^{(1)}[u,i], \ldots, \sigma_{n,k}^{(\ell_{\max})}[u,i]\right),$$

where the semi-unitary matrices $\mathbf{U}_{n,k}[u,i] \in \mathbb{C}^{N_{\mathrm{R}}[u,i]\times\ell_{\max}}$ and $\mathbf{V}_{n,k}[u,i] \in \mathbb{C}^{N_{\mathrm{T}}[i]\times\ell_{\max}}$ denote the matrices of left singular vectors and right singular vectors, respectively. The νth diagonal element $\sigma_{n,k}^{(\nu)}[u,i]$ of the singular value matrix $\boldsymbol{\Sigma}_{n,k}[u,i]$ is equal to the νth largest singular value of $\mathbf{H}_{n,k}[u,i]$. Assuming a full-rank channel, the maximum number of streams is $\ell_{\max} = \min\left(N_{\mathrm{T}}[i], N_{\mathrm{R}}[u,i]\right)$. By setting the precoder and the receive filter according to

$$\mathbf{F}_{n,k}[u,i] = \mathbf{V}_{n,k}[u,i]\mathbf{P}_{n,k}[u,i]^{1/2}, \tag{5.3}$$
$$\mathbf{G}_{n,k}[u,i] = \mathbf{U}_{n,k}[u,i], \tag{5.4}$$

with $\mathbf{P}_{n,k}[u,i] = \operatorname{diag}\left(p_1[n,k], \ldots, p_{\ell_{\max}}[n,k]\right)$ being a diagonal power loading matrix, the effective channel is decomposed into $\ell_{\max}$ parallel non-interfering AWGN SISO channels, denoted as spatial modes. The input-output relationship simplifies to

$$\left[\mathbf{y}_{n,k}[u,i]\right]_{\nu} = \sqrt{p_{\nu}[n,k]}\sigma_{n,k}^{(\nu)}[u,i]\left[\mathbf{x}_{n,k}[u,i]\right]_{\nu} + \tilde{z}_{n,k}[u,i], \quad \nu \in \{1,\ldots,\ell_{\max}\}, \tag{5.5}$$

with $\tilde{z}_{n,k}[u,i] \sim \mathcal{N}_{\mathbb{C}}\left(0, P_{\tilde{z}}\right)$ because the receive filter is semi-unitary. To achieve the highest transmission rate, the diagonal elements of $\mathbf{P}_{n,k}[u,i]$ have to be set according to the water-filling power allocation policy over the squared singular values normalized by the effective noise variance. The water level is equal to the transmit power P_i [8]. Assuming equal and spatially uncorrelated out-of-cell interference on all receive antennas, the effective noise variance is calculated as

$$\tilde{\sigma}_z^2 = \sigma_z^2 + \frac{1}{N_{\mathrm{R}}[u,i]} \sum_{j=0, j\neq i}^{I} \left\| \mathbf{H}_{n,k}^{(j)}[u,i]\mathbf{F}_{n,k}[j] \right\|^2, \tag{5.6}$$

with $\mathbf{H}_{n,k}^{(j)}[u,i]$ denoting the channel of user u at base station i with respect to base station j, and $\mathbf{F}_{n,k}[j]$ being the precoder applied at base station j. Notice that knowledge of the out-of-cell channel matrices and the precoders applied in the other cells is not required in practice to estimate the effective noise variance [28]. Due to the applied water level, the total power constraint

$$\left\|\mathbf{F}_{n,k}[u,i]\right\|^2 = P_i \tag{5.7}$$

is satisfied by the precoder. In this derivation independent per RE power constraints are assumed. If power loading over REs is considered, the water-filling power allocation is calculated over spatial streams and REs; see [13, 19].

Codebook Based Precoding with Linear Equalization:

In the previous paragraph it is shown that the capacity achieving transmit strategy of the point-to-point AWGN MIMO channel involves linear precoding and reception employing semi-unitary matrices. To calculate the precoder at the base station, knowledge of the left singular matrix together with the SNR obtained on each spatial mode is required. This suggests that with limited feedback these values should be quantized and fed back by the users [29].

Nonetheless, with codebook based precoding a different approach is taken, which is shown to outperform direct quantization of the channel matrix in [6]. Instead of quantizing the channel, the optimal precoder is determined by the user from a given precoder quantization codebook. In [6], criteria for the selection of the optimal precoder and corresponding quantization codebook constructions are proposed for single carrier systems. It is shown that the optimal quantization codebooks for the transmission of ℓ spatial streams are maximally spaced subspace packings in the Grassmann manifold of ℓ-dimensional subspaces in the $N_{\mathrm{T}}[i]$-dimensional Euclidean space $\mathscr{G}(N_{\mathrm{T}}[i], \ell)$. For a short introduction of the Grassmann manifold see [15, Appendix C]. Depending on the considered performance criterion, different distance metrics are employed for the construction of the Grassmannian codebooks, e.g., the projection two-norm, the Fubini-Study distance or the chordal distance [6]. The corresponding codebooks can be represented by sets of semi-unitary matrices, i.e., orthonormal bases, spanning the ℓ-dimensional subspaces. Notice that this approach can also be interpreted as a high SNR approximation of direct channel quantization. At high SNR it is known that the water-filling power allocation converges to equal power allocation over all spatial modes [30]. Therefore, quantization of the left singular matrix only is sufficient, which can be efficiently achieved using a Grassmannian codebook. Alternative codebook designs are based on vector quantization [31], discrete Fourier-transform matrices [32], QAM [33] and other concepts [34].

Motivated by the gains promised by these investigations, codebook based precoding has been implemented in commercial cellular technology, e.g., in LTE's Closed Loop Spatial Multiplexing (CLSM) transmission mode [35]. Hence, the precoder codebook employed for SU-MIMO transmission is assumed as given. Specifically, the precoder codebook for the transmission of ℓ streams over $N_{\mathrm{T}}[i]$ transmit antennas, consisting of semi-unitary precoders, is denoted

$$\mathcal{Q}_\ell^{(N_\mathrm{T}[i])} \subset \left\{\mathbf{Q} \in \mathbb{C}^{N_\mathrm{T}[i]\times \ell} \middle| \mathbf{Q}^\mathrm{H}\mathbf{Q} = \mathbf{I}_\ell\right\}. \tag{5.8}$$

Such precoder codebooks are defined for all possible numbers of data streams $\ell \in \{1, \ldots, N_\mathrm{T}[i]\}$. Assuming that precoder $\mathbf{Q}_{n,k}[u,i] \in \mathcal{Q}_{\ell_k[u,i]}^{(N_\mathrm{T}[i])}$ is selected as the preferred precoder of user u on RE $[n,k]$, the precoder applied for transmission in cell i is

$$\mathbf{F}_{n,k}[u,i] = \sqrt{\frac{P_i}{\ell_k[u,i]}}\mathbf{Q}_{n,k}[u,i], \tag{5.9}$$

accounting for the power constraint P_i of the base station. The power is equally distributed over all $\ell_k[u,i]$ spatial streams. Power allocation among spatial streams is restricted to an on-off switching of spatial modes, by selecting the preferred transmission rank $\ell_k[u,i]$. Hence, the continuous trade-off between MIMO beamforming and spatial multiplexing, which is achieved with the water-filling power allocation employed with SVD based precoding, is coarsely approximated in this case. The capacity loss of SU-MIMO with limited feedback is known to decrease exponentially in the number of feedback bits b [29]. The number of feedback bits b is related to the codebook size as

$$b = \log_2\left(Q_\ell^{(N_\mathrm{T}[i])}\right), \quad Q_\ell^{(N_\mathrm{T}[i])} = \left|\mathcal{Q}_\ell^{(N_\mathrm{T}[i])}\right|. \tag{5.10}$$

To separate the spatial data streams at the receiver, a linear equalizer filter $\mathbf{G}_{n,k}[u,i]$ is employed. Specifically, in the presented simulations Zero Forcing (ZF) and Minimum Mean Square Error (MMSE) equalization are considered, with the corresponding receive filtering matrices

$$\mathbf{G}_{n,k}^{(\mathrm{ZF})}[u,i] = \left((\mathbf{H}_{n,k}[u,i]\mathbf{F}_{n,k}[u,i])^\mathrm{H}\mathbf{H}_{n,k}[u,i]\mathbf{F}_{n,k}[u,i]\right)^{-1}(\mathbf{H}_{n,k}[u,i]\mathbf{F}_{n,k}[u,i])^\mathrm{H}, \tag{5.11}$$

$$\mathbf{G}_{n,k}^{(\mathrm{MMSE})}[u,i] = \left((\mathbf{H}_{n,k}[u,i]\mathbf{F}_{n,k}[u,i])^\mathrm{H}\mathbf{H}_{n,k}[u,i]\mathbf{F}_{n,k}[u,i] + P_{\tilde{z}}\mathbf{I}_{\ell_k[u,i]}\right)^{-1} \times (\mathbf{H}_{n,k}[u,i]\mathbf{F}_{n,k}[u,i])^\mathrm{H}. \tag{5.12}$$

In the presented simulations, the precoder codebooks proposed in the LTE specification [22] are employed. Instead of applying maximally spaced Grassmannian subspace packings, these codebooks are designed having computational and implementation complexity in mind, e.g., by minimizing the amount of complex multiplications involved in precoding or by changing only the phase of the transmit signal and not its amplitude, reducing the requirements posed on amplifier linearity [36]. In case of two transmit antennas, only seven possible precoders are defined by the LTE standard. With four and eight transmit antennas, 64 and 621 precoders are supported, respectively [22]. Other transmit antenna configurations are currently not considered by the standard.

5.2 Implicit CSI Feedback Algorithms

The aim of the CSI feedback selection algorithms proposed in this section is to maximize the instantaneous user throughput, given the CSI available at the receiver as obtained from channel estimation using, e.g., training symbols [37]. Naturally, instantaneous CSI feedback is reasonable only if the current channel estimate at the receiver, employed for calculating the feedback, is representative for the time when the feedback is utilized to adapt the data transmission. Hence, the delay experienced in the feedback path must be sufficiently small compared to the coherence time of the channel to ensure similar channel conditions; see Sect. 5.3 for a more detailed investigation on the impact of a feedback delay on the proposed algorithms. If this cannot be guaranteed, statistical (or long-term) CSI feedback should be employed instead, as proposed, e.g., in [38].

5.2.1 Feedback Clustering

Before going into the details of the proposed feedback selection algorithms, some useful notation is introduced in this section to capture the concept of feedback clustering. As mentioned above, feedback clustering is a technique for reducing the CSI feedback overhead, by exploiting the correlation of the wireless channel in the time-frequency domain. With feedback clustering, the optimal CSI feedback is determined not for each RE individually, but for a set of consecutive time instants and subcarriers. The intuition behind this approach is that the best precoder stays constant over a specific time-frequency interval, due to the restricted size of the precoder codebook $\mathcal{Q}_{\ell}^{(N_{\mathrm{T}}[i])}$ and the correlation of the channel. Therefore, the size of this interval is determined by the coherence time and bandwidth of the channel, but also by the size of the codebook; see [39] for an overview of different clustering approaches for precoder feedback. Similarly, due to the restricted set of supported MCSs, the achievable transmission rate does not change arbitrarily fast. But even if this assumption is not fulfilled and significant variations of the channel occur over the cluster size, optimal feedback indicators can still be determined using appropriate averaging as explained in Sect. 5.2.2, though for the price of a reduced throughput as investigated in Sect. 5.3.3.

The idea of clustering is illustrated in Fig. 5.3. This figure shows the time-frequency resource grid spanned by the OFDM subcarriers and symbols. The pair $[n, k]$ of subcarrier index and symbol-time index is denoted as RE, as already introduced before. With clustering, a set of $N_{\mathrm{clust}}^{(f)}$ subcarriers and $N_{\mathrm{clust}}^{(t)}$ time instants is combined to a so called RB.[1] The number of REs per cluster is $N_{\mathrm{RE}} = N_{\mathrm{clust}}^{(f)} N_{\mathrm{clust}}^{(t)}$.

[1]Notice the slight abuse of LTE notation; the RBs considered can have variable size, while those of LTE consist of exactly 12 subcarriers and 7 OFDM symbols.

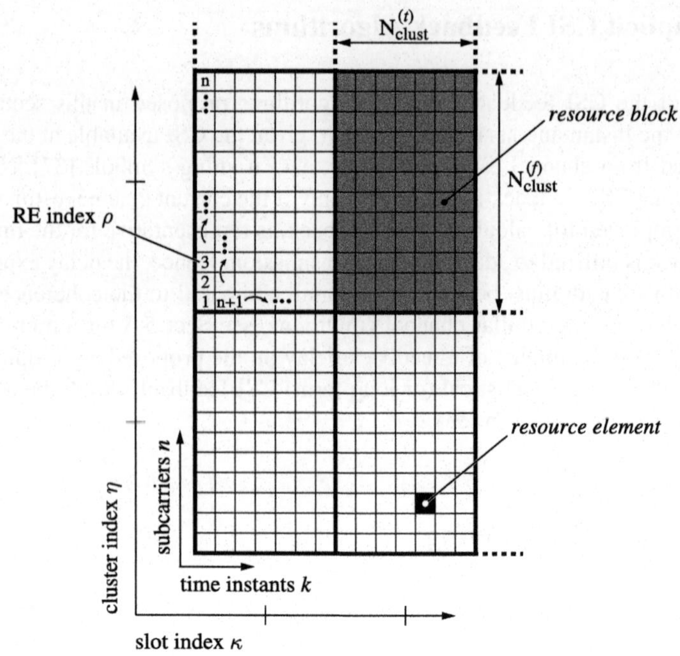

Fig. 5.3 OFDM time-frequency resource grid and clustering of REs into RBs

The time axis is divided into slots by the time domain clustering. The feedback is calculated for each slot κ individually. Similarly, the frequency axis is divided into clusters by the frequency domain clustering. A specific RB is indexed with the pair $[\eta, \kappa]$ of cluster index and slot index. The number of RBs per time slot is $N_{\mathrm{RB}} = \frac{N_{\mathrm{tot}}}{N_{\mathrm{clust}}^{(f)}}$, with N_{tot} denoting the total number of subcarriers. To index the REs of a specific RB $[\eta, \kappa]$, the single RB specific RE index $\rho \in \{1, \ldots, N_{\mathrm{RE}}\}$ is employed, which relates to the index pair $[n, k]$ as

$$k = (\kappa - 1)N_{\mathrm{clust}}^{(t)} + \left\lceil \frac{\rho}{N_{\mathrm{clust}}^{(f)}} \right\rceil, \; n = (\eta - 1)N_{\mathrm{clust}}^{(f)} + \rho - (k-1)N_{\mathrm{clust}}^{(f)}, \tag{5.13}$$

$$\eta = \left\lceil \frac{n}{N_{\mathrm{clust}}^{(f)}} \right\rceil, \; \kappa = \left\lceil \frac{k}{N_{\mathrm{clust}}^{(t)}} \right\rceil, \tag{5.14}$$

$$\rho = \left(n - N_{\mathrm{clust}}^{(f)}(\eta - 1)\right) + \left(k - N_{\mathrm{clust}}^{(t)}(\kappa - 1) - 1\right) N_{\mathrm{clust}}^{(f)}. \tag{5.15}$$

This is equivalent to indexing the REs within an RB along subcarriers one time instant after the other, as illustrated in Fig. 5.3.

With clustering, the same precoder is applied for all REs within an RB. Hence, the RB index η, κ is employed in that case to index the precoders instead of the

RE index n, k, i.e., $\mathbf{Q}_{n,k}[u,i]$ is replaced with $\mathbf{Q}_{\eta,\kappa}[u,i]$. Similarly, the transmission rank is indexed with the time slot index κ instead of the OFDM symbol index k, i.e., $\ell_\kappa[u,i]$.

5.2.2 Feedback Selection Algorithm

With the support of AMC and codebook based precoding with a variable transmission rank, the assumed transceiver architecture is able to adapt the parameters of the SU-MIMO transmission to the current channel conditions such as to maximize the MIMO gain and to ensure a reliable data transmission. This is achieved with the provision of CSIT for each time slot κ using the following three feedback indicators: Channel Quality Indicator (CQI)

- The *preferred MCSs* $\mathbf{m}_{\eta,\kappa}[u,i] \in \mathcal{M}^{C_\ell}$, $\forall \eta \in \{1,\ldots,N_{\mathrm{RB}}\}$ are signaled with the *CQIs*. The set of supported MCSs of the considered technology is denoted $\mathcal{M}$. As the MCS determines the spectral efficiency of the data transmission, the CQI is equivalent to the achievable transmission rate over $\mathrm{RB}_{\eta,\kappa}$. RB and codeword specific CQI feedback is considered to enable exploitation of the multi-user diversity during scheduling.
- The *preferred transmission rank* $\ell_\kappa[u,i]$ is signaled with the *Rank Indicator (RI)*. With rank adaptive transmission, a trade-off between the SNR gain provided by beamforming and the spatial multiplexing gain of the MIMO system is enabled, adjusting the transmission to the current SINR experienced by a user.
- The *Precoding Matrix Indicator (PMI)* is employed for MIMO precoding, selecting the *favored precoders* $\mathbf{Q}_{\eta,\kappa}[u,i]$, $\forall \eta \in \{1,\ldots,N_{\mathrm{RB}}\}$ from the quantization codebooks $\mathcal{Q}_\ell^{(N_{\mathrm{T}}[i])}$, $\ell \in \{1,\ldots,\ell_{\max}\}$. The maximum number of spatial streams is $\ell_{\max} = \min(N_{\mathrm{T}}[i], N_{\mathrm{R}}[u,i])$.

The selection of the preferred feedback indicators is based on maximizing the achievable throughput. Due to the finite block-length of the codewords $c_\kappa[u,i] \in \{1,\ldots,C_{\ell_\kappa[u,i]}\}$ and other imperfections of the channel code, a vanishing BLER is in general not achieved. The system is rather designed to operate below a target BLER $P_b^{(t)}$ that is commonly determined by the application driving the data transmission. This target BLER is considered in the proposed feedback selection algorithm as an implicit constraint of the optimization problem.

In the following, feedback clustering with a cluster size of $N_{\mathrm{clust}}^{(f)}$ in the frequency domain and $N_{\mathrm{clust}}^{(t)}$ in the time domain is assumed. The same cluster size is applied for CQI and PMI feedback to simplify the exposition, although different cluster sizes can be accommodated with an extended notation; see [18]. The RI is a wide band feedback value that is valid for all REs of a time slot κ. The CSI feedback is determined for each slot κ individually; hence, κ is assumed as fixed in the following.

Assuming that precoder $\mathbf{Q}_\eta \in \mathcal{Q}_\ell^{(N_\mathrm{T}[i])}$ is employed during transmission on $\mathrm{RB}_{\eta,\kappa}$, the post-equalization SINR of data stream $\nu \in \{1, \ldots, \ell\}$ on $\mathrm{RE}_{n,k}$ equals

$$\beta_{n,k}[\nu, u, i] = \frac{\frac{P_i}{\ell} \left|\mathbf{g}_{n,k}[\nu, u, i]^\mathrm{H} \mathbf{H}_{n,k}[u, i]^\mathrm{H} \mathbf{q}_\eta[\nu]\right|^2}{\frac{P_i}{\ell} \sum_{\mu=1,\mu\neq\nu}^{\ell} \left|\mathbf{g}_{n,k}[\nu, u, i]^\mathrm{H} \mathbf{H}_{n,k}[u, i]^\mathrm{H} \mathbf{q}_\eta[\mu]\right|^2 + P_{\tilde{z}} \left\|\mathbf{g}_{n,k}[\nu, u, i]\right\|^2}, \tag{5.16}$$

with $\mathbf{g}_{n,k}[\nu, u, i] = \left[\mathbf{G}_{n,k}[u, i]^\mathrm{H}\right]_{:,\nu}$, $\mathbf{q}_\eta[\nu] = \left[\mathbf{Q}_\eta\right]_{:,\nu}$ and $P_{\tilde{z}}$ from Eq. (5.6). Due to SU-MIMO transmission, the in-cell interference term of the more general instantaneous SINR defined in (1.44) does not appear in Eq. (5.16). Also, the semi-unitary precoder $\mathbf{Q}$ is employed in (5.16) instead of its scaled version $\mathbf{F}$ from (5.9) as in Eq. (1.44), to express the dependency of the SINR on the precoder codebook more clearly.

In order to determine the achievable transmission rate over an RB comprising N_RE REs, Mutual Information Effective Signal to Interference and Noise Ratio Mapping (MIESM) is employed in this work to average the corresponding post-equalization SINRs [40]. The idea of MIESM is to identify the time-frequency selective channel experienced over the N_RE REs, with an AWGN channel that achieves the same average spectral efficiency in terms of the BICM capacity. Assuming transmission with MCS $m_\eta \in \mathcal{M}$ on $\mathrm{RB}_{\eta,\kappa}$, this is achieved with the following averaging function

$$\beta_{\eta,\kappa}^{(m_\eta)}[\nu, u, i] = B_{m_\eta}^{-1}\left(\frac{1}{N_\mathrm{RE}} \sum_{\rho=1}^{N_\mathrm{RE}} B_{m_\eta}\left(\beta_\rho[\nu, u, i]\right)\right), \tag{5.17}$$

where the absolute RE index n, k has been replaced with the implicit RB specific RE index ρ using Eq. (5.15). In Eq. (5.17), the function $B_{m_\eta}(\beta)$ is defined as

$$B_{m_\eta}(\beta) = B_{A_{m_\eta}}\left(\frac{\beta}{\phi_{m_\eta}}\right), \tag{5.18}$$

with $B_{A_{m_\eta}}(\beta)$ denoting the BICM capacity of the modulation alphabet $A_{m_\eta} \in \mathcal{A}$ that is associated with the MCS m_η. The scalar ϕ_{m_η} is employed for calibration purposes, to adapt the MIESM averaging to the performance of the different MCSs. The inverse of the bijective function $B_{m_\eta}(\beta)$ is denoted $B_{m_\eta}^{-1}(\cdot)$.

The RB specific AWGN equivalent effective SNR $\beta_{\eta,\kappa}^{(m_\eta)}[\nu, u, i]$ is dependent on the spatial stream index ν. As mentioned in Sect. 5.1.1, in LTE the data of multiple spatial streams can be jointly coded, that is, a codeword $c \in \{1, \ldots, C_\ell\}$ is mapped onto several streams $\nu \in \{1, \ldots, \ell\}$. If this is the case, it is sufficient to provide CQI feedback for each codeword only, instead of for each stream. To accommodate this case, MIESM averaging is applied to determine the average SNR not only over REs, but also over spatial streams. Denoting the set of spatial streams associated with codeword c as $\mathcal{L}_c \subseteq \{1, \ldots, \ell\}$, the codeword dependent effective SNR is defined as

$$\beta_{c,u,i}^{(m_\eta)}[\eta,\kappa] = B_{m_\eta}^{-1}\left(\frac{1}{N_{\text{RE}}\,|\mathscr{L}_c|}\sum_{\nu\in\mathscr{L}_c}\sum_{\rho=1}^{N_{\text{RE}}} B_{m_\eta}\left(\beta_{\nu,u,i}[\rho]\right)\right). \tag{5.19}$$

The AWGN equivalent SNR $\beta_{\eta,\kappa}^{(m_\eta)}[c,u,i]$ is employed to estimate the BLER achieved with MCS m_η on $\text{RB}_{\eta,\kappa}$, using precomputed AWGN look-up tables that quantify the performance of the considered technology. For the AWGN BLERs attained with the MCSs of LTE, which are employed as look-up tables in the simulations, as well as for details on MIESM, see [15, Appendix E]. Denoting the relationship between the SNR β of an AWGN channel and the corresponding BLER of MCS m_η as $g_{m_\eta}(\beta)$, the estimated BLER of codeword c over $\text{RB}_{\eta,\kappa}$ equals

$$P_{\eta,\kappa}^{(m_\eta)}[c,u,i] = g_{m_\eta}\left(\beta_{\eta,\kappa}^{(m_\eta)}[c,u,i]\right). \tag{5.20}$$

To account for the target BLER $P_b^{(t)}$, a function $h_{m_\eta}(P, P_b^{(t)})$ is defined that outputs the spectral efficiency e_{m_η} of MCS m_η if the BLER P is less than the target BLER and zero otherwise

$$h_{m_\eta}(P, P_b^{(t)}) = \begin{cases} e_{m_\eta}, & P \le P_b^{(t)} \\ 0, & P > P_b^{(t)} \end{cases}. \tag{5.21}$$

With this function, the spectral efficiency of codeword c using MCS m_η for the transmission over RB $[\eta,\kappa]$ is estimated as

$$e_{c,u,i}^{(m_\eta)}[\eta,\kappa] = \left(1 - P_{c,u,i}^{(m_\eta)}[\eta,\kappa]\right) h_{m_\eta}\left(P_{c,u,i}^{(m_\eta)}[\eta,\kappa], P_b^{(t)}\right). \tag{5.22}$$

In this equation, the success probability of the data transmission is taken into account in the first term, and the achieved spectral efficiency in case of successful transmission is quantified with the second term. Hence, $e_{c,u,i}^{(m_\eta)}[\eta,\kappa]$ corresponds to the expected value of the spectral efficiency. Notice though that zero spectral efficiency is output in case that the BLER target is not satisfied, preventing these MCSs to be selected by the optimization problem defined below. The target BLER is hence implicitly considered as a hard constraint of the optimization problem.

The optimal feedback values at time slot κ are jointly determined from the following optimization problem maximizing the achievable throughput, i.e., the sum spectral efficiency over all RBs and codewords

$$\left\{\ell_{u,i}[\kappa], \mathbf{Q}_{u,i}[\eta,\kappa], \mathbf{m}_{u,i}[\eta,\kappa]\right\} = \underset{\ell,\mathbf{Q}[\eta],\mathbf{m}[\eta]}{\operatorname{argmax}} \sum_{\eta=1}^{N_{\text{RB}}}\sum_{c=1}^{C_\ell} e_{c,u,i}^{(m_c[\eta])}[\eta,\kappa] \tag{5.23}$$

$$\begin{aligned} \text{subject to: } & \ell \le \min\left(N_{\text{T}}[i], N_{\text{R}}[u,i]\right), \\ & \mathbf{Q}[\eta] \in \mathscr{Q}_\ell^{(N_{\text{T}}[i])}, \ \forall \eta \in \{1,\ldots,N_{\text{RB}}\} \\ & \mathbf{m}[\eta] = \left[m_1[\eta],\ldots,m_{C_\ell}[\eta]\right]^{\text{T}} \in \mathscr{M}^{C_\ell}, \ \forall \eta \in \{1,\ldots,N_{\text{RB}}\}. \end{aligned}$$

In general, a joint optimization over all N_{RB} RBs is required to solve this problem, because the individual summands of the sum over η are coupled over the transmission rank ℓ. Only if ℓ is fixed, the summands in (5.23) decouple and the optimal precoder and MCSs for each RB can be determined independently. Based on this observation, the most efficient way for solving the joint optimization problem (5.23) is

- Fix the transmission rank ℓ and solve the decoupled optimization problem for each η

$$\left\{\mathbf{Q}_{u,i}^{(\ell)}[\eta,\kappa], \mathbf{m}_{u,i}^{(\ell)}[\eta,\kappa], e_{u,i}^{(\ell)}[\eta,\kappa]\right\} = \underset{\mathbf{Q}[\eta],\mathbf{m}[\eta]}{\operatorname{argmax}} \sum_{c=1}^{C_\ell} e_{c,u,i}^{(m_c[\eta])}[\eta,\kappa] \tag{5.24}$$
$$\text{subject to: } \mathbf{Q}[\eta] \in \mathcal{Q}_\ell^{(N_{\mathrm{T}}[i])},$$
$$\mathbf{m}[\eta] \in \mathcal{M}^{C_\ell}.$$

  Here, the optimized efficiency on RB $[\eta,\kappa]$ when transmitting ℓ streams is denoted $e_{u,i}^{(\ell)}[\eta,\kappa]$.
- Maximize the sum of the optimized efficiencies with respect to the transmission rank ℓ

$$\ell_{u,i}[\kappa] = \underset{\ell\in\{1,\ldots,\ell_{\max}\}}{\operatorname{argmax}} \sum_{\eta=1}^{N_{\mathrm{RB}}} e_{u,i}^{(\ell)}[\eta,\kappa], \tag{5.25}$$
$$\mathbf{Q}_{u,i}[\eta,\kappa] = \mathbf{Q}_{u,i}^{(\ell_{u,i}[\kappa])}[\eta,\kappa],$$
$$\mathbf{m}_{u,i}[\eta,\kappa] = \mathbf{m}_{u,i}^{(\ell_{u,i}[\kappa])}[\eta,\kappa].$$

A closed-form solution of this optimization problem is not possible, because the variables are confined to pre-specified codebooks and are hence discrete. As soon as the optimization problem is solved, the feedback indicators are obtained as the codebook indices corresponding to the optimal solutions, i.e., the PMIs are the indices of the precoders $\mathbf{Q}_{u,i}[\eta,\kappa]$ in the codebook $\mathcal{Q}_{\ell_{u,i}[\kappa]}^{(N_{\mathrm{T}}[i])}$, the RI is equal to the transmission rank $\ell_{u,i}[\kappa]$ and the CQIs are determined by the indices of $\mathbf{m}_{u,i}[\eta,\kappa]$ within $\mathcal{M}$.

5.2.3 *Approximate Sequential Solution*

The complexity involved in the exhaustive search required to obtain the optimal solution of Eq. (5.23) may often be too high for practical implementations. For example, finding the optimum of only one RB in an LTE compliant system with eight transmit antennas and four receive antennas already requires a search over almost 10,000 options, which may not be feasible within the strict delay requirements of instantaneous CSI feedback.

To reduce the complexity of the exhaustive search, an approximate sequential solution is proposed, selecting the precoders independently of the MCSs by employing a coarser estimation of the achievable throughput. This estimation is obtained from the BICM capacity. To make the calculation independent of the modulation alphabet $A \in \mathscr{A}$ and hence of the MCS, the *BICM system capacity* is defined as the envelope of the modulation alphabet dependent capacities

$$B(\beta) = \max_{A \in \mathscr{A}} B_A(\beta). \tag{5.26}$$

The RB specific estimated achievable throughput using precoder $\mathbf{Q}[\eta] \in \mathscr{Q}_\ell^{(N_\mathrm{T}[i])}$ is

$$B_{u,i}^{(\mathbf{Q}[\eta])}[\eta, \kappa] = \sum_{\nu=1}^{\ell} \sum_{\rho=1}^{N_\mathrm{RE}} B\left(\beta_{\nu,u,i}[\rho]\right), \tag{5.27}$$

with $\beta_{\nu,u,i}[\rho]$ from Eq. (5.16). Maximizing this value with respect to the precoders, the transmission rank dependent optimal precoders are obtained

$$\left\{\mathbf{Q}_{u,i}^{(\ell)}[\eta, \kappa], B_{u,i}^{(\ell)}[\eta, \kappa]\right\} = \underset{\mathbf{Q}[\eta] \in \mathscr{Q}_\ell^{(N_\mathrm{T}[i])}}{\operatorname{argmax}} \; B_{u,i}^{(\mathbf{Q}[\eta])}[\eta, \kappa], \tag{5.28}$$

with $B_{u,i}^{(\ell)}[\eta, \kappa]$ denoting the transmission rank dependent optimized achievable throughput. The solutions for the precoders can then be employed as a-priori knowledge in Eq. (5.23) to obtain the corresponding transmission rank and MCSs.

If this is still too complex, the transmission rank ℓ can also be selected independently of the MCSs by maximizing the sum of $B_{u,i}^{(\ell)}[\eta, \kappa]$ over the RBs

$$\ell_{u,i}[\kappa] = \underset{\ell \in \{1,\ldots,\ell_\mathrm{max}\}}{\operatorname{argmax}} \sum_{\eta=1}^{N_\mathrm{RB}} B_{u,i}^{(\ell)}[\eta, \kappa], \tag{5.29}$$
$$\mathbf{Q}_{u,i}[\eta, \kappa] = \mathbf{Q}_{u,i}^{(\ell_{u,i}[\kappa])}[\eta, \kappa].$$

Utilizing these solutions as a-priori knowledge in Eq. (5.23) the corresponding MCSs are obtained. The optimization of the MCSs cannot be further simplified to ensure that the upper bound on the BLER specified by $P_b^{(t)}$ is satisfied.

The most expensive step involved in solving these optimization problems is the calculation of the post-equalization SINR in (5.16) for every precoder, because it involves matrix inversions to determine the equalizer $\mathbf{G}_{n,k}[u, i]$. This can be avoided by employing the *pre-equalization mutual information* achieved with Gaussian signaling to estimate the achievable throughput according to [41]

$$I_{u,i}[n, k] = \log_2 \det\left(\mathbf{I}_{N_\mathrm{R}[u,i]} + \frac{P_i}{\tilde{\sigma}_z^2 \ell} \mathbf{H}_{n,k}[u, i]\mathbf{Q}[\eta]\mathbf{Q}[\eta]^\mathrm{H}\mathbf{H}_{n,k}[u, i]^\mathrm{H}\right). \tag{5.30}$$

Replacing $\sum_{\nu=1}^{\ell} B\left(\beta_{\nu,u,i}[\rho]\right)$ in Eq. (5.27) with $I_{u,i}[n,k]$, the same optimization steps as in Eqs. (5.28) and (5.29) can again be performed to calculate suboptimal solutions for the precoders and the transmission rank. Then, the post-equalization SINR hast to be calculated only once, for the optimization of the MCSs. The precoder selection based on the mutual information is also proposed in [6]. The performance of the different methods is compared in Sect. 5.3.2.

5.2.4 *Antenna Subset Selection for DASs*

The available transmit power P_i of cell i is distributed uniformly over all $N_\mathrm{T}[i]$ transmit antennas by the precoders of the considered semi-unitary codebook (5.8). In case of distributed transmit antennas, when pathloss differences between the Remote Radio Units (RRUs) become significant, it is beneficial to concentrate the transmit power on those antenna arrays that experience good channel quality, thus employing only a subset of the antennas for transmission. To transmit the data from $\tilde{N}_i \le N_\mathrm{T}[i]$ antennas, the reduced size codebook $\mathscr{Q}_\ell^{(\tilde{N}_i)}$ is employed to distribute the signal onto the $\tilde{N}_i$ activated antennas, while the other antenna elements are deactivated.

The preferred antenna subset can be determined by solving the optimization problem (5.23) for all possible antenna subset choices, amounting in

$$\sum_{\tilde{N}_i=1}^{N_\mathrm{T}[i]} \binom{N_\mathrm{T}[i]}{\tilde{N}_i} = 2^{N_\mathrm{T}[i]} - 1 \tag{5.31}$$

possibilities and thus in an additional feedback overhead of roughly $N_\mathrm{T}[i]$ bits.

To enable efficient transmit antenna subset selection, with minimal extra CSI feedback overhead, it is assumed that the macroscopic in-cell channel gains with respect to all distributed antenna arrays are known by both, the base station and the user. This is a reasonable, often made assumption in DASs, because the large-scale statistics change only very slowly over time and frequency, and can thus be learned either from the uplink [42], provided the duplex distance is not too large, or via a very low rate feedback link (e.g., during connection setup). The CSI feedback overhead can then be reduced by considering only the antenna subsets with the largest channel gains as possible candidates for antenna subset selection. Thus, e.g., if $\tilde{N}_i = 4$, the optimization problem (5.23) is only solved for the four strongest distributed antennas, reducing the additional CSI feedback overhead to $\log_2\left(N_\mathrm{T}[i]\right)$ bits.

As an alternative to antenna subset selection, power loading over transmit antennas can be employed [43], which requires accurate knowledge of the instantaneous channel gain at the transmitter, thus increasing the CSI feedback overhead significantly.

5.3 Performance Investigation

In this section, the performance of the proposed CSI feedback selection algorithms is investigated by means of Monte-Carlo simulations. Due to the involved non-linear functions (MIESM, BICM capacity) and the discrete nature of the proposed optimization problem, an analytic performance investigation was not successful. Instead, the theoretical throughput bounds developed in [15, Appendix D] are applied to evaluate the performance of LTE using the proposed CSI feedback algorithms, and to identify the dominant sources of the observed throughput loss with respect to the Shannon capacity. In Sect. 5.3.2, the performance of the individual feedback indicators, i.e., the RI, CQI and PMI, is scrutinized and the impact of a delay in the feedback path is evaluated. Finally, in Sect. 5.3.3, the impact of the CSI feedback granularity applied with feedback clustering on the achieved throughput is examined.

The simulations presented here are mostly obtained with the Vienna Long Term Evolution-Advanced (LTE-A) link level simulator [26, 44]. This open source *MATLAB* based LTE compliant link level simulator is publicly available for download, pursuing the goal of facilitating reproducibility in wireless communications research. The link level simulator is augmented with a system level simulator enabling the simulation of large cellular networks with reasonable complexity. The results are presented in dependence of the equivalent average transmit SNR defined as

$$\mathrm{SNR} = \frac{P_i}{\sigma_z^2} = \frac{E_i}{N_0}, \tag{5.32}$$

with E_i denoting the symbol energy and N_0 being the noise power spectral density.

Centralized antenna systems, without RRUs, are studied in the simulations presented in this section. Pathloss and shadow fading are not considered, the instantaneous SNR observed by a user is rather determined by the microscopic channel fading realization and the variance of the receiver noise. In case of multi-user simulations, the average SNR of all users is equal. A block fading channel model is assumed with a temporally constant channel for the duration of one LTE Transmission Time Interval (TTI), denoted as subframe (1 ms or 14 OFDM symbols). The channel realizations of consecutive subframes are either statistically independent or correlated according to Clarke's model. CSI feedback is provided for each TTI. In most simulations a feedback delay of zero is assumed, meaning that the CSIT is acausally available before the transmission. The impact of a feedback delay is separately investigated. Proportional fair multi-user scheduling is applied [45] utilizing the LTE optimized scheduling framework of [46]. Important simulation parameters are summarized in Table 5.1. Notice, parameter sets are specified by curly brackets $\{\cdot\}$; intervals are defined by square brackets $[\cdot]$.

Table 5.1 Simulation parameters of the SUMIMO performance investigation

Parameter	Value
Carrier bandwidth	{1.4, 10} MHz
Carrier frequency	$f_c = 2\,\text{GHz}$
Signal to noise ratio	SNR $\in [-10, 40]$ dB
Number of subcarriers	$N_{\text{tot}} \in \{72, 600\}$
Channel models	Power delay profile based: Vehicular A, pedestrian B [47] Typical urban [48]
Number of receive antennas	$N_{\text{R}}[u, i] \in \{4, 8\}$
Number of transmit antennas	$N_{\text{T}}[i] \in \{4, 8\}$
Number of users	$U_i \in \{1, 2, 4, \ldots, 64\}$
Spatial correlation parameter	$\alpha_{\text{corr}} \in \{0, 0.9\}$
Maximum channel doppler frequency	$f_d \in [10, 500]$ Hz
CSI feedback delay	$\tau_d \in [0, 8]$ TTIs
Multi-user scheduling	Proportional fair
MIMO receiver	ZF

5.3.1 Comparison of LTE to Theoretical Throughput Bounds

In this section, a single user scenario employing the largest LTE compliant antenna configuration of $N_{\text{T}}[i] = 8$ transmit antennas and $N_{\text{R}}[u, i] = 8$ receive antennas is assumed, transmitting over a bandwidth of 1.4 MHz (72 subcarriers). The spatial correlation parameter of the receive antenna array is optimistically assumed as $\alpha_{\text{corr}} = 0$. The performance of LTE's CLSM transmission mode, utilizing PMI, RI and CQI feedback, is compared to the throughput bounds derived in [15, Appendix D]. CSI feedback clustering is applied with $N^{(t)}_{\text{clust}} = 14$ and $N^{(f)}_{\text{clust}} = 12$. The performance with perfect and estimated channel knowledge at the receiver is simulated, employing a least-squares channel estimator [37]. The additional noise caused by the channel estimator has to be considered in the post-equalization SINR $\beta_{\nu,u,i}[n, k]$ used for the calculation of the feedback indicators to ensure an accurate CQI estimation [17]. This is achieved by employing the first-order Taylor approximation proposed in [49] together with the analytic expression of the channel estimation error variance of the least-squares channel estimator derived in [50].

The results of the investigation are shown in Fig. 5.4 in terms of the absolute throughput and the relative throughput with respect to channel capacity. As detailed in [15, Appendix D], the proposed throughput upper bounds shown in the figure account for the throughput loss due to the system overhead for guard bands and reference symbols (achievable capacity), the restriction to codebook based precoding (Closed Loop Mutual Information (CLMI)), the application of a linear receiver (Closed Loop Mutual Information with Linear Receiver (CLMI LR)) and the BICM

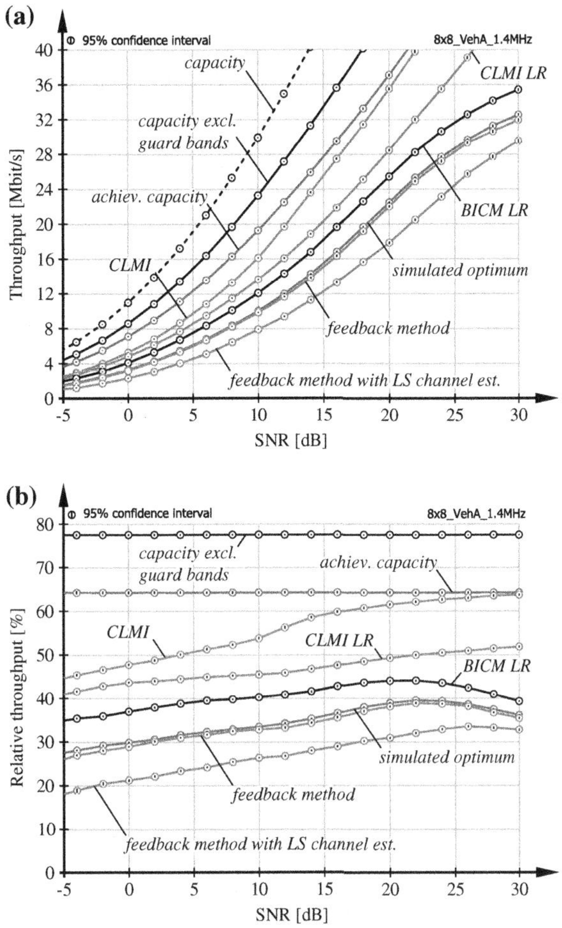

Fig. 5.4 Comparison of the achieved throughput to channel capacity and the proposed throughput bounds with $N_{\mathrm{T}}[i] \times N_{\mathrm{R}}[u, i] = 8 \times 8$. **a** Absolute throughput. **b** Relative throughput

architecture (Bit Interleaved Coded Modulation with Linear Receiver (BICM LR)). It can be seen in Fig. 5.4b that each of these factors entails a throughput loss in the order of 10–20 % of channel capacity, amounting in an achievable throughput of approximately 35–45 % of channel capacity (BICMLR). The simulated optimum performance of LTE with perfect channel knowledge at the receiver, obtained by exhaustive search, is found another 3–7 % below the BICM LR bound. With the proposed feedback selection algorithms close to optimal performance is attained (feedback method). The loss of approximately 1 % can be explained by the non-zero BLER achieved with the proposed CQI selection. With imperfect channel knowledge at the receiver a further throughput reduction to 20–30 % of channel capacity is experienced (feedback method with LS channel est.).

Notice that the performance of LTE is pessimistically represented in these results due to some assumptions of the simulation. With all other LTE compliant carrier bandwidths (3/5/10/20 MHz), the guard band overhead is reduced from 23 to 10 %. The bandwidth of 1.4 MHz was chosen for complexity reasons to be able to determine the optimal performance of LTE with an exhaustive search. The channel estimator performance can be improved, e.g., by means of linear MMSE channel estimation providing close to perfect performance [37]. Also, significantly improved MIMO detection is possible, e.g., with successive interference cancellation [8]. Still, it is questionable whether such highly complex algorithms are feasible in practice in the near future, particularly in mobile phones.

5.3.2 Evaluation of the Feedback Algorithms

The performance of the individual feedback indicators is evaluated in the following in more detail. In the first scenario, an $N_\mathrm{T}[i] \times N_\mathrm{R}[u,i] = 4 \times 4$ antenna configuration is investigated assuming strongly correlated receive antennas, i.e., $\alpha_\mathrm{corr} = 0.9$ in Eq. (5.33). The correlation matrices are generated according to the 3rd Generation Partnership Project (3GPP) document [51, Appendix B.2.3] assuming uniform linear arrays; e.g., the user-side correlation matrices for two and four receive antennas are

$$\mathbf{R}_\mathrm{RX} = \begin{bmatrix} 1 & \alpha_\mathrm{corr} \\ \alpha_\mathrm{corr} & 1 \end{bmatrix}, \quad \mathbf{R}_\mathrm{RX} = \begin{bmatrix} 1 & \alpha_\mathrm{corr}^{\frac{1}{9}} & \alpha_\mathrm{corr}^{\frac{4}{9}} & \alpha_\mathrm{corr} \\ \alpha_\mathrm{corr}^{\frac{1}{9}} & 1 & \alpha_\mathrm{corr}^{\frac{1}{9}} & \alpha_\mathrm{corr}^{\frac{4}{9}} \\ \alpha_\mathrm{corr}^{\frac{4}{9}} & \alpha_\mathrm{corr}^{\frac{1}{9}} & 1 & \alpha_\mathrm{corr}^{\frac{1}{9}} \\ \alpha_\mathrm{corr} & \alpha_\mathrm{corr}^{\frac{4}{9}} & \alpha_\mathrm{corr}^{\frac{1}{9}} & 1 \end{bmatrix}, \qquad (5.33)$$

with $\alpha_\mathrm{corr} \in [0, 1]$ determining the strength of the correlation.

Other simulation parameters are set as in the previous section considering perfect channel estimation at the receiver. The corresponding throughput versus SNR is shown in Fig. 5.5a, comparing the performance of fixed rank transmission ($\ell_{u,i} \in \{1, 2, 3, 4\}$) to rank adaptation by means of the proposed joint feedback selection algorithm of Eq. (5.23). These results show that the proposed feedback selection algorithm is able to determine the optimal transmission rank in dependency of the channel conditions, and automatically trades-off the MIMO beamforming gain for the MIMO multiplexing gain to achieve the best performance. The corresponding BLER of the two codewords is shown in Fig. 5.5b demonstrating that the target BLER $P_b^{(t)} = 0.1$ is observed by the proposed CQI selection. Due to the limited set of only 15 MCSs supported in LTE the achieved BLER is markedly below the constraint $P_b^{(t)} = 0.1$. Notice that the second codeword is activated just in case $\ell_{u,i} \geq 2$; hence the BLER curve starts only at 14 dB SNR.

The same scenario is considered in the results presented in Fig. 5.6a to investigate the performance of the approximate sequential solutions (5.28) and (5.29) of the joint CSI feedback selection problem (5.23). The figure shows that the best performance is

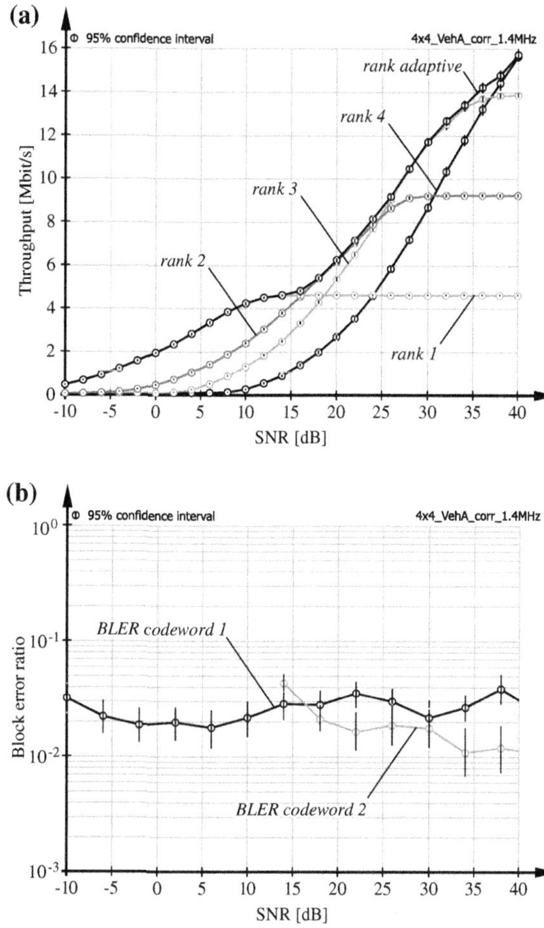

Fig. 5.5 Comparison of the proposed rank adaptive scheme to fixed rank transmission and simulated BLER achieved with the proposed feedback algorithms. **a** Rank adaptation versus fixed rank transmission. **b** BLER achieved with the proposed feedback algorithms

achieved with the accurate rate estimation considered in the joint selection algorithm. If the PMI and RI are selected from the BICM capacity as in Sect. 5.2.3 almost equally good performance is achieved, with a complexity that is approximately a factor of ten below the joint selection in case of LTE. Similar throughput is attained if the pre-equalization mutual information of Eq. (5.30) is employed instead of the BICM capacity for the PMI selection to avoid the calculation of the post-equalization SINR. Notice though that the RI selection based on the pre-equalization mutual information leads to a significant throughput reduction, because the performance of higher rank transmission is obviously overestimated. For comparison the achieved transmission rate with random precoder selection is also shown in Fig. 5.6a. It can be seen that a gain of approximately 5–10 dB in SNR is obtained with optimal precoder selection. Similar behavior is observed in Fig. 5.6b for the case of eight transmit antennas. Notice that the LTE precoder codebook for eight antennas is four times larger than the four antenna codebook.

Fig. 5.6 Performance of approximate sequential CSI feedback selection schemes compared to the joint optimization. **a** $N_i \times M_{u,i} = 4 \times 4$. **b** $N_i \times M_{u,i} = 8 \times 4$

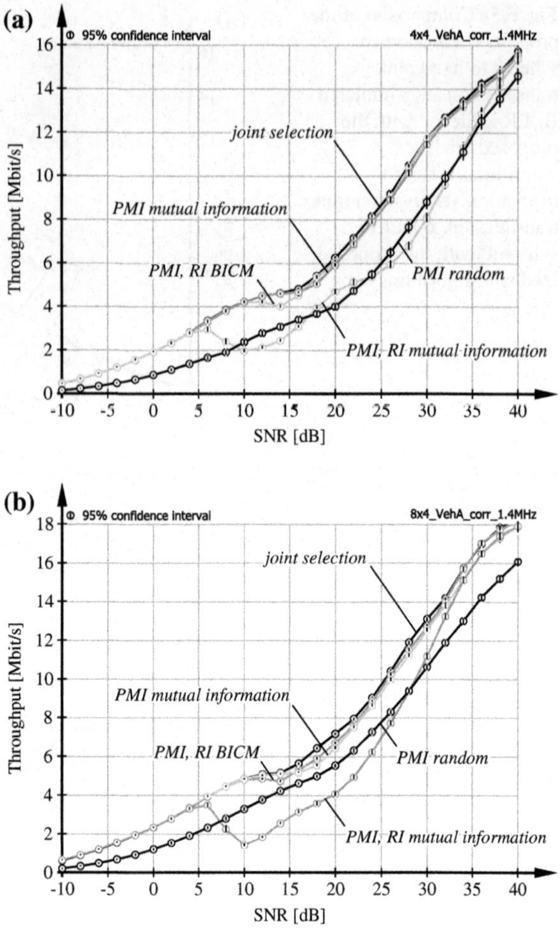

Next, the impact of a delay in the feedback path on the transmission rate obtained with the proposed instantaneous CSI feedback algorithms is investigated. Again, the $N_\mathrm{T}[i] \times N_\mathrm{R}[u,i] = 4 \times 4$ configuration is assumed, this time with uncorrelated receive antennas, i.e., $\alpha_\mathrm{corr} = 0$. The effect of the feedback delay is determined by the strength of the temporal correlation of the wireless channel. Hence, the performance of the system is evaluated in terms of the maximum channel Doppler frequency f_d, defining the channel coherence time according to Eq. (5.34). The 50 % coherence time of the channel, assuming Clarke's model [52], is defined as [53]

$$T_C = \frac{9}{16\pi f_d}, \tag{5.34}$$

where f_d is the maximum channel Doppler frequency. The results are plotted in dependency of the normalized channel Doppler frequency

$$\nu_d = f_d \, T_s, \tag{5.35}$$

with $T_s = 1$ ms being the temporal sampling rate of the channel, i.e., the CSI feedback interval. The throughput degradation due to the feedback delay is shown in Fig. 5.7a. With a delay of 1 TTI, i.e., $\tau_d = 1$ ms, optimal performance is achieved up to a normalized Doppler frequency of $\nu_d^{(c)} = 0.025$, corresponding to a user speed of

$$v = \frac{\nu_d^{(c)}}{T_s \, f_c} c \approx 15 \,\text{km/h}, \tag{5.36}$$

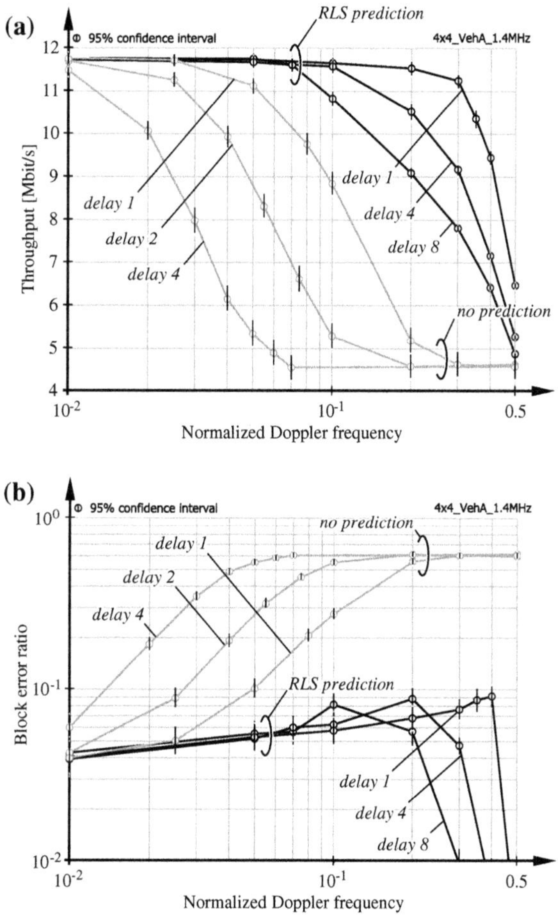

Fig. 5.7 Sensitivity of the proposed instantaneous CSI feedback algorithms with respect to a delay in the feedback path. **a** Throughput versus channel Doppler frequency. **b** Block error ratio versus channel Doppler frequency

with $f_c = 2\,\text{GHz}$ denoting the carrier frequency and c being the speed of light. With $\tau_d = 4\,\text{ms}$ a further reduction to 5 km/h is incurred. Hence, without any delay-compensation, the proposed algorithms are suitable for quasi-stationary situations. The reason for the performance degradation can be seen in Fig. 5.7b. Due to the mismatch between the channel observed during CSI feedback calculation and the channel for which the outdated CSIT is employed for rate adaptation, the BLER is severely increased and the target BLER is not satisfied. A simple means to overcome this problem is the inclusion of an adaptive link margin to account for the channel mismatch [54].

Alternatively, channel prediction can be employed to compensate for the feedback delay. To demonstrate the capabilities of this approach, Finite Impulse Response (FIR) filter based channel prediction is implemented. For simplicity, each column of the channel matrix $\mathbf{H}_{n,k}[u, i]$ is independently predicted. Also, each RB $[\eta, \kappa]$ is represented by the single channel matrix in the center of the RB for CSI feedback calculation, utilizing the channel sub sampling approach of [16]. The FIR filter coefficients are trained by means of the Recursive Least Squares (RLS) algorithm [55]. The corresponding results, labeled with RLS prediction, are shown in Fig. 5.7. The delay sensitivity is significantly reduced with channel prediction, enabling optimal performance up to user speeds of 100 and 40 km/h at delays of 1 TTI and 8 TTIs, respectively. Interestingly, the throughput reduction at high Doppler frequencies is not caused by an increased BLER. It is rather a consequence of the reduction of the gain of the predicted channel matrix, due to diminishing magnitudes of the FIR filter coefficients with decreasing channel correlation. Hence, the SINR is automatically reduced with decreasing channel correlation, entailing a more conservative CQI estimation.

5.3.3 Impact of CSI Feedback Granularity

The SU-MIMO CSI feedback indicators investigated in this chapter can be exploited in a twofold way to improve the downlink data rate of cellular networks. Firstly, the single user throughput is increased through the optimized selection of the precoders, transmission rank and MCSs enabled by the proposed feedback selection algorithm of Eq. (5.23). Secondly, the CQI feedback is useful in multi-user scheduling to achieve a multi-user diversity gain and hence to improve the sum data rate in a multi-user scenario [46, 56]. In both cases, the potential gain is larger with a smaller CSI feedback cluster size. The downside of reducing the cluster size is an increase in the CSI feedback overhead. This trade-off is investigated below by means of a simulation based study.

In the first scenario the improvement of the single user throughput with reducing PMI feedback cluster size and with increasing precoder codebook size is evaluated. To enable a variable precoder codebook size, which is not considered in LTE, a random codebook of independent and isotropically distributed semi-unitary matrices is generated [57]. Such random codebooks are known to be asymptotically optimal in the case of Identically Distributed (I.I.D.) channel matrices; hence α_{corr} is set equal to zero. The results are averaged over random codebook realizations. An $N_{\mathrm{T}}[i] \times N_{\mathrm{R}}[u, i] = 8 \times 1$ antenna configuration is considered. The typical urban channel model [48] is employed, having a 50 % coherence bandwidth of $B_C = 400$ kHz. A carrier bandwidth of 10 MHz is assumed. The precoder codebook size is varied from one element to 2^{12} elements corresponding to a feedback overhead of 0–12 bits per cluster. The number of feedback clusters per TTI is varied from 1–50. The results of the simulation are shown in Fig. 5.8a. A significant throughput improvement from 14 Mbit/s without precoder feedback to 34 Mbit/s with a feedback overhead of $12 \cdot 50$ bit/ms = 600 kbit/s is obtained. Such investigations are also useful to determine the optimal trade-off between cluster size and codebook size for a fixed amount of CSI feedback bits. For example, the best performance with 10 bits per TTI is obtained with a single cluster using a precoder codebook of size 2^{10}, amounting in 22 Mbit/s downlink throughput. On the contrary, with 10 clusters each using a codebook of

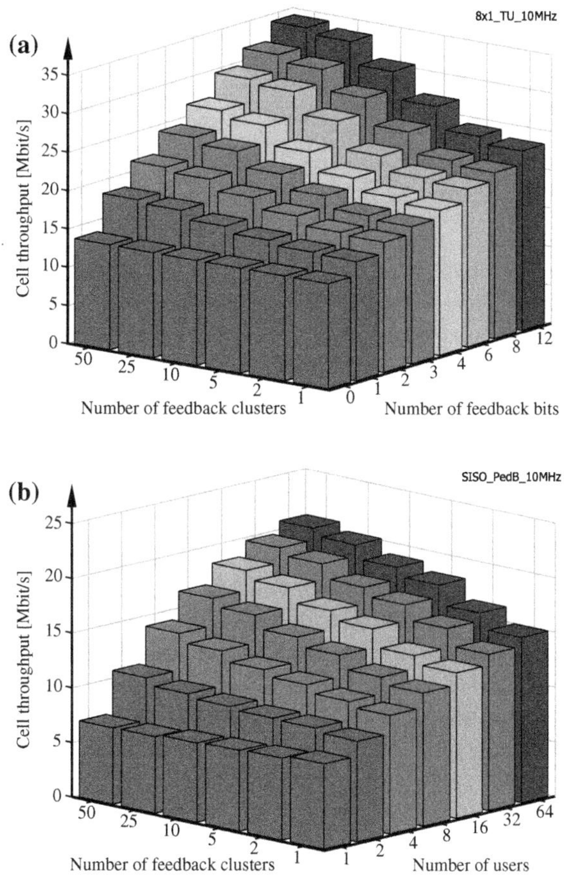

Fig. 5.8 Cell throughput with proportional fair scheduling versus CSI feedback granularity, codebook size and number of users. **a** Variation of PMI feedback granularity and codebook size. **b** Variation of CQI feedback granularity and user number

size two only 17 Mbit/s are achieved. Naturally, these results depend on the channel coherence bandwidth.

In the second scenario the impact of the CQI cluster size on the achievable multi-user scheduling gain is investigated. A 10 MHz SISO system is assumed to eliminate the effect of the precoder feedback. The pedestrian B channel model [47] with a coherence bandwidth of $B_C = 267$ kHz is employed. The number of users U_i is varied from one to 64 and the number of CSI feedback clusters is increased from one to 50. The simulation results are plotted in Fig. 5.8b, visualizing the improvement of the multi-user gain with an increasing number of feedback clusters. With one feedback cluster, only temporal multi-user diversity can be exploited, leading to a throughput improvement by a factor of two when increasing the number of users form one to 64. With 50 feedback clusters a threefold throughput gain is obtained as the frequency diversity can additionally be exploited during scheduling. With 64 users the cell throughput grows by 6 Mbit/s when increasing the number of clusters from one to 50. Notice though that this downlink improvement is outweighed by the increased CQI feedback overhead, scaling from $4 \cdot 64$ bit/ms $= 256$ kbit/s up to $4 \cdot 64 \cdot 50$ bit/ms $=$ 12.8 Mbit/s, if 4 bits per CQI are employed as in LTE. Solutions for this problem have been published, e.g., K-significant scheduling [58] and thresholding, i.e., providing feedback only in case a threshold is exceeded [59].

5.4 Summary

In this chapter, limited feedback SU-MIMO transmission in wireless communications is investigated. Following the approach that is currently implemented in most commercial systems, the transmission parameters of the single user link are adapted by means of AMC and codebook based precoding to ensure reliable and spectrally efficient data transmission. This is achieved by the provision of CSIT using feedback indicators for the selection of the optimal transmission rate, transmission rank and precoders. The best performance is achieved with instantaneous CSIT enabling the exploitation of the space, time, frequency and multi-user diversity inherent in wireless communication systems.

By deriving an accurate estimate of the achievable user data rate, based on the post-equalization SINR, the optimization problem for the joint-selection of the optimal CSI feedback indicators is derived in this chapter. The performance of this method is investigated by means of simulations, demonstrating that close to optimal throughput is achieved with the proposed algorithm. Several simplifying approximations are considered to reduce the complexity of the exhaustive search. A promising trade-off between complexity and performance is obtained by selecting the precoders based on the pre-equalization mutual information and considering the post-equalization SINR only for transmission rank and rate adaptation, significantly reducing the amount of matrix inversion.

The sensitivity of instantaneous CSI feedback with respect to a delay in the feedback path is investigated. The obtained results show that this kind of feedback is restricted to quasi-stationary situations if no delay compensation is incorporated. The scope of the algorithms can be extended to moderate mobility (~50 km/h) if delay compensation by means of channel prediction is implemented. The achievable throughput with limited feedback is strongly dependent on the granularity of the feedback indicators in the OFDM time-frequency grid, requiring a careful choice of the feedback parameters to achieve a good trade-off between the uplink overhead and the downlink throughput.

References

1. A. Paulraj, T. Kailath, Increasing capacity in wireless broadcast systems using distributed transmission/directional reception (DTDR), U.S. patent 5,345,599, Sept. 1994
2. G.J. Foschini, M.J. Gans, On limits of wireless communications in a fading environment when using multiple antennas. Wirel. Pers. Commun. **6**, 311–335 (1998)
3. L. Zheng, D.N.C. Tse, Diversity and multiplexing: a fundamental tradeoff in multiple antenna channels. IEEE Trans. Inf. Theory **49**, 1073–1096 (2002)
4. A. Paulraj, D. Gore, R. Nabar, H. Bolcskei, An overview of MIMO communications—a key to Gigabit wireless. Proc. IEEE **92**(2), 198–218 (2004)
5. A. Goldsmith, S. Jafar, N. Jindal, S. Vishwanath, Capacity limits of MIMO channels. IEEE J. Sel. Areas Commun. **21**(5), 684–702 (2003)
6. D. Love, R. Heath Jr., Limited feedback unitary precoding for spatial multiplexing systems. IEEE Trans. Inf. Theory **51**(8), 2967–2976 (2005)
7. A. Dabbagh, D. Love, Feedback rate-capacity loss tradeoff for limited feedback MIMO systems. IEEE Trans. Inf. Theory **52**(5), 2190–2202 (2006)
8. D. Tse, P. Viswanath, *Fundamentals of Wireless Communication*, Wiley series in telecommunications (Cambridge University Press, Cambridge, 2005)
9. T. Pande, D. Love, J. Krogmeier, Reduced feedback MIMO-OFDM precoding and antenna selection. IEEE Trans. Signal Proc. **55**(5), 2284–2293 (2007)
10. J. Choi, R. Heath Jr., Interpolation based transmit beamforming for MIMO-OFDM with limited feedback. IEEE Trans. Signal Proc. **53**(11), 4125–4135 (2005)
11. G. Caire, G. Taricco, E. Biglieri, Capacity of bit-interleaved channels. Electron. Lett. **32**(12) (1996)
12. A. Guillen i Fabregas, A. Martinez, G. Caire, Bit-interleaved coded modulation. Found. Trends Commun. Inf. Theory **5**, 1–153 (2008)
13. S. Caban, C. Mehlführer, M. Rupp, M. Wrulich, *Evaluation of HSDPA and LTE: From Testbed Measurements to System Level Performance* (Wiley, UK, 2012)
14. P. Viswanath, Opportunistic communication: a system view, *Space-Time Wireless Systems* (Cambridge University Press, Cambridge, 2006), pp. 426–442 (Cambridge Books Online)
15. S. Schwarz, Limited feedback transceiver design for downlink MIMO OFDM cellular networks, Ph.D. dissertation, Technische Universität Wien, 2013, http://theses.eurasip.org/theses/514/limited-feedback-transceiver-design-for-downlink/
16. S. Schwarz, M. Wrulich, M. Rupp, Mutual information based calculation of the precoding matrix indicator for 3GPP UMTS/LTE, in *International ITG Workshop on Smart Antennas* (Bremen, Germany, Feb. 2010) pp. 52–58
17. S. Schwarz, C. Mehlführer, M. Rupp, Calculation of the spatial preprocessing and link adaption feedback for 3GPP UMTS/LTE, in *6th Conference on Wireless Advanced* (London, UK, June 2010) pp. 1–6
18. S. Schwarz, M. Rupp, Throughput maximizing feedback for MIMO OFDM based wireless communication systems, in *IEEE 12th International Workshop on Signal Processing Advances in Wireless Communications*, (San Francisco, CA, June 2011) pp. 316–320

19. S. Schwarz, M. Simko, M. Rupp, On performance bounds for MIMO OFDM based wireless communication systems, in *IEEE 12th International Workshop on Signal Processing Advances in Wireless Communications*, (San Francisco, CA, June 2011) pp. 311–315
20. S. Schwarz, R. Heath Jr., M. Rupp, Single-user MIMO versus multi-user MIMO in distributed antenna systems with limited feedback. EURASIP J. Adv. Signal Proc. **2013**(54), 1–20 (2013)
21. E. Armanious, D. Falconer, H. Yanikomeroglu, Adaptive modulation, adaptive coding, and power control for fixed cellular broadband wireless systems: some new insights. IEEE Wirel. Commun. Netw. **1**, 238–242 (2003)
22. Technical Specification Group RAN, E-UTRA; physical channels and modulation, 3GPP, Technical Report, TS 36.211 Version 10.0.0 (2010)
23. 3GPP, Technical specification group radio access network; evolved universal terrestrial radio access (E-UTRA); multiplexing and channel coding (release 10), http://www.3gpp.org/ftp/Specs/html-info/36212.htm (2010)
24. L.-J. Chen, T. Sun, M. Y. Sanadidi, M. Gerla, *Improving wireless link throughput via interleaved FEC*, in *Proceedings of the 9th International Symposium on Computers and Communications*, (Washington, DC, USA, 2004) pp. 539–544
25. A. Chindapol, J. Ritcey, Design, analysis, and performance evaluation for BICM-ID with square QAM constellations in Rayleigh fading channels. IEEE J. Sel. Areas Commun. **19**(5), 944–957 (2001)
26. C. Mehlführer, J.C. Ikuno, M. Simko, S. Schwarz, M. Rupp, The vienna LTE simulators—enabling reproducibility in wireless communications research. EURASIP J. Adv. Signal Proc. Spec. Issue Rep. Res. **2011**(29), 1–14 (2011)
27. M. Wrulich, C. Mehlführer, M. Rupp, Managing the interference structure of MIMO HSDPA: a multi-user interference aware MMSE receiver with moderate complexity. IEEE Trans. Wirel. Commun. **9**(4), 1472–1482 (2010)
28. T. Yucek, H. Arslan, MMSE noise power and SNR estimation for OFDM systems, in *IEEE Sarnoff Symposium* (2006) pp. 1–4,
29. J. Zheng, B. Rao, Capacity analysis of MIMO systems using limited feedback transmit precoding schemes. IEEE Trans. Signal Proc. **56**(7), 2886–2901 (2008)
30. A. Paulraj, R. Nabar, D. Gore, *Introduction to Space-Time Wireless Communications* (Cambridge University Press, Cambridge, 2003)
31. J. Roh, B. Rao, Design and analysis of MIMO spatial multiplexing systems with quantized feedback. IEEE Trans. Signal Proc. **54**(8), 2874–2886 (2006)
32. D. Yang, L.-L. Yang, L. Hanzo, DFT-based beamforming weight-vector codebook design for spatially correlated channels in the unitary precoding aided multiuser downlink, in *IEEE International Conference on Communications*, (Cape Town, South Africa, 2010) pp. 1–5
33. D. Ryan, I. Vaughan, L. Clarkson, I. Collings, D. Guo, M. Honig, QAM codebooks for low-complexity limited feedback MIMO beamforming, in *IEEE International Conference on Communications*, (Glasgow, Scotland, 2007) pp. 4162–4167
34. T. Inoue, R. Heath, Jr., Kerdock codes for limited feedback MIMO systems, in *IEEE International Conference on Acoustics, Speech and Signal Processing*, (Las Vegas, NV, 2008) pp. 3113–3116
35. 3GPP, Technical specification group radio access network; evolved universal terrestrial radio access (E-UTRA); physical layer procedures (release 10) (2010). http://www.3gpp.org/ftp/Specs/html-info/36213.htm
36. S. Sesia, I. Toufik, M. Baker, *LTE, the UMTS Long Term Evolution: From Theory to Practice, ser* (Wiley InterScience online books, Wiley, 2009)
37. M. Simko, C. Mehlfuhrer, M. Wrulich, M. Rupp, Doubly dispersive channel estimation with scalable complexity, in *International ITG Workshop on Smart Antennas*, (Feb. 2010) pp. 251–256
38. H. Zhang, S. Wei, G. Ananthaswamy, D. Goeckel, Adaptive signaling based on statistical characterizations of outdated feedback in wireless communications. Proc. IEEE **95**(12), 2337–2353 (2007)

39. M. Wu, C. Shen, Z. Qiu, Feedback reduction based on clustering in MIMO-OFDM beamforming systems, in *5th International Conference on Wireless Communications, Networking and Mobile Computing (WiCom)* (2009) pp. 1–4
40. L. Wan, S. Tsai, M. Almgren, A fading-insensitive performance metric for a unified link quality model. IEEE Wirel. Commun. Netw. Conf. **4**, 2110–2114 (2006)
41. Z. Wang, G. Giannakis, Outage mutual information of space-time MIMO channels. IEEE Trans. Inf. Theory **50**(4), 657–662 (2004)
42. M. Jordan, A. Dimofte, X. Gong, G. Ascheid, Conversion from uplink to downlink spatio-temporal correlation with cubic splines, in *IEEE 69th Vehicular Technology Conference*, (April 2009) pp. 1–5
43. D. Lim, K. Choi, H. Liu, Optimum power allocation for distributed antenna systems with large-scale fading-only feedback, in *Sixth International Conference on Information Technology* (New Generations) (April 2009)
44. S. Schwarz, J. Ikuno, M. Simko, M. Taranetz, Q. Wang, M. Rupp, Pushing the limits of LTE: a survey on research enhancing the standard. IEEE Access **1**, 51–62 (2013)
45. R. Jain, D. Chiu, W. Hawe, A Quantitative Measure of Fairness and Discrimination for Resource Allocation in Shared Computer Systems, DEC, Technical Report TR-301, Sept. 1984
46. S. Schwarz, C. Mehlführer, M. Rupp, Low complexity approximate maximum throughput scheduling for LTE, in *Conference Record of the Forty Fourth Asilomar Conference on Signals, Systems, and Computers*, (Pacific Grove, California, Nov. 2010) pp. 1563–1569
47. ITU, Recommendation ITU-R M.1225: Guidelines for Evaluation of Radio Transmission Technologies for IMT-2000, ITU, Technical Report (1997)
48. 3GPP, Technical specification group radio access networks; Deployment aspects (Release 8) (2008). http://www.3gpp.org/ftp/Specs/html-info/25943.htm
49. T. Weber, A. Sklavos, M. Meurer, Imperfect channel-state information in MIMO transmission. IEEE Trans. Commun. **54**(3), 543–552 (2006)
50. M. Simko, S. Pendl, S. Schwarz, Q. Wang, J. Ikuno, M. Rupp, Optimal pilot symbol power allocation in LTE, in *IEEE 74th Vehicular Technology Conference*, (San Francsico, CA, 2011) pp. 1–5
51. Technical Specification Group Radio Access Network, Evolved universal terrestrial radio access (E-UTRA); user equipment (UE) radio transmission and reception, 3GPP, Technical Report TS 36.101 Version 8.5.1 (2009)
52. R.H. Clarke, A statistical theory of mobile radio reception. Bell Syst. Tech. J. **47**, 957–1000 (1968)
53. T. Rappaport, *Wireless Communications: Principles and Practice*, (Prentice Hall communications engineering and emerging technologies series, Dorling Kindersley, 2009)
54. T. Cui, F. Lu, V. Sethuraman, A. Goteti, S. Rao, P. Subrahmanya, Throughput optimization in high speed downlink packet access (HSDPA). IEEE Trans. Wirel. Commun. **10**(2), 474–483 (2011)
55. S. Haykin, *Adaptive Filter Theory*, 4th edn. (Prentice Hall, Upper Saddle River, 2002)
56. S. Schwarz, C. Mehlführer, M. Rupp, Throughput maximizing multiuser scheduling with adjustable fairness, in *International Conference on Communications*,(Kyoto, Japan, June 2011) pp. 1–5
57. N. Ravindran, N. Jindal, Limited feedback-based block diagonalization for the MIMO broadcast channel. IEEE J. Sel. Areas Commun. **26**(8), 1473–1482 (2008)
58. D. Gesbert, M.-S. Alouini, Selective multi-user diversity, in *Proceedings of the 3rd IEEE International Symposium on Signal Processing and Information Technology (ISSPIT)* (2003), pp. 162–165
59. S. Sanayei, A. Nosratinia, Exploiting multiuser diversity with only 1-bit feedback. IEEE Wirel. Commun. Netw. Conf. **2**, 978–983 (2005)

[illegible]

Chapter 6
Multi User and Multi Cell Simulations

Markus Rupp, Stefan Schwarz and Martin Taranetz

Spatial multiplexing of multiple users, i.e., Multi User Multiple-Input Multiple-Output (MU-MIMO), is considered as a promising technique in multi-antenna broadcast systems to achieve high spectral efficiencies by serving multiple users in parallel over the same time-frequency resources [1, 2]. In contrast to Single User Multiple-Input Multiple-Output (SU-MIMO), the potential multiplexing gain of MU-MIMO is only confined by the capabilities of the transmitter. Hence, with MU-MIMO the need for multiple antennas at the users is eliminated, facilitating the development of small and cheap user equipments. Another important advantage of MU-MIMO over SU-MIMO is the reduced sensitivity to channel impairments such as line-of-sight propagation and antenna correlation, which cause an increase in the singular value imbalance of the single-user channel and hence effectively limit the transmission rank of SU-MIMO [3, 4]. Unfortunately, these advantages come at the cost of significantly increased susceptibility of the achievable multi-user throughput with respect to the accuracy of the Channel State Information at the Transmitter (CSIT), due to residual multi-user interference incurred with imperfect channel knowledge [5–8].

The literature on MU-MIMO precoding can be coarsely partitioned into two main concepts, that is, linear precoding and non-linear precoding. While non-linear techniques based on Tomlinson-Harashima precoding [7] and vector perturbation [8, 9] provide an advantage in terms of the achievable throughput, linear strategies such as Zero Forcing (ZF) beamforming [10] and Block Diagonalization (BD) precoding [11] are considered as practically important for complexity reasons. In addition to their complexity advantage, ZF and BD precoding have the benefit that partial CSIT, specifically channel subspace information, is sufficient for the calculation of the precoders, reducing the burden on the feedback channel. Notice that BD precoding

M. Rupp (✉) · S. Schwarz · M. Taranetz
Institute of Telecommunications, TU Wien, Vienna, Austria
e-mail: mrupp@nt.tuwien.ac.at

S. Schwarz
e-mail: sschwarz@nt.tuwien.ac.at

M. Taranetz
e-mail: martin.taranetz@tuwien.ac.at

M. Rupp et al., *The Vienna LTE-Advanced Simulators*,
Signals and Communication Technology, DOI 10.1007/978-981-10-0617-3_6

can be viewed as a generalization of ZF beamforming to multiple data stream transmission per user; hence in the sequel mostly BD precoding is used to refer to both methods.

For the reasons mentioned above, MU-MIMO transmission based on BD precoding is investigated. Although codebook based precoding as in SU-MIMO is an option for MU-MIMO as well [12], the precoders employed with BD precoding are not confined to a codebook but are calculated during operation from explicit channel knowledge at the base station. Correspondingly, the implicit Channel State Information (CSI) feedback of SU-MIMO is replaced with direct quantization of the channel matrix in MU-MIMO. To this end, memoryless and predictive CSI quantization algorithms are proposed in this chapter, building upon the concepts of Grassmannian quantization [13, 14]. The temporal correlation of the wireless channel can be exploited by the proposed predictive quantizer to achieve high fidelity quantization in low to moderate mobility scenarios, i.e., when the channel coherence time is large. In case of distributed antennas, knowledge of the channel gain matrix is utilized during quantization to further improve the accuracy.

Similar to SU-MIMO, Channel Quality Indicator (CQI) feedback is employed in MU-MIMO as well, for transmission rate adaptation and multi-user scheduling. In contrast to single-user transmission, the achievable rate with MU-MIMO cannot be estimated accurately by the users ahead of scheduling due to the unknown precoders. Exploiting the precoder construction of BD precoding, a lower bound on the expected Signal to Interference and Noise Ratio (SINR) of a user with imperfect CSIT is derived, facilitating estimation of the achievable transmission rate. This bound is proposed as CQI feedback for multi-user scheduling.

In case the number of data streams per user is less than the number of receive antennas, interference cancellation by means of BD is only achieved over a subspace of a users' channel matrix. Ideally, joint user subspace selection and scheduling must be performed at the transmitter to maximize the achievable throughput, requiring full channel knowledge for all users at the base station. To reduce the CSI feedback overhead, it is, however, proposed to perform a pre-selection of the preferred channel subspace by the users based on selfish arguments, effectively decreasing the dimensionality of the Grassmannian quantization problem. To filter-out the interference-free subspace at the users, semi-unitary antenna combiners are employed. It is shown how the Degrees of Freedom (DoF) provided by the excess antennas can be utilized to trade-off the residual multi-user interference, due to imperfect CSI, for the effective channel gain experienced by a user.

The proposed CSI feedback algorithms are extended to frequency selective Orthogonal Frequency Division Multiplexing (OFDM) systems employing the same concepts as in SU-MIMO, i.e., by means of CSI feedback clustering and interpolation. The performance of these two methods is evaluated in this chapter using simulations, with the result that interpolation is only reasonable if the distance between CSI pilots is small compared to the channel coherence bandwidth, challenging its value for practical implementations.

This chapter is organized as follows: In Sect. 6.1, the ZF beamforming and BD precoding constructions are introduced, providing the motivation for the proposed

Grassmannian quantization algorithms. The memoryless and predictive CSI quantizers are detailed in Sects. 6.2.1 and 6.2.2, respectively. In Sect. 7.1, the proposed antenna combiners are derived.

6.1 Zero-Forcing and Block-Diagonalization Precoding

In this section, an overview of ZF and BD precoding is provided and the relevant CSIT for precoder calculation is identified. It is shown that channel subspace information is sufficient if uniform power allocation over spatial streams and users is considered, which is known to be a asymptotically optimal in the limit of high Signal to Noise Ratio (SNR) [15]. Although not considered here, optimal power allocation by means of water-filling can be facilitated with additional channel magnitude feedback [11].

6.1.1 Transmit Strategy

The input-output relationship of user u in cell i on Resource Element (RE) $[n, k]$, assuming MU-MIMO transmission and employing the description with conjugate-transpose channel matrices as introduced in Sect. 1.5, is

$$\begin{aligned}\mathbf{y}_{n,k}[u,i] &= \mathbf{H}_{n,k}^{(\mathrm{ect})}[u,i]^{\mathrm{H}}\mathbf{F}_{n,k}[u,i]\mathbf{x}_{n,k}[u,i] \\ &\quad + \mathbf{H}_{n,k}^{(\mathrm{ect})}[u,i]^{\mathrm{H}} \sum_{\substack{s\in\mathscr{S}_{n,k}[i] \\ s\neq u}} \mathbf{F}_{n,k}[s,i]\mathbf{x}_{n,k}[s,i] + \mathbf{G}_{n,k}^{(\mathrm{ct})}[u,i]^{\mathrm{H}}\,\tilde{\mathbf{z}}_{n,k}[u,i],\end{aligned} \tag{6.1}$$

with the effective channel $\mathbf{H}_{n,k}^{(\mathrm{ect})}[u,i] \in \mathbb{C}^{N_{\mathrm{T}}[i]\times\ell_k[u,i]}$ being specified in Eq. (1.42) and $\tilde{\mathbf{z}}_{n,k}[u,i] \sim \mathscr{N}_{\mathbb{C}}\left(\mathbf{0}, \tilde{\sigma}_z^2\,\mathbf{I}_{N_{\mathrm{R}}[u,i]}\right)$, hence the out-of-cell interference is treated as additional Gaussian noise. To simplify notations, the user set $\mathscr{U}_i$ is re-ordered after scheduling such that the served users $\mathscr{S}_{n,k}[i]$ can be indexed as $\mathscr{S}_{n,k}[i] = \{1,\ldots,S_{n,k}[i]\}$. The goal of BD precoding is to calculate the precoders $\mathbf{F}_{n,k}[s,i], \forall s$ such that the interference at each user due to all other users is perfectly eliminated. BD was originally proposed for the case that each user is served over $\ell_k[u,i] = N_{\mathrm{R}}[u,i]$ streams. Then no antenna combiner $\mathbf{G}_{n,k}^{(\mathrm{ct})}[u,i]$ is required at the users. If the number of data streams $\ell_k[u,i]$ is less than the number of receive antennas $N_{\mathrm{R}}[u,i]$, the BD precoder is calculated for the effective channel $\mathbf{H}_{n,k}^{(\mathrm{ect})}[u,i]$. Hence the multi-user interference is only cancelled in the $\ell_k[u,i]$-dimensional subspace of span $\left(\mathbf{H}_{n,k}^{(\mathrm{ct})}[u,i]\right)$ that is spanned by $\mathbf{H}_{n,k}^{(\mathrm{ect})}[u,i]$. The precoders are obtained from the following conditions

$$\mathbf{H}_{n,k}^{(\mathrm{ect})}[u,i]^{\mathrm{H}}\mathbf{F}_{n,k}[s,i] = \mathbf{0}_{\ell_k[u,i]\times\ell_k[s,i]},\ \forall u,s\in\mathscr{S}_{n,k}[i] \text{ and } s\neq u, \tag{6.2}$$

$$\operatorname{rank}\left(\mathbf{H}_{n,k}^{(\mathrm{ect})}[u,i]^{\mathrm{H}}\mathbf{F}_{n,k}[u,i]\right) = \ell_k[u,i],\ \forall u\in\mathscr{S}_{n,k}[i]. \tag{6.3}$$

Notice the similarity of this formulation of BD to interference alignment [16, 17], with the difference that there is only a single transmitter in BD that has access to the data of all users. When applied for Coordinated Multi-Point (CoMP) transmission, BD is hence classified as a joint transmission strategy. Certain feasibility conditions on the number of streams per user and the total number of spatial streams $\ell_{n,k}[i]$ have to be fulfilled to assure that a solution to this problem exists, such as $\ell_k[u,i] \leq N_\mathrm{R}[u,i]$ and $\ell_{n,k}[i] \leq N_\mathrm{T}[i]$ [11]. Provided these feasibility conditions are satisfied, the solution is obtained from

$$\mathbf{F}_{n,k}[u,i] \in \operatorname{null}\left(\overline{\mathbf{H}}_{n,k}[u,i]\right), \ \operatorname{rank}\left(\mathbf{F}_{n,k}[u,i]\right) = \ell_k[u,i], \ \forall u \in \mathscr{S}_{n,k}[i], \quad (6.4)$$

$$\overline{\mathbf{H}}_{n,k}[u,i] = \left[\tilde{\mathbf{H}}_{n,k}[1,i], \ldots, \tilde{\mathbf{H}}_{n,k}[u-1,i], \tilde{\mathbf{H}}_{n,k}[u+1,i], \ldots, \tilde{\mathbf{H}}_{n,k}[S_{n,k}[i],i]\right]^\mathrm{H}, \quad (6.5)$$

$$\operatorname{span}\left(\tilde{\mathbf{H}}_{n,k}[s,i]\right) = \operatorname{span}\left(\mathbf{H}_{n,k}^{(\mathrm{ect})}[s,i]\right), \quad (6.6)$$

$$\tilde{\mathbf{H}}_{n,k}[s,i]^\mathrm{H}\tilde{\mathbf{H}}_{n,k}[s,i] = \mathbf{I}_{\ell_k[s,i]}, \ \tilde{\mathbf{H}}_{n,k}[s,i] \in \mathbb{C}^{N_\mathrm{T}[i] \times \ell_k[s,i]}, \ \forall s \in \mathscr{S}_{n,k}[i]. \quad (6.7)$$

Here, the matrices $\tilde{\mathbf{H}}_{n,k}[s,i]$ form orthonormal bases for the respective effective channels $\mathbf{H}_{n,k}^{(\mathrm{ect})}[s,i]$ and can, e.g., be obtained from a QR or a Singular Value Decomposition (SVD) of $\mathbf{H}_{n,k}^{(\mathrm{ect})}[s,i]$. The precoder of user u lies in the left null space of all other users' effective channels. Hence, after antenna combining, the transmission to user u does not interfere with the transmission to any of the other users. As the same holds true for every $u \in \mathscr{S}_{n,k}[i]$, interference-free transmission to all users is achieved. To satisfy the rank constraint (6.3), $\operatorname{rank}\left(\mathbf{F}_{n,k}[u,i]\right) = \ell_k[u,i]$ must be fulfilled.

As mentioned above, to maximize the SINR of the users, a joint optimization of the precoders $\mathbf{F}_{n,k}[s,i]$ and the antenna combiners $\mathbf{G}_{n,k}^{(\mathrm{ct})}[s,i]$ at the base station, based on perfect knowledge of $\mathbf{H}_{n,k}^{(\mathrm{ct})}[s,i]$, $\forall s$, is required; see, e.g., [18]. With the proposed approach, however, it is sufficient to feed back the subspace $\operatorname{span}\left(\mathbf{H}_{n,k}^{(\mathrm{ect})}[s,i]\right)$, $\forall s$, effectively reducing the feedback overhead, facilitating independent design of the transmit- and receive-filters and enabling the application of Grassmannian quantization, as detailed Sect. 6.1.2. Useful selections of $\mathbf{H}_{n,k}^{(\mathrm{ect})}[s,i]$ are proposed in Sect. 7.1.

In case of ZF beamforming, transmitting a single stream per user $\ell_k[u,i] = 1$, $\forall u$, the effective channel matrix $\mathbf{H}_{n,k}^{(\mathrm{ect})}[u,i]$ of user u is reduced to the effective channel vector $\mathbf{h}_{n,k}^{(\mathrm{ect})}[u,i]$ and the precoding matrix $\mathbf{F}_{n,k}[u,i]$ is obtained as the column-vector $\mathbf{f}_{n,k}[u,i] \in \mathbb{C}^{N_\mathrm{T}[i] \times 1}$. Combining the normalized effective channel vectors (channel directions)

$$\tilde{\mathbf{h}}_{n,k}[u,i] = \frac{\mathbf{h}_{n,k}^{(\mathrm{ect})}[u,i]}{\left\|\mathbf{h}_{n,k}^{(\mathrm{ect})}[u,i]\right\|} \quad (6.8)$$

of all served users in the so-called cell channel direction matrix

$$\tilde{\mathbf{H}}_{n,k}[i] = \left[\tilde{\mathbf{h}}_{n,k}[1,i], \ldots, \tilde{\mathbf{h}}_{n,k}[S_{n,k}[i],i]\right]^{\mathrm{H}} \in \mathbb{C}^{S_{n,k}[i] \times N_{\mathrm{T}}[i]}, \tag{6.9}$$

a closed-form solution of the *cell ZF precoding matrix* $\mathbf{F}_{n,k}[i]$ of all users is obtained as

$$\mathbf{F}_{n,k}[i] = \tilde{\mathbf{H}}_{n,k}[i]^{\mathrm{H}} \left(\tilde{\mathbf{H}}_{n,k}[i]\tilde{\mathbf{H}}_{n,k}[i]^{\mathrm{H}}\right)^{-1} \mathbf{P}_{n,k}[i]^{1/2} = \tilde{\mathbf{F}}_{n,k}[i]\mathbf{P}_{n,k}[i]^{1/2}, \tag{6.10}$$
$$\mathbf{F}_{n,k}[i] = \left[\mathbf{f}_{n,k}[1,i], \ldots, \mathbf{f}_{n,k}[S_{n,k}[i],i]\right], \ \mathbf{P}_{n,k}[i] = \mathrm{diag}\left(p_{n,k}[1,i], \ldots, p_{n,k}[S_{n,k}[i],i]\right),$$

where the precoding vectors $\mathbf{f}_{n,k}[u,i]$ are obtained as the columns of $\mathbf{F}_{n,k}[i]$. Considering equal power allocation, the powers $p_{n,k}[u,i]$ are calculated as

$$p_{n,k}[u,i] = \frac{P_i}{S_{n,k}[i] \left\| [\tilde{\mathbf{F}}_{n,k}[i]]_{:,u} \right\|^2}, \ \forall u \in \{1, \ldots, S_i[n,k]\}, \tag{6.11}$$

with P_i denoting the instantaneous total power constraint of the base station.

With the BD condition (6.2), the input-output relationship with perfect CSIT simplifies to

$$\mathbf{y}_{n,k}[u,i] = \mathbf{H}_{n,k}^{(\mathrm{ect})}[u,i]^{\mathrm{H}} \, \mathbf{F}_{n,k}[u,i] \, \mathbf{x}_{n,k}[u,i] + \mathbf{G}_{n,k}^{(\mathrm{ct})}[u,i]^{\mathrm{H}} \, \tilde{\mathbf{z}}_{n,k}[u,i]. \tag{6.12}$$

The precoder $\mathbf{F}_{n,k}[u,i]$ obtained from (6.4) is unique only up to right-multiplication with any full-rank $\ell_k[u,i] \times \ell_k[u,i]$ matrix, as this multiplication has no impact on $\mathrm{span}\left(\mathbf{F}_{n,k}[u,i]\right)$.

In [11], this ambiguity is removed by treating the interference-free input-output relationship (6.12) as a SU-MIMO system (c.f. Eq. (5.1)), and additionally performing SVD based precoding with water-filling power allocation over the obtained single-user channel. Hence, if $\tilde{\mathbf{F}}_{n,k}[u,i]$ denotes an arbitrary semi-unitary solution of (6.4), the unambiguous precoder according to [11] is obtained as

$$\mathbf{H}_{n,k}^{(\mathrm{ect})}[u,i]^{\mathrm{H}} \, \tilde{\mathbf{F}}_{n,k}[u,i] = \mathbf{U}_{n,k}[u,i] \boldsymbol{\Sigma}_{n,k}[u,i] \mathbf{V}_{n,k}[u,i]^{\mathrm{H}}, \tag{6.13}$$
$$\mathbf{F}_{n,k}[u,i] = \tilde{\mathbf{F}}_{n,k}[u,i] \mathbf{V}_{n,k}[u,i] \mathbf{P}_{n,k}[u,i]^{1/2}, \tag{6.14}$$

with (6.13) denoting an SVD of $\left(\mathbf{H}_{n,k}^{(\mathrm{ect})}[u,i]^{\mathrm{H}} \, \tilde{\mathbf{F}}_{n,k}[u,i]\right)$ and $\mathbf{P}_{n,k}[u,i]$ being the diagonal power allocation matrix obtained form the water-filling solution; see Sect. 5.1.2.

This, however, is not a reasonable approach with the subspace CSI feedback proposed below, because the feedback only provides an arbitrary orthonormal basis for $\mathbf{H}_{n,k}^{(\mathrm{ect})}[u,i]$ that neither matches the individual orientations of the eigenmodes of the effective channel nor contains information on their magnitude. Instead, semi-unitary precoding with equal power allocation is considered, enforcing the following additional precoder constraint

$$\mathbf{F}_{n,k}[u,i]^{\mathrm{H}}\mathbf{F}_{n,k}[u,i] = \frac{P_i}{S_{n,k}[i]\,\ell_k[u,i]}\tilde{\mathbf{F}}_{n,k}[u,i]^{\mathrm{H}}\tilde{\mathbf{F}}_{n,k}[u,i] = P_{u,i}[n,k]\mathbf{I}_{\ell_k[u,i]}. \tag{6.15}$$

The corresponding solution is unique up to right-multiplication with any unitary matrix, which has no impact on the *achievable per user transmission rate with perfect CSIT and equal power allocation*

$$R_{\mathrm{BD}} = \mathbb{E}\log_2\det\left(\mathbf{I}_{\ell_k[u,i]} + \rho\,\mathbf{H}_{n,k}^{(\mathrm{ect})}[u,i]^{\mathrm{H}}\tilde{\mathbf{F}}_{n,k}[u,i]\tilde{\mathbf{F}}_{n,k}[u,i]^{\mathrm{H}}\mathbf{H}_{n,k}^{(\mathrm{ect})}[u,i]\right), \quad \rho = \frac{P_{u,i}[n,k]}{\bar{\sigma}_z^2}. \tag{6.16}$$

6.1.2 Limited Feedback Model

With the assumption of equal power allocation, BD precoders can be calculated at the base station if knowledge of the $\ell_k[u,i]$-dimensional subspace span $\left(\mathbf{H}_{n,k}^{(\mathrm{ect})}[u,i]\right)$ is available for every $u \in \mathcal{S}_{n,k}[i]$. This subspace can be interpreted as a point on the Grassmann manifold of $\ell_k[u,i]$-dimensional subspaces in the $N_{\mathrm{T}}[i]$-dimensional Euclidean space, denoted as $\mathcal{G}\,(N_{\mathrm{T}}[i], \ell_k[u,i])$. In general it can be represented equivalently by any matrix $\mathbf{S}$ spanning the subspace

$$\mathbf{S} \equiv \mathbf{H}_{n,k}^{(\mathrm{ect})}[u,i] \Leftrightarrow \mathrm{span}\,(\mathbf{S}) = \mathrm{span}\left(\mathbf{H}_{n,k}^{(\mathrm{ect})}[u,i]\right). \tag{6.17}$$

The most common approach to represent points on the Grassmann manifold is employed, using orthonormal bases as in Eqs. (6.6) and (6.7). For a short introduction of the Grassmann manifold see [19, Appendix C].

Perfect knowledge of $\tilde{\mathbf{H}}_{n,k}[u,i]$, $\forall u \in \mathcal{S}_{n,k}[i]$ from (6.6) is required at the base station to achieve zero multi-user interference. With limited feedback, however, a quantized version $\hat{\mathbf{H}}_{n,k}[u,i]$ of the subspace is fed back to the transmitter. To convey the CSI to the transmitter, a quantization codebook $\mathcal{Q}_{n,k}[u,i]$ of size 2^b is employed by user u, with b denoting the number of feedback bits

$$\mathcal{Q}_{n,k}[u,i] = \left\{\mathbf{Q}_j \in \mathbb{C}^{N_{\mathrm{T}}[i]\times\ell_k[u,i]} \middle| \mathbf{Q}_j^{\mathrm{H}}\mathbf{Q}_j = \mathbf{I}_{\ell_k[u,i]},\ j \in \{1,\ldots,2^b\}\right\}. \tag{6.18}$$

The codebook is known by the base station. The quantized channel subspace is obtained by minimizing the subspace chordal distance [6] between $\mathbf{H}_{n,k}^{(\mathrm{ect})}[u,i]$ and the elements of the codebook

$$\hat{\mathbf{H}}_{n,k}[u,i] = \underset{\mathbf{Q}_j\in\mathcal{Q}_{n,k}[u,i]}{\mathrm{argmin}}\ \mathrm{d}_{\mathrm{c}}^2\left(\mathbf{H}_{n,k}^{(\mathrm{ect})}[u,i], \mathbf{Q}_j\right), \tag{6.19}$$

$$\mathrm{d}_{\mathrm{c}}^2\left(\mathbf{H}_{n,k}^{(\mathrm{ect})}[u,i], \mathbf{Q}_j\right) = \ell_k[u,i] - \mathrm{tr}\left(\tilde{\mathbf{H}}_{n,k}[u,i]^{\mathrm{H}}\mathbf{Q}_j\mathbf{Q}_j^{\mathrm{H}}\tilde{\mathbf{H}}_{n,k}[u,i]\right), \tag{6.20}$$

$$\mathrm{d}^2_{\mathrm{c,min}} = \mathrm{d}^2_{\mathrm{c}}\left(\mathbf{H}^{(\mathrm{ect})}_{n,k}[u,i], \hat{\mathbf{H}}_{n,k}[u,i]\right) = \ell_k[u,i] - \sum_{\ell=1}^{\ell_k[u,i]} \cos\left(\varphi^{(q)}_{\ell}\right)^2. \tag{6.21}$$

In this equation, the variables $\varphi^{(q)}_{\ell}$ denote the $\ell_k[u,i]$ principal angles between $\mathrm{span}\left(\mathbf{H}^{(\mathrm{ect})}_{n,k}[u,i]\right) = \mathrm{span}\left(\tilde{\mathbf{H}}_{n,k}[u,i]\right)$ and $\mathrm{span}\left(\hat{\mathbf{H}}_{n,k}[u,i]\right)$. When the chordal distance is employed as quantization metric the expected value of the residual multi-user interference, due to the CSI quantization error, is minimized. This is investigated in more detail in Sects. 7.1 and 7.2.2, when considering the theoretical performance of BD precoding with limited feedback and deriving the lower bound on the expected SINR. The quantization problem (6.19) is well known as Grassmannian quantization on the Grassmann manifold $\mathcal{G}\left(N_{\mathrm{T}}[i], \ell_k[u,i]\right)$ [13].

It is assumed that the base station treats the quantized subspaces $\hat{\mathbf{H}}_{n,k}[u,i]$ as the actual subspaces $\tilde{\mathbf{H}}_{n,k}[u,i]$, and calculates the precoders from (6.4), replacing $\overline{\mathbf{H}}_{n,k}[u,i]$ with

$$\hat{\overline{\mathbf{H}}}_{n,k}[u,i] = \left[\hat{\mathbf{H}}_{n,k}[1,i], \ldots, \hat{\mathbf{H}}_{n,k}[u-1,i], \hat{\mathbf{H}}_{n,k}[u+1,i], \ldots, \hat{\mathbf{H}}_{n,k}[S_{n,k}[i],i]\right]^{\mathrm{H}}. \tag{6.22}$$

The effective channel $\mathbf{H}^{(\mathrm{ect})}_{n,k}[u,i]$ can be decomposed into its range space component $\mathbf{H}^{(\mathrm{ect,r})}_{n,k}[u,i]$ and its left null space component $\mathbf{H}^{(\mathrm{ect,n})}_{n,k}[u,i]$, with respect to the orthonormal basis $\hat{\mathbf{H}}_{n,k}[u,i]$

$$\begin{aligned}\mathbf{H}^{(\mathrm{ect})}_{n,k}[u,i] &= \hat{\mathbf{H}}_{n,k}[u,i]\hat{\mathbf{H}}_{n,k}[u,i]^{\mathrm{H}}\,\mathbf{H}^{(\mathrm{ect})}_{n,k}[u,i] \\ &\quad + \left(\mathbf{I}_{N_{\mathrm{T}}[i]} - \hat{\mathbf{H}}_{n,k}[u,i]\hat{\mathbf{H}}_{n,k}[u,i]^{\mathrm{H}}\right)\mathbf{H}^{(\mathrm{ect})}_{n,k}[u,i] \\ &= \underbrace{\hat{\mathbf{H}}_{n,k}[u,i]\hat{\mathbf{H}}_{n,k}[u,i]^{\mathrm{H}}\,\mathbf{H}^{(\mathrm{ect})}_{n,k}[u,i]}_{\mathbf{H}^{(\mathrm{ect,r})}_{n,k}[u,i]} + \underbrace{\hat{\mathbf{H}}^{\perp}_{n,k}[u,i]\hat{\mathbf{H}}^{\perp}_{n,k}[u,i]^{\mathrm{H}}\,\mathbf{H}^{(\mathrm{ect})}_{n,k}[u,i]}_{\mathbf{H}^{(\mathrm{ect,n})}_{n,k}[u,i]},\end{aligned} \tag{6.23}$$

with $\hat{\mathbf{H}}^{\perp}_{n,k}[u,i] \in \mathbb{C}^{N_{\mathrm{T}}[i]\times N_{\mathrm{T}}[i]-\ell_k[u,i]}$ being an orthonormal basis for the orthogonal complement of $\mathrm{span}\left(\hat{\mathbf{H}}_{n,k}[u,i]\right)$. Using this notation and $\hat{\mathbf{H}}_{n,k}[u,i]^{\mathrm{H}}\mathbf{F}_{n,k}[s,i] = 0,\ \forall s \neq u$, due to the BD construction, the input-output relationship with quantized CSIT is

$$\begin{aligned}\mathbf{y}_{n,k}[u,i] &= \mathbf{H}^{(\mathrm{ect})}_{n,k}[u,i]^{\mathrm{H}}\mathbf{F}_{n,k}[u,i]\mathbf{x}_{n,k}[u,i] \\ &\quad + \mathbf{H}^{(\mathrm{ect,n})}_{n,k}[u,i]^{\mathrm{H}} \sum_{s=1,s\neq u}^{S_{n,k}[i]} \mathbf{F}_{n,k}[s,i]\mathbf{x}_{n,k}[s,i] + \mathbf{G}^{(\mathrm{ct})}_{n,k}[u,i]^{\mathrm{H}}\,\tilde{\mathbf{z}}_{n,k}[u,i].\end{aligned} \tag{6.24}$$

Thus, with quantized CSIT perfect interference cancellation is not achieved. The residual interference is determined by the quantization error of $\tilde{\mathbf{H}}_{n,k}[u,i]$, captured by the null space component $\mathbf{H}_{n,k}^{(\mathrm{ect,n})}[u,i]$. The *per-user transmission rate with quantized CSIT* is [6]

$$R_{\text{BD-Quant}} = \mathbb{E}\log_2\det\left(\mathbf{I}_{\ell_k[u,i]} + \sum_{s=1}^{S_{n,k}[i]} \mathbf{H}_{n,k}^{(\mathrm{ect})}[u,i]^{\mathrm{H}}\mathbf{F}_{n,k}[s,i]\mathbf{F}_{n,k}[s,i]^{\mathrm{H}}\mathbf{H}_{n,k}^{(\mathrm{ect})}[u,i]\right) \\ - \mathbb{E}\log_2\det\left(\mathbf{I}_{\ell_k[u,i]} + \sum_{s=1,s\neq u}^{S_{n,k}[i]} \mathbf{H}_{n,k}^{(\mathrm{ect})}[u,i]^{\mathrm{H}}\mathbf{F}_{n,k}[s,i]\mathbf{F}_{n,k}[s,i]^{\mathrm{H}}\mathbf{H}_{n,k}^{(\mathrm{ect})}[u,i]\right), \tag{6.25}$$

where the expectation is taken with respect to channel and corresponding precoder realizations.

6.2 Explicit CSI Feedback Algorithms

In interference limited multi-user precoding systems, such as MU-MIMO and interference alignment, strict requirements are placed on the CSIT accuracy, because CSI imperfections directly impact the residual multi-user interference and strongly deteriorate the throughput performance [5, 6, 20]. Therefore, efficient CSI quantization is central in such systems to obtain sufficiently accurate CSIT with a reasonable feedback overhead. As mentioned above, for the calculation of the BD precoders with equal power allocation channel subspace information is sufficient. The same holds true for other concepts as well, e.g., interference alignment [21] and signal to leakage and noise ratio beamforming [22]. This information can efficiently be conveyed by the Grassmann manifold, putting Grassmannian quantization at the center of this section. In Sect. 6.2.1, memoryless quantization of channel subspaces is studied, where the CSI at each Transmission Time Interval (TTI) is independently quantized, i.e., without considering the past. Efficient memoryless quantization in Distributed Antennas Systems (DASs) is achieved by taking into account the channel gain with respect to the individual antenna arrays during construction of the quantization codebook, to match the spatial distribution of the compound channel matrix. For this purpose, we employ the following decomposition of the channel matrix into a macroscopic and a microscopic fading part

$$\mathbf{H}_{n,k}^{(\mathrm{ct})}[u,i] = \mathbf{C}[u,i]\bar{\mathbf{H}}_{n,k}^{(\mathrm{ct})}[u,i], \tag{6.26}$$

$$\mathbf{C}[u,i] = \operatorname{diag}\left(\gamma_1[u,i],\ldots,\gamma_{N_{\mathrm{T}}[i]}[u,i]\right), \tag{6.27}$$

with $\gamma_j[u,i]$ denoting the large-scale channel gain between transmit antenna j of cell i and all receive antennas of user u in cell i.

Often sufficient CSIT accuracy cannot be obtained with an acceptable feedback overhead employing memoryless techniques. To circumvent this problem, predictive quantization is proposed in Sect. 6.2.2, exploiting the temporal channel correlation to improve the quantizer's fidelity. The performance of the quantizers is investigated numerically by evaluating the quantization Mean Square Error (MSE) in Sect. 6.2.3. The derivation of channel quality feedback for transmission rate adaptation and multi-user scheduling is postponed until Sect. 7.2.2 when frequency selective systems are treated.

The methods proposed in this section are designed for CSI quantization on a single frequency flat subcarrier. Efficient extensions to multi-carrier OFDM are proposed in Sect. 7.2, considering feedback clustering and interpolation. Also, antenna combining is not considered in Sects. 6.2.1 and 6.2.2; hence, $\ell_k[u,i] = N_\mathrm{R}[u,i]$ and $\mathbf{G}_{n,k}^{(\mathrm{ct})}[u,i] = \mathbf{I}_{N_\mathrm{R}[u,i]}$ is assumed. Excess receive antennas, i.e., the case $\ell_k[u,i] \leq N_\mathrm{R}[u,i]$, and antenna combining are covered in Sect. 7.1. Notice that it is possible to apply the predictive quantization proposed in Sect. 6.2.2 in the frequency domain as well, i.e., over OFDM subcarriers. This is not treated here, although it has been investigated in [23, 24]. With sufficiently dense placement of CSI pilots in the frequency domain, a substantial prediction gain can be achieved, especially if large system bandwidths are considered. The interpolation and clustering methods proposed here, however, facilitate a significant reduction of the required CSI pilots in the frequency domain.

6.2.1 Memoryless Grassmannian Quantization

With memoryless quantization, the temporal correlation of the source to be quantized is neglected and the quantization is performed independently at each time instant k, employing a pre-specified quantization codebook $\mathcal{Q}_{n,k}[u,i]$. Possibly different codebooks are used by different users to avoid situations in which users feed back the same CSI, excluding them from being spatially multiplexed with each other. The average quantization distortion in terms of the chordal distance MSE

$$D = \mathbb{E}\left(\mathrm{d}_{\mathrm{c,min}}^2\right) \tag{6.28}$$

is critically determined by the construction of the quantization codebook.

Codebooks for Isotropic Subspaces

If the channel subspace span $\left(\tilde{\mathbf{H}}_{n,k}[u,i]\right)$ is uniformly distributed on the Grassmann manifold $\mathcal{G}\left(N_\mathrm{T}[i], N_\mathrm{R}[u,i]\right)$, i.e., the channel matrix $\mathbf{H}_{n,k}^{(\mathrm{ct})}[u,i]$ is isotropically distributed (e.g., independent and Identically Distributed (I.I.D.) Rayleigh fading), the minimum average quantization distortion is achieved by maximally spaced Grassmannian subspace packings with respect to the chordal distance [25]. These packings are essentially uniform on the Grassmann manifold in terms of the chordal distance.

Unfortunately, finding good codebooks is in general hard except for special cases; an algorithm for obtaining good codebooks is provided, e.g., in [26].

More suitable for analytical investigations, however, are random quantization codebooks, consisting of elements $\mathbf{Q}_j \in \mathcal{Q}_{n,k}[u,i]$ that are chosen independently and uniformly distributed on $\mathcal{G}(N_\mathrm{T}[i], N_\mathrm{R}[u,i])$ [6]. The elements of such codebooks can be obtained as the compact left singular matrices of I.I.D. Gaussian matrices $\mathbf{H} \in \mathbb{C}^{N_\mathrm{T}[i]\times N_\mathrm{R}[u,i]}$, $[\mathbf{H}]_{m,n} \sim \mathcal{N}_\mathbb{C}(0,1)$. Such a codebook construction, denoted as Random Vector Quantization (RVQ), is applied in here whenever memoryless quantization is evaluated. In that case, the results are averaged over codebook realizations by employing independent quantization codebooks at each time instant k. Bounds on the average distortion achieved with random isotropically distributed Grassmannian quantization codebooks are derived in [14], and the codebooks are shown to perform asymptotically optimal in a number of applications, e.g., [27].

Codebooks for Correlated Subspaces

In case the channel subspace to be quantized is not obtained from an isotropically distributed channel matrix, better quantization performance can be achieved if knowledge of the spatial channel correlation is exploited. An efficient heuristic codebook construction is proposed in [28] for single-user multiple-input single-output beamforming systems, which effectively "colors" a Grassmannian line packing according to the channel correlation matrix. This approach is extended here for Grassmannian subspace packings to exploit the pathloss differences experienced in DASs. Considering a DAS model, a random Grassmannian codebook that is statistically matched to the distribution of the channel subspace can be obtained as

$$\mathcal{Q}_{n,k}^{(\mathrm{corr})}[u,i] = \left\{ \mathbf{Q}_j^{(\mathrm{corr})} \middle| \mathbf{Q}_j^{(\mathrm{corr})} \mathbf{\Sigma} \mathbf{V}^\mathrm{H} = \mathbf{C}[u,i]^{1/2} \bar{\mathbf{H}} \in \mathbb{C}^{N_\mathrm{T}[i]\times N_\mathrm{R}[u,i]},\ [\bar{\mathbf{H}}]_{m,n} \sim \mathcal{N}_\mathbb{C}(0,1) \right\}. \tag{6.29}$$

Here, $\mathbf{Q}_j^{(\mathrm{corr})} \mathbf{\Sigma} \mathbf{V}^\mathrm{H}$ represents a compact SVD of the matrix $\mathbf{C}[u,i]^{1/2}\bar{\mathbf{H}}$. The codebook construction is denoted as correlated RVQ. An analytic performance characterization of this construction was not yet successful. Simulations demonstrating the performance gain achieved in DASs with this codebook are provided in [29, 30]. The method can be applied to account for the spatial correlation in-between different entries of the channel matrix as well. Alternative approaches to consider the pathloss differences in DASs or the spatial channel correlation exist in literature, e.g., in [31] a quantization codebook for ZF beamforming based on estimated channel statistics is proposed, and in [32] a feedback bit allocation scheme for DASs is derived.

6.2.2 Predictive Grassmannian Quantization

MU-MIMO techniques are mostly considered for situations in which users are either static or moving slowly through the cell. If this is the case, the temporal correlation of the wireless channel is often high, which can be exploited to improve the CSI quantization accuracy and to reduce the feedback overhead. A common approach in quantization and source coding for exploiting the correlation of the source signal is vector quantization [33], where multiple consecutive samples of the source signal are combined to a vector and are jointly coded/quantized. Although such techniques are rate-distortion optimal, they cannot be applied for CSI quantization in wireless communications, because the throughput performance of MU-MIMO with outdated CSI degrades in a similar way as with a quantization error [34]. Hence, the delay of vector quantization caused by gathering and processing the data cannot be tolerated. Instead, predictive quantization on the Grassmann manifold is proposed in this section to achieve efficient subspace quantization. Predictive quantizers are able to leverage the temporal correlation of the source signal at consecutive time instants to provide higher fidelity at a given quantization rate. The prediction of points on the Grassmannian is realized by translating the problem to the tangent space associated with the manifold. The statistics of the prediction error in the tangent space are approximated with a Gaussian distribution, which is valid for sufficiently small prediction errors. Based on this approximation, a Grassmannian quantization codebook is generated to quantize the prediction error. The performance of the proposed algorithm is evaluated by means of Monte Carlo simulations of the quantization MSE in Sect. 6.2.3. An application of the proposed quantizer for CSI quantization in limited feedback MU-MIMO is investigated in Sect. 7.2.3. For a short introduction of the Grassmann manifold see [19, Appendix C]. To shorten notations, the RE index $[n, k]$ is partly omitted for intermediate steps and auxiliary variables of the derivation.

Predictive Quantization Model

The predictive Grassmannian quantizer derived below explicitly considers the effect of distributing antenna arrays over the cell area, by employing the decomposition of the channel matrix into a temporally constant channel gain matrix $\mathbf{C}[u, i]$ and a small-scale fading matrix $\bar{\mathbf{H}}_{u,i}[n, k]$. The channel gain matrix $\mathbf{C}[u, i]$ is supposed to be known by the transmitter and the receiver. The temporal variation of the channel due to the movement of obstacles and/or users is hence captured in $\bar{\mathbf{H}}_{u,i}[n, k]$. In the derivation of the quantizer, a general model for the temporal evolution of $\bar{\mathbf{H}}_{u,i}[n, k]$ is considered

$$\bar{\mathbf{H}}^{(\mathrm{ct})}_{n,k}[u, i] = G\left(\bar{\mathbf{H}}^{(\mathrm{ct})}_{n,k-1}[u, i], \bar{\mathbf{H}}^{(\mathrm{ct})}_{n,k-2}[u, i], \ldots\right) + \bar{\mathbf{J}}_{n,k}[u, i] = \bar{\mathbf{H}}^{(\mathrm{d})}_{n,k}[u, i] + \bar{\mathbf{J}}_{n,k}[u, i]. \tag{6.30}$$

Here, the deterministic dependence of the current channel on the past is described with $\bar{\mathbf{H}}^{(\mathrm{d})}_{n,k}[u, i]$ and the random innovation is captured in the matrix $\bar{\mathbf{J}}_{n,k}[u, i]$, which is assumed as I.I.D. Gaussian

$$\mathrm{vec}\left(\bar{\mathbf{J}}_{n,k}[u, i]\right) \sim \mathcal{N}_{\mathbb{C}}\left(\mathbf{0}, \sigma_j^2[n, k]\,\mathbf{I}_{N_{\mathrm{T}}[i]\cdot N_{\mathrm{R}}[u,i]}\right). \tag{6.31}$$

No specific assumptions are made about the nature of the deterministic function $G(\cdot)$ in (6.30), because the behavior of the wireless channel is strongly dependent on the surrounding environment [35]. In [36] it is shown that Rayleigh fading processes can be accurately modeled using auto-regressive relations, fitting well to the considered decomposition in (6.30).

In the proposed quantizer, a prediction algorithm is implemented to provide an estimate of the deterministic evolution $\bar{\mathbf{H}}^{(\mathrm{d})}_{n,k}[u,i]$, based on previously quantized channel observations. Assuming that a prediction $\bar{\mathbf{H}}^{(\mathrm{p})}_{n,k}[u,i]$ of $\bar{\mathbf{H}}^{(\mathrm{d})}_{n,k}[u,i]$ is available, the channel is written as

$$\begin{aligned}\mathbf{H}^{(\mathrm{ct})}_{n,k}[u,i] &= \mathbf{C}[u,i]^{1/2}\Big(\bar{\mathbf{H}}^{(\mathrm{p})}_{n,k}[u,i] + \underbrace{\bar{\mathbf{E}}^{(\mathrm{p})}_{n,k}[u,i] + \bar{\mathbf{J}}_{n,k}[u,i]}_{\bar{\mathbf{E}}_{n,k}[u,i]}\Big)\\ &= \bar{\mathbf{H}}^{(\mathrm{p})}_{n,k}[u,i] + \mathbf{C}[u,i]^{1/2}\bar{\mathbf{E}}_{n,k}[u,i],\end{aligned} \tag{6.32}$$

with $\bar{\mathbf{E}}^{(\mathrm{p})}_{n,k}[u,i]$ denoting the prediction error. The prediction error is assumed as independent of the innovation and I.I.D. Gaussian:

$$\mathrm{vec}\big(\bar{\mathbf{E}}^{(\mathrm{p})}_{n,k}[u,i]\big) \sim \mathcal{N}_{\mathbb{C}}\big(\mathbf{0}, \sigma_p^2[n,k]\,\mathbf{I}_{N_{\mathrm{T}}[i]\cdot N_{\mathrm{R}}[u,i]}\big).$$

Thus, the total error $\bar{\mathbf{E}}_{n,k}[u,i]$ is distributed as

$$\mathrm{vec}\left(\bar{\mathbf{E}}_{n,k}[u,i]\right) \sim \mathcal{N}_{\mathbb{C}}\left(\mathbf{0}, \sigma_e^2[n,k]\,\mathbf{I}_{N_{\mathrm{T}}[i]\cdot N_{\mathrm{R}}[u,i]}\right), \quad \sigma_e^2[n,k] = \sigma_p^2[n,k] + \sigma_j^2[n,k]. \tag{6.33}$$

Given the prediction $\bar{\mathbf{H}}^{(\mathrm{p})}_{n,k}[u,i]$ at both ends of the CSI feedback link, the channel can be reproduced from knowledge of the error $\bar{\mathbf{E}}_{n,k}[u,i]$. With an accurate prediction ($\sigma_p^2[n,k] \ll 1$) and a smoothly varying channel ($\sigma_j^2[n,k] \ll 1$), the variance of the signal to be quantized can thus be reduced by quantizing $\bar{\mathbf{E}}_{n,k}[u,i]$ instead of $\bar{\mathbf{H}}^{(\mathrm{ct})}_{n,k}[u,i]$.

By taking advantage of the CSIT invariances introduced in Sect. 6.1.2, the required feedback information can further be reduced. Using orthonormal bases $\tilde{\mathbf{H}}_{n,k}[u,i]$ and $\tilde{\mathbf{H}}^{(\mathrm{p})}_{n,k}[u,i]$ for the actual channel $\mathbf{H}^{(\mathrm{ct})}_{n,k}[u,i]$ and the predicted channel $\bar{\mathbf{H}}^{(\mathrm{p})}_{n,k}[u,i]$, as defined by Equation (6.7), the channel matrix can be written as

$$\mathbf{H}^{(\mathrm{ct})}_{n,k}[u,i] = \tilde{\mathbf{H}}_{n,k}[u,i]\mathbf{D}_{n,k}[u,i] = \tilde{\mathbf{H}}^{(\mathrm{p})}_{n,k}[u,i]\mathbf{D}^{(\mathrm{p})}_{n,k}[u,i] + \mathbf{C}[u,i]^{1/2}\bar{\mathbf{E}}_{n,k}[u,i], \tag{6.34}$$

$$\mathbf{D}_{n,k}[u,i] = \tilde{\mathbf{H}}_{n,k}[u,i]^{\mathrm{H}}\mathbf{H}^{(\mathrm{ct})}_{n,k}[u,i], \quad \mathbf{D}^{(\mathrm{p})}_{n,k}[u,i] = \tilde{\mathbf{H}}^{(\mathrm{p})}_{n,k}[u,i]^{\mathrm{H}}\bar{\mathbf{H}}^{(\mathrm{p})}_{n,k}[u,i]. \tag{6.35}$$

The channel subspace, represented by the orthonormal basis $\tilde{\mathbf{H}}_{n,k}[u,i]$, is thus obtained as

$$\tilde{\mathbf{H}}_{n,k}[u,i] = \left(\tilde{\mathbf{H}}^{(\mathrm{p})}_{n,k}[u,i]\mathbf{D}^{(\mathrm{p})}_{n,k}[u,i] + \mathbf{C}[u,i]^{1/2}\bar{\mathbf{E}}_{n,k}[u,i]\right)^{-1}\mathbf{D}_{n,k}[u,i]. \tag{6.36}$$

Exploiting the differential geometry associated with the Grassmannian, it is possible to describe the error between the predicted subspace $\text{span}\left(\tilde{\mathbf{H}}_{n,k}^{(\text{p})}[u,i]\right)$ and the observed subspace $\text{span}\left(\tilde{\mathbf{H}}_{n,k}[u,i]\right)$ with the error tangent

$$\mathbf{T}_{n,k}[u,i] = T\left(\tilde{\mathbf{H}}_{n,k}^{(\text{p})}[u,i], \tilde{\mathbf{H}}_{n,k}[u,i]\right) \in \mathscr{T}\left(\tilde{\mathbf{H}}_{n,k}^{(\text{p})}[u,i]\right) \subset \mathbb{C}^{N_{\text{T}}[i]\times N_{\text{R}}[u,i]}, \quad (6.37)$$

where $\mathscr{T}\left(\tilde{\mathbf{H}}_{n,k}^{(\text{p})}[u,i]\right)$ denotes the tangent space associated with the prediction $\tilde{\mathbf{H}}_{n,k}^{(\text{p})}[u,i]$. The shortest path on the manifold between the predicted subspace and the actual subspace, which is denoted as the geodesic $\Gamma\left(\tilde{\mathbf{H}}_{n,k}^{(\text{p})}[u,i], \mathbf{T}_{n,k}[u,i], p\right)$, is specified by the tangent $\mathbf{T}_{n,k}[u,i]$ and parametrized by variable $p \in [0,1]$. Having knowledge of $\tilde{\mathbf{H}}_{n,k}^{(\text{p})}[u,i]$ and $\mathbf{T}_{n,k}[u,i]$, it is hence possible to obtain an equivalent representation of $\text{span}\left(\tilde{\mathbf{H}}_{n,k}[u,i]\right)$ from

$$\tilde{\mathbf{H}}_{n,k}[u,i] \equiv \tilde{\mathbf{H}}_{n,k}^{(\text{e})}[u,i] = \Gamma\left(\tilde{\mathbf{H}}_{n,k}^{(\text{p})}[u,i], \mathbf{T}_{n,k}[u,i], 1\right), \quad (6.38)$$

$$\text{span}\left(\tilde{\mathbf{H}}_{n,k}[u,i]\right) = \text{span}\left(\tilde{\mathbf{H}}_{n,k}^{(\text{e})}[u,i]\right), \quad \tilde{\mathbf{H}}_{n,k}^{(\text{e})}[u,i]^{\text{H}}\tilde{\mathbf{H}}_{n,k}^{(\text{e})}[u,i] = \mathbf{I}_{N_{\text{R}}[u,i]}. \quad (6.39)$$

In general, the two orthonormal bases $\tilde{\mathbf{H}}_{n,k}[u,i]$ and $\tilde{\mathbf{H}}_{n,k}^{(\text{e})}[u,i]$ are not equal but equivalent representatives of the same point on the Grassmann manifold $\mathscr{G}\left(N_{\text{T}}[i], N_{\text{R}}[u,i]\right)$, since they span the same subspace. For the purpose of precoder calculation for BD MU-MIMO either matrix can be employed; an explicit distinction between $\tilde{\mathbf{H}}_{n,k}[u,i]$ and $\tilde{\mathbf{H}}_{n,k}^{(\text{e})}[u,i]$ is hence omitted in the following. To calculate the subspace $\text{span}\left(\tilde{\mathbf{H}}_{n,k}[u,i]\right)$ at the base station, it is sufficient to feed back the error tangent $\mathbf{T}_{n,k}[u,i]$. Notice that this tangent lies in the $(N_{\text{T}}[i] - N_{\text{R}}[u,i])$-dimensional tangent space associated with $\tilde{\mathbf{H}}_{n,k}^{(\text{p})}[u,i]$. Therefore, compared to quantization of $\bar{\mathbf{E}}_{n,k}[u,i]$, the dimensionality of the quantization problem is reduced from $N_{\text{T}}[i]$ to $N_{\text{T}}[i] - N_{\text{R}}[u,i]$.

Approximation of the Error Tangent Statistics

To derive an efficient quantizer for the error tangent, knowledge about the statistics of $\mathbf{T}_{n,k}[u,i]$ is required. Due to the nonlinear operations involved in the calculation of the tangent, obtaining a closed-form statistical description is hard. Therefore, an approximation is employed in the following that is valid if the channel correlation is sufficiently large.

The tangent $\mathbf{T}_{n,k}[u,i]$ is obtained as

$$\mathbf{T}_{n,k}[u,i] = \mathbf{U}\boldsymbol{\Phi}\mathbf{V}^{\text{H}}, \quad \mathbf{U}\tan\left(\boldsymbol{\Phi}\right)\mathbf{V}^{\text{H}} = \boldsymbol{\Theta}_{n,k}[u,i], \quad (6.40)$$

$$\boldsymbol{\Theta}_{n,k}[u,i] = \Big(\underbrace{\mathbf{I}_{N_{\text{T}}[i]} - \tilde{\mathbf{H}}_{n,k}^{(\text{p})}[u,i]\tilde{\mathbf{H}}_{n,k}^{(\text{p})}[u,i]^{\text{H}}}_{\mathbf{P}_{n,k}[u,i]}\Big)\tilde{\mathbf{H}}_{n,k}[u,i]\left(\tilde{\mathbf{H}}_{n,k}^{(\text{p})}[u,i]^{\text{H}}\tilde{\mathbf{H}}_{n,k}[u,i]\right)^{-1}. \quad (6.41)$$

Here, the diagonal matrix $\boldsymbol{\Phi} = \mathrm{diag}\big(\varphi_1^{(\mathrm{p})}, \ldots, \varphi_{N_\mathrm{R}[u,i]}^{(\mathrm{p})}\big)$ is composed of the principal angles between the subspaces spanned by $\tilde{\mathbf{H}}_{n,k}^{(\mathrm{p})}[u,i]$ and $\tilde{\mathbf{H}}_{n,k}[u,i]$, which are obtained from the arctangent of the singular values of $\boldsymbol{\Theta}_{n,k}[u,i]$. Under the assumption $\varphi_i^{(\mathrm{p})} \leq 0.30 \mathrel{\hat{=}} 17°$ an error that is below 3 % is caused by the approximation

$$\tan\left(\varphi_i^{(\mathrm{p})}\right) \approx \varphi_i^{(\mathrm{p})}. \tag{6.42}$$

With this approximation, the tangent simplifies to

$$\mathbf{T}_{n,k}[u,i] \approx \boldsymbol{\Theta}_{n,k}[u,i] = \mathbf{P}_{n,k}[u,i]\tilde{\mathbf{H}}_{n,k}[u,i]\left(\tilde{\mathbf{H}}_{n,k}^{(\mathrm{p})}[u,i]^\mathrm{H}\tilde{\mathbf{H}}_{n,k}[u,i]\right)^{-1}, \tag{6.43}$$

where the matrix $\mathbf{P}_{n,k}[u,i]$, which is defined in Eq. (6.41), can be identified as a projection onto the orthogonal complement of $\tilde{\mathbf{H}}_{n,k}^{(\mathrm{p})}[u,i]$. Substituting $\tilde{\mathbf{H}}_{n,k}[u,i]$ from Eq. (6.36) the approximated tangent can be written as

$$\mathbf{T}_{n,k}[u,i] \approx \mathbf{P}_{n,k}[u,i]\mathbf{C}[u,i]^{1/2}\bar{\mathbf{E}}_{n,k}[u,i]\left(\tilde{\mathbf{H}}_{n,k}^{(\mathrm{p})}[u,i]^\mathrm{H}\mathbf{H}_{n,k}^{(\mathrm{ct})}[u,i]\right)^{-1}, \tag{6.44}$$

where the first term from (6.36) vanishes due to the projection $\mathbf{P}_{n,k}[u,i]$ and the matrix $\mathbf{D}_{n,k}[u,i]^{-1}$ is pulled inside the brackets. The tangent matrix is hence obtained as the product of the left null space component of the channel matrix with respect to the predicted subspace $\mathrm{span}\big(\tilde{\mathbf{H}}_{n,k}^{(\mathrm{p})}[u,i]\big)$, and the inverse of the term $\tilde{\mathbf{H}}_{n,k}^{(\mathrm{p})}[u,i]^\mathrm{H}\mathbf{H}_{n,k}^{(\mathrm{ct})}[u,i]$ that is determined by the range space component. Applying an SVD to the term in brackets

$$\begin{gathered}\tilde{\mathbf{H}}_{n,k}^{(\mathrm{p})}[u,i]^\mathrm{H}\mathbf{H}_{n,k}^{(\mathrm{ct})}[u,i] = \mathbf{Y}_{n,k}[u,i]\boldsymbol{\Lambda}_{n,k}[u,i]\mathbf{W}_{n,k}[u,i]^\mathrm{H}, \\ \mathbf{Y}_{n,k}[u,i],\ \boldsymbol{\Lambda}_{n,k}[u,i],\ \mathbf{W}_{n,k}[u,i] \in \mathbb{C}^{N_\mathrm{R}[u,i]\times N_\mathrm{R}[u,i]},\end{gathered} \tag{6.45}$$

the approximated tangent can be formulated as

$$\mathbf{T}_{n,k}[u,i] \approx \mathbf{P}_{n,k}[u,i]\mathbf{C}[u,i]^{1/2}\bar{\mathbf{E}}_{n,k}[u,i]\left(\mathbf{W}_{n,k}[u,i]^{-1}\boldsymbol{\Lambda}_{n,k}[u,i]\mathbf{Y}_{n,k}[u,i]^\mathrm{H}\right). \tag{6.46}$$

Assuming $\tilde{\mathbf{H}}_{n,k}^{(\mathrm{p})}[u,i]^\mathrm{H}\mathbf{H}_{n,k}^{(\mathrm{ct})}[u,i]$ as fixed/observed, the distribution of the tangent is determined by the null space component $\mathbf{P}_{n,k}[u,i]\mathbf{C}[u,i]^{1/2}\bar{\mathbf{E}}_{n,k}[u,i]$. This term is obtained as the projection of the zero-mean Gaussian error $\mathbf{C}[u,i]^{1/2}\bar{\mathbf{E}}_{n,k}[u,i]$ onto the orthogonal complement of $\mathrm{span}\big(\tilde{\mathbf{H}}_{n,k}^{(\mathrm{p})}[u,i]\big)$ and is therefore itself Gaussian distributed. The distribution of $\mathbf{T}_{n,k}[u,i]$ is hence obtained as

$$\mathrm{vec}\left(\mathbf{T}_{n,k}[u,i]\right) \sim \mathcal{N}_\mathbb{C}\left(\mathbf{0}, \mathbf{C}_{n,k}^{(\mathrm{t})}[u,i]\right), \tag{6.47}$$

$$\mathbf{C}_{n,k}^{(\mathrm{t})}[u,i] = \sigma_e^2[n,k]\Big(\underbrace{\mathbf{Y}_{n,k}[u,i]\boldsymbol{\Lambda}_{n,k}[u,i]^{-2}\mathbf{Y}_{n,k}[u,i]^{\mathrm{H}}}_{\mathbf{C}_M \in \mathbb{C}^{N_\mathrm{R}[u,i]\times N_\mathrm{R}[u,i]}} \otimes \underbrace{\mathbf{P}_{n,k}[u,i]\mathbf{C}[u,i]\mathbf{P}_{n,k}[u,i]}_{\mathbf{C}_N \in \mathbb{C}^{N_i \times N_i}}\Big). \tag{6.48}$$

Notice that the correlation matrix (6.48) is obtained from a Kronecker product. It is therefore possible to derive a statistically matched quantization codebook for the error tangent $\mathbf{T}_{n,k}[u,i]$ by correlating I.I.D. Gaussian matrices

$$\mathcal{Q}_{n,k}^{(\mathrm{t,corr})}[u,i] = \left\{\mathbf{C}_N^{1/2}\mathbf{Q}_j\mathbf{C}_M^{1/2}\,\middle|\,\mathbf{Q}_j \in \mathbb{C}^{N_\mathrm{T}[i]\times N_\mathrm{R}[u,i]},\ \left[\mathbf{Q}_j\right]_{l,m} \sim \mathcal{N}_\mathbb{C}\left(0,\sigma_e^2[n,k]\right)\right\}. \tag{6.49}$$

The correlation matrix $\mathbf{C}_N$ can be calculated by the transmitter and the receiver, because both have knowledge of the predicted subspace and of the channel gain matrix. Unfortunately, $\mathbf{C}_M$ is not known a-priori and can therefore not be exploited to improve the quantization efficiency. Hence, the codebook is designed under the assumption that $\mathbf{C}_M$ is a scaled identity, which is the case if $\boldsymbol{\Lambda}_{n,k}[u,i]$ in (6.48) is a scaled identity. The *error tangent correlation matrix* is then obtained as

$$\mathbf{C}_{n,k}^{(\mathrm{t})}[u,i] \approx \left(\frac{\sigma_e[n,k]}{\lambda_{n,k}[u,i]}\right)^2 \left(\mathbf{I}_{N_\mathrm{R}[u,i]} \otimes \mathbf{C}_N\right). \tag{6.50}$$

To further motivate this approach, two special cases are considered below that enable a more detailed investigation of the factor $\lambda_{n,k}[u,i]$.

Error Tangent Statistics with $N_\mathrm{R}[u,i] = 1$

When the users are equipped with single receive antennas, the channel matrix is reduced to the channel vector $\mathbf{h}_{n,k}^{(\mathrm{ct})}[u,i]$ and $\mathbf{C}_M$ is obtained as

$$\mathbf{C}_M = \frac{1}{\left\|\mathbf{h}_{n,k}^{(\mathrm{ct})}[u,i]\right\|^2 \cos\left(\varphi^{(\mathrm{p})}\right)^2},\quad \cos\left(\varphi^{(\mathrm{p})}\right)^2 = \left|\tilde{\mathbf{h}}_{n,k}^{(\mathrm{p})}[u,i]^{\mathrm{H}}\tilde{\mathbf{h}}_{n,k}[u,i]\right|^2, \tag{6.51}$$

with $0 \le \varphi^{(\mathrm{p})} \le \pi/2$ denoting the principal angle of the subspace prediction error. In this case, $\mathbf{C}_M$ is reduced to a scalar and has an impact only on the variance of the error tangent. Notice that the variance is increased in fading dips, i.e., when $\left\|\mathbf{h}_{n,k}^{(\mathrm{ct})}[u,i]\right\|^2$ is small, and in case the subspace prediction error is large.

Error Tangent Statistics for I.I.D. Gaussian Channels

When the channel matrix is I.I.D. Gaussian distributed $\left[\mathbf{H}_{n,k}^{(\mathrm{ct})}[u,i]\right]_{l,m} \sim \mathcal{N}_\mathbb{C}\left(0,\gamma_{u,i}\right)$, it is possible to determine the expected value of the matrix $\mathbf{C}_M$ with respect to the unknown range space component of $\mathbf{H}_{n,k}^{(\mathrm{ct})}[u,i]$ within $\mathrm{span}\left(\tilde{\mathbf{H}}_{n,k}^{(\mathrm{p})}[u,i]\right)$. This is achieved by considering the inverse of $\mathbf{C}_M$

$$\mathbf{C}_M^{-1} = \mathbf{Y}_{n,k}[u,i]\boldsymbol{\Lambda}_{n,k}[u,i]^2\mathbf{Y}_{n,k}[u,i]^{\mathrm{H}} = \tilde{\mathbf{H}}_{n,k}^{(\mathrm{p})}[u,i]^{\mathrm{H}}\mathbf{H}_{n,k}^{(\mathrm{ct})}[u,i]\mathbf{H}_{n,k}^{(\mathrm{ct})}[u,i]^{\mathrm{H}}\tilde{\mathbf{H}}_{n,k}^{(\mathrm{p})}[u,i]. \tag{6.52}$$

With a compact SVD of the channel matrix $\mathbf{H}_{n,k}^{(\mathrm{ct})}[u,i] = \mathbf{U}\boldsymbol{\Sigma}\mathbf{V}^{\mathrm{H}}$, this product is written as

$$\mathbf{C}_M^{-1} = \tilde{\mathbf{H}}_{n,k}^{(\mathrm{p})}[u,i]^{\mathrm{H}}\mathbf{U}\boldsymbol{\Sigma}^2\mathbf{U}^{\mathrm{H}}\tilde{\mathbf{H}}_{n,k}^{(\mathrm{p})}[u,i]. \tag{6.53}$$

Notice that $\mathbf{U}$ is an isotropically distributed unitary matrix that is statistically independent of the singular value matrix $\boldsymbol{\Sigma}$, due to the assumption that the channel is I.I.D. Gaussian and hence also isotropic [37, Theorem 1]. Applying an SVD to the product $\tilde{\mathbf{H}}_{n,k}^{(\mathrm{p})}[u,i]^{\mathrm{H}}\mathbf{U} = \mathbf{Q}\cos(\boldsymbol{\Phi})\,\mathbf{W}^{\mathrm{H}}$, the inverse of $\mathbf{C}_M$ is obtained as

$$\mathbf{C}_M^{-1} = \mathbf{Q}\cos(\boldsymbol{\Phi})\left(\mathbf{W}^{\mathrm{H}}\boldsymbol{\Sigma}^2\mathbf{W}\right)\cos(\boldsymbol{\Phi})\,\mathbf{Q}^{\mathrm{H}}. \tag{6.54}$$

Due to the isotropy of $\mathbf{U}$, its projection onto $\mathrm{span}\left(\tilde{\mathbf{H}}_{n,k}^{(\mathrm{p})}[u,i]\right)$ is isotropically distributed within the subspace. Hence, the $N_{\mathrm{R}}[u,i] \times N_{\mathrm{R}}[u,i]$ matrices $\mathbf{Q}$, $\cos(\boldsymbol{\Phi})$ and $\mathbf{W}$ are statistically independent, and $\mathbf{Q}$ and $\mathbf{W}$ are isotropically distributed unitary matrices [37, Theorem 1]. Furthermore, $\cos(\boldsymbol{\Phi})$ is composed of the cosines of the principal angles between $\mathrm{span}\left(\tilde{\mathbf{H}}_{n,k}^{(\mathrm{p})}[u,i]\right)$ and $\mathrm{span}\left(\tilde{\mathbf{H}}_{n,k}[u,i]\right)$

$$\cos(\boldsymbol{\Phi}) = \mathrm{diag}\left(\cos\left(\varphi_1^{(\mathrm{p})}\right),\ldots,\cos\left(\varphi_{N_{\mathrm{R}}[u,i]}^{(\mathrm{p})}\right)\right). \tag{6.55}$$

Because of the isotropy of $\mathbf{W}$ and its statistical independence of $\boldsymbol{\Sigma}$, it follows that

$$\mathbf{W}^{\mathrm{H}}\boldsymbol{\Sigma}^2\mathbf{W} \sim \mathscr{W}_{N_{\mathrm{R}}[u,i]}^{\mathbb{C}}\left(N_{\mathrm{T}}[i],\gamma_{u,i}\,\mathbf{I}_{N_{\mathrm{R}}[u,i]}\right), \tag{6.56}$$

with $\mathscr{W}_{N_{\mathrm{R}}[u,i]}^{\mathbb{C}}\left(N_{\mathrm{T}}[i],\gamma_{u,i}\,\mathbf{I}_{N_{\mathrm{R}}[u,i]}\right)$ denoting a central complex-valued Wishart distribution of dimension $N_{\mathrm{R}}[u,i]$, having $N_{\mathrm{T}}[i]$ DoF. With this notation, the correlation matrix $\mathbf{C}_M$ is obtained as

$$\mathbf{C}_M = \mathbf{Q}\cos(\boldsymbol{\Phi})^{-1}\left(\mathbf{W}^{\mathrm{H}}\boldsymbol{\Sigma}^2\mathbf{W}\right)^{-1}\cos(\boldsymbol{\Phi})^{-1}\,\mathbf{Q}^{\mathrm{H}}, \tag{6.57}$$

where the inverse matrix in the center is distributed according to the inverse Wishart distribution $\mathscr{W}_{N_{\mathrm{R}}[u,i]}^{-\mathbb{C}}\left(N_{\mathrm{T}}[i],\gamma_{u,i}^{-1}\,\mathbf{I}_{N_{\mathrm{R}}[u,i]}\right)$. Correspondingly, the expected value of $\mathbf{C}_M$ can be calculated as [38]

$$\begin{aligned}\mathbb{E}\left(\mathbf{C}_M\right) &= \frac{\mathbb{E}\left(\mathbf{Q}\cos(\boldsymbol{\Phi})^{-2}\,\mathbf{Q}^{\mathrm{H}}\right)}{\gamma_{u,i}\left(N_{\mathrm{T}}[i]-N_{\mathrm{R}}[u,i]\right)} \\ &= \frac{1}{\gamma_{u,i}\left(N_{\mathrm{T}}[i]-N_{\mathrm{R}}[u,i]\right)}\left(\frac{1}{N_{\mathrm{R}}[u,i]}\sum_{j=1}^{N_{\mathrm{R}}[u,i]}\frac{1}{\cos\left(\varphi_j^{(\mathrm{p})}\right)^2}\right)\mathbf{I}_{N_{\mathrm{R}}[u,i]},\end{aligned} \tag{6.58}$$

where the second equality follows from the isotropy of $\mathbf{Q}$. Hence, averaging out the unknown orientation of the range space component of $\mathbf{H}_{n,k}^{(\text{ct})}[u,i]$ within the unknown singular values and span $\left(\tilde{\mathbf{H}}_{n,k}^{(\text{p})}[u,i]\right)$, a scaled identity matrix is obtained. The exact scaling is determined by the distribution of the chordal distance prediction error among the subspace dimensions. Notice that the expected value of (6.51) in case of I.I.D. Rayleigh fading is obtained from (6.58) by setting $N_{\text{R}}[u,i]=1$.

Quantization Algorithm

Based on the observations of the previous paragraphs, a predictive Grassmannian quantizer is proposed that generates at each time instant k a quantization codebook, according to the tangent statistics derived in Eqs. (6.47) and (6.50). To determine the correlation matrix $\mathbf{C}_{n,k}^{(\text{t})}[u,i]$ in (6.50), a subspace prediction $\tilde{\mathbf{H}}_{n,k}^{(\text{p})}[u,i]$ is required at the CSI encoder and the decoder. In this paragraph the prediction is assumed to be given; subspace prediction algorithms are proposed in the next paragraph. In general the scaling $\left(\frac{\sigma_e[n,k]}{\lambda_{n,k}[u,i]}\right)^2$ of $\mathbf{C}_{n,k}^{(\text{t})}[u,i]$ is time-dependent; a tracking algorithm is proposed to adapt the scaling of the quantization codebook correspondingly. As described in Sect. 6.1.2, the chordal distance is employed as quantization metric. The chordal distance cannot be evaluated directly in the tangent space associated with the Grassmannian. It is therefore necessary to project the tangent codebook onto the manifold to obtain an equivalent Grassmannian codebook that enables the quantization. The quantization algorithm is described in the following steps:

1. Initialize the codebook scale parameter $s_{n,0}[u,i]=0$ and the scale growth rate $g>1$. These parameters are used to track the scaling of $\mathbf{C}_{n,k}^{(\text{t})}[u,i]$.
2. Initialize a Gaussian quantization codebook $\mathcal{Q}_{u,i}^{(0)}$ of size 2^b for a standard normal random matrix of size $N_{\text{T}}[i]\times N_{\text{R}}[u,i]$. Here this codebook is randomly generated
$$\mathcal{Q}_{u,i}^{(0)}=\left\{\mathbf{Q}_j^{(0)}\middle|\left[\mathbf{Q}_j^{(0)}\right]_{l,m}\sim\mathcal{N}_{\mathbb{C}}(0,1)\right\}. \tag{6.59}$$
3. Correlate the elements of $\mathcal{Q}_{u,i}^{(0)}$ to match the correlation of the prediction error $\mathbf{C}[u,i]^{1/2}\bar{\mathbf{E}}_{n,k}[u,i]$
$$\mathcal{Q}_{u,i}^{(\text{corr})}=\left\{\mathbf{C}[u,i]^{1/2}\mathbf{Q}_j^{(0)}\middle|\,\forall\mathbf{Q}_j^{(0)}\in\mathcal{Q}_{u,i}^{(0)}\right\}. \tag{6.60}$$

Repeat the following steps for each $k>0$:

4. Predict the current subspace and calculate the projection matrix $\mathbf{P}_{n,k}[u,i]$ in Eq. (6.41).
5. Project the correlated codebook onto the orthogonal complement of span$\left(\tilde{\mathbf{H}}_{n,k}^{(\text{p})}[u,i]\right)$ to determine the tangent codebook matching the structure of the error tangent correlation matrix $\mathbf{C}_{n,k}^{(\text{t})}[u,i]$ defined in (6.50)

$$\mathscr{Q}^{(t)}_{n,k}[u,i] = \left\{ \mathbf{P}_{n,k}[u,i]\mathbf{Q}_j \middle| \forall \mathbf{Q}_j \in \mathscr{Q}^{(\text{corr})}_{u,i} \right\} \subset \mathscr{T}\left(\tilde{\mathbf{H}}^{(p)}_{n,k}[u,i]\right). \tag{6.61}$$

6. Calculate two scaled codebooks to track the scaling of $\mathbf{C}^{(t)}_{n,k}[u,i]$

$$\mathscr{Q}^{(t)}_{-,n,k}[u,i] = \left\{ g^{s_-}\, \mathbf{Q}^{(t)}_j \middle| \forall \mathbf{Q}^{(t)}_j \in \mathscr{Q}^{(t)}_{n,k}[u,i] \right\}, \tag{6.62}$$

$$\mathscr{Q}^{(t)}_{+,n,k}[u,i] = \left\{ g^{s_+}\, \mathbf{Q}^{(t)}_j \middle| \forall \mathbf{Q}^{(t)}_j \in \mathscr{Q}^{(t)}_{n,k}[u,i] \right\}, \tag{6.63}$$

$$s_- = s_{n,k-1}[u,i] - 1, \quad s_+ = \min\left(s_{n,k-1}[u,i] + 1, 0\right). \tag{6.64}$$

7. Project the obtained tangent codebooks onto the manifold to obtain two Grassmannian codebooks $\mathscr{Q}_{-,u,i}[n,k]$ and $\mathscr{Q}_{+,u,i}[n,k]$. This is achieved with the geodesic $\Gamma(\cdot)$

$$\mathscr{Q}_{-,n,k}[u,i] = \left\{ \Gamma\left(\tilde{\mathbf{H}}^{(p)}_{n,k}[u,i], \mathbf{Q}^{(t)}_j, 1\right) \middle| \forall \mathbf{Q}^{(t)}_j \in \mathscr{Q}^{(t)}_{-,n,k}[u,i] \right\}, \tag{6.65}$$

$$\mathscr{Q}_{+,n,k}[u,i] = \left\{ \Gamma\left(\tilde{\mathbf{H}}^{(p)}_{n,k}[u,i], \mathbf{Q}^{(t)}_j, 1\right) \middle| \forall \mathbf{Q}^{(t)}_j \in \mathscr{Q}^{(t)}_{+,n,k}[u,i] \right\}. \tag{6.66}$$

8. Quantize the observed subspace $\tilde{\mathbf{H}}_{n,k}[u,i]$ with respect to both Grassmannian codebooks

$$\mathbf{Q}_- = \underset{\mathbf{Q}_j \in \mathscr{Q}_{-,n,k}[u,i]}{\text{argmin}}\ \mathrm{d}_\mathrm{c}^2\left(\tilde{\mathbf{H}}_{n,k}[u,i], \mathbf{Q}_j\right), \ d_- = \mathrm{d}_\mathrm{c}^2\left(\tilde{\mathbf{H}}_{n,k}[u,i], \mathbf{Q}_-\right), \tag{6.67}$$

$$\mathbf{Q}_+ = \underset{\mathbf{Q}_j \in \mathscr{Q}_{+,n,k}[u,i]}{\text{argmin}}\ \mathrm{d}_\mathrm{c}^2\left(\tilde{\mathbf{H}}_{n,k}[u,i], \mathbf{Q}_j\right), \ d_+ = \mathrm{d}_\mathrm{c}^2\left(\tilde{\mathbf{H}}_{n,k}[u,i], \mathbf{Q}_+\right). \tag{6.68}$$

9. Determine the "winning" codebook, i.e., the one with minimum quantization error

$$w = \underset{i \in \{-,+\}}{\text{argmin}}\, d_i. \tag{6.69}$$

10. Determine the quantized channel subspace and update the codebook scale parameter

$$\hat{\mathbf{H}}_{n,k}[u,i] = \mathbf{Q}_w \in \mathscr{Q}_{w,n,k}[u,i], \quad s_{n,k}[u,i] = s_w. \tag{6.70}$$

11. Feedback the indices of $\mathbf{Q}_w$ and s_w.

The tangent codebook obtained in step 2 of the algorithm is equal in distribution to the error tangent specified by Eqs. (6.47) and (6.50), despite a scaling factor. The codebook scale parameter $s_{n,k}[u,i]$ is employed to track the scaling $\left(\frac{\sigma_e[n,k]}{\lambda_{n,k}[u,i]}\right)^2$ of $\mathbf{C}^{(t)}_{n,k}[u,i]$, by determining the appropriate up- or down-scaling of the codebook scale parameter in steps 3 and 6 of the algorithm. The ratio of the scaling between two consecutive time instants is given by

$$\frac{g^{s_{n,k}[u,i]}}{g^{s_{n,k-1}[u,i]}} = g^{\pm 1}. \tag{6.71}$$

Thus, the tracking speed of the algorithm is determined by the variance growth rate g; increasing g causes a faster tracking speed, but also a larger steady state tracking error. The optimal value of g is determined by means of simulations in my corresponding publication [39], where it is shown that the optimum depends on the subspace dimension, but the sensitivity of the algorithm to the calibration of g is not strongly pronounced. In this publication also the accuracy of the approximation of the error tangent statistics in (6.47) is evaluated, demonstrating that the estimated tangent variance fits well to the observed variance up to very high channel Doppler frequencies.

The feedback overhead of the proposed quantization algorithm is $b + 1$ bits per quantization instant. The 1 bit overhead is caused by the codebook scale parameter that must be signaled to the decoder as additional side-information. The same prediction algorithm is employed at the encoder and at the decoder, such that both sides are able to calculate the two scaled codebooks $\mathcal{Q}_{-,n,k}[u,i]$ and $\mathcal{Q}_{+,n,k}[u,i]$. With the feedback information, i.e., the indices of $\mathbf{Q}_w$ and s_w, the quantized channel subspace $\hat{\mathbf{H}}_{n,k}[u,i]$ and the scale parameter $s_{n,k}[u,i]$ can therefore be reproduced by the decoder.

Prediction Algorithm

The Grassmannian quantization algorithm proposed in the previous paragraph is based on the availability of a prediction $\tilde{\mathbf{H}}^{(\mathrm{p})}_{n,k}[u,i]$ of the current channel subspace, to calculate a quantization codebook that is matched to the temporal evolution of the subspace. The prediction algorithm must be based on quantized CSI to enable the prediction at the encoder and at the decoder. Trivially, the previously quantized subspace can be employed as a prediction $\tilde{\mathbf{H}}^{(\mathrm{p})}_{n,k}[u,i] = \hat{\mathbf{H}}_{n,k-1}[u,i]$, leading to a differential quantizer as proposed in [40]. Better performance can be achieved with a more sophisticated prediction. One possibility is to implement the prediction directly on the Grassmann manifold, e.g., by means of a linear predictor

$$\tilde{\mathbf{H}}^{(\mathrm{p})}_{n,k}[u,i] = \sum_{j=1}^{N_p} \hat{\mathbf{H}}_{n,k-j}[u,i]\, a_j. \tag{6.72}$$

In general, the obtained prediction $\tilde{\mathbf{H}}^{(\mathrm{p})}_{n,k}[u,i]$ is not compatible with the considered semi-unitary matrix representation of points on $\mathcal{G}\left(N_{\mathrm{T}}[i], N_{\mathrm{R}}[u,i]\right)$, which could be resolved by applying an SVD decomposition after linear prediction to determine the corresponding subspace. With this approach, however, an optimal solution for the filter coefficients is hard to obtain.

This difficulty can be avoided by predicting the tangent $T\left(\hat{\mathbf{H}}_{n,k-1}[u,i], \tilde{\mathbf{H}}_{n,k}[u,i]\right)$ between the quantized subspace $\hat{\mathbf{H}}_{n,k-1}[u,i]$ at time instant $k-1$ and the observed subspace $\tilde{\mathbf{H}}_{n,k}[u,i]$ at time instant k. With the geodesic defined by this predicted tangent, a subspace prediction $\tilde{\mathbf{H}}^{(\mathrm{p})}_{n,k}[u,i]$ can indirectly be obtained. In the tangent space well known prediction algorithms from Euclidean geometry can be reused. Although statistical models for the temporal evolution of the channel matrix exist, there are no corresponding models for the temporal evolution of tangents available. Due to the highly nonlinear relationship between subspaces and tangents, it is difficult to derive such a model. Instead, predictors that are not based on an underlying channel model are proposed here. For one-dimensional subspaces adaptive Finite Impulse Response (FIR) filters are employed, while a regression model is considered for higher dimensional subspaces. The adaptive filtering approach is not used for higher dimensional subspaces, due to the large number of filter coefficients required to achieve an accurate prediction and the corresponding slow filter convergence speed.

The basic idea of regression based prediction is proposed by Zhang et.al. for one-dimensional subspaces in [41] under the name *robust Grassmannian prediction*. Its extension to higher dimensional subspaces is straightforward and is conducted in my publication [42]; the interested reader is referred to this publication for details. The basic idea is to apply a linear regression to the tangents observed over multiple time instants and to predict the current tangent based on this regression. Compared to the adaptive filter based approach proposed below, the regression has the advantage that the filter convergence phase of the adaptive filter is omitted, despite a slightly reduced prediction performance as shown in Sect. 6.2.3.

Adaptive Filter Based Prediction of One-Dimensional Subspaces

The following predictor is derived for a prediction on the Grassmann manifold $\mathscr{G}(N_{\mathrm{T}}[i], 1)$. An auto-regressive model of order N_p is considered to describe the temporal evolution of the tangent random process. The prediction is based on quantized CSI. The tangent describing the geodesic between the quantized channel subspaces $\hat{\mathbf{h}}_{n,k-1}[u,i]$ and $\hat{\mathbf{h}}_{n,k}[u,i]$ at consecutive time instants is denoted

$$\mathbf{t}^{(q)}_{n,k}[u,i] = T\left(\hat{\mathbf{h}}_{n,k-1}[u,i], \hat{\mathbf{h}}_{n,k}[u,i]\right) \in \mathscr{T}\left(\hat{\mathbf{h}}_{n,k-1}[u,i]\right). \tag{6.73}$$

At time instant k, a prediction $\tilde{\mathbf{h}}^{(p)}_{n,k}[u,i]$ of the subspace $\tilde{\mathbf{h}}_{n,k}[u,i]$ is required by the quantizer to calculate the quantization codebook. This prediction is obtained indirectly by first predicting the tangent $\mathbf{t}_{n,k}[u,i] = T\left(\hat{\mathbf{h}}_{n,k-1}[u,i], \tilde{\mathbf{h}}_{n,k}[u,i]\right)$ using a linear predictor of order N_p

$$\mathbf{t}^{(p)}_{n,k}[u,i] = \sum_{j=1}^{N_p} \bar{\mathbf{t}}^{(q)}_{n,k-j}[u,i]\, a^{(j)}_{n,k}[u,i] = \mathbf{T}^{(q)}_{n,k}[u,i]\mathbf{a}_{n,k}[u,i], \tag{6.74}$$

$$\mathbf{T}^{(q)}_{n,k}[u,i] = \left[\bar{\mathbf{t}}^{(q)}_{n,k-1}[u,i], \ldots, \bar{\mathbf{t}}^{(q)}_{n,k-N_p}[n,k-N_p]\right] \in \mathscr{T}\left(\hat{\mathbf{h}}_{n,k-1}[u,i]\right), \tag{6.75}$$

$$\mathbf{a}_{n,k}[u,i] = \left[a^{(1)}_{n,k}[u,i], \ldots, a^{(N_p)}_{n,k}[u,i]\right]^{\mathrm{T}}. \tag{6.76}$$

A linear combination of tangent vectors as in (6.74) is only meaningful if the tangents are defined in the same tangent space. Hence, the tangents $\mathbf{t}_{n,k-j}^{(q)}[u,i]$, $j \in \{1,\ldots,N_p\}$ cannot directly be employed in (6.74). Instead, it is necessary to transport the geometric information contained in $\mathbf{t}_{n,k-j}^{(q)}[u,i]$ from the respective tangent space $\mathscr{T}\left(\hat{\mathbf{h}}_{n,k-j-1}[u,i]\right)$ to the current position $\mathscr{T}\left(\hat{\mathbf{h}}_{n,k-1}[u,i]\right)$. This is enabled by means of parallel transport. Thus, parallel transported versions $\bar{\mathbf{t}}_{n,k-j}^{(q)}[u,i]$ of the previously observed tangents $\mathbf{t}_{n,k-j}^{(q)}[u,i]$ are employed in (6.74). This is achieved by updating the matrix $\mathbf{T}_{n,k}^{(q)}[u,i]$ at each time instant k, as soon as $\hat{\mathbf{h}}_{n,k}[u,i]$ is observed, as follows:

1. Circularly shift the columns of $\mathbf{T}_{n,k}^{(q)}[u,i]$ by one to the right.
2. Calculate the tangent $\mathbf{t}_{n,k}^{(q)}[u,i]$ and replace the first column of $\mathbf{T}_{n,k}^{(q)}[u,i]$ with $\mathbf{t}_{n,k}^{(q)}[u,i]$.
3. Parallel transport the columns of $\mathbf{T}_{n,k}^{(q)}[u,i]$ along the geodesic defined by $\mathbf{t}_{n,k}^{(q)}[u,i]$ from $\mathscr{T}\left(\hat{\mathbf{h}}_{n,k-1}[u,i]\right)$ to $\mathscr{T}\left(\hat{\mathbf{h}}_{n,k}[u,i]\right)$.

$$\left[\mathbf{T}_{n,k+1}^{(q)}[u,i]\right]_{:,j} = \Pi\left(\hat{\mathbf{h}}_{n,k-1}[u,i], \hat{\mathbf{h}}_{n,k}[u,i], \left[\mathbf{T}_{n,k}^{(q)}[u,i]\right]_{:,j}\right), \tag{6.77}$$

with $\Pi(\cdot)$ denoting the parallel transport operator.

The predictor coefficients $\mathbf{a}_{n,k}[u,i]$ are trained with a stochastic gradient algorithm, namely the normalized least mean squares algorithm [43]. Defining the tangent prediction error at time k as

$$\mathbf{e}_{n,k}^{(p)}[u,i] = \mathbf{t}_{n,k}^{(q)}[u,i] - \mathbf{t}_{n,k}^{(p)}[u,i] = \mathbf{t}_{n,k}^{(q)}[u,i] - \mathbf{T}_{n,k}^{(q)}[u,i]\mathbf{a}_{n,k}[u,i], \tag{6.78}$$

the filter coefficient update rule is obtained as

$$\mathbf{a}_{n,k+1}[u,i] = \mathbf{a}_{n,k}[u,i] + \mu \frac{\mathbf{T}_{n,k}^{(q)}[u,i]^{\mathrm{H}}}{\left\|\mathbf{T}_{n,k}^{(q)}[u,i]\right\|^2} \mathbf{e}_{n,k}^{(p)}[u,i]. \tag{6.79}$$

The step size μ, determining the trade-off between the filter convergence speed and the steady-state MSE, has to satisfy $0 < \mu < 2$ for convergence [43]. Notice that the error (6.78) is calculated with the tangent $\mathbf{t}_{n,k}^{(q)}[u,i]$ obtained from the quantized subspace $\hat{\mathbf{h}}_{n,k}[u,i]$ and not from the actual subspace $\tilde{\mathbf{h}}_{n,k}[u,i]$. Therefore, the accuracy of the prediction of $\tilde{\mathbf{h}}_{n,k}[u,i]$ is impacted by the quantization error. This approach is necessary to enable the prediction at the decoder, which has only access to quantized CSI.

In Fig. 6.1, a schematic of the concatenation of the proposed quantization, prediction and codebook generation methods is illustrated, visualizing the interplay between the different components. The quantized subspace is passed from the

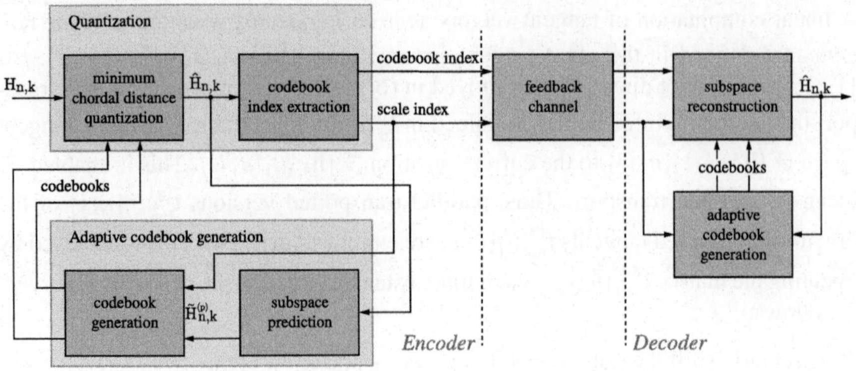

Fig. 6.1 Structure of the Grassmannian subspace quantizer visualizing the different components

quantizer to the predictor at both ends of the feedback link to enable the prediction of the next subspace. Based on this prediction and on information about the previously employed codebook scaling, a new quantization codebook is calculated that spans a certain volume on the Grassmann manifold around the predicted subspace. Over a dedicated feedback channel the quantized CSI is passed from the encoder to the decoder in form of the codebook index and the scale index. This information is sufficient for the decoder to reconstruct the quantized channel subspace and to prepare the codebook for the next time instant.

6.2.3 Evaluation of the Quantization MSE

In this section, the performance of the proposed Grassmannian quantization algorithms is investigated by means of Monte-Carlo simulations. The chordal distance MSE as defined in Eq. (6.28), providing a natural distance measure for points on the Grassmannian, is employed as performance metric. The chordal distance also determines the achievable rate loss of BD based MU-MIMO with quantized CSIT compared to perfect CSIT; see Sect. 7.1 for details.

In the first simulation, predictive Grassmannian quantization on $\mathcal{G}(4, 1)$ is considered using the quantization codebook sizes $Q \in \{8, 32, 128\}$ which translates to a CSI feedback overhead of $b \in \{3, 5, 7\} + 1$ bit, due to the 1 bit overhead caused by the codebook scale index. The performance of differential quantization [40], robust Grassmannian prediction [41] and the adaptive filter based prediction proposed in the previous section is compared. The temporal correlation is determined by Clarke's model. The results are plotted in dependency of the normalized channel Doppler frequency ν_d defined in Eq. (5.35). The appropriate length N_p of the prediction filter is determined by the channel correlation according to Clarke's model. The length is set as

$$N_p = \min\left(\left\lceil \frac{1.5}{2\pi \nu_d} \right\rceil, 20\right), \tag{6.80}$$

employing past values with a correlation of ≥ 0.5 for prediction, where the argument of the Bessel function determining the correlation in Clarke's model is approximately 1.5 [44]. The maximum filter length is restricted to 20 taps to achieve moderate complexity and a reasonable filter convergence speed. The channel gain matrix is assumed as an identity matrix.

The results of the simulation are shown in Fig. 6.2a. The performance of the quantizer is strongly dependent on the Doppler frequency, which determines the channel correlation according to (5.34). With increasing Doppler frequency the channel correlation is reduced, leading to a larger prediction error and thus to an increased variance of the error tangent $\mathbf{T}_{n,k}[u, i]$. It can be seen that the slope of the MSE achieved with

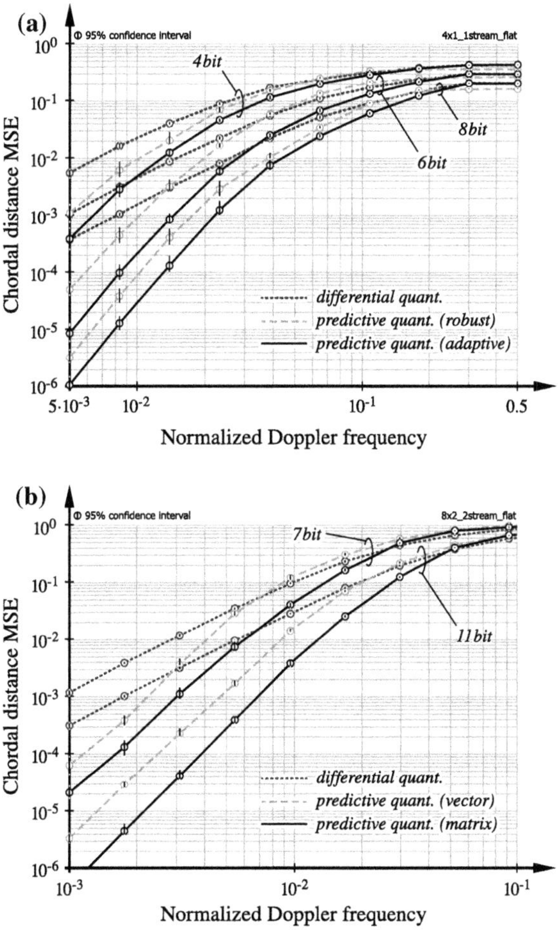

Fig. 6.2 MSE performance comparison of the proposed predictive Grassmannian quantizer using different prediction algorithms. **a** Predictive quantization on $\mathcal{G}(4, 1)$. **b** Predictive quantization on $\mathcal{G}(8, 2)$.

the adaptive filter based predictor and the robust predictor is similar. Differential quantization is significantly outperformed by predictive quantization if the channel correlation is sufficiently large, i.e., at low Doppler frequencies. Notice though that this observation depends on the considered channel model. In [40] it is observed that quantizers using higher order prediction do not achieve a gain over differential quantizers in case the temporal channel evolution is determined by an autoregressive model of order 1. This is evident, because in this case the current channel is obtained by adding I.I.D. noise to the previous channel realization; as the I.I.D. noise cannot be predicted, the best approach is to use the previous quantized observation as prediction. Comparing adaptive prediction and robust prediction, it can be seen that the adaptive filter is able to extract a prediction gain already at higher Doppler frequencies. With a similar or even slightly improved slope, this gain is maintained over the full range of considered Doppler frequencies. The gain is achieved because the adaptive filter adjusts to the temporal statistics of the tangents, in contrast to the tangent regression model employed by [41], which does not exploit any statistical information.

Similar observations are obtained if quantization of higher dimensional subspaces is investigated. In Fig. 6.2b, the quantization performance on $\mathcal{G}(8, 2)$ is shown. In this case, a CSI feedback overhead of 7 bit and 11 bit is considered. A comparison of differential quantization and predictive quantization is conducted. The predictive quantizers employ the robust prediction algorithm of [42]. Again it is observed that a larger MSE slope is achieved with predictive quantization compared to differential quantization, leading to a significantly reduced quantization error at low Doppler frequencies. The gain obtained by jointly quantizing all subspace dimensions (denoted as matrix quantization) is investigated by comparing to an individual quantization of the subspace dimensions (vector quantization). In case of matrix quantization, a matrix codebook of size $Q_m \in \{64, 1024\}$ is employed, while with vector quantization the codebook size has to be reduced to $Q_v \in \{6, 23\}$ to achieve the same feedback overhead of $\log_2(64) + 1 \approx 2(\log_2(6) + 1) \approx 7$ bit respectively 11 bit. Due to the reduced codebook size, individual quantization of the subspace dimensions is outperformed by joint quantization. With a growing subspace dimension the performance difference is also increased.

In the next simulation, the effect of different channel gains experienced in a DAS on the quantization performance is considered, by assuming a channel gain matrix of the form

$$\mathbf{C}[u, i] = \operatorname{diag}(1, 1, 1, 1, \gamma_s, \gamma_s, \gamma_s, \gamma_s),$$

with $\gamma_s \in [0, 1]$. Effectively this causes a gradual switching from the quantization on $\mathcal{G}(4, 1)$ to $\mathcal{G}(8, 1)$. A codebook size of $Q = 16$ is assumed. The performance of memoryless quantization, using the correlated RVQ codebook proposed in Sect. (6.2.1), is compared to the MSE of predictive quantization, employing adaptive filter based prediction. The obtained quantization MSE is shown in Fig. 6.3. It can be seen how the quantizers adapt to the varying channel gains. With decreasing γ_s the quantization MSE is improved from the pure 8×1 performance at $\gamma_s = 1$ to

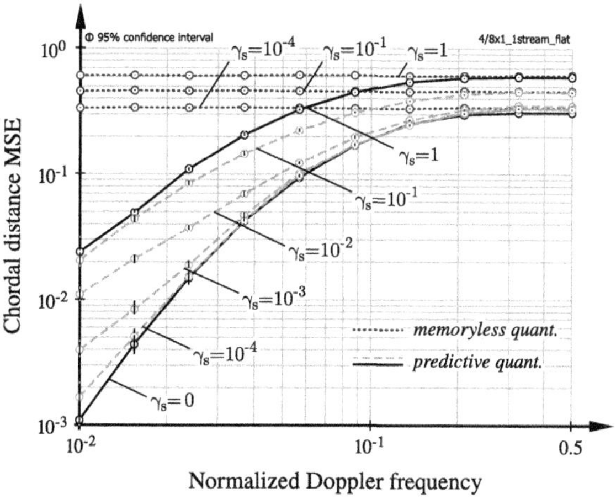

Fig. 6.3 Quantization MSE achieved with memoryless and predictive quantization in a DAS with varying channel gain differences

the pure 4×1 curve at $\gamma_s = 0$. Also, the performance improvement of predictive quantization over memoryless quantization at low to moderate Doppler frequencies is demonstrated in Figure 6.3.

References

1. H. Weingarten, Y. Steinberg, S. Shamai, The capacity region of the Gaussian multiple-input multiple-output broadcast channel. IEEE Trans. Inf. Theor. **52**(9), 3936–3964 (2006)
2. D. Gesbert, M. Kountouris, R. Heath Jr., C. Chae, T. Slzer, From single user to multiuser communications: shifting the MIMO paradigm. IEEE Signal Process. Mag. **24**(5), 36 (2007)
3. D. Shiu, G. Foschini, M. Gans, J. Kahn, Fading correlation and its effect on the capacity of multielement antenna systems. IEEE Trans. Commun. **48**(3), 502–513 (2000)
4. A. Molisch, A generic model for MIMO wireless propagation channels in macro- and micro-cells. IEEE Trans. Signal Process. **52**(1), 61–71 (2004)
5. N. Jindal, MIMO broadcast channels with finite-rate feedback. IEEE Trans. Inf. Theor. **52**(11), 5 (2006)
6. N. Ravindran, N. Jindal, Limited feedback-based block diagonalization for the MIMO broadcast channel. IEEE J. Sel. Areas Commun. **26**(8), 1473–1482 (2008)
7. H. Bizaki, A. Falahati, Tomlinson-harashima precoding with imperfect channel state information. IET Commun. **2**(1), 151–158 (2008)
8. D. Ryan, I. Collings, I.V.L. Clarkson, R. Heath Jr., Performance of vector perturbation multiuser MIMO systems with limited feedback. IEEE Trans. Commun. **57**(9), 2633–2644 (2009)
9. A. Razi, D. Ryan, I. Collings, J. Yuan, Sum rates, rate allocation, and user scheduling for multi-user MIMO vector perturbation precoding. IEEE Trans. Wirel. Commun. **9**(1), 356–365 (2010)
10. C. Peel, B. Hochwald, A. Swindlehurst, A vector-perturbation technique for near-capacity multiantenna multiuser communication-part I: channel inversion and regularization. IEEE Trans. Commun. **53**(1), 195–202 (2005)
11. Q. Spencer, A. Swindlehurst, M. Haardt, Zero-forcing methods for downlink spatial multiplexing in multiuser MIMO channels. IEEE Trans. Signal Process. **52**(2), 461–471 (2004)

12. J.S. Kim, H. Kim, C.S. Park, K.B. Lee, On the performance of multiuser MIMO systems in WCDMA/HSDPA: beamforming, feedback and user diversity. IEICE Trans. **89**, 2161–2169 (2006)
13. B. Mondal, S. Dutta, R. Heath Jr., Quantization on the Grassmann manifold. IEEE Trans. Signal Process. **55**(8), 4208–4216 (2007)
14. W. Dai, Y. Liu, B. Rider, Quantization bounds on Grassmann manifolds and applications to MIMO communications. IEEE Trans. Inf. Theor. **54**(3), 1108–1123 (2008)
15. J. Lee, N. Jindal, High SNR analysis for MIMO broadcast channels: dirty paper coding versus linear precoding. IEEE Trans. Inf. Theor. **53**(12), 4787–4792 (2007)
16. V. Cadambe, S. Jafar, Interference alignment and degrees of freedom of the K-user interference channel. IEEE Trans. Inf. Theor. **54**(8), 3425–3441 (2008)
17. M. Maddah-Ali, A. Motahari, A. Khandani, Communication over MIMO X channels: Interference alignment, decomposition, and performance analysis. IEEE Trans. Inform. Theor. **54**(8), 3457–3470 (2008)
18. J. Park, B. Lee, B. Shim, A MMSE vector precoding with block diagonalization for multiuser MIMO downlink. IEEE Trans. Commun. **60**(2), 569–577 (2012)
19. S. Schwarz, Limited feedback transceiver design for downlink MIMO OFDM cellular networks, Ph.D. Dissertation, Technische Universität Wien (2013). http://theses.eurasip.org/theses/514/limited-feedback-transceiver-design-for-downlink/
20. B. Nosrat-Makouei, J. Andrews, R. Heath Jr., MIMO interference alignment over correlated channels with imperfect CSI. IEEE Trans. Signal Process. **59**(6), 2783–2794 (2011)
21. M. Rezaeekheirabadi, M. Guillaud, Limited feedback for interference alignment in the k-user MIMO interference channel, in *Proceedings of the Information Theory Workshop* (Lausanne, 2012), pp. 1–5
22. L. Qiang, Y. Yang, F. Shu, W. Gang, SLNR precoding based on QBC with limited feedback in downlink CoMP system," in *International Conference on Wireless Communications and Signal Processing* (2010), pp. 1–5
23. S. Schwarz, M. Rupp, Adaptive channel direction quantization for frequency selective channels, in *20th European Signal Processing Conference* (Bucarest, 2012), pp. 2536–2540
24. T. Inoue, R. Heath, Jr., Grassmannian predictive frequency domain compression for limited feedback beamforming, in *Information Theory and Applications Workshop* (2010), pp. 173–177
25. D. Love, R. Heath Jr., Limited feedback unitary precoding for spatial multiplexing systems. IEEE Trans. Inf. Theor. **51**(8), 2967–2976 (2005)
26. I.S. Dhillon, R. Heath Jr., T. Strohmer, J.A. Tropp, Constructing packings in Grassmannian manifolds via alternating projection, ArXiv e-prints (2007)
27. W. Santipach, M. Honig, Capacity of a multiple-antenna fading channel with a quantized precoding matrix. IEEE Trans. Inf. Theor. **55**(3), 1218–1234 (2009)
28. D. Love, R. Heath Jr., Limited feedback diversity techniques for correlated channels. IEEE Trans. Veh. Tech. **55**(2), 718–722 (2006)
29. S. Schwarz, R. Heath Jr., M. Rupp, Multiuser MIMO in distributed antenna systems with limited feedback, in *IEEE 4th International Workshop on Heterogeneous and Small Cell Networks, GLOBECOM* (Anaheim, 2012)
30. S. Schwarz, R. Heath Jr., M. Rupp, Single-user MIMO versus multi-user MIMO in distributed antenna systems with limited feedback. EURASIP J. Adv. Signal Process. **2013**(54), 1–20 (2013)
31. B. Clerckx, G. Kim, S. Kim, MU-MIMO with channel statistics-based codebooks in spatially correlated channels, in *IEEE Global Telecommunications Conference* (2008) pp. 1–5
32. E. Park, H. Kim, H. Park, I. Lee, Feedback bit allocation schemes for multi-user distributed antenna systems. IEEE Commun. Lett. **17**(1), 99–102 (2013)
33. A. Gersho, R. Gray, *Vector Quantization and Signal Compression*, The Kluwer international series in engineering and computer science : communications and information theory (Kluwer Academic Publishers, Boston, 1992)
34. B. Zhou, L. Jiang, S. Zhao, C. He, BER analysis of TDD downlink multiuser MIMO systems with imperfect channel state information. EURASIP J. Adv. Signal Process. **2011**(1), 104 (2011)

35. D. McNamara, M. Beach, P. Fletcher, Experimental investigation of the temporal variation of MIMO channels, in *IEEE 54th Vehicular Technology Conference*, vol. 2 (2001), pp. 1063–1067
36. K. Baddour, N. Beaulieu, Autoregressive modeling for fading channel simulation. IEEE Trans. Wirel. Commun. **4**(4), 1650–1662 (2005)
37. B. Hassibi, Random matrices, integrals and space-time systems, in *DIMACS Workshop on Algebraic Coding Theory and Information Theory* (Piscataway, 2003)
38. D. Maiwald, D. Kraus, Calculation of moments of complex Wishart and complex inverse Wishart distributed matrices. IEE Proc. Radar Sonar Navig. **147**(4), 162–168 (2000)
39. S. Schwarz, R. Heath Jr., M. Rupp, Adaptive quantization on a Grassmann-manifold for limited feedback beamforming systems. IEEE Trans. Signal Process. **61**(18), 4450–4462 (2013)
40. O. El Ayach, R. Heath Jr., Grassmannian differential limited feedback for interference alignment. IEEE Trans. Signal Process. **60**(12), 6481–6494 (2012)
41. Y. Zhang, M. Lei, Robust Grassmannian prediction for limited feedback multiuser MIMO systems, in *IEEE Wireless Communications and Networking Conference* (2012), pp. 863–867
42. S. Schwarz, R. Heath, Jr., M. Rupp, Adaptive quantization on the Grassmann-manifold for limited feedback multi-user MIMO systems, in *38th International Conference on Acoustics, Speech and Signal Processing* (Vancouver, 2013)
43. S. Haykin, *Adaptive filter theory*, 4th edn. (Prentice Hall, New Jersey, 2002)
44. R.H. Clarke, A statistical theory of mobile radio reception. Bell Syst. Tech. J. **47**, 957–1000 (1968)

Chapter 7
Advanced Multi User MIMO Concepts

Markus Rupp, Stefan Schwarz and Martin Taranetz

We start with extensions to systems with excess antennas in Sect. 7.1. In the following Sect. 7.2, the feedback methods are extended to frequency selective systems by means of interpolation and clustering, and the lower bound on the Signal to Interference and Noise Ratio (SINR) for Channel Quality Indicator (CQI) feedback is presented in Sect. 7.2.2. The performance of the methods and algorithms is investigated in Sect. 7.2.3 through extensive Monte-Carlo simulations. Relevant background information on the Grassmann manifold, useful for the understanding of the proposed Grassmannian quantizer, is provided in [1, Appendix C]. Proofs and derivations related to the antenna combiners and the SINR lower bound proposed in this chapter are presented in [2, 3]. The Multiple-Input Multiple-Output (MIMO) Minimum Mean Square Error (MMSE) equalizer that is employed in some of the simulations presented in Sect. 7.2.3 is derived in [4]. The impact of the residual multi-user interference due to Channel State Information at the Transmitter (CSIT) inaccuracies is effectively reduced by this equalizer by exploiting the Block Diagonalization (BD) construction to estimate the multi-user interference. More details are in [2–11].

7.1 Extension to Systems with Excess Antennas

The performance of Zero Forcing (ZF) beamforming and BD precoding based Multi User Multiple-Input Multiple-Output (MU-MIMO) with quantized CSIT has been studied thoroughly in the literature, for the case that the number of data streams per

M. Rupp (✉) · S. Schwarz · M. Taranetz
Institute of Telecommunications, TU Wien, Vienna, Austria
e-mail: mrupp@nt.tuwien.ac.at

S. Schwarz
e-mail: sschwarz@nt.tuwien.ac.at

M. Taranetz
e-mail: martin.taranetz@tuwien.ac.at

M. Rupp et al., *The Vienna LTE-Advanced Simulators*,
Signals and Communication Technology, DOI 10.1007/978-981-10-0617-3_7

user $\ell[u, i][k]$ is equal to the number of receive antennas $N_R[u, i]$. It is shown in [12] that the Channel State Information (CSI) feedback overhead of ZF beamforming must be scaled linearly with the logarithmic Signal to Noise Ratio (SNR) (the SNR in [dB]) to achieve the same multiplexing gain as in a system with perfect CSIT. A similar bit-scaling law is determined in [13] for BD precoding to multiple users, if the number of data streams per user is equal to the number of receive antennas. In [14], the results of [12] on ZF beamforming are extended to the case that the users are equipped with multiple receive antennas. An efficient antenna combining algorithm denoted as Quantization Based Combining (QBC) is proposed, which exploits the excess antennas to minimize the CSI quantization error. With this strategy a significant reduction of the residual multi-user interference is achieved, implying a reduced slope of the feedback bit-scaling law.

In this section, the QBC algorithm of [14] is extended to multiple data-streams per user via BD precoding, for the case that the number of data streams per user is less than or equal to the number of receive antennas, i.e., $\ell[u, i][k] \leq N_R[u, i]$. The performance of the obtained Subspace Quantization Based Combining (SQBC) algorithm is investigated analytically by deriving the statistics of the Gramian of the effective channel when including the antenna combiner. An upper bound on the rate loss of a BD system employing SQBC with quantized CSIT compared to a BD system with perfect CSIT is derived. It is shown that this bound generalizes the previous results [12–14]. The corresponding scaling law of the feedback overhead to achieve the same multiplexing gain as a system with perfect CSIT is calculated. Depending on the number of data streams per user, the slope of the bit-scaling law is significantly reduced compared to BD without antenna combining.

To set the basis for the proposed antenna combining algorithm, previous results on the rate loss of ZF and BD precoding systems with quantized CSIT are summarized in Sect. 7.1.1. The SQBC strategy, derived and analyzed in Sect. 7.1.2, is compared to a conventional antenna combining method, namely Maximum Eigenmode Transmission (MET), which considers maximization of the effective channel gain of a user. The performance of MET with quantized CSIT is evaluated in Sect. 7.1.3 by deriving an upper bound on the rate loss with respect to perfect CSIT. Although MET provides an advantage in case of perfect CSIT, a significant throughput improvement is obtained with SQBC in interference-limited scenarios if a reasonable CSI feedback overhead is considered. The RE index $[n, k]$ is partly omitted for intermediate steps and auxiliary variables to shorten notations.

7.1.1 Summary of Previous Results

Important results that assess the sensitivity of special cases of BD based MU-MIMO systems with respect to the CSI quantization error are summarized below under the assumption of independent and Identically Distributed (I.I.D.) Rayleigh fading channels, i.e., $\left[\mathbf{H}_{n,k}^{(\mathrm{ct})}[u, i]\right]_{n,m} \sim \mathcal{N}_{\mathbb{C}}(0, \gamma[u, i])$, with $\gamma[u, i]$ denoting the macroscopic fading loss. A symmetric system is investigated where all users have the same number of antennas and receive an equal number of data streams $\ell_k[u, i] = N_\mathrm{L}, \forall u$. The schedule $\mathscr{S}_{n,k}[i]$ is supposed to be time independent, constantly serving the same set of $S[i] = \frac{N_\mathrm{T}[i]}{N_\mathrm{L}}$ users in parallel (supposed to be integer-valued). CSI quantization is achieved with an Random Vector Quantization (RVQ) quantization codebook.

ZF Beamforming with $N_\mathrm{L} = N_\mathrm{R}[u, i] = 1$

The gap between the per-user transmission rate achieved with ZF beamforming based on perfect CSIT (see Eq. (6.16)) and the rate of ZF beamforming with quantized CSIT (see Eq. (6.25)) employing RVQ is upper-bounded by [12]

$$R_\mathrm{ZF} - R_\mathrm{ZF\text{-}Quant} \leq \log_2\left(1 + \frac{P_i\,\gamma[u, i]}{\tilde{\sigma}_z^2}D\right), \quad D = 2^{-\frac{b}{N_\mathrm{T}[i]-1}}, \tag{7.1}$$

with D being the average distortion achieved with RVQ. To maintain a bounded rate gap, the number of feedback bits b must be scaled linearly with the logarithmic SNR, with a slope that is determined by the number of transmit antennas $N_\mathrm{T}[i]$.

BD Precoding with $N_\mathrm{L} = N_\mathrm{R}[u, i] \geq 1$

The performance of "pure" BD precoding without receive antenna combining is investigated in [13]. The number of streams N_L per user is here equal to the number of receive antennas $N_\mathrm{R}[u, i]$. The per-user rate gap between BD precoding with perfect CSIT and BD precoding with quantized CSIT is upper-bounded by

$$R_\mathrm{BD} - R_\mathrm{BD\text{-}Quant} \leq N_\mathrm{R}[u, i]\log_2\left(1 + \frac{P_i\,\gamma[u, i]}{\tilde{\sigma}_z^2\,N_\mathrm{R}[u, i]}D\right), \tag{7.2}$$

$$D = C_\mathrm{BD}\,2^{-\frac{b}{N_\mathrm{R}[u,i](N_\mathrm{T}[i]-N_\mathrm{R}[u,i])}}. \tag{7.3}$$

As before it is observed that the number of feedback bits must be scaled linearly with the SNR in [dB] to maintain a bounded rate gap, with a slope that depends on both $N_\mathrm{T}[i]$ and $N_\mathrm{R}[u, i]$. The constant C_BD is specified in [15]. Setting $N_\mathrm{L} = N_\mathrm{R}[u, i] = 1$ this bound reduces to the result of [12].

ZF Beamforming with $N_\mathrm{L} = 1, N_\mathrm{R}[u, i] \geq 1$ and Receive Antenna Combining

ZF beamforming with multiple receive antennas is considered in [14]. The QBC antenna combiner proposed in [14] results in an effective channel that can be quantized with minimal quantization error, given the quantization codebook $\mathscr{Q}[u, i]$. In that way, the expected residual multi-user interference is minimized without requiring knowledge about the interference statistics. On the downside, the gain of the effective channel is reduced when applying QBC, causing a loss in the received power of the intended signal. Still, in interference limited scenarios this power-loss is outweighed by the reduction of the multi-user interference. The rate gap between ZF beamforming with $N_\mathrm{R}^{(\mathrm{ZF})}[u, i] = 1$ and perfect CSIT and ZF beamforming with $N_\mathrm{R}[u, i] > 1$ employing QBC and quantized CSIT is bounded by

$$R_{\mathrm{ZF}} - R_{\mathrm{QBC}}^{(N_{\mathrm{R}}[u,i])} \leq \log_2\left(1 + \frac{P_i\,\gamma[u,i]}{\tilde{\sigma}_z^2}\frac{N_{\mathrm{T}}[i] - N_{\mathrm{R}}[u,i] + 1}{N_{\mathrm{T}}[i]}D\right)$$
$$+\log_2(e)\sum_{\ell=N_{\mathrm{T}}[i]-N_{\mathrm{R}}[u,i]+1}^{N_{\mathrm{T}}[i]-1}\frac{1}{\ell}, \tag{7.4}$$

with D being proportional to $2^{-\frac{b}{N_{\mathrm{T}}[i]-N_{\mathrm{R}}[u,i]}}$ as specified in [14]. Hence, with $N_{\mathrm{R}}[u,i] > 1$, the distortion is reduced compared to Eq. (7.1). The superscript $^{(N_{\mathrm{R}}[u,i])}$ is employed to highlight the important dependency of the performance of QBC on the number of available receive antennas. In contrast to the other bounds considered above, a constant residual rate loss (the second summand) is caused by the application of QBC, which does not depend on the quantization accuracy and thus cannot be reduced by increasing the feedback overhead. This loss is caused by the reduced channel gain of QBC compared to the single receive antenna system. With $N_{\mathrm{R}}[u,i] = 1$ the result of [12] is recovered.

7.1.2 Subspace Quantization Based Combining

Considering the list of previous work on theoretical performance bounds for ZF and BD MU-MIMO with limited feedback presented in the previous section, it is noticed that the general case of BD precoding with $1 \leq \ell_k[u,i] \leq N_{\mathrm{R}}[u,i]$ and receive antenna combining is not evaluated. This general case is investigated below by first extending the QBC method to multi-stream transmission per user and then deriving an upper-bound on the rate loss compared to perfect CSIT, generalizing the results of [12–14] presented above.

SQBC Algorithm

The proposed antenna combiner is designed such as to generate an effective channel that can be quantized with minimal subspace quantization error, given the quantization codebook $\mathcal{Q}_{n,k}[u,i]$ and the number of streams $\ell_k[u,i]$, that is,

$$\left\{\mathbf{G}_{n,k}^{(\mathrm{SQBC})}[u,i],\ \hat{\mathbf{H}}_{n,k}^{(\mathrm{SQBC})}[u,i]\right\} = \underset{\mathbf{G},\mathbf{Q}_j}{\operatorname{argmin}}\ \mathrm{d}_{\mathrm{c}}^2\left(\mathbf{H}_{n,k}^{(\mathrm{ect})}[u,i], \mathbf{Q}_j\right)$$
$$= \underset{\mathbf{G},\mathbf{Q}_j}{\operatorname{argmin}}\ \mathrm{d}_{\mathrm{c}}^2\left(\mathbf{H}_{n,k}^{(\mathrm{ct})}[u,i]\mathbf{G}, \mathbf{Q}_j\right), \tag{7.5}$$

$$\text{subject to: } \mathbf{G} \in \mathbb{C}^{N_{\mathrm{R}}[u,i]\times\ell_k[u,i]},\ \mathbf{G}^{\mathrm{H}}\mathbf{G} = \mathbf{I}_{\ell_k[u,i]},$$
$$\mathbf{Q}_j \in \mathcal{Q}_{n,k}[u,i] = \left\{\mathbf{Q}_j \in \mathbb{C}^{N_{\mathrm{T}}[i]\times\ell_k[u,i]} | \mathbf{Q}_j^{\mathrm{H}}\mathbf{Q}_j = \mathbf{I}_{\ell_k[u,i]}, j \in \{1,\ldots,2^b\}\right\}.$$

Considering any $\mathbf{Q}_j \in \mathcal{Q}_{n,k}[u,i]$, a decomposition into its range space and left null space components with respect to $\mathbf{H}_{n,k}^{(\mathrm{ct})}[u,i]$ can be applied, using an orthonormal basis $\mathbf{B}[u,i]$ for span $\left(\mathbf{H}_{n,k}^{(\mathrm{ct})}[u,i]\right)$

$$\mathbf{Q}_j = \mathbf{Q}_j^{(\mathrm{R})} + \mathbf{Q}_j^{(\mathrm{N})}, \tag{7.6}$$

$$\begin{aligned}\mathbf{Q}_j^{(\mathrm{R})} &= \mathbf{H}_{n,k}^{(\mathrm{ct})}[u,i]\left(\mathbf{H}_{n,k}^{(\mathrm{ct})}[u,i]^{\mathrm{H}}\mathbf{H}_{n,k}^{(\mathrm{ct})}[u,i]\right)^{-1}\mathbf{H}_{n,k}^{(\mathrm{ct})}[u,i]^{\mathrm{H}}\mathbf{Q}_j \\ &= \mathbf{B}[u,i]\mathbf{B}[u,i]^{\mathrm{H}}\mathbf{Q}_j, \end{aligned}\tag{7.7}$$

$$\mathbf{Q}_j^{(\mathrm{N})} = \left(\mathbf{I}_{N_{\mathrm{T}}[i]} - \mathbf{B}[u,i]\mathbf{B}[u,i]^{\mathrm{H}}\right)\mathbf{Q}_j, \tag{7.8}$$

$$\mathrm{span}\,(\mathbf{B}[u,i]) = \mathrm{span}\left(\mathbf{H}_{n,k}^{(\mathrm{ct})}[u,i]\right), \tag{7.9}$$

$$\mathbf{B}[u,i] \in \mathbb{C}^{N_{\mathrm{T}}[i]\times N_{\mathrm{R}}[u,i]},\ \mathbf{B}[u,i]^{\mathrm{H}}\mathbf{B}[u,i] = \mathbf{I}_{N_{\mathrm{R}}[u,i]}. \tag{7.10}$$

As $\mathbf{Q}_j^{(\mathrm{R})}$ is in the range space of $\mathbf{H}_{n,k}^{(\mathrm{ct})}[u,i]$, it is possible to find an antenna combiner $\mathbf{G}_{n,k}^{(\mathrm{ct})}[u,i]$ such that the effective channel spans the same space as $\mathbf{Q}_j^{(\mathrm{R})}$, i.e., $\mathrm{span}\left(\mathbf{H}_{n,k}^{(\mathrm{ct})}[u,i]\mathbf{G}_{n,k}^{(\mathrm{ct})}[u,i]\right) = \mathrm{span}\left(\mathbf{Q}_j^{(\mathrm{R})}\right)$. On the other hand, the subspace distance to the component $\mathbf{Q}_j^{(\mathrm{N})}$ in the orthogonal complement of $\mathrm{span}\left(\mathbf{H}_{n,k}^{(\mathrm{ct})}[u,i]\right)$ cannot be reduced with antenna combining. Hence, with appropriate antenna combining, the subspace quantization error is only determined by $\mathbf{Q}_j^{(\mathrm{N})}$, and the minimal quantization error is obtained with that $\mathbf{Q}_j$ that is closest to the range space of $\mathbf{H}_{n,k}^{(\mathrm{ct})}[u,i]$.

Theorem 7.1 (Subspace Quantization Based Combining) *An $\ell_k[u,i] \le N_{\mathrm{R}}[u,i]$ dimensional subspace of a channel matrix $\mathbf{H}_{n,k}^{(\mathrm{ct})}[u,i] \in \mathbb{C}^{N_{\mathrm{T}}[i]\times N_{\mathrm{R}}[u,i]}$ is to be quantized with a given quantization codebook*

$$\mathcal{Q}_{n,k}[u,i] = \left\{\mathbf{Q}_j \in \mathbb{C}^{N_{\mathrm{T}}[i]\times \ell_k[u,i]} | \mathbf{Q}_j^{\mathrm{H}}\mathbf{Q}_j = \mathbf{I}_{\ell_k[u,i]}, j \in \{1,\ldots,2^b\}\right\}. \tag{7.11}$$

Applying an antenna combiner to generate an effective channel of dimension $N_{\mathrm{T}}[i] \times \ell_k[u,i]$, the minimal quantization error that can be obtained and the corresponding quantized channel subspace are

$$\begin{aligned}\mathrm{d}^2_{\mathrm{c,SQBC}}[n,k] &= \min_{\mathbf{Q}_j\in\mathcal{Q}_{n,k}[u,i]} \left\|\mathbf{Q}_j^{(\mathrm{N})}\right\|^2 = \min_{\mathbf{Q}_j\in\mathcal{Q}_{n,k}[u,i]} \mathrm{tr}\left((\mathbf{Q}_j^{(\mathrm{N})})^{\mathrm{H}}\mathbf{Q}_j^{(\mathrm{N})}\right) \\ &= \min_{\mathbf{Q}_j\in\mathcal{Q}_{n,k}[u,i]} \ell_k[u,i] - \mathrm{tr}\left(\mathbf{B}[u,i]^{\mathrm{H}}\mathbf{Q}_j\mathbf{Q}_j^{\mathrm{H}}\mathbf{B}[u,i]\right) \\ &= \min_{\mathbf{Q}_j\in\mathcal{Q}_{n,k}[u,i]} \mathrm{d}^2_{\mathrm{c}}\left(\mathbf{H}_{n,k}^{(\mathrm{ct})}[u,i], \mathbf{Q}_j\right), \end{aligned}\tag{7.12}$$

$$\hat{\mathbf{H}}_{n,k}^{(\mathrm{SQBC})}[u,i] = \operatorname*{argmin}_{\mathbf{Q}_j\in\mathcal{Q}_{n,k}[u,i]} \mathrm{d}^2_{\mathrm{c}}\left(\mathbf{H}_{n,k}^{(\mathrm{ct})}[u,i], \mathbf{Q}_j\right), \tag{7.13}$$

with $\mathrm{span}\,(\mathbf{B}[u,i]) = \mathrm{span}\left(\mathbf{H}_{n,k}^{(\mathrm{ct})}[u,i]\right)$, $\mathbf{B}[u,i]^{\mathrm{H}}\mathbf{B}[u,i] = \mathbf{I}_{N_{\mathrm{R}}[u,i]}$. *The corresponding receive antenna combiner, generating the effective channel that achieves this error, is obtained from the conditions*

$$\mathrm{span}\left(\mathbf{H}_{n,k}^{(\mathrm{ct})}[u,i]\mathbf{G}_{n,k}^{(\mathrm{SQBC})}[u,i]\right) \stackrel{!}{=} \mathrm{span}\left((\mathbf{B}[u,i]\mathbf{B}[u,i]^{\mathrm{H}})\hat{\mathbf{H}}_{n,k}^{(\mathrm{SQBC})}[u,i]\right), \tag{7.14}$$

$$(\mathbf{G}_{n,k}^{(\mathrm{SQBC})}[u,i])^{\mathrm{H}}\mathbf{G}_{n,k}^{(\mathrm{SQBC})}[u,i] \stackrel{!}{=} \mathbf{I}_{\ell_k[u,i]}. \tag{7.15}$$

A formal proof of this theorem is provided in [1, Appendix F]. Notice that the quantization metric (7.13) is independent of the antenna combiner. Hence, it is not necessary for CSI quantization to calculate the combiner for each $\mathbf{Q}_j$, providing an advantage in terms of computational complexity.

A solution for the antenna combiner obtained from the conditions stated in Theorem 7.1 is unique only up to right-multiplication with any unitary $\ell_k[u,i] \times \ell_k[u,i]$ matrix. In the following a specific solution is derived that enables further investigations on the channel statistics; an equivalent alternative solution is provided in [4]. The condition (7.14) can be written as

$$\mathbf{H}_{n,k}^{(\mathrm{ct})}[u,i]\mathbf{G}_{n,k}^{(\mathrm{SQBC})}[u,i] = (\mathbf{B}[u,i]\mathbf{B}[u,i]^{\mathrm{H}})\hat{\mathbf{H}}_{n,k}^{(\mathrm{SQBC})}[u,i]\tilde{\mathbf{K}}[u,i], \tag{7.16}$$

with an appropriate full-rank matrix $\tilde{\mathbf{K}}[u,i] \in \mathbb{C}^{\ell_k[u,i]\times\ell_k[u,i]}$. The orthonormal basis $\mathbf{B}[u,i]$ can be chosen such that the first $\ell_k[u,i]$ columns of $\mathbf{B}[u,i]$ correspond to the range space component of $\hat{\mathbf{H}}_{n,k}^{(\mathrm{SQBC})}[u,i]$ with respect to $\mathbf{H}_{n,k}^{(\mathrm{ct})}[u,i]$ and the remaining $N_\mathrm{R}[u,i] - \ell_k[u,i]$ columns are orthogonal to $\hat{\mathbf{H}}_{n,k}^{(\mathrm{SQBC})}[u,i]$

$$\mathbf{B}[u,i]^{\mathrm{H}}\hat{\mathbf{H}}_{n,k}^{(\mathrm{SQBC})}[u,i] = \mathbf{W}[u,i]\mathbf{R}[u,i] = \begin{bmatrix} \mathbf{I}_{\ell_k[u,i]} \\ \mathbf{0} \end{bmatrix}\mathbf{R}[u,i], \tag{7.17}$$
$$\mathbf{W}[u,i] \in \mathbb{C}^{N_\mathrm{R}[u,i]\times\ell_k[u,i]},\ \mathbf{R}[u,i] \in \mathbb{C}^{\ell_k[u,i]\times\ell_k[u,i]}.$$

The channel $\mathbf{H}_{n,k}^{(\mathrm{ct})}[u,i]$ is decomposed in terms of $\mathbf{B}[u,i]$, resulting in $\mathbf{H}_{n,k}^{(\mathrm{ct})}[u,i] = \mathbf{B}[u,i]\mathbf{D}[u,i]$ with $\mathbf{D}[u,i] = \mathbf{B}[u,i]^{\mathrm{H}}\mathbf{H}_{n,k}^{(\mathrm{ct})}[u,i] \in \mathbb{C}^{N_\mathrm{R}[u,i]\times N_\mathrm{R}[u,i]}$. With these decompositions the solution for $\mathbf{G}_{n,k}^{(\mathrm{SQBC})}[u,i]$ is obtained as

$$\mathbf{G}_{n,k}^{(\mathrm{SQBC})}[u,i] = \mathbf{H}_{n,k}^{(\mathrm{ct})}[u,i]^{\dagger}\mathbf{B}[u,i]\mathbf{W}[u,i]\underbrace{\mathbf{R}[u,i]\tilde{\mathbf{K}}[u,i]}_{\mathbf{K}[u,i]} = \mathbf{D}[u,i]^{-1}\mathbf{W}[u,i]\mathbf{K}[u,i]. \tag{7.18}$$

The undetermined matrix $\mathbf{K}[u,i]$ of size $\ell_k[u,i] \times \ell_k[u,i]$ is obtained by invoking condition (7.15)

$$\begin{aligned}(\mathbf{G}_{n,k}^{(\mathrm{SQBC})}[u,i])^{\mathrm{H}}\mathbf{G}_{n,k}^{(\mathrm{SQBC})}[u,i] \\ = \mathbf{K}[u,i]^{\mathrm{H}}\left(\mathbf{W}[u,i]^{\mathrm{H}}\left(\mathbf{D}[u,i]\mathbf{D}[u,i]^{\mathrm{H}}\right)^{-1}\mathbf{W}[u,i]\right)\mathbf{K}[u,i] \stackrel{!}{=} \mathbf{I}_{\ell_k[u,i]},\end{aligned} \tag{7.19}$$

$$\Rightarrow \mathbf{K}[u,i] = \left(\mathbf{W}[u,i]^{\mathrm{H}}\left(\mathbf{B}[u,i]^{\mathrm{H}}\mathbf{H}_{n,k}^{(\mathrm{ct})}[u,i]\mathbf{H}_{n,k}^{(\mathrm{ct})}[u,i]^{\mathrm{H}}\mathbf{B}[u,i]\right)^{-1}\mathbf{W}[u,i]\right)^{-1/2}. \tag{7.20}$$

With this solution, the effective channel employing SQBC is given by

$$\mathbf{H}_{n,k}^{(\mathrm{ect})}[u,i] = \mathbf{H}_{n,k}^{(\mathrm{ct})}[u,i]\mathbf{G}_{n,k}^{(\mathrm{SQBC})}[u,i] = \mathbf{B}[u,i]\mathbf{W}[u,i]\mathbf{K}[u,i] = \tilde{\mathbf{H}}_{n,k}[u,i]\mathbf{K}[u,i]. \tag{7.21}$$

Channel Statistics

To derive statements about the statistics of the effective channel obtained with SQBC, and to develop the throughput bound of the rate loss incurred with quantized CSIT compared to perfect CSIT, it is necessary to impose additional assumptions on the channel $\mathbf{H}_{n,k}^{(\mathrm{ct})}[u,i]$ and the quantization codebook $\mathscr{Q}_{n,k}[u,i]$. Specifically, in the following two paragraphs the channel matrix is assumed as I.I.D. Rayleigh fading, i.e., $[\mathbf{H}_{n,k}^{(\mathrm{ct})}[u,i]]_{n,m} \sim \mathscr{N}_{\mathbb{C}}(0, \gamma[u,i])$, and RVQ is employed for channel subspace quantization. Also, the number of streams per user is assumed as time independent $\ell_k[u,i] = \ell[u,i]$.

Lemma 7.1 *The subspaces spanned by the effective channels* $\mathbf{H}_{n,k}^{(\mathrm{ect})}[u,i], \forall u$ *are statistically independent and isotropically (uniformly) distributed on* $\mathscr{G}(N_{\mathrm{T}}[i], \ell[u,i])$.

Proof From (7.14) it is known that the subspace spanned by $\mathbf{H}_{n,k}^{(\mathrm{ect})}[u,i]$ is determined by the projection of the best quantization matrix $\hat{\mathbf{H}}_{n,k}[u,i]$ onto span $\left(\mathbf{H}_{n,k}^{(\mathrm{ct})}[u,i]\right)$. Since the quantization matrices are isotropically distributed on $\mathscr{G}(N_{\mathrm{T}}[i], \ell[u,i])$, their projections onto span $\left(\mathbf{H}_{n,k}^{(\mathrm{ct})}[u,i]\right)$ are isotropically distributed within this subspace. This holds also true for the best quantization matrix, since it is chosen based solely on the Frobenius norm of the null space component. Furthermore, the subspace span $\left(\mathbf{H}_{n,k}^{(\mathrm{ct})}[u,i]\right)$ itself is isotropically distributed on $\mathscr{G}(N_{\mathrm{T}}[i], N_{\mathrm{R}}[u,i])$, since we assume i.i.d. Rayleigh fading. Thus, span $\left(\mathbf{H}_{n,k}^{(\mathrm{ect})}[u,i]\right)$ is isotropically distributed on $\mathscr{G}(N_{\mathrm{T}}[i], \ell[u,i])$. Finally, the quantization codebooks and channels of different users are statistically independent, implying statistical independence of $\mathbf{H}_{1,i}^{\mathrm{eff}}[n,k], \ldots, \mathbf{H}_{S[i],i}^{\mathrm{eff}}[n,k]$.

Lemma 7.2 *The Gramian of the effective channel* $(\mathbf{H}_{n,k}^{(\mathrm{ect})}[u,i])^{\mathrm{H}}\mathbf{H}_{n,k}^{(\mathrm{ect})}[u,i]$ *is complex-valued Wishart distributed of dimension* $\ell[u,i]$, *with* $N_{\mathrm{T}}[i] - N_{\mathrm{R}}[u,i] + \ell[u,i]$ *degrees of freedom and identity scale matrix*

$$(\mathbf{H}_{n,k}^{(\mathrm{ect})}[u,i])^{\mathrm{H}}\mathbf{H}_{n,k}^{(\mathrm{ect})}[u,i] \sim \mathscr{W}_{\ell[u,i]}^{\mathbb{C}}\left(N_{\mathrm{T}}[i] - N_{\mathrm{R}}[u,i] + \ell[u,i], \gamma[u,i]\,\mathbf{I}_{\ell[u,i]}\right). \tag{7.22}$$

The proof of this lemma is provided in [1, Appendix F].

Throughput Analysis

Utilizing the results of the previous paragraph, it is possible to derive an upper bound on the rate loss of BD precoding with quantized CSIT and excess receive antennas $N_{\mathrm{R}}[u,i] \geq \ell[u,i]$, with respect to BD precoding with perfect CSIT but having no excess antennas at the receivers, i.e., $N_{\mathrm{R}}^{(\mathrm{BD})}[u,i] = \ell[u,i]$. In case of excess antennas, it is assumed that the users employ SQBC antenna combining to determine the channel subspace feedback according to the quantization metric in Theorem 7.1.

The quantized CSIT is used by the base station to calculate the precoders from (6.4) and (6.22), respectively. The same symmetric scenario as in Sect. 7.1.1 is considered, i.e., all $S[i]$ users are equipped with the same number of $N_{\mathrm{R}}[u,i]$ receive antennas and are served over the same number of $\ell[u,i] = N_{\mathrm{L}}$ data streams. An RVQ quantization codebook is employed for quantization of the effective channel subspace.

According to (6.16), the achievable user rate of BD with perfect CSIT and no excess antennas is

$$R_{\mathrm{BD}}^{(N_{\mathrm{L}})} = \mathbb{E}\log_2\det\left(\mathbf{I}_{N_{\mathrm{L}}} + \rho\,\mathbf{H}_{n,k}^{(\mathrm{ct})}[u,i]^{\mathrm{H}}\tilde{\mathbf{F}}_{n,k}[u,i]\tilde{\mathbf{F}}_{n,k}[u,i]^{\mathrm{H}}\mathbf{H}_{n,k}^{(\mathrm{ct})}[u,i]\right), \tag{7.23}$$

$$\rho = \frac{P_i}{\tilde{\sigma}_z^2\,S[i]\,N_{\mathrm{L}}}, \tag{7.24}$$

with $\mathbf{H}_{n,k}^{(\mathrm{ct})}[u,i] = \mathbf{H}_{n,k}^{(\mathrm{ect})}[u,i]$ because $N_{\mathrm{R}}^{(\mathrm{BD})}[u,i] = N_{\mathrm{L}}$. The expected value is calculated with respect to the channel and the precoder. Similarly, with quantized CSIT, $N_{\mathrm{L}} \leq N_{\mathrm{R}}[u,i]$ data streams and application of SQBC antenna combining, the achievable rate is obtained as

$$\begin{aligned} R_{\mathrm{SQBC}}^{(N_{\mathrm{L}},N_{\mathrm{R}}[u,i])} &= \mathbb{E}\log_2\det\left(\mathbf{I}_{N_{\mathrm{L}}} + \rho\sum_{s=1}^{S[i]}\mathbf{H}_{n,k}^{(\mathrm{ect})}[u,i]^{\mathrm{H}}\tilde{\mathbf{F}}_{n,k}[s,i]\tilde{\mathbf{F}}_{n,k}[s,i]^{\mathrm{H}}\mathbf{H}_{n,k}^{(\mathrm{ect})}[u,i]\right) \\ &- \mathbb{E}\log_2\det\left(\mathbf{I}_{N_{\mathrm{L}}} + \rho\sum_{s=1,s\neq u}^{S[i]}\mathbf{H}_{n,k}^{(\mathrm{ect})}[u,i]^{\mathrm{H}}\tilde{\mathbf{F}}_{n,k}[s,i]\tilde{\mathbf{F}}_{n,k}[s,i]^{\mathrm{H}}\mathbf{H}_{n,k}^{(\mathrm{ect})}[u,i]\right), \end{aligned} \tag{7.25}$$

$$\rho = \frac{P_i}{\tilde{\sigma}_z^2\,S[i]\,N_{\mathrm{L}}}, \tag{7.26}$$

with $\mathbf{H}_{n,k}^{(\mathrm{ect})}[u,i]$ being determined by Eq. (7.21). Here, the expected value is additionally calculated over quantization codebook realizations. Similar to the bounds proposed in [12–14], i.e., Eqs. (7.1), (7.3), and (7.4), the throughput loss $R_{\mathrm{BD}}^{(N_{\mathrm{L}})} - R_{\mathrm{SQBC}}^{(N_{\mathrm{L}},N_{\mathrm{R}}[u,i])}$ can be upper bounded:

Theorem 7.2 (SQBC rate loss) *Consider a broadcast system with $N_{\mathrm{T}}[i]$ transmit antennas, $N_{\mathrm{R}}[u,i]$ receive antennas per user and transmit power P_i. The system serves $S[i]$ users with $N_{\mathrm{L}} \leq N_{\mathrm{R}}[u,i]$ streams each, over I.I.D. Rayleigh fading channels with additive Gaussian receiver noise of variance $\tilde{\sigma}_z^2$. The per-user throughput loss of BD precoding employing SQBC with quantized CSIT compared to BD precoding with perfect CSIT, but having only $N_{\mathrm{R}}[u,i]^{(BD)} = N_{\mathrm{L}}$ receive antennas, is upper bounded by*

$$R_{\mathrm{BD}}^{(N_{\mathrm{L}})} - R_{\mathrm{SQBC}}^{(N_{\mathrm{L}},N_{\mathrm{R}}[u,i])} \leq N_{\mathrm{L}}\log_2\left(1 + \rho\,\gamma[u,i]\,\frac{N_{\mathrm{T}}[i] - N_{\mathrm{R}}[u,i] + N_{\mathrm{L}}}{N_{\mathrm{T}}[i] - N_{\mathrm{L}}}\,(S[i]-1)\,D\right)$$

$$+\log_2(e)\sum_{k=0}^{N_\mathrm{L}-1}\sum_{\ell=N_\mathrm{T}[i]-N_\mathrm{R}[u,i]+N_\mathrm{L}}^{N_\mathrm{T}[i]-1}\frac{1}{\ell-k},\quad \rho=\frac{P_i}{\tilde{\sigma}_z^2\,S[i]\,N_\mathrm{L}}. \tag{7.27}$$

Here, the average quantization distortion in terms of subspace chordal distance achieved with RVQ is denoted D and the macroscopic pathloss is considered in $\gamma[u,i]$.

The proof of this theorem is provided in Appendix [1, Appendix F]. As can be seen from (7.12), the relevant distortion D for SQBC is the distortion achieved when quantizing subspaces from $\mathscr{G}(N_\mathrm{T}[i], N_\mathrm{R}[u,i])$ using a quantization codebook with entries from $\mathscr{G}(N_\mathrm{T}[i], N_\mathrm{L})$. This quantization problem is considered in [15]. It is shown in [15], that the average distortion with random isotropically distributed quantization codebooks, i.e., RVQ, is obtained as

$$D=\mathbb{E}\left(\mathrm{d}_{\mathrm{c,SQBC}}^2\right)\approx C_\mathrm{SQBC}\,2^{-\frac{b}{N_\mathrm{L}(N_\mathrm{T}[i]-N_\mathrm{R}[u,i])}}, \tag{7.28}$$

with C_SQBC being a function of N_L, $N_\mathrm{T}[i]$ and $N_\mathrm{R}[u,i]$ as specified in [15, Eqs. (8) and (11)], and 2^b being the size of the codebook. Notice, for random quantization the upper bound of [15, Eq. (11)] is relevant; the $\mathscr{O}(1)$ term is omitted. By setting appropriate values for N_L, $N_\mathrm{T}[i]$ and $N_\mathrm{R}[u,i]$ and considering the case $S[i]=\frac{N_\mathrm{T}[i]}{N_\mathrm{L}}$ it can be verified that (7.1), (7.3) and (7.4) are contained in (7.27).

Discussion of the Bound

The upper bound on the rate loss in (7.27) is composed of two summands capturing distinct effects of the SQBC antenna combiner. The second summand, which is abbreviated by Δa, is independent of the quantization accuracy D. From the argumentation in the proof in [1, Appendix F], it follows that Δa gives a tight upper bound on the rate loss $R_\mathrm{BD}^{(N_\mathrm{L})}-R_\mathrm{SQBC}^{(N_\mathrm{L},N_\mathrm{R}[u,i])}$ for the case of perfect CSIT ($b\to\infty, D\to 0$) and $\rho\,\gamma[u,i]\to\infty$, i.e., at high SINR. This high SINR rate loss is caused by the reduction of the effective channel gain due to the antenna combiner. Specifically, for BD with $N_\mathrm{R}[u,i]^{(\mathrm{BD})}=N_\mathrm{L}$, the term $\mathbf{H}_{n,k}^{(\mathrm{ect})}[u,i]^\mathrm{H}\mathbf{H}_{n,k}^{(\mathrm{ect})}[u,i]$ is distributed according to $\mathscr{W}_{N_\mathrm{L}}^{\mathbb{C}}\left(N_\mathrm{T}[i],\gamma[u,i]\,\mathbf{I}_{N_\mathrm{L}}\right)$, while it is shown in Lemma 7.2 that it is distributed as $\mathscr{W}_{N_\mathrm{L}}^{\mathbb{C}}\left(N_\mathrm{T}[i]-N_\mathrm{R}[u,i]+N_\mathrm{L},\gamma[u,i]\,\mathbf{I}_{N_\mathrm{L}}\right)$ for SQBC. This loss in Degrees of Freedom (DoF) of the Wishart distribution is the cause for the constant rate offset Δa. Hence, SQBC is disadvantageous if the CSI quantization accuracy is already very high. In this case, it is better to invest the available DoF provided by the $N_\mathrm{R}[u,i]\geq N_\mathrm{L}$ antennas to obtain a large effective channel gain, instead of further reducing the quantization error.

The more important insights for limited feedback systems are captured in the first summand of (7.27) (abbreviated by Δb). If b is fixed, this term grows to infinity as the SNR increases, meaning that the system becomes interference limited. If b is scaled with the SNR, however, the rate gap can be kept constant and a multiplexing gain of $N_T[i]$ can be achieved. Assuming $S[i] = \frac{N_T[i]}{N_L}$ the necessary feedback bit scaling is determined by setting Δb equal to a constant rate loss ΔR in [bits/s/Hz]

$$b = N_L \left(N_T[i] - N_R[u,i]\right)\left(\log_2\left(\frac{P_i}{\tilde{\sigma}_z^2}\frac{\gamma[u,i]}{N_T[i]}\right)\right.$$
$$\left.+ \log_2\left(\frac{N_T[i] - N_R[u,i] + N_L}{N_L} C_{SQBC}\right) - \log_2\left(2^{\frac{\Delta R}{N_L}} - 1\right)\right). \tag{7.29}$$

The important insight is the growth-rate of the number of feedback bits with the SNR β_{dB} in [dB]

$$\frac{db}{d\beta_{dB}} = N_L \left(N_T[i] - N_R[u,i]\right)\frac{\log_2(10)}{10} \approx \frac{N_L \left(N_T[i] - N_R[u,i]\right)}{3}, \tag{7.30}$$
$$\beta_{dB} = 10\log_{10}\left(\frac{P_i}{\tilde{\sigma}_z^2}\right).$$

It can be seen that the slope of the feedback overhead with SNR grows linearly with the number of streams N_L, and reduces with the number of receive antennas $N_R[u,i]$. Hence, having more receive antennas with a fixed N_L the CSI feedback overhead can be decreased. Still, this does not give the full picture, because varying N_L and/or $N_R[u,i]$ also impacts the absolute achievable throughput.

To investigate this influence, the high SINR sum rate difference between two SQBC systems with the same $N_T[i]$ and N_L, but different $N_R[u,i] \in \{M_1, M_2\}$ is determined. This rate difference is obtained from (7.27) by considering the bound on $R_{BD}^{(N_L)} - R_{SQBC}^{(N_L,M_2)} - \left(R_{BD}^{(L)} - R_{SQBC}^{(N_L,M_1)}\right)$ at high SINR

$$\Delta R(M_1,M_2) = \frac{N_T[i]}{N_L}\left(R_{SQBC}^{(N_L,M_1)} - R_{SQBC}^{(N_L,M_2)}\right)$$
$$= \frac{N_T[i]}{N_L}\log_2(e)\sum_{k=0}^{N_L-1}\sum_{\ell=N_T[i]-M_2+N_L}^{N_T[i]-M_1+N_L-1}\frac{1}{\ell-k}, \tag{7.31}$$

which results in a positive rate loss if $M_2 > M_1$. Thus, in contrast to most conventional antenna combining strategies, if SQBC is employed with perfect CSIT, a rate reduction is incurred at high SINR with growing number of receive antennas $N_R[u,i]$. Similarly, it can be shown that the sum rate of SQBC is improved if the number of streams per user is increased and the number of users is correspondingly decreased; see [2] for details.

7.1.3 Maximum Eigenmode Transmission

In this section, the performance of an alternative interference-unaware receive antenna combining algorithm, namely MET, is investigated. With MET the effective channel generated by a user is composed of the N_{L} maximum eigenmodes of the channel matrix. Hence, with perfect CSIT the N_{L}-dimensional dominant subspace of each users' channel is kept free of interference, providing a potentially large channel gain. On the other hand, the CSI quantization error achieved with MET is significantly larger than with SQBC, and thus the residual multi-user interference has a much stronger impact on the performance of MET. The main reason for considering MET as an alternative strategy to SQBC is that it enables an instructive investigation on the trade off between investing the provided DoF of having $N_{\mathrm{R}}[u, i] \geq N_{\mathrm{L}}$ excess antennas to maximizing the signal power by means of MET, in contrast to minimizing the CSI quantization error (and hence the expected interference power) utilizing SQBC. Also, the complexity and CSI feedback requirements (subspace information) of MET are very similar to those of SQBC. In [16], MET is combined with a coordinated eigenmode selection by the base station, such that users with close to orthogonal channels are served in parallel. In the following investigation, scheduling is not explicitly considered.

MET Algorithm

The goal of MET is to generate an N_{L}-dimensional effective channel that maximizes the achievable transmission rate of a user in the absence of multi-user interference, by applying a semi-unitary antenna combiner. This is achieved, if the transmission to the user takes place over the N_{L}-dimensional dominant subspace of the channel matrix $\mathbf{H}_{n,k}^{(\mathrm{ct})}[u, i]$. Consider an Singular Value Decomposition (SVD) of the channel matrix $\mathbf{H}_{n,k}^{(\mathrm{ct})}[u, i]$ in compact form

$$\mathbf{H}_{n,k}^{(\mathrm{ct})}[u, i] = \mathbf{U}_{n,k}[u, i]\boldsymbol{\Sigma}_{n,k}[u, i]\mathbf{V}_{n,k}[u, i]^{\mathrm{H}} \tag{7.32}$$
$$\mathbf{U}_{n,k}[u, i] \in \mathbb{C}^{N_{\mathrm{T}}[i]\times N_{\mathrm{R}}[u,i]},\ \boldsymbol{\Sigma}_{n,k}[u, i] \in \mathbb{C}^{N_{\mathrm{R}}[u,i]\times N_{\mathrm{R}}[u,i]},\ \mathbf{V}_{n,k}[u, i] \in \mathbb{C}^{N_{\mathrm{R}}[u,i]\times N_{\mathrm{R}}[u,i]}.$$

Notice that $\mathbf{U}_{n,k}[u, i]$, $\boldsymbol{\Sigma}_{n,k}[u, i]$, $\mathbf{V}_{n,k}[u, i]$ are statistically independent, and $\mathbf{U}_{n,k}[u, i]$ and $\mathbf{V}_{n,k}[u, i]$ are isotropic for I.I.D. Rayleigh fading. The channel subspace to be quantized is chosen as the first N_{L} columns of $\mathbf{U}_{n,k}[u, i]$, corresponding to the N_{L} largest singular values of $\mathbf{H}_{n,k}^{(\mathrm{ct})}[u, i]$. The quantized subspace is obtained as

$$\hat{\mathbf{H}}_{n,k}^{(\mathrm{MET})}[u, i] = \underset{\mathbf{Q}_j \in \mathcal{Q}_{n,k}[u,i]}{\operatorname{argmin}}\ \mathrm{d}_{\mathrm{c}}^2\left(\left[\mathbf{U}_{n,k}[u, i]\right]_{:,1:N_{\mathrm{L}}}, \mathbf{Q}_j\right). \tag{7.33}$$

As with SQBC, the quantized channel subspace can be obtained without having to calculate the antenna combiner for each $\mathbf{Q}_j$. The codebook index of the quantized channel subspace is fed back to the base station by the users, and the BD precoder is calculated from quantized CSIT. With the provided subspace feedback power loading over the eigenmodes is not reasonable, because no information about the magnitude of the singular values is available at the base station. The corresponding MET antenna combiner and the effective channel are given by

$$\mathbf{G}_{n,k}^{(\mathrm{MET})}[u,i] = \left[\mathbf{V}_{n,k}[u,i]\right]_{:,1:N_\mathrm{L}}, \tag{7.34}$$

$$\begin{aligned}\mathbf{H}_{n,k}^{(\mathrm{ect})}[u,i] &= \left[\mathbf{U}_{n,k}[u,i]\right]_{:,1:N_\mathrm{L}}\left[\boldsymbol{\Sigma}_{n,k}[u,i]\right]_{1:N_\mathrm{L},1:N_\mathrm{L}}\\ &= \mathbf{U}_{n,k}^{(N_\mathrm{L})}[u,i]\boldsymbol{\Sigma}_{n,k}^{(N_\mathrm{L})}[u,i] = \mathbf{H}_{n,k}^{(N_\mathrm{L})}[u,i].\end{aligned} \tag{7.35}$$

Throughput Analysis

The performance of BD precoding with MET antenna combining and quantized CSIT can be evaluated in a similar way as the performance of SQBC under the same assumptions (symmetric scenario, I.I.D. Rayleigh fading, RVQ), by deriving an upper bound on the rate loss with respect to perfect CSIT. With the effective channel from Eq. (7.35), the achievable per-user rate with perfect CSIT according to Eq. (6.16) is

$$R_\mathrm{MET} = \mathbb{E}\log_2\det\left(\mathbf{I}_{N_\mathrm{L}} + \rho\,\mathbf{H}_{n,k}^{(N_\mathrm{L})}[u,i]^\mathrm{H}\tilde{\mathbf{F}}_{n,k}[u,i]\tilde{\mathbf{F}}_{n,k}[u,i]^\mathrm{H}\mathbf{H}_{n,k}^{(N_\mathrm{L})}[u,i]\right). \tag{7.36}$$

Similarly to (7.25), with quantized CSIT the achievable per-user rate is obtained as

$$\begin{aligned}R_\text{MET-Quant} = {}&\mathbb{E}\log_2\det\left(\mathbf{I}_{N_\mathrm{L}} + \rho\sum_{s=1}^{S[i]}\mathbf{H}_{n,k}^{(N_\mathrm{L})}[u,i]^\mathrm{H}\tilde{\mathbf{F}}_{n,k}[s,i]\tilde{\mathbf{F}}_{n,k}[s,i]^\mathrm{H}\mathbf{H}_{n,k}^{(N_\mathrm{L})}[u,i]\right)\\ &-\mathbb{E}\log_2\det\left(\mathbf{I}_{N_\mathrm{L}} + \rho\sum_{s=1,s\neq u}^{S[i]}\mathbf{H}_{n,k}^{(N_\mathrm{L})}[u,i]^\mathrm{H}\tilde{\mathbf{F}}_{n,k}[s,i]\tilde{\mathbf{F}}_{n,k}[s,i]^\mathrm{H}\mathbf{H}_{n,k}^{(N_\mathrm{L})}[u,i]\right).\end{aligned} \tag{7.37}$$

Theorem 7.3 (MET rate loss) *Consider a multi-user broadcast system with $N_\mathrm{T}[i]$ transmit antennas, $N_\mathrm{R}[u,i]$ receive antennas per user and transmit power P_i. The system serves $S[i]$ users with $N_\mathrm{L} \le N_\mathrm{R}[u,i]$ spatial streams each, over I.I.D. Rayleigh fading channels with additive Gaussian receiver noise of variance $\tilde{\sigma}_z^2$. The per-user throughput loss of BD precoding employing MET antenna combining with quantized CSIT compared to perfect CSIT is upper bounded by*

$$R_\mathrm{MET} - R_\text{MET-Quant} \le \sum_{\ell=1}^{N_\mathrm{L}}\log_2\left(1 + \rho\,\bar{\sigma}_{\ell,u,i}^2\frac{S[i]-1}{N_\mathrm{T}[i]-N_\mathrm{L}}D\right),\quad \rho = \frac{P_i}{\tilde{\sigma}_z^2\,S[i]\,N_\mathrm{L}}. \tag{7.38}$$

Here, the average quantization distortion in terms of subspace chordal distance achieved with RVQ is denoted D. The set $\left\{\bar{\sigma}_1^2[u,i]\ldots,\bar{\sigma}_{N_\mathrm{L}}^2[u,i]\right\}$ is composed of *the expected values of the N_{L} largest squared singular values of the channel matrix $\mathbf{H}_{n,k}^{(\mathrm{ct})}[u,i]$. Hence, $\bar{\sigma}_{\ell}^{2}[u,i]$ is equal to the expected value of the ℓ-th largest eigenvalue of $\mathbf{H}_{n,k}^{(\mathrm{ct})}[u,i]^{\mathrm{H}}\mathbf{H}_{n,k}^{(\mathrm{ct})}[u,i] \sim \mathscr{W}_{N_{\mathrm{R}}[u,i]}^{\mathbb{C}}\left(N_{\mathrm{T}}[i], \gamma[u,i]\,\mathbf{I}_{N_{\mathrm{R}}[u,i]}\right)$, with $\gamma[u,i]$ denoting the macroscopic pathloss.*

The proof of this theorem is provided in [1, Appendix F]. Notice that the macroscopic pathloss $\gamma[u,i]$ does not appear explicitly in (7.38) but is captured in the squared singular values. Closed form expressions for the expected eigenvalues of $\mathbf{H}_{n,k}^{(\mathrm{ct})}[u,i]^{\mathrm{H}}\mathbf{H}_{n,k}^{(\mathrm{ct})}[u,i]$ are known [17], but the expressions are involved and do not provide further analytical insights. In contrast to the distortion D in (7.28) achieved with SQBC, the average distortion D with MET is determined by quantizing an isotropically distributed subspace from $\mathscr{G}\left(N_{\mathrm{T}}[i], N_{\mathrm{L}}\right)$ with a random codebook of isotropically distributed entries from $\mathscr{G}\left(N_{\mathrm{T}}[i], N_{\mathrm{L}}\right)$ [15]

$$D \approx C_{\mathrm{MET}}\, 2^{-\frac{b}{N_{\mathrm{L}}(N_{\mathrm{T}}[i]-N_{\mathrm{L}})}}. \tag{7.39}$$

It is hard to derive a closed form solution for b in dependency of the rate loss as in (7.29), but the required number of bits can be evaluated numerically. The important fact to note is that with MET the exponent of the distortion scales inversely proportional to $N_{\mathrm{L}}(N_{\mathrm{T}}[i]-N_{\mathrm{L}})$ in contrast to $N_{\mathrm{L}}(N_{\mathrm{T}}[i]-N_{\mathrm{R}}[u,i])$, which is achieved with SQBC. Hence, with SQBC having $N_{\mathrm{R}}[u,i] > N_{\mathrm{L}}$ leads to a reduction of the average quantization error, which is not the case with MET.

7.1.4 Achievable Rate Comparison of SQBC and MET

Whether SQBC or MET is considered as the preferred receive antenna combining strategy in combination with BD precoding is determined by the available CSI feedback resources. This is investigated and explained in more detail in this section. For that purpose, the achievable transmission rates of the two strategies in the limiting case of perfect CSIT, i.e., $b \to \infty$ is evaluated. The rates are calculated from the corresponding achievable rate Eqs. (7.23) and (7.25) by estimating the expected value by means of Monte-Carlo simulations. The achievable rate is calculated for a single Orthogonal Frequency Division Multiplexing (OFDM) subcarrier, assuming frequency flat Rayleigh fading with $\left[\mathbf{H}_{n,k}^{(\mathrm{ct})}[u,i]\right]_{l,m} \sim \mathscr{N}_{\mathbb{C}}(0,1)$. The result is shown in Fig. 7.1a for the case $N_{\mathrm{T}}[i]=6, N_{\mathrm{L}}=2$ and $N_{\mathrm{R}}[u,i] \in \{2,3,4,5\}$. As scheduling is not considered in this investigation, it is supposed that the set of scheduled users $\mathscr{S}[i]$ is composed of all $U_i = 3$ users in the cell (hence also $S[i]=3$).

It is observed in Fig. 7.1a that the achievable sum rate of SQBC decreases with increasing number of receive antennas, while the throughput of MET improves. This is in conformance with our theoretical investigation of Sect. 7.1.2. More specifically, the throughput loss of SQBC at high SNR is given by the value $\Delta R(M_1, M_2)$ calculated in (7.31), e.g., $\Delta R(2,5) = 8.08$ bits/s/Hz according to (7.31). Hence, with perfect CSIT, SQBC is not a reasonable choice. This behavior can be explained by

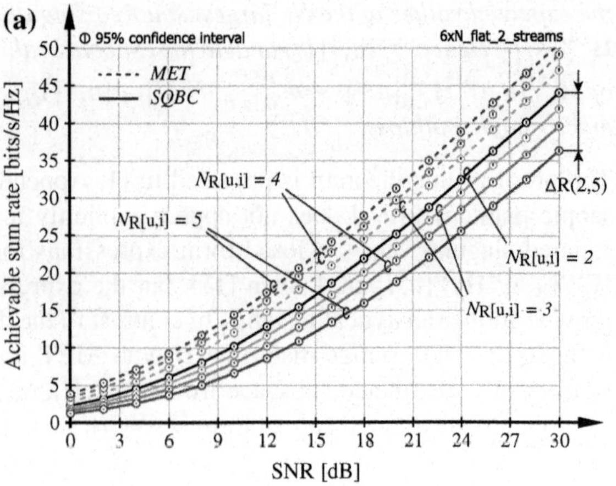

Fig. 7.1 Evaluation of the achievable rate performance and CSI feedback overhead requirement of BD precoding with SQBC and MET antenna combining. Three users are served over six transmit antennas via two streams per user. **a** Achievable sum rate with perfect CSIT. **b** Feedback bit-scaling for a per-user rate loss of 1 bits/s/Hz

considering the channel statistics provided in Lemma 7.2. According to this lemma, the DoF of the Wishart distribution defining the statistics of $\mathbf{H}_{n,k}^{(\mathrm{ect})}[u,i]^{\mathrm{H}}\mathbf{H}_{n,k}^{(\mathrm{ect})}[u,i]$ are reduced with a growing number of receive antennas $N_{\mathrm{R}}[u,i]$. Correspondingly, the eigenvalues of this Wishart matrix are decreased on average, causing a reduction of the average effective channel gain. On the other hand, with MET the channel gain is determined by the N_{L} maximum eigenvalues of a matrix with Wishart distribution $\mathscr{W}_{N_{\mathrm{R}}[u,i]}^{\mathbb{C}}\left(N_{\mathrm{T}}[i], \mathbf{I}_{N_{\mathrm{R}}[u,i]}\right)$ according to Theorem 7.3. It is well known that these eigenvalues increase on average with growing dimension $N_{\mathrm{R}}[u,i]$.

The advantage of SQBC is visualized in Fig. 7.1b. In this figure the sufficient number of CSI feedback bits to achieve a per-user rate loss of 1 bits/s/Hz compared to perfect CSIT is shown. With MET, the slope of the feedback overhead versus the SNR for a given number of data streams N_{L} per user is independent of the number of receive antennas $N_{\mathrm{R}}[u,i]$ and the required feedback overhead is significantly larger than in case of SQBC. With SQBC, the feedback overhead can be substantially reduced by increasing the number of receive antennas $N_R[u, i]$, for the cost of a moderate SNR offset in the achievable throughput. This reduction is due to the decreasing exponent in the average distortion (7.28) with increasing $N_R[u, i]$, obtained from the degree of freedom to select the best N_L-dimensional subspace within an $N_R[u, i]$-dimensional space during CSI quantization.

In Fig. 7.2, the validity of the bit-scaling law derived in (7.29) is investigated. In the results shown in Fig. 7.2a the base station is equipped with $N_T[i] = 6$ transmit antennas. $S[i] = 3$ users are served in parallel over $N_L = 2$ streams each. The CSI feedback overhead is scaled such as to achieve a sum rate loss of 1.5 bits/s/Hz for the case of SQBC with $N_R[u, i] = 5$. At an SNR of $\beta_{dB} = 0$ dB, a feedback overhead of $b = 0$ bits is required to maintain the intended rate loss, which is increased to $b = 17$ bits at $\beta_{dB} = 30$ dB. The other configurations considered in Fig. 7.2b are simulated with the same number of feedback bits. It is observed that the actually achieved rate loss is equal to 1.3 bits/s/Hz, which is close to the desired value. With the same

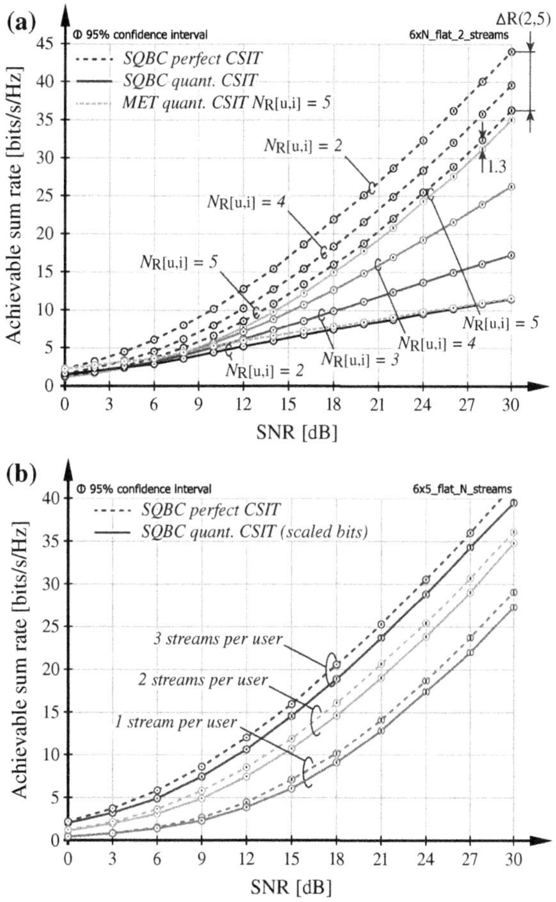

Fig. 7.2 Achievable sum rate of BD-based MU-MIMO systems employing SQBC with quantized and perfect CSIT and $N_T[i] = 6$. **a** $N_L = 2$, three users and $N_R \in \{2, \ldots, 5\}$. **b** $N_R = 5$, $N_L \in \{1, 2, 3\}$ and $S = U = \frac{N_T[i]}{N_L}$

feedback overhead a significant reduction in throughput is incurred when $N_R[u, i]$ is decreased. This reduction is caused by the increased residual multi-user interference due to the CSI quantization error. If MET is employed with $N_R[u, i] = 5$ and the same feedback overhead, a much worse performance compared to SQBC is observed if $\beta_{dB} > 10$ dB. At low SNR, however, SQBC is outperformed by MET, because the transmission rate is limited by the noise rather than the multi-user interference. In this case, the interference reduction capabilities of SQBC are outweighed by the channel gain improvement of MET.

In Fig. 7.2b, the performance of SQBC with $N_T[i] = 6$, $N_R[u, i] = 5$ and $N_L \in \{1, 2, 3\}$ is shown. The feedback overhead is scaled such as to achieve a sum rate loss of 2 bits/s/Hz with respect to perfect CSIT. This is obtained with an overhead of $b \in [0, 8]$ bits/user for $N_L = 1$, $b \in [0, 16.1]$ bits/user for $N_L = 2$ and $b \in [0, 18.3]$ bits/user for $N_L = 3$. With decreasing number of streams N_L per user a sum rate reduction is observed. This is because the interference between a larger number of users must be canceled by the BD precoder when the number of streams per user is reduced and the total number of streams is kept constant, leading to an SNR loss of the effective user channel [18].

Further simulation results investigating the tightness of the proposed bounds on the rate loss are provided in [2]. A trade-off between SQBC and MET can be achieved by restricting the SQBC algorithm to finding the best subspace within only a subset of the eigenmodes of the channel. This is explained in Sect. 7.2 when extending the method to frequency selective channels.

7.1.5 Adjustment of the Grassmannian CSI Feedback

In Sect. 6.2, memoryless and predictive Grassmannian quantization codebooks are proposed under the assumption that the number of data streams per user $\ell_k[u, i]$ is equal to the number of receive antennas $N_R[u, i]$, rendering the application of an antenna combiner obsolete. In this case, the quantization metric is given by the chordal distance between the $N_R[u, i]$-dimensional subspace defined by the channel matrix $\mathbf{H}_{n,k}^{(ct)}[u, i]$ and the $N_R[u, i]$-dimensional subspaces spanned by the elements $\mathbf{Q}_j$ of the codebook $\mathcal{Q}_{n,k}[u, i]$; see Eq. (6.19).

With the antenna combiners of the previous section the quantization metric is modified, which can be considered in correlated RVQ to improve the quantizer efficiency. Specifically, with SQBC the chordal distance between the $N_R[u, i]$-dimensional channel subspace and each of the $\ell_k[u, i]$-dimensional subspaces defined by the elements of the codebook is minimized according to Eq. (7.13). The minimum is achieved with that element of the codebook, that has the smallest left null space component with respect to the channel matrix. As the null space component of the codebook elements is the decisive factor during quantization, there exists no preferred subspace orientation within the range space of the channel matrix. Hence, efficient memoryless quantization is enabled if the quantization codebook is matched to the distribution of any $\ell_k[u, i]$-dimensional subspace of the channel matrix. With correlated RVQ, the quantization codebook can be generated according to

$$\mathscr{Q}_{n,k}^{(\mathrm{SQBC})}[u,i] = \Big\{ \mathbf{Q}_j^{(\mathrm{SQBC})} = \mathbf{Q}_j^{(\mathrm{corr})}\mathbf{U} | \mathbf{Q}_j^{(\mathrm{corr})} \in \mathscr{Q}_{n,k}^{(\mathrm{corr})}[u,i], \\ \mathbf{U}\boldsymbol{\Sigma}\mathbf{V}^{\mathrm{H}} = \bar{\mathbf{H}} \in \mathbb{C}^{N_{\mathrm{R}}[u,i]\times \ell_k[u,i]},\ [\bar{\mathbf{H}}]_{m,n} \sim \mathscr{N}_{\mathbb{C}}(0,1) \Big\}. \tag{7.40}$$

Here, the codebook defined in (6.29) is employed to obtain semi-unitary matrices $\mathbf{Q}_j^{(\mathrm{corr})}$ of size $N_{\mathrm{T}}[i] \times N_{\mathrm{R}}[u,i]$ whose span is matched in distribution to the subspace span$\left(\mathbf{H}_{n,k}^{(\mathrm{ct})}[u,i]\right)$. The multiplication with the isotropic semi-unitary matrix $\mathbf{U} \in \mathbb{C}^{N_{\mathrm{R}}[u,i]\times \ell_k[u,i]}$ then generates an orthonormal basis $\mathbf{Q}_j^{(\mathrm{SQBC})}$ that spans a uniformly distributed subspace within span$\left(\mathbf{Q}_j^{(\mathrm{corr})}\right)$.

With MET, the quantization metric (7.33) is determined by the chordal distance between the subspace spanned by the $\ell_k[u,i]$ maximum eigenmodes of the channel matrix and the elements of the codebook. Efficient memoryless quantization is hence enabled by matching the quantization codebook to the distribution of these eigenmodes. This can, e.g., be achieved by modifying the codebook construction in Sect. 6.2.1 to select only the first $\ell_k[u,i]$ columns of the matrix of left singular vectors

$$\mathscr{Q}_{n,k}^{(\mathrm{MET})}[u,i] = \left\{ \mathbf{Q}_j^{(\mathrm{MET})} = \left[\mathbf{Q}_j^{(\mathrm{corr})}\right]_{:,1:\ell_k[u,i]} \middle| \mathbf{Q}_j^{(\mathrm{corr})} \in \mathscr{Q}_{n,k}^{(\mathrm{corr})}[u,i] \right\}. \tag{7.41}$$

In case of predictive quantization, deriving the tangent statistics when MET or SQBC is applied was not successful. The tangent codebook is therefore generated assuming the channel subspace to be uniformly distributed on $\mathscr{G}(N_{\mathrm{T}}[i], \ell_k[u,i])$. This is not a critical issue, because the predictive quantizer automatically adjusts to the statistics of the subspace spanned by the effective channel using the adaptive codebook construction described in Sect. 6.2.2.

7.2 Extension to Frequency-Selective Systems

The CSI feedback algorithms and channel subspace selection methods proposed in Sects. 6.2 and 7.1 are derived for frequency flat channels and are hence applicable to OFDM on a per-subcarrier basis. Providing CSI feedback for each Resource Element (RE), however, implies a large feedback overhead, which cannot be sustained in practical systems. The CSI feedback overhead can be reduced by employing the same approaches as in Single User Multiple-Input Multiple-Output (SU-MIMO), i.e., CSI feedback clustering and interpolation. CSI feedback interpolation for ZF and BD based MU-MIMO is considered, e.g., in [7, 19–21]. Linear subspace interpolation on the Grassmannian is possible by sampling the geodesic between neighboring CSI pilots. Geodesic interpolation is reviewed in Sect. 7.2.1 for completeness. Higher order spline interpolation on the unit-sphere has been proposed in [22] in the context of graphics and animations. These algorithms are applicable to CSI interpolation for ZF beamforming, but suffer from high computational complexity. An issue of interpolation based methods is the high density of CSI pilots required to achieve sufficiently accurate results; see Sect. 7.2.3. If this cannot be sustained due to limitations on the feedback overhead, better performance is possible with suitable feedback clustering. To this end, a clustering approach is proposed in Sect. 7.2.1, in which a single representative subspace is determined for each Resource Block (RB) such that the average chordal distance is minimized. This method can be viewed as a natural extension of the SQBC idea to multiple REs, providing a significant performance improvement especially in systems with excess antennas. With the SQBC clustering approach, it is possible to exploit the time-frequency channel correlation within each RB separately. The residual interference in-between RBs, on the other hand, is utilized by the predictive quantizer proposed in Sect. 6.2.2. Combining the two methods, efficient CSI quantization is achieved, as demonstrated in Sect. 7.2.3.

In Sect. 7.2.2, channel quality feedback for transmission rate adaptation and multi-user scheduling in the space, time and frequency domain is considered. A combination of the Semi-Orthogonal User Selection (SUS) algorithm [23] with proportional fair scheduling [24] is employed to determine the multi-user resource allocation. The scheduling is based on a proposed estimate of the achievable user data rate with BD precoding. The performance of the proposed techniques is evaluated by means of simulations in Sect. 7.2.3. Notice that the notation introduced for feedback clustering in Sect. 5.2.1 is reused in this section.

7.2.1 Grassmannian Interpolation and Clustering

Geodesic Interpolation

Interpolation on the Grassmannian can be achieved by exploiting the differential geometry associated with the manifold. The concept of a straight line in the Euclidean space is generalized to curved spaces and manifolds with the geodesic; see, e.g., [1, Appendix C]. Linear interpolation on the manifold is possible by equidistantly sampling the geodesic between neighbouring CSI pilots.

When CSI interpolation is employed, it is assumed that the OFDM time-frequency resource grid is partitioned into RBs as visualized in Fig. 5.3. With the notation introduced in Sect. 5.2.1, the index pair $[\eta, \kappa]$ is employed to indicate the RBs. The channel matrix observed on the RE in the center of the RB is quantized and fed back by the users, employing the Grassmannian quantizers of Sect. 6.2. The corresponding subcarrier index and symbol-time index are denoted as

$$n_\eta = (\eta - 1)N_{\text{clust}}^{(f)} + \left\lfloor \frac{N_{\text{clust}}^{(f)}}{2} \right\rfloor, \quad k_\kappa = (\kappa - 1)N_{\text{clust}}^{(t)} + \left\lceil \frac{N_{\text{clust}}^{(t)}}{2} \right\rceil. \tag{7.42}$$

The quantized subspace on RB $[\eta, \kappa]$ is thus obtained as

$$\hat{\mathbf{H}}_{\eta,\kappa}[u,i] = \underset{\mathbf{Q}_j \in \mathscr{Q}_{\eta,\kappa}[u,i]}{\operatorname{argmin}} \ \mathrm{d}_{\mathrm{c}}^2 \left(\tilde{\mathbf{H}}_{n_\eta,k_\kappa}[u,i], \mathbf{Q}_j \right). \tag{7.43}$$

Because CSI feedback is provided only once per RB, the quantization codebook $\mathscr{Q}_{\eta,\kappa}[u,i]$ is adapted on a per-RB basis. To simplify the exposition, block fading is considered in this section; hence, the channel is assumed as temporally constant within each RB. Then 1D interpolation in the frequency domain is sufficient. A trivial extension to time-frequency selective channels is possible by consecutive 1D interpolation in time and frequency (or vice versa); more sophisticated multi-dimensional manifold interpolators are proposed in [25].

Considering two neighbouring quantized channel subspaces $\hat{\mathbf{H}}_{\eta,\kappa}[u,i]$ as well as $\hat{\mathbf{H}}_{\eta+1,\kappa}[u,i]$, the tangent defining the geodesic is denoted as

$$\mathbf{T}_{\eta,\kappa}[u,i] = T\left(\hat{\mathbf{H}}_{\eta,\kappa}[u,i], \hat{\mathbf{H}}_{\eta+1,\kappa}[u,i]\right) \in \mathscr{T}\left(\hat{\mathbf{H}}_{\eta,\kappa}[u,i]\right). \tag{7.44}$$

We define the linearly interpolated subspace between the subspaces $\operatorname{span}\left(\hat{\mathbf{H}}_{\eta,\kappa}[u,i]\right)$ and $\operatorname{span}\left(\hat{\mathbf{H}}_{\eta+1,\kappa}[u,i]\right)$ at distance $\Delta f = N_{\mathrm{clust}}^{(f)} p,\ p \in [0,1]$ from $\operatorname{span}\left(\hat{\mathbf{H}}_{\eta,\kappa}[u,i]\right)$ via the geodesic $\Gamma(\cdot)$

$$\hat{\mathbf{H}}_{n_\eta+\Delta f,k_\kappa}[u,i] = \Gamma\left(\hat{\mathbf{H}}_{\eta,\kappa}[u,i], \mathbf{T}_{\eta,\kappa}[u,i], p\right), \tag{7.45}$$

with $N_{\mathrm{clust}}^{(f)}$ denoting the number of subcarriers in-between $\hat{\mathbf{H}}_{\eta,\kappa}[u,i]$ and $\hat{\mathbf{H}}_{\eta+1,\kappa}[u,i]$. At the boundaries of the system bandwidth CSI extrapolation is necessary. Considering, e.g., the upper boundary, CSI extrapolation is achieved by extending the tangent between $\hat{\mathbf{H}}_{N_{\mathrm{RB}}-1,\kappa}[u,i]$ and $\hat{\mathbf{H}}_{N_{\mathrm{RB}},\kappa}[u,i]$ beyond the subcarrier $n_{N_{\mathrm{RB}}}$, using (7.45) with $p > 1$.

SQBC Clustering

As an alternative to CSI interpolation, feedback clustering is considered, avoiding the need for an interpolator at the base station. The clustering approach is suitable when the distance $N_{\mathrm{clust}}^{(f)}$ between CSI pilots is large compared to the channel coherence bandwidth, entailing unsatisfactory performance of linear interpolation because of significant channel variations within the RBs. For such situations it is proposed that the users determine the best $\ell_\kappa[u,i]$-dimensional subspace representation for each RB, assuming zeroth-order interpolation. The best subspace representation $\tilde{\mathbf{H}}_{\eta,\kappa}[u,i]$ of RB $[\eta, \kappa]$ is defined by minimizing the average chordal distance over the RB

$$\tilde{\mathbf{H}}_{\eta,\kappa}[u,i] = \underset{\tilde{\mathbf{H}}}{\operatorname{argmin}} \ \frac{1}{N_{\mathrm{RE}}} \sum_{\rho=1}^{N_{\mathrm{RE}}} \mathrm{d}_{\mathrm{c}}^2 \left(\mathbf{U}_{n,k}[u,i], \tilde{\mathbf{H}} \right)$$

$$= \underset{\tilde{\mathbf{H}}}{\operatorname{argmin}} \; \frac{1}{N_{\mathrm{RE}}} \sum_{\rho=1}^{N_{\mathrm{RE}}} \ell_\kappa[u,i] - \operatorname{tr}\left(\tilde{\mathbf{H}}^{\mathrm{H}}\left(\mathbf{U}_{n,k}[u,i]\mathbf{U}_{n,k}[u,i]^{\mathrm{H}}\right)\tilde{\mathbf{H}}\right), \quad (7.46)$$

$$\text{subject to: } \tilde{\mathbf{H}} \in \mathbb{C}^{N_{\mathrm{T}}[i]\times \ell_\kappa[u,i]}, \quad \tilde{\mathbf{H}}^{\mathrm{H}}\tilde{\mathbf{H}} = \mathbf{I}_{\ell_\kappa[u,i]},$$

with $\mathbf{U}_{n,k}[u,i]$ being obtained from an SVD of the channel matrix as in (7.32), and the RE index $[n,k]$ being implicitly determined by $[\eta,\kappa]$ and ρ as specified in (5.13). Notice the similarity of this optimization problem to the SQBC optimization in Eq. (7.13). In fact, (7.46) can be combined with SQBC to not only finding the best subspace representation in terms of zeroth-order interpolation, but minimizing both, the interpolation and the quantization error

$$\hat{\mathbf{H}}_{\eta,\kappa}[u,i] = \underset{\mathbf{Q}_j}{\operatorname{argmin}} \; \frac{1}{N_{\mathrm{RE}}} \sum_{\rho=1}^{N_{\mathrm{RE}}} \mathrm{d}_{\mathrm{c}}^2\left(\mathbf{U}_{n,k}[u,i], \mathbf{Q}_j\right), \quad (7.47)$$

$$\text{subject to: } \mathbf{Q}_j \in \mathcal{Q}_{\eta,\kappa}[u,i] = \left\{\mathbf{Q}_j \in \mathbb{C}^{N_{\mathrm{T}}[i]\times\ell_\kappa[u,i]} \middle| \mathbf{Q}_j^{\mathrm{H}}\mathbf{Q}_j = \mathbf{I}_{\ell_\kappa[u,i]}\right\}.$$

The solution to problem (7.46) is obtained according to

$$\tilde{\mathbf{H}}_{\eta,\kappa}[u,i] = \left[\bar{\mathbf{U}}_{\eta,\kappa}[u,i]\right]_{:,1:\ell_\kappa[u,i]} = \bar{\mathbf{U}}_{\eta,\kappa}^{(\ell_\kappa[u,i])}[u,i], \quad (7.48)$$

$$\bar{\mathbf{U}}_{\eta,\kappa}[u,i]\bar{\boldsymbol{\Lambda}}_{\eta,\kappa}[u,i]\bar{\mathbf{U}}_{\eta,\kappa}[u,i]^{\mathrm{H}} = \bar{\mathbf{R}}_{\eta,\kappa}[u,i], \quad (7.49)$$

$$\bar{\mathbf{R}}_{\eta,\kappa}[u,i] = \frac{1}{N_{\mathrm{RE}}} \sum_{\rho=1}^{N_{\mathrm{RE}}} \mathbf{U}_{n,k}[u,i]\mathbf{U}_{n,k}[u,i]^{\mathrm{H}}, \quad (7.50)$$

with (7.49) denoting an eigendecomposition of $\bar{\mathbf{R}}_{\eta,\kappa}[u,i]$. Matrix $\bar{\mathbf{R}}_{\eta,\kappa}[u,i]$ can be interpreted as a *subspace correlation matrix*. Notice that the eigenvalues in $\bar{\boldsymbol{\Lambda}}_{\eta,\kappa}[u,i]$ are assumed in decreasing order. Similarly, the solution of (7.47) is determined by the quantization metric

$$\hat{\mathbf{H}}_{\eta,\kappa}^{(\mathrm{SQBC})}[u,i] = \underset{\mathbf{Q}_j\in\mathcal{Q}_{\eta,\kappa}[u,i]}{\operatorname{argmin}} \; \ell_\kappa[u,i] - \operatorname{tr}\left(\bar{\boldsymbol{\Lambda}}_{\eta,\kappa}[u,i]\left(\bar{\mathbf{U}}_{\eta,\kappa}[u,i]^{\mathrm{H}}\mathbf{Q}_j\mathbf{Q}_j^{\mathrm{H}}\bar{\mathbf{U}}_{\eta,\kappa}[u,i]\right)\right),$$

$$\triangleq \underset{\mathbf{Q}_j\in\mathcal{Q}_{\eta,\kappa}[u,i]}{\operatorname{argmin}} \; \mathrm{d}_{\mathrm{c,w}}^2\left(\bar{\mathbf{U}}_{\eta,\kappa}[u,i], \mathbf{Q}_j, \bar{\boldsymbol{\Lambda}}_{\eta,\kappa}[u,i]\right), \quad (7.51)$$

$$\mathrm{d}_{\mathrm{c,SQBC}}^2[\eta,\kappa] = \min_{\mathbf{Q}_j\in\mathcal{Q}_{\eta,\kappa}[u,i]} \mathrm{d}_{\mathrm{c,w}}^2\left(\bar{\mathbf{U}}_{\eta,\kappa}[u,i], \mathbf{Q}_j, \bar{\boldsymbol{\Lambda}}_{\eta,\kappa}[u,i]\right). \quad (7.52)$$

Here, $\mathrm{d}_{\mathrm{c,w}}^2\left(\bar{\mathbf{U}}_{\eta,\kappa}[u,i], \mathbf{Q}_j, \bar{\boldsymbol{\Lambda}}_{\eta,\kappa}[u,i]\right)$ can be viewed as a *weighted chordal distance* with weighting matrix $\bar{\boldsymbol{\Lambda}}_{\eta,\kappa}[u,i]$. The importance of the individual eigenmodes of the subspace correlation matrix in the quantization metric (7.51) is specified by the diagonal weighting matrix.

By applying the SQBC antenna combiner of Sect. 7.1.2 to generate the effective channel on RE $[n,k]$ based on the quantized subspace $\hat{\mathbf{H}}_{\eta,\kappa}^{(\mathrm{SQBC})}[u,i]$, the quantization error (7.52) is achieved.

Dimensionality Adaptation

As observed in Sect. 7.1.4, SQBC is subject to the problem that all $N_{\mathrm{R}}[u,i]$ modes of the channel matrix $\mathbf{H}_{n,k}^{(\mathrm{ct})}[u,i]$ are equally treated in the quantization metric (7.13), potentially causing a weak channel gain of the effective channel matrix. This is also the case with SQBC clustering and its corresponding quantization metric (7.51). When the transmission is noise limited, rather than interference limited, the effective channel gain is of greater importance than the CSI quantization error. A trade-off between MET and SQBC can be achieved by considering only the subset of d maximum eigenmodes of the channel during quantization, with $\ell_\kappa[u,i] \leq d \leq N_{\mathrm{R}}[u,i]$. Then, the weighted chordal distance quantization metric (7.51) is replaced with

$$\hat{\mathbf{H}}_{\eta,\kappa}^{(d)}[u,i] = \underset{\mathbf{Q}_j \in \mathscr{Q}_{\eta,\kappa}[u,i]}{\mathrm{argmin}}\ \mathrm{d}_{\mathrm{c,w}}^2\left(\bar{\mathbf{U}}_{\eta,\kappa}^{(d)}[u,i], \mathbf{Q}_j, \bar{\boldsymbol{\Lambda}}_{\eta,\kappa}^{(d)}[u,i]\right), \tag{7.53}$$

$$\bar{\mathbf{U}}_{\eta,\kappa}^{(d)}[u,i]\bar{\boldsymbol{\Lambda}}_{\eta,\kappa}^{(d)}[u,i]\bar{\mathbf{U}}_{\eta,\kappa}^{(d)}[u,i]^{\mathrm{H}} = \bar{\mathbf{R}}_{\eta,\kappa}^{(d)}[u,i], \tag{7.54}$$

$$\bar{\mathbf{R}}_{\eta,\kappa}^{(d)}[u,i] = \frac{1}{N_{\mathrm{RE}}}\sum_{\rho=1}^{N_{\mathrm{RE}}} \mathbf{U}_{n,k}^{(d)}[u,i]\mathbf{U}_{n,k}^{(d)}[u,i]^{\mathrm{H}}, \tag{7.55}$$

$$\mathbf{U}_{n,k}^{(d)}[u,i] = \left[\mathbf{U}_{n,k}[u,i]\right]_{:,1:d}. \tag{7.56}$$

With this metric, the best $\ell_\kappa[u,i]$-dimensional subspace representation in the quantization codebook $\mathscr{Q}_{\eta,\kappa}[u,i]$ is determined, with respect to the average chordal distance to the d maximum eigenmodes of the channel matrices within the RB $[\eta,\kappa]$.

The natural question that arises when considering the quantization metric (7.53) is how to select the dimension d. To answer this question, an estimate of the pre-equalization achievable data rate with BD precoding and quantized CSIT is employed. The corresponding pre-equalization input-output relationship is given in Eq. (1.37). According to [26], the instantaneous mutual information between the channel input and output, determining the achievable transmission rate, is

$$R_{n,k}^{(d)}[u,i] = \log_2 \det\Big(\mathbf{I}_{N_{\mathrm{R}}[u,i]} + \mathbf{H}_{n,k}^{(\mathrm{ct})}[u,i]^{\mathrm{H}}\mathbf{S}_{n,k}[u,i]\mathbf{H}_{n,k}^{(\mathrm{ct})}[u,i]\left(\tilde{\sigma}_z^2\mathbf{I}_{N_{\mathrm{R}}[u,i]} + \mathbf{R}_{n,k}^{(d)}[u,i]\right)^{-1}\Big), \tag{7.57}$$

with $\mathbf{S}_{n,k}[u,i]$ denoting the covariance matrix of the channel input and $\mathbf{R}_{n,k}^{(d)}[u,i]$ being the interference covariance matrix. During the calculation of the CSI feedback, the precoders are unknown to the users. To determine an estimate of the mutual information, the precoders are hence considered as random and are taken into account in the covariance matrices

$$\mathbf{S}_{n,k}[u,i] = \mathbb{E}\left(\mathbf{F}_{n,k}[u,i]\mathbf{x}_{n,k}[u,i](\mathbf{F}_{n,k}[u,i]\mathbf{x}_{n,k}[u,i])^{\mathrm{H}}\right) = \mathbb{E}\left(\mathbf{F}_{n,k}[u,i]\mathbf{F}_{n,k}[u,i]^{\mathrm{H}}\right), \tag{7.58}$$

$$\mathbf{R}^{(d)}_{n,k}[u,i] = \sum_{s=1,s\neq u}^{S_{n,k}[i]} \mathbf{H}^{(\mathrm{ct})}_{n,k}[u,i]^{\mathrm{H}} \mathbb{E}\left(\mathbf{F}_{s,i}[n,k]\mathbf{F}_{s,i}[n,k]^{\mathrm{H}}\right) \mathbf{H}^{(\mathrm{ct})}_{n,k}[u,i], \tag{7.59}$$

with the expectation being taken with respect to the transmit signals and the precoders and considering the statistical independence of transmit signals corresponding to different streams and users. The precoder $\mathbf{F}_{n,k}[u,i]$ is determined by the channels of the other users according to Eq. (6.4). As these channels are unknown to user u, $\mathbf{F}_{n,k}[u,i]$ is assumed as isotropically distributed, implying

$$\begin{aligned} \mathbf{S}_{n,k}[u,i] &= P_{n,k}[u,i]\,\mathbb{E}\left(\tilde{\mathbf{F}}_{n,k}[u,i]\tilde{\mathbf{F}}_{n,k}[u,i]^{\mathrm{H}}\right) \\ &= P_{n,k}[u,i]\,\frac{\ell_\kappa[u,i]}{N_{\mathrm{T}}[i]}\mathbf{I}_{\ell_\kappa[u,i]} = \frac{P_i}{N_{\mathrm{T}}[i]\,S_{n,k}[i]}\mathbf{I}_{\ell_\kappa[u,i]}, \end{aligned} \tag{7.60}$$

with $\tilde{\mathbf{F}}_{n,k}[u,i]$ as defined in Eq. (6.15). Due to the BD construction, the precoders $\mathbf{F}_{s,i}[n,k]$ of the other users are restricted to the left null space of $\hat{\mathbf{H}}^{(d)}_{\eta,\kappa}[u,i]$. Considering this knowledge in the calculation of $\mathbf{R}^{(d)}_{n,k}[u,i]$ and assuming that $S_{n,k}[i] = \frac{N_{\mathrm{T}}[i]}{\ell_\kappa[u,i]}$ users are served in parallel each over $\ell_\kappa[u,i]$ streams, the interference covariance matrix is obtained as

$$\mathbf{R}^{(d)}_{n,k}[u,i] = \frac{P_i}{N_{\mathrm{T}}[i]}\mathbf{H}^{(\mathrm{ct})}_{n,k}[u,i]^{\mathrm{H}}\left(\mathbf{I}_{N_{\mathrm{R}}[u,i]} - \hat{\mathbf{H}}^{(d)}_{\eta,\kappa}[u,i]\hat{\mathbf{H}}^{(d)}_{\eta,\kappa}[u,i]^{\mathrm{H}}\right)\mathbf{H}^{(\mathrm{ct})}_{n,k}[u,i]. \tag{7.61}$$

The *preferred dimensionality* $d_{\eta,\kappa}[u,i]$ is selected by maximizing the sum rate

$$d_{\eta,\kappa}[u,i] = \underset{\ell_\kappa[u,i]\leq d\leq N_{\mathrm{R}}[u,i]}{\operatorname{argmax}} \sum_{\rho=1}^{N_{\mathrm{RE}}} R^{(d)}_{n,k}[u,i], \tag{7.62}$$

where the RE index $[n,k]$ is implicitly determined by ρ and $[\eta,\kappa]$, according to Eq. (5.13). Notice that knowledge of $d_{\eta,\kappa}[u,i]$ is not required at the base station; hence dimensionality adaptation does not imply an additional feedback overhead.

The sum rate in (7.62) could potentially also be considered as quantization metric for the channel subspace. This is not followed up here for two reasons. Firstly, calculating (7.62) for all quantization matrices is computationally expensive. The second reason is only relevant for predictive quantization. The performance of the prediction algorithm in the proposed quantizer of Sect. 6.2.2 is strongly impacted by the chordal distance quantization error, because the prediction is based on quantized CSI. Not choosing the quantized subspace according to the minimal chordal distance can entail a significant degradation of the prediction accuracy.

7.2.2 Channel Quality Feedback and Multi-user Scheduling

CQI Feedback for Limited Feedback BD

To determine the multi-user resource allocation in the time, frequency and spatial domain, it is necessary to calculate the achievable data rate of a given schedule at the base station. In the spatial domain this rate is dependent on the number of users that is served in parallel, because the available transmit power is equally distributed among users and spatial streams according to Eq. (6.15). Therefore, it must be possible to update the CQI feedback at the base station to account for the number of users served in parallel. For this reason, the pre-equalization mutual information estimated in the previous paragraph in Eq. (7.57) is not employed as channel quality feedback, because the power allocation is hidden within the $\log_2 \det(\cdot)$ and cannot be updated subsequently.

The achievable data rate of a realistic communication system is determined by the post-equalization SINR, as argued in Sect. 5.2.2. It is therefore proposed to employ an estimate of the post-equalization SINR, achieved with limited feedback based BD precoding, as CQI feedback. An accurate SINR estimate cannot be obtained during feedback calculation, because the applied precoders are unknown at that time, in contrast to SU-MIMO with codebook based precoding. To circumvent this problem, a lower bound on the expected value of the SINR is utilized instead, similar to the proposal in [27] for ZF beamforming. The derivation of this lower bound is provided in [1, Appendix G]; the *lower bound on the expected SINR* of stream ν is expressed as

$$\tilde{\beta}_{\nu,n,k}[u,i] = \frac{c_{S,n,k}\, \sigma^{\text{eff}}_{\nu,n,k}[u,i]^2}{\tilde{\sigma}_z^2 + c_{I,n,k}\, \sigma^{\text{eff}}_{\nu,n,k}[u,i]^2}, \tag{7.63}$$

$$\boldsymbol{\Sigma}^{\text{eff}}_{u,i}[n,k] = \operatorname{diag}\left(\sigma^{\text{eff}}_{1,n,k}[u,i], \ldots, \sigma^{\text{eff}}_{\ell[u,i],n,k}[u,i]\right),$$

$$c_{S,n,k} = \frac{P_i}{S_{n,k}[i](N_{\mathrm{T}}[i] - \bar{\ell}_i[n,k])} \cdot \left(1 - \frac{\mathrm{d}_{\mathrm{c}}^2\left(\hat{\mathbf{H}}_{n,k}[u,i], \mathbf{B}_{n,k}[u,i]\right)}{\ell_k[u,i]}\right)\left(1 - \frac{\mathrm{d}_{\mathrm{c}}^2\left(\tilde{\mathbf{H}}_{n,k}[u,i], \hat{\mathbf{H}}_{n,k}[u,i]\right)}{\ell_k[u,i]}\right),$$

$$c_{I,n,k} = \frac{P_i}{N_{\mathrm{T}}[i]\,\ell_k[u,i]}\, \mathrm{d}_{\mathrm{c}}^2\left(\tilde{\mathbf{H}}_{n,k}[u,i], \hat{\mathbf{H}}_{n,k}[u,i]\right).$$

with $\boldsymbol{\Sigma}^{\text{eff}}_{n,k}[u,i]$ denoting the matrix of singular values of the effective user channel $\mathbf{H}^{(\text{ect})}_{n,k}[u,i]$, $\tilde{\mathbf{H}}_{n,k}[u,i]$ as defined in (6.6), $\hat{\mathbf{H}}_{n,k}[u,i]$ being the quantized channel subspace, $\mathbf{B}_{n,k}[u,i]$ denoting an orthonormal basis for the orthogonal complement of the other served users' effective channels and $\bar{\ell}_{n,k}[i] = \sum_{s \in \mathscr{S}_{n,k}[i], s \neq u} \ell_k[s,i]$. The residual multi-user interference due to imperfect CSI feedback is determined by the chordal distance quantization error in $c_{I,n,k}$, justifying the application of the chordal distance as CSI quantization metric.

Considering the constant $c_{S,n,k}$, the first two factors cannot be determined by the user u during feedback calculation, because neither the number of served users

$S_{n,k}[i]$ nor the number of interfering streams $\bar{\ell}_{n,k}[i]$ is known, and also the other users' quantized effective channels are unknown. The last term, however, depending on the quantization error, can be calculated by the user. Also, the constant $c_{I,n,k}$ and the singular values in $\boldsymbol{\Sigma}^{\text{eff}}_{n,k}[u,i]$ are available at the user, because they depend on local CSI only. It is therefore proposed to employ the following value as *per-stream CQI feedback*

$$\text{CQI}_{\nu,n,k}[u,i] = \frac{\sigma^{\text{eff}}_{\nu,n,k}[u,i]^2 \left(1 - \frac{\text{d}_\text{c}^2\left(\tilde{\mathbf{H}}_{n,k}[u,i], \hat{\mathbf{H}}_{n,k}[u,i]\right)}{\ell_k[u,i]}\right)}{\tilde{\sigma}_z^2 + \frac{P_i}{N_\text{T}[i]\,\ell_k[u,i]} \text{d}_\text{c}^2\left(\tilde{\mathbf{H}}_{n,k}[u,i], \hat{\mathbf{H}}_{n,k}[u,i]\right) \sigma^{\text{eff}}_{\nu,n,k}[u,i]^2}. \tag{7.64}$$

With this CQI feedback, it is possible to obtain an estimate of the achievable user rate $R_{n,k}[u,i]$ for a given schedule $\mathscr{S}_{n,k}[i]$, because the remaining two factors of $c_{S,n,k}$ can be calculated by the base station, enabling the calculation of the SINR

$$\tilde{\beta}_{\nu,n,k}[u,i] = \frac{P_i}{S_{n,k}[i](N_\text{T}[i] - \bar{\ell}_{n,k}[i])} \left(1 - \frac{\text{d}_\text{c}^2(\hat{\mathbf{H}}_{n,k}[u,i], \mathbf{B}_{n,k}[u,i])}{\ell_k[u,i]}\right) \text{CQI}_{\nu,n,k}[u,i], \tag{7.65}$$

$$R_{n,k}[u,i] \approx \sum_{\nu=1}^{\ell_k[u,i]} \log_2\left(1 + \tilde{\beta}_{\nu,n,k}[u,i]\right). \tag{7.66}$$

In case of feedback clustering or interpolation, CQI feedback is provided only once per RB. In the proposed SU-MIMO feedback algorithms of Sect. 5.2.2, Mutual Information Effective Signal to Interference and Noise Ratio Mapping (MIESM) is applied to obtain a single SNR that represent the average channel quality of an RB. This approach is not applicable here, because the CQI in (7.64) represents only a scaled version of the SINR (7.63), and MIESM is not a linear function; hence, without the correct scaling the application of MIESM is pointless. Instead, it is proposed to calculate an average CQI for RB $[\eta,\kappa]$, by linearly averaging the chordal distance quantization error and the squared singular values

$$\text{CQI}_{\nu,\eta,\kappa}[u,i] = \frac{\bar{\sigma}^{\text{eff}}_{\nu,\eta,\kappa}[u,i]^2 \left(1 - \frac{\bar{\text{d}}^2_{c,\eta,\kappa}}{\ell_\kappa[u,i]}\right)}{\tilde{\sigma}_z^2 + \frac{P_i}{N_\text{T}[i]\,\ell_\kappa[u,i]} \bar{\text{d}}^2_{c,\eta,\kappa}\, \bar{\sigma}^{\text{eff}}_{\nu,\eta,\kappa}[u,i]^2}, \tag{7.67}$$

$$\bar{\sigma}^{\text{eff}}_{\nu,\eta,\kappa}[u,i]^2 = \frac{1}{N_\text{RE}} \sum_{\rho=1}^{N_\text{RE}} \sigma^{\text{eff}}_{\nu,n,k}[u,i]^2, \quad \bar{\text{d}}^2_{c,\eta,\kappa} = \frac{1}{N_\text{RE}} \sum_{\rho=1}^{N_\text{RE}} \text{d}_\text{c}^2\left(\tilde{\mathbf{H}}_{n,k}[u,i], \hat{\mathbf{H}}_{n,k}[u,i]\right),$$

with $[n,k]$ being determined by $[\eta,\kappa]$ and ρ as specified in Eq. (5.13).

The performance of the proposed CQI feedback for multi-user scheduling is evaluated in Sect. 7.2.3, demonstrating reasonably close to optimal results with quantized CSIT. Notice though that the CQI is not sufficiently accurate for transmission rate adaptation. As the CQI is based on a lower bound on the expected value of the SINR, the instantaneous SINR achieved with a given schedule can be considerably above or below this value. It is therefore proposed to employ (7.64) only as an initial CQI for multi-user scheduling. As soon as the schedule is fixed, the users are able to estimate the instantaneous SINR defined in (1.44). Using this instantaneous SINR in combination with the CQI feedback algorithm of Sect. 5.2.2, accurate transmission rate adaptation can be achieved. Naturally this approach is only useful if the schedule is kept fixed for several Transmission Time Intervals (TTIs), implying a loss of temporal multi-user diversity. However, accounting for the downlink signaling overhead involved in changing the multi-user resource allocation, the performance loss is negligible in the considered low to moderate mobility scenarios.

Multi-user Scheduling

The sum throughput achieved with ZF beamforming and BD precoding is very much dependent on the selected set of users that is served in parallel [23]. If the channels of the selected users are orthogonal, the signal energy of the user can be steered into the user's channel subspace without causing any interference. This effect is also observable in the term $c_{S,n,k}$ of the proposed SINR lower bound in Eq. (7.63): when the channels of the selected users are orthogonal, the chordal distance $\mathrm{d}_{\mathrm{c}}^2\left(\hat{\mathbf{H}}_{n,k}[u,i], \mathbf{B}_{n,k}[u,i]\right)$ is equal to zero, implying no reduction of the channel gain by the precoder. In principal, the schedule can hence be obtained from an exhaustive search employing the achievable rate (7.66) as the scheduling metric. To avoid the complexity of an exhaustive search, greedy scheduling based on the SUS algorithm is instead considered here. To achieve some level of fairness among the users in the cell and to avoid user starvation, proportional fairness is utilized as scheduling metric [24]. Scheduling is applied on an RB basis, due to the availability of RB-specific CQI feedback. The scheduling procedure is summarized below for RB $[\eta, \kappa]$; the same approach is applied on all RBs. To simplify the presentation, it is assumed that $\ell_{\eta,\kappa}[s,i] = N_{\mathrm{L}}, \forall s \in \mathscr{U}_i$ and that $\frac{N_{\mathrm{T}}[i]}{N_{\mathrm{L}}}$ is integer-valued. Otherwise, an explicit validation of the feasibility conditions of BD precoding must be performed by the scheduler.

1. Initialize the weighted sum rate $R_{\eta,\kappa}[i] = 0$, the set of scheduled users $\mathscr{S}_{\eta,\kappa}[i] = \{\}$, the set of potential users $\mathscr{P}_{\eta,\kappa}[i] = \mathscr{U}_i$ and the number of scheduled users $S_{\eta,\kappa}[i] = 0$.
   Repeat the following steps until the maximum number of $S_{\eta,\kappa}[i] = \frac{N_{\mathrm{T}}[i]}{N_{\mathrm{L}}}$ users is served:
2. Calculate an orthonormal basis $\mathbf{B}_{\eta,\kappa}[i]$ for the space spanned by the channels of the served users in $\mathscr{S}_{\eta,\kappa}[i]$.
3. Find the semi-orthogonal user set $\tilde{\mathscr{S}}_{\eta,\kappa}[i]$, by determining all users $s \in \mathscr{P}_{\eta,\kappa}[i]$ for which the SUS condition is fulfilled:

$$\mathrm{tr}\left(\hat{\mathbf{H}}_{\eta,\kappa}[s,i]^{\mathrm{H}}\mathbf{B}_{\eta,\kappa}[i]\mathbf{B}_{\eta,\kappa}[i]^{\mathrm{H}}\hat{\mathbf{H}}_{\eta,\kappa}[s,i]\right) \le \alpha_{\mathrm{SUS}} N_{\mathrm{L}}. \quad (7.68)$$

4. If the SUS user set is empty, stop the algorithm and serve the users in $\mathscr{S}_{\eta,\kappa}[i]$.

5. For all users $s \in \tilde{\mathscr{S}}_{\eta,\kappa}[i]$, calculate the estimated achievable rate $R_{\eta,\kappa}[s,i]$ in (7.66) (replacing $\tilde{\beta}_{\nu,n,j}[s,i]$ with $\tilde{\beta}_{\nu,\eta,\kappa}[s,i]$ as obtained from $\mathrm{CQI}_{\nu,\eta,\kappa}[s,i]$) under the assumption that user s is served in parallel with the users in $\mathscr{S}_{\eta,\kappa}[i]$.
6. Determine the user $s \in \tilde{\mathscr{S}}_{\eta,\kappa}[i]$ that achieves the largest weighted rate

$$\hat{s} = \underset{s \in \tilde{\mathscr{S}}_{\eta,\kappa}[i]}{\mathrm{argmax}} \frac{R_{\eta,\kappa}[s,i]}{T[s,i]}, \tag{7.69}$$

with $T[s,i]$ denoting the average throughput of user s achieved over the past.
7. Update the estimated achievable rates $R_{\eta,\kappa}[s,i]$ of the users $s \in \mathscr{S}_{\eta,\kappa}[i]$, assuming that $\hat{s}$ is served in addition to the users in $\mathscr{S}_{\eta,\kappa}[i]$.
8. Calculate the weighted sum rate of the schedule $\{\mathscr{S}_{\eta,\kappa}[i], \hat{s}\}$

$$R = \sum_{s \in \{\mathscr{S}_{\eta,\kappa}[i], \hat{s}\}} \frac{R_{\eta,\kappa}[s,i]}{T[s,i]}. \tag{7.70}$$

9. If $R \geq R_{\eta,\kappa}[i]$, add user $\hat{s}$ to $\mathscr{S}_{\eta,\kappa}[i]$ and update $S_{\eta,\kappa}[i]$, set $R_{\eta,\kappa}[i] = R$ and remove $\hat{s}$ from $\mathscr{P}_{\eta,\kappa}[i]$. Otherwise, stop the algorithm and serve the users in $\mathscr{S}_{\eta,\kappa}[i]$

In step 2 of the scheduling algorithm, a pre-selection of users is performed, based on their subspace distance to the already served users. Only if a user is close to orthogonal to the previously served users, he is considered as a potential additional user. The exact meaning of "close to orthogonal" is determined by the choice of the SUS parameter α_{SUS}; see [23] for details on the selection of this parameter. In the presented simulations, the parameter $\alpha_{\mathrm{SUS}} = 0.35$ turned out as a good choice.

7.2.3 Performance Investigation

In this section, the performance of CSI feedback using subspace clustering and interpolation for BD precoding based MU-MIMO transmission over a frequency-selective OFDM broadcast channel is investigated. Monte-Carlo simulations of the quantization Mean Square Error (MSE) are conducted and the corresponding throughput achieved in the downlink of a single Long Term Evolution (LTE) compliant cell is evaluated. The frequency-selectivity of the wireless channel is characterized by the coherence bandwidth or equivalently by the Root Mean Square (RMS) delay spread τ_{RMS} of the channel multipath power delay profile. When the sensitivity of the proposed methods with respect to the frequency-selectivity of the channel is investigated, the simulation results are presented in dependence of the normalized sampling bandwidth

$$B_s = \Delta f_s \, N_{\mathrm{clust}}^{(f)} \, \tau_{\mathrm{RMS}}, \tag{7.71}$$

with Δf_s [Hz] being the OFDM subcarrier spacing ($\Delta f_s = 15$ kHz in case of LTE) and $N_{\text{clust}}^{(f)}$ denoting the size of a feedback cluster in the frequency domain in multiples of subcarriers. The advantage of SQBC clustering in terms of the chordal distance quantization MSE is demonstrated, but also its downside, i.e., the implied channel gain reduction, is highlighted, which can cause a significant throughput degradation, especially in case of strongly correlated receive antennas. In order to enable efficient CSI feedback operation, channel prediction, subspace clustering and Grassmannian quantization are combined at the user. The MSE contribution of the individual components is investigated to reveal the corresponding limitations and operating regimes of the proposed feedback algorithms. Finally, the performance achieved with the proposed SUS based multi-user scheduler, utilizing the derived CQI feedback, is evaluated by comparing to the throughput attained with the optimal schedule, as determined from an exhaustive search.

Chordal Distance MSE Evaluation

In the first simulation, the chordal distance MSE obtained with MET subspace selection, according to Sect. 7.1.3, in combination with zeroth- and first-order interpolation, as detailed in Sect. 7.2.1, is compared to SQBC clustering. The channel matrix is of size $N_T[i] \times N_R[u, i] = 8 \times 4$, and an $\ell[u, i] = 1$-dimensional subspace is selected as CSI feedback. Unquantized feedback of this representative subspace is considered; hence the error is caused only by the subspace clustering and the interpolation, respectively. In case of MET, the maximum eigenmode of the channel matrix experienced in the center of each RB is used as feedback information, while with SQBC clustering the subspace feedback is determined by solving the optimization problem (7.46). The MSE is estimated by means of Monte-Carlo simulations as

$$D \approx \frac{1}{N_{\text{tot}}} \frac{1}{K} \sum_{n=1}^{N_{\text{tot}}} \sum_{k=1}^{K} \mathrm{d}_{\mathrm{c}}^2 \left(\hat{\mathbf{H}}_{n,k}[u, i], \mathbf{H}_{n,k}^{(\text{ct})}[u, i] \right), \tag{7.72}$$

where the interpolated subspace $\hat{\mathbf{H}}_{n,k}[u, i] \in \mathbb{C}^{N_T[i] \times \ell[u,i]}$ is obtained from Eq. (7.45) in case of first-order interpolation, while $\hat{\mathbf{H}}_{n,k}[u, i] = \tilde{\mathbf{H}}_{\eta,\kappa}[u, i]$ is employed with zeroth-order interpolation and SQBC clustering ($[\eta, \kappa]$ is defined in Eq. (5.14).

The effect of dimensionality adaptation, as defined in Eq. (7.53), on the MSE performance of SQBC clustering is investigated. A system bandwidth of 5 MHz is assumed and the power delay profile of the channel is determined by the SCME *urban micro* channel model [28], having a coherence bandwidth of $B_C = 680$ kHz.

The results of the simulation are shown in Fig. 7.3a. Considering MET subspace selection with CSI interpolation, it is observed that zeroth-order interpolation is outperformed by first-order (linear) geodesic interpolation, as soon as the distance between the CSI pilots is sufficiently small. With SQBC clustering, a substantial MSE reduction is achieved by increasing the dimensionality d of the optimization problem (7.53), which determines the subspace selection. Notice that the dimensionality d has no impact on the dimension $\ell[u, i]$ of the subspace feedback; it is rather

Fig. 7.3 Chordal distance MSE investigation of the proposed CSI feedback methods for frequency-selective channels. **a** Comparison of feedback clustering and interpolation. **b** Contribution of CSI prediction, clustering and quantization

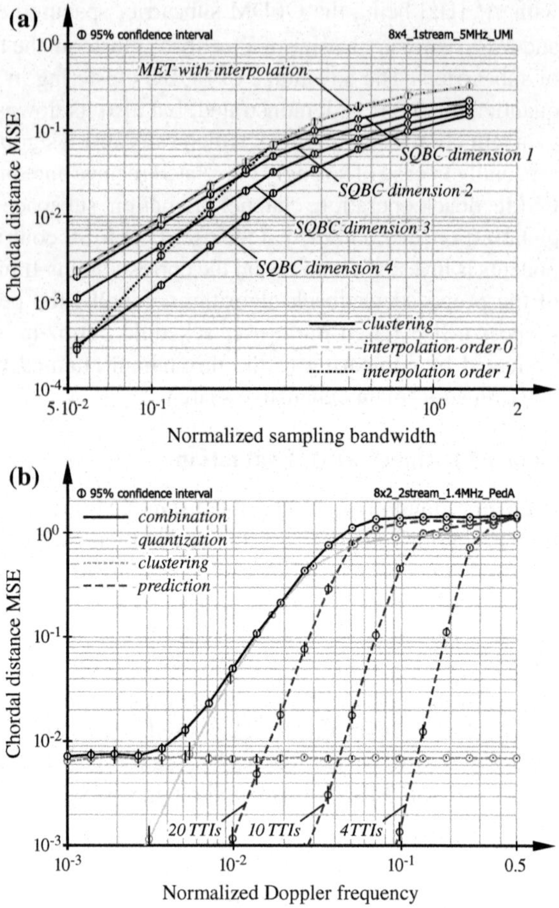

the dimension of the *search space* in the optimization problem (7.53) that is specified with d. Comparing CSI interpolation and clustering, a significant MSE improvement is observed with clustering when a small density of CSI pilots, i.e., a large sampling bandwidth B_s, is applied. This MSE improvement, however, is obtained at the cost of a channel gain reduction of the effective channel, causing a throughput loss at low SNR, as demonstrated below.

In the next simulation, predictive CSI quantization and Grassmannian subspace clustering are combined to enable efficient limited feedback operation. Additionally, the effect of a delay in the feedback path is investigated. To compensate for this delay, Finite Impulse Response (FIR) based channel prediction, as described in Sect. 5.3.2, is employed by the user. Notice that this channel predictor is only required at the user and not at the base station and it can therefore be based on perfect CSI. Channel prediction for delay compensation is not related to the operation of the predictive quantizer; the quantizer is working independently on top of the channel predictor output. Feedback of an $\ell[u, i] = 2$-dimensional subspace of an $N_T[i] \times N_R[u, i] = 8 \times 2$ dimensional channel matrix is considered. The *PedA* channel model [29], with $B_C = 4.4$ MHz, is employed and a single feedback cluster is applied to represent the 1.4 MHz system bandwidth, resulting in a feedback overhead of 7 bit per TTI (7 kbit/s assuming the LTE subframe duration of 1 ms).

The results of the investigation are shown in Fig. 7.3b. The individual MSE contributions of the channel prediction, the subspace clustering and the quantization, as well as the performance of the concatenated system are separately plotted. At low normalized Doppler frequencies (see (5.35)), i.e., when the channel variation over time is slow, the overall MSE is dominated by the subspace representation error, due to feedback clustering. In this case, the adaptive codebook construction of the predictive quantizer is able to track the channel variation very well, causing a negligible quantization error. A reduction of the observed error floor is only possible by reducing the cluster size, implying an increased feedback overhead. At intermediate Doppler frequencies, the overall MSE is determined by the quantization error. This error can be reduced by enlarging the quantization codebook, for the cost of increasing the feedback rate. Only at high Doppler frequencies and with a large feedback delay, the error caused by the channel prediction comes into play. In the region of interest for predictive quantization ($\nu_d \approx 10^{-2}$), even a feedback delay of 20 TTIs is irrelevant; hence, the feedback delay is not further considered in the remaining simulations of this chapter.

Throughput Evaluation

Next, the throughput performance of BD based MU-MIMO is investigated, when the proposed limited feedback algorithms are applied to provide CSIT. In the first simulation, the transmission rate reduction with respect to perfect CSIT, due to the subspace representation error caused by SQBC clustering, is evaluated, demonstrating the effect of dimensionality adaptation on the achieved throughput in dependence of the equivalent average transmit SNR (5.32).

The base station is equipped with $N_T[i] = 8$ transmit antennas; $S[i] = 8$ users, having $N_R[u, i] = 4$ receive antennas, are served in parallel, each over a single spatial stream $\ell[u, i] = 1$. The power delay profile of the channel is specified by the *rural area* model [30], which has a coherence bandwidth of $B_C = 2$ MHz. The carrier bandwidth of 1.4 MHz is represented with a single CSI feedback cluster, using unquantized CSI feedback. Correlated receive antennas are assumed according to Eq. (5.33) with $\alpha_{\text{corr}} \in \{0, 0.9\}$. The interference-averaged MMSE equalizer proposed in [4] is applied by the users to detect their data streams. The throughput simulations are conducted with the standard compliant Vienna LTE link level simulator [31].

The obtained simulations results are shown in Fig. 7.4. At low SNR, the best performance is achieved with a dimensionality of $d = 1$. With $d = 1$, only the maximum eigenmodes of the channels experienced over the subcarriers of an RB are taken into account for the calculation of the subspace representation according to Eq. (7.53). The calculated subspace is kept free of interference by the BD precoder; hence, the interference-free subspace has a potentially large channel gain. Due to clustering, however, the single subspace representation is imperfect for a given RE, implying

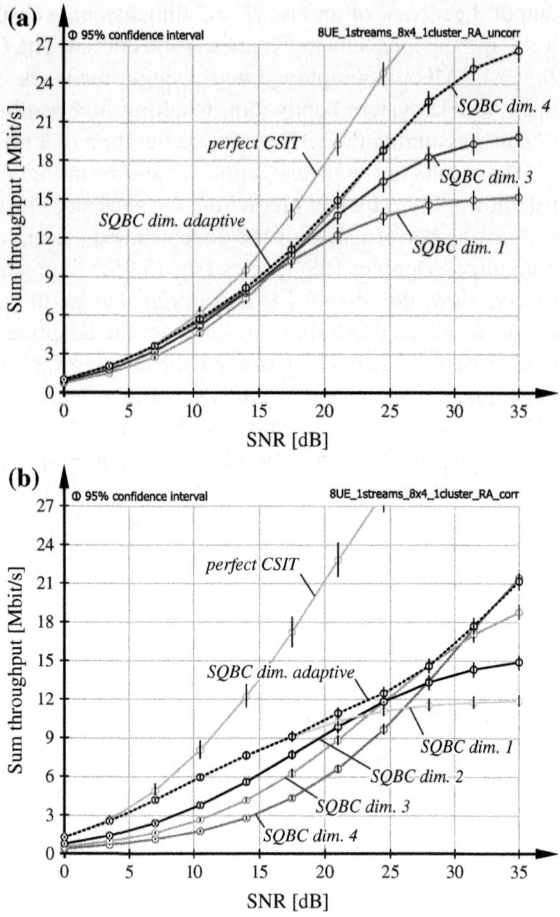

Fig. 7.4 Sum throughput achieved with ZF beamforming and SQBC feedback clustering in an $N_{\mathrm{T}}[i] \times N_{\mathrm{R}}[u, i] = 8 \times 4$ system, serving $S[i] = 8$ users over $\ell[u, i] = 1$ spatial stream each. The impact of dimensionality adaptation is investigated. **a** No receive antenna correlation $\alpha_{\mathrm{corr}} = 0$. **b** Strong receive antenna correlation $\alpha_{\mathrm{corr}} = 0.9$

residual multi-user interference after precoding. Therefore, an interference-limitation is observed at high SNR. The strength of the residual multi-user interference is impacted by the dimensionality d of the feedback clustering. By increasing d the *search space* for the subspace representation in (7.53) is extended, enabling a reduction of the chordal distance error, as shown in Fig. 7.3a, which implies reduced residual multi-user interference according to (7.63). The optimal dimensionality is dependent on the SNR. The proposed dimensionality adaptation (7.62) is able to identify the optimal dimensionality, as demonstrated in Fig. 7.4. Comparing Fig. 7.4a, b, it is observed that the throughput difference between different dimensionalities is increased with growing antenna correlation. This behavior is due to the increased singular value spread of the channel matrix with larger α_{corr}. Notice the similarity of the performance observed with dimensionality adaptation and with transmission rank adaptation, as shown in Fig. 5.5a. In the current simulations, however, the transmission rank is not changed; the total number of data streams is always equal to eight.

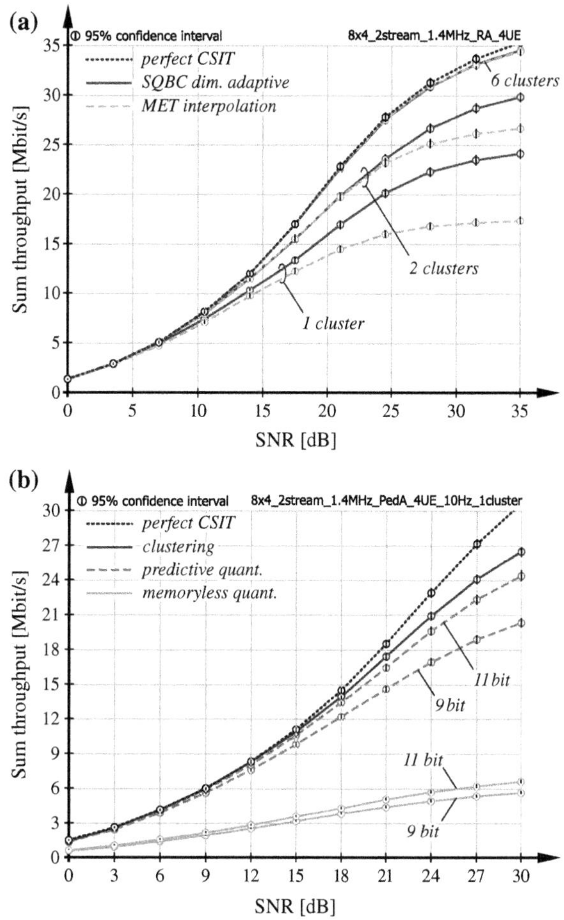

Fig. 7.5 Sum throughput achieved with BD precoding and the proposed limited feedback clustering and quantization algorithms in an $N_T[i] \times N_R[u, i] = 8 \times 4$ system, serving $S[i] = 4$ users over $\ell[u, i] = 2$ spatial streams each. **a** MET feedback with interpolation versus SQBC clustering. **b** Predictive versus memoryless quantization

In Fig. 7.5a, a similar scenario is evaluated where four users are served over $\ell[u, i] = 2$ streams each. The throughput of SQBC clustering with dimensionality adaptation is compared to MET subspace selection with linear geodesic interpolation. The number of feedback clusters is varied from $N_{RB} = 1$ to $N_{RB} = 6$ and uncorrelated receive antennas, $\alpha_{corr} = 0$, are assumed. It is observed that MET is outperformed by SQBC at high SNR if the feedback cluster size is large, because the residual multi-user interference achieved with SQBC is smaller due to the improved subspace representation. SQBC clustering is never surpassed by MET feedback, not even at low SNR when the throughput is determined by the effective channel gain, due to the application of dimensionality adaptation, which trades-off the effective channel gain for the residual multi-user interference depending on the SNR.

Finally, in Fig. 7.5b, the performance of the combination of feedback clustering and quantization is evaluated. The 1.4 MHz channel generated with the *PedA* model [29], assuming strongly correlated receive antennas, $\alpha_{corr} = 0.9$, is quantized with a single feedback cluster. Nine and eleven bit of feedback per TTI are considered, corresponding to a feedback rate of 9 kbit/s and 11 kbit/s, respectively. A maximum normalized Doppler frequency of $\nu_d = 0.01$ is assumed, representing, at a center frequency of 2 GHz, a walking user with a speed of approximately 5 km/h.

With these parameters, a negligible throughput degradation is caused by the subspace clustering, compared to perfect CSIT, up to an SNR of approximately 15 dB. Above this value, a throughput loss is observed, due to the residual multi-user interference induced by the frequency selectivity of the channel. When the SQBC subspace selection is conveyed to the base station by means of predictive quantization, close to optimal performance is achieved with 11 bit of feedback per TTI over a large SNR range. Naturally, a throughput deterioration is incurred when the quantization codebook size is reduced. Considering memoryless quantization, however, the obtained CSIT accuracy is insufficient to ensure reliable transmission to four users in parallel, each being served over $\ell[u,i] = 2$ streams. In this case, better performance is possible by employing a multi-user scheduler to select only a subset of the users for transmission, implying reduced multi-user interference.

Multi-user Scheduling

In this section, the efficiency of the SUS based greedy multi-user scheduler, proposed in Sect. 7.2.2, is investigated by comparing the achieved sum throughput to the optimal schedule, as obtained from an exhaustive search. Two different antenna configurations are considered, i.e., $N_\mathrm{T}[i] \times N_\mathrm{R}[u,i] = 4 \times 1$ and $N_\mathrm{T}[i] \times N_\mathrm{R}[u,i] = 8 \times 4$. In the 4×1 system, a single stream is transmitted per user, $\ell[u,i] = 1$, while with 8×4 antennas $\ell[u,i] = 2$ streams per user are employed. At most four users can thus be spatially multiplexed in both cases. The *rural area* channel model [30] is used and the antenna correlation parameter α_corr is set equal to zero. In case of imperfect CSIT, unquantized feedback of a single feedback cluster is considered for the total system bandwidth of 1.4 MHz using SQBC clustering. All users are simulated with the same SNR of 20 dB.

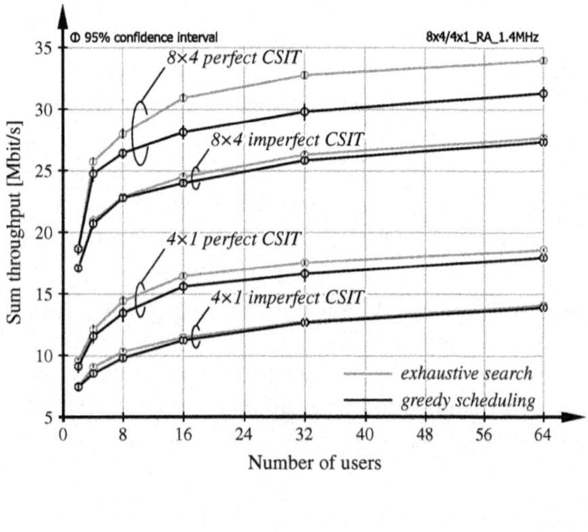

Fig. 7.6 Performance of the proposed SUS scheduling algorithm with perfect and imperfect CSIT compared to an exhaustive search. Two configurations are considered: $N_\mathrm{T}[i] \times N_\mathrm{R}[u,i] = 4 \times 1$ with $\ell[u,i] = 1$ and $N_\mathrm{T}[i] \times N_\mathrm{R}[u,i] = 8 \times 4$ with $\ell[u,i] = 2$

The simulation results are presented in Fig. 7.6, in terms of the sum throughput achieved in the downlink of the cell versus the number of served users. It can be seen that a similar multi-user diversity is attained with the exhaustive search scheduler and the proposed greedy scheduling algorithm, albeit an approximately constant rate loss that is dependent on the settings of the system. Especially in the realistic situation of imperfect CSIT, the throughput obtained with the greedy algorithm is close to the rate achieved with the optimal schedule.

References

1. S. Schwarz, Limited feedback transceiver design for downlink MIMO OFDM cellular networks. Ph.D. dissertation, Technische Universität Wien (2013). http://theses.eurasip.org/theses/514/limited-feedback-transceiver-design-for-downlink/
2. S. Schwarz, M. Rupp, Subspace quantization based combining for limited feedback block-diagonalization. IEEE Trans. Wirel. Commun. **12**(11), 5868–5879 (2013). doi:10.1109/TWC.2013.093013.130403
3. S. Schwarz, M. Rupp, Evaluation of distributed downlink multi-user MIMO-OFDM with limited feedback. IEEE Trans. Wirel. Commun. **13**(11), 6081–6094 (2014)
4. S. Schwarz, M. Rupp, Antenna combiners for block-diagonalization based multi-user MIMO with limited feedback, in *Proceedings of the IEEE International Conference on Communications, Workshop Beyond LTE-A*. Budapest, Hungary, June 2013
5. S. Schwarz, M. Rupp, Adaptive channel direction quantization based on spherical prediction, in *Proceedings of the IEEE International Conference on Communications*, pp. 3757–3762. Ottawa, Canada, June 2012
6. S. Schwarz, M. Rupp, Adaptive channel direction quantization— enabling multi user MIMO gains in practice, in *Proceedings of the IEEE International Conference on Communications*, pp. 6947–6952. Ottawa, Canada, June 2012
7. S. Schwarz, M. Rupp, Adaptive channel direction quantization for frequency selective channels, in *Proceedings of the 20th European Signal Processing Conference*, pp. 2536–2540. Bucarest, Romania, Aug 2012
8. S. Schwarz, R. Heath Jr., M. Rupp, Multiuser MIMO in distributed antenna systems with limited feedback, in *Proceedings of the IEEE 4th International Workshop on Heterogeneous and Small Cell Networks, GLOBECOM*. Anaheim, CA, Dec 2012
9. S. Schwarz, R. Heath Jr., M. Rupp, Adaptive quantization on the Grassmann-manifold for limited feedback multi-user MIMO systems, in *Proceedings of the 38th International Conference on Acoustics, Speech and Signal Processing*. Vancouver, Canada, May 2013
10. S. Schwarz, R. Heath Jr., M. Rupp, Adaptive quantization on a Grassmann-manifold for limited feedback beamforming systems. IEEE Trans. Signal Process. **61**(18), 4450–4462 (2013)
11. S. Schwarz, R. Heath Jr., M. Rupp, Single-user MIMO versus multi-user MIMO in distributed antenna systems with limited feedback. EURASIP J. Adv. Signal Process. **2013**(54), 1–20 (2013)
12. N. Jindal, MIMO broadcast channels with finite-rate feedback. IEEE Trans. Inf. Theory **52**(11), 5 (2006)
13. N. Ravindran, N. Jindal, Limited feedback-based block diagonalization for the MIMO broadcast channel. IEEE J. Sel. Areas Commun. **26**(8), 1473–1482 (2008)
14. N. Jindal, Antenna combining for the MIMO downlink channel. IEEE Trans. Wirel. Commun. **7**(10), 3834–3844 (2008)
15. W. Dai, Y. Liu, B. Rider, Quantization bounds on Grassmann manifolds and applications to MIMO communications. IEEE Trans. Inf. Theory **54**(3), 1108–1123 (2008)

16. F. Boccardi, H. Huang, A near-optimum technique using linear precoding for the MIMO broadcast channel, in *Proceedings of the IEEE International Conference on Acoustics, Speech and Signal Processing*, vol. 3, pp. III-17–III-20, April 2007
17. A. Zanella, M. Chiani, The PDF of the lth largest eigenvalue of central Wishart matrices and its application to the performance analysis of MIMO systems, in *Proceedings of the IEEE Global Telecommunications Conference*, pp. 1–6. New Orleans, LA, Dec 2008
18. J. Lee, N. Jindal, High SNR analysis for MIMO broadcast channels: dirty paper coding versus linear precoding. IEEE Trans. Inf. Theory **53**(12), 4787–4792 (2007)
19. T. Li, L. Yang, Z. He, Interpolation-based multiuser precoding for MIMO-OFDM system with limited feedback, in *Proceedings of the International Conference on Neural Networks and Signal Processing*, pp. 205–209 (2008)
20. T. Inoue, R. Heath Jr., Grassmannian predictive frequency domain compression for limited feedback beamforming, in *Proceedings of the Information Theory and Applications Workshop*, pp. 173–177, Feb 2010
21. J. Chang, I. Lu, Y. Li, Adaptive codebook-based channel prediction and interpolation for multiuser multiple-input multiple-output orthogonal frequency division multiplexing systems. IET Commun. **6**(3), 281–288 (2012)
22. S.R. Buss, J.P. Fillmore, Spherical averages and applications to spherical splines and interpolation. ACM Trans. Graph. **20**, 95–126 (2001)
23. T. Yoo, A. Goldsmith, On the optimality of multiantenna broadcast scheduling using zero-forcing beamforming. IEEE J. Sel. Areas Commun. **24**(3), 528–541 (2006)
24. R. Jain, D. Chiu, W. Hawe, A quantitative measure of fairness and discrimination for resource allocation in shared computer systems. DEC, Technical report TR-301, Sept 1984
25. D. Amsallem, Interpolation on manifolds of CFD-based fluid and finite element-based structural reduced-order models for on-line aeroelastic predictions, Dissertation, Standford University (2010)
26. R. Blum, MIMO capacity with interference. IEEE J. Sel. Areas Commun. **21**(5), 793–801 (2003)
27. M. Trivellato, F. Boccardi, F. Tosate, User selection schemes for MIMO broadcast channels with limited feedback, in *Proceedings of the IEEE 65th Vehicular Technology Conference*, pp. 2089–2093. Dublin, Ireland, April 2007
28. D. Baum, J. Hansen, J. Salo, An interim channel model for beyond-3G systems: extending the 3GPP spatial channel model (SCM), in *Proceedings of the IEEE 61st Vehicular Technology Conference*, vol. 5, pp. 3132–3136. Stockholm, Sweden (2005)
29. ITU, Recommendation ITU-R M.1225: Guidelines for Evaluation of Radio Transmission Technologies for IMT-2000, ITU, Technical report (1997)
30. 3GPP, Technical specification group radio access networks; Deployment aspects (Release 8), Dec 2008. http://www.3gpp.org/ftp/Specs/html-info/25943.htm
31. C. Mehlführer, J.C. Ikuno, M. Simko, S. Schwarz, M. Rupp, The Vienna LTE simulators—enabling reproducibility in wireless communications research. EURASIP J. Adv. Signal Process., Special issue on reproducible research, **2011**(29), 1–14 (2011)

Chapter 8
LTE-Advanced Uplink Transmissions

Stefan Pratschner and Erich Zöchmann

In Long Term Evolution (LTE) uplink transmissions Single-Carrier Frequency Division Multiplexing (SC-FDM) is employed, mainly to achieve a lower Peak-to-Average Power Ratio (PAPR) at the transmitting mobile device, when compared to Orthogonal Frequency Division Multiplexing (OFDM) transmissions. SC-FDM, however, is essentially OFDM with an additional Discrete Fourier Transform (DFT) spreading at the transmitter and a de-spreading at the receiver. Employing such modulation scheme, subcarriers are still orthogonal due to the OFDM processing. Unlike in pure OFDM transmissions, individual subcarriers cannot be treated independently, since each data symbol is spread over all subcarriers by means of a DFT. Therefore, we present a Multiple-Input Multiple-Output (MIMO) SC-FDM system model, suitable to describe the Long Term Evolution-Advanced (LTE-A) uplink link level, together with appropriate Signal to Interference and Noise Ratio (SINR) expressions. Already at this point, severe differences to the downlink occur in the post-equalization SINR for Zero Forcing (ZF) and Minimum Mean Square Error (MMSE) receiver. Due to the spreading of symbols across all subcarriers, the system performance is dominated by the subcarrier with the lowest channel gain. Employing an MMSE receiver in SC-FDM is thus even more beneficial.

Similar to downlink transmissions the Modulation and Coding Scheme (MCS) is adapted to the current channel state by the post-equalization SINR. However, in uplink transmissions,it turns out that the post-spreading SINR is the relevant metric to select the optimal Adaptive Modulation and Coding (AMC).

Within this framework, we compare SC-FDM performance to OFDM in terms of Bit Error Ratio (BER) and PAPR by replacing the spreading and de-spreading with the identity function. This even allows to compare the achieved throughput as the system's overhead is not influenced. We show that the reduction in PAPR of

S. Pratschner (✉) · E. Zöchmann
Institute of Telecommunications, TU Wien, Vienna, Austria
e-mail: spratsch@nt.tuwien.ac.at

E. Zöchmann
e-mail: ezoechma@nt.tuwien.ac.at

M. Rupp et al., *The Vienna LTE-Advanced Simulators*,
Signals and Communication Technology, DOI 10.1007/978-981-10-0617-3_8

SC-FDM compared to OFDM transmissions is significant, even at a low bandwidth of 1.4 MHz. This, however, comes at the cost of a performance loss, as in an OFDM transmissions the coded BER is in general lower.

Another aspect of differences between uplink and downlink transmissions is observed when it comes to Channel Estimation (CE). In both cases, CE for coherent detection is pilot aided. While the allocation of pilot symbols in the downlink shows a diamond pattern (see Fig. 2.2 in Chap. 2), the uplink pilot symbols (see Fig. 8.7 further ahead) are allocated at symbol time $n = 3$ in each slot, spanning the entire scheduled bandwidth range. Also, user specific reference symbols, required for channel estimation in the downlink, are multiplexed such that reference symbols of different spatial transmission layers do not overlap. This allows to estimate all MIMO channel components separately. The uplink reference symbols, of different transmission streams, on the other hand, overlap entirely and have to be separated by exploiting their Frequency Domain Code Division Multiplexing (FD-CDM) orthogonality. Further, the non-diamond shaped pilot allocation in the uplink hampers sophisticated interpolation between the reference symbols. We present well known DFT based CE, as well as novel estimation methods with improved performance together with time interpolation methods in the last part of this chapter.

8.1 LTE-Specific System Model and Post-equalization SINR

We focus one those details, necessary to describe our system model at time n. In an SC-FDM model, OFDM tranmissions can be considered a special case. The major difference is an additional spreading and de-spreading stage at the transmitter and receiver, highlighted via dashed boxes in Fig. 8.1. The common parts of the system model will be described from left to right. An equivalent mathematical model is given in Eqs. (8.1) and (8.2).

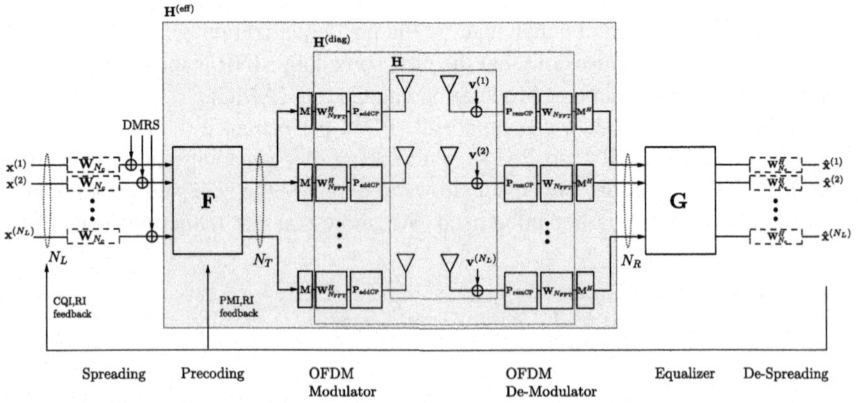

Fig. 8.1 The LTE-A uplink transceiver

$$\begin{aligned}
\hat{\mathbf{x}} = & \left(\mathbf{I}_{N_\mathrm{L}} \otimes \mathbf{W}_{N_\mathrm{c}}^H\right) \mathbf{G} \left(\mathbf{I}_{N_\mathrm{R}} \otimes \mathbf{M}^H \mathbf{W}_{N_\mathrm{FFT}} \mathbf{P}_\mathrm{remCP}\right) && \text{(receiver)} \\
& \times \mathbf{H} && \text{(channel)} \\
& \times \left(\mathbf{I}_{N_\mathrm{T}} \otimes \mathbf{P}_\mathrm{addCP} \mathbf{W}_{N_\mathrm{FFT}}^H \mathbf{M}\right) \left(\mathbf{F} \otimes \mathbf{I}_{N_\mathrm{c}}\right) \left(\mathbf{I}_{N_\mathrm{L}} \otimes \mathbf{W}_{N_\mathrm{c}}\right) \mathbf{x} && \text{(transmitter)} \\
& + \underbrace{\left(\mathbf{I}_{N_\mathrm{L}} \otimes \mathbf{W}_{N_\mathrm{c}}^H\right) \mathbf{G} \left(\mathbf{I}_{N_\mathrm{R}} \otimes \mathbf{M}^H \mathbf{W}_{N_\mathrm{FFT}} \mathbf{P}_\mathrm{remCP}\right) \mathbf{v}}_{\tilde{\mathbf{v}}} && \text{(transformed noise)} \\
= & \underbrace{\left(\mathbf{I}_{N_\mathrm{L}} \otimes \mathbf{W}_{N_\mathrm{c}}^H\right)}_{\substack{\mathbf{I}_{N_\mathrm{L} N_\mathrm{c}} \\ \text{for OFDM}}} \mathbf{G}\mathbf{H}^{(\mathrm{eff})} \underbrace{\left(\mathbf{I}_{N_\mathrm{L}} \otimes \mathbf{W}_{N_\mathrm{c}}\right)}_{\substack{\mathbf{I}_{N_\mathrm{L} N_\mathrm{c}} \\ \text{for OFDM}}} \mathbf{x} + \tilde{\mathbf{v}} && (8.1) \\
= & \,\mathbf{K}\mathbf{x} + \tilde{\mathbf{v}} = \underbrace{\mathbf{I} \odot \mathbf{K}\mathbf{x}}_{\text{desired signal}} + \underbrace{(\mathbf{K} - \mathbf{I} \odot \mathbf{K})\,\mathbf{x}}_{\text{intra- and interlayer interference}} + \tilde{\mathbf{v}}. && (8.2)
\end{aligned}$$

Right after the DFT spreading, the Demodulation Reference Signals (DMRS) are inserted. DMRS will be considered later for the purpose of CE. Next, MIMO precoding is carried out, exploiting a set of semi-unitary precoding matrices $\mathbf{F}$, pooled in the precoder codebook $\mathscr{Q}$, as defined in [1]. For LTE-A uplink transmission, the precoding matrix applied for a given user is equal for all Resource Block (RB) assigned to this user. In case of spatial multiplexing, each spatial layer is transmitted with equal power.

Each antenna is equipped with its own OFDM modulator, consisting of subcarrier mapping, Inverse Fast Fourier Transform (IFFT) and an Cyclic Prefix (CP) addition.

At the transmitter, processing occurs in reversed order. First, the OFDM demodulation by an Fast Fourier Transform (FFT) takes place to get back into the frequency domain. The immunity to multipath propagation (stemming from the CP) allows to employ one-tap frequency domain equalizers $\mathbf{G} \in \mathbb{C}^{N_\mathrm{L} N_\mathrm{c} \times N_\mathrm{R} N_\mathrm{c}}$ without performance loss. Eventually, de-spreading delivers the data estimates.

All this previously informally described processing is linear and we are able to formulate a matrix-vector input–output relationship between a (stacked) data-vector $\mathbf{x}$ and its estimate $\hat{\mathbf{x}}$. For simplicity we assume that the channel stays constant during one OFDM symbol. A more detailed system description can be found in [2, 3].

In order to adapt the data transmission to the current channel state, LTE-A applies limited feedback. Limited feedback is depicted via the feedback arrow in Fig. 8.1. The data vector $\mathbf{x}^{(l)} \in \mathbb{C}^{N_\mathrm{c} \times 1}$ of layer $l \in \{1, \ldots, N_\mathrm{L}\}$ contains modulated and coded symbols for each of the N_c subcarriers; the modulation format and code rate is adapted via the Channel Quality Indicator (CQI). The number of transmit layers depends on the LTE-A specific Rank Indicator (RI) feedback. All $\mathbf{x}^{(l)}$ are stacked into one vector $\mathbf{x} \in \mathbb{C}^{N_\mathrm{c} N_\mathrm{L} \times 1}$ on which layer-wise spreading and joint precoding—according to the Precoding Matrix Indicator (PMI)—of all subcarriers takes place. The subsequent OFDM modulator consists of the localized subcarrier mapping $\mathbf{M}$, mapping N_c subcarriers to the center of an N_FFT point IFFT, and the addition of the CP.

Depending on the level of abstraction, our system model can be described via different channel matrices. The physical baseband time domain channel is described by a block-wise Toeplitz matrix $\mathbf{H} \in \mathbb{C}^{(N_{\text{FFT}}+N_{\text{cp}})N_{\text{R}}\times(N_{\text{FFT}}+N_{\text{cp}})N_{\text{T}}}$, with N_{T} transmit and N_{R} receive antennas, which turns block-wise circulant after addition ($\mathbf{P}_{\text{addCP}}$) and removal ($\mathbf{P}_{\text{remCP}}$) of an appropriately chosen CP of length N_{cp}. Finally, it turns diagonal after the IFFT and FFT on the transmitter and receiver, respectively.

$$\mathbf{H}^{(\text{diag})} = \left(\mathbf{I}_{N_{\text{R}}} \otimes \mathbf{W}_{N_{\text{FFT}}}\mathbf{P}_{\text{remCP}}\right)\mathbf{H}\left(\mathbf{I}_{N_{\text{T}}} \otimes \mathbf{P}_{\text{addCP}}\mathbf{W}_{N_{\text{FFT}}}^{H}\right). \tag{8.3}$$

The last step of the OFDM de-modulator is the reversal of the localized subcarrier mapping $\mathbf{M}^{H}$. The effective MIMO channel from N_{L} transmit layers to N_{R} receive antennas, incorporating the precoder, the OFDM modulator, the time-domain MIMO channel $\mathbf{H}$ and the OFDM de-modulator, is abstracted to one block matrix $\mathbf{H}^{(\text{eff})}$. This greatly facilitates the readability of all formulas later on.

$$\mathbf{H}^{(\text{eff})} = \left(\mathbf{I}_{N_{\text{R}}} \otimes \mathbf{M}^{H}\right)\mathbf{H}^{(\text{diag})}\left(\mathbf{I}_{N_{\text{T}}} \otimes \mathbf{M}\right)\left(\mathbf{F} \otimes \mathbf{I}_{N_{\text{c}}}\right) \tag{8.4}$$

The additive noise is assumed independent across antennas and is distributed zero mean, white Gaussian $\mathbf{v}^{(i)} \sim \mathcal{N}_{\mathbb{C}}\left(\mathbf{0}, P_{\text{V}}\mathbf{I}\right), \quad i \in \{1, \ldots, N_{\text{R}}\}$. The stacked noise vector $\mathbf{v} = \left((\mathbf{v}^{(1)})^{T}, \ldots, (\mathbf{v}^{(N_{\text{R}})})^{T}\right)^{T}$ is thus zero mean, white Gaussian as well.

The frequency domain one-tap equalizer $\mathbf{G}$ is chosen conforming to different criteria, either the ZF criterion, which removes all channel distortions at risk of noise enhancement, or the MMSE criterion, that tries to minimize the effects of noise enhancement and channel distortion.

After the de-spreading operation the data estimates $\hat{\mathbf{x}}$ of the noisy, received signal are given in Eq. (8.2), with the before mentioned convenient abbreviation (8.4) and $\mathbf{W}_{N_{\text{FFT}}}$ is the DFT matrix of size N_{FFT}.

8.1.1 *Post-spreading SINR in SC-FDM Transmissions*

The special structure of Eq. (8.2), due to the frequency domain one tap equalizer and the DFT spreading, yields a block-wise circulant input–output matrix,

$$\mathbf{K} = \left(\mathbf{I}_{N_{\text{L}}} \otimes \mathbf{W}_{N_{\text{c}}}^{H}\right)\mathbf{G}\mathbf{H}^{(\text{eff})}\left(\mathbf{I}_{N_{\text{L}}} \otimes \mathbf{W}_{N_{\text{c}}}\right). \tag{8.5}$$

This block-wise circulant structure produces a constant post equalization and post spreading SINR over all subcarriers within one layer $l = 1, 2, \ldots, N_{\text{L}}$ [3, 4]. The detailed derivation is provided in the appendix of [5]. We obtain for the SINR

$$\mathbf{SINR}^{\mathrm{SC-FDM},\,(l)} = \frac{\frac{P_\mathrm{S}}{N_\mathrm{c}}\left|\mathbb{1}_{N_\mathrm{c}}^T \mathbf{S}^{(l)}\mathrm{diag}\left(\mathbf{GH}^{(\mathrm{eff})}\right)\right|^2}{P_\mathrm{S}\|\mathbf{S}^{(l)}\mathbf{GH}^{(\mathrm{eff})}\|_F^2 - \frac{P_\mathrm{S}}{N_\mathrm{c}}\left|\mathbb{1}_{N_\mathrm{c}}^T \mathbf{S}^{(l)}\mathrm{diag}\left(\mathbf{GH}^{(\mathrm{eff})}\right)\right|^2 + P_\mathrm{V}\|\mathbf{S}^{(l)}\mathbf{G}\|_F^2},$$

where

$$\mathbf{S}^{(l)} = \left(\mathbf{0},\quad \mathbf{I}_{N_\mathrm{c}},\quad \mathbf{0}\right), \tag{8.6}$$

selects that part of $\mathbf{GH}^{(\mathrm{eff})}$, that effects the lth layer. The second moment of the zero-mean data symbols equals the baseband transmit power P_S as LTE-A has standardized semi-unitary precoders $\mathbf{F}$, so that the overall transmitter (spreading, precoding and OFDM modulation) is unitary.

8.1.2 Post-equalization SINR in OFDM Transmissions

In contrast to SC-FDM, no spreading takes place for OFDM. The dashed boxes in Fig. 8.1 are replaced by identity matrices; they are thus simply omitted. Different subcarriers k are orthogonal/independent and the equalizer treats the corresponding subcarrier channel $\mathbf{H}_k$ only. Subscript k denotes the relevant part of the effective channel matrix $\mathbf{H}_k^{(\mathrm{eff})}$ for the kth subcarrier. The corresponding indexes within diagonal matrix $\mathbf{H}^{(\mathrm{diag})}$ are $\mathbb{1}_{N_\mathrm{R}\times N_\mathrm{T}} \otimes \mathrm{Diag}\left(\mathbf{e}_k\right)$, with the canonical base vectors $\mathbf{e}_k$. With this notation, the effective subcarrier channel $\mathbf{H}_k^{(\mathrm{eff})} \in \mathbb{C}^{N_\mathrm{R}\times N_\mathrm{L}}$ is

$$\mathbf{H}_k^{(\mathrm{eff})} = \left[\mathbf{H}^{(\mathrm{diag})}\right]_{\mathbb{1}_{N_\mathrm{R}\times N_\mathrm{T}}\otimes\mathrm{Diag}(\mathbf{e}_k)}\mathbf{F}, \tag{8.7}$$

and $\mathbf{G}_k$ reduces to its linear one tap equalizer. Precoding matrix $\mathbf{F}$ remains constant for all subcarriers. The SINR formula is quite similar to the SC-FDM case, except that the SINR shows subcarrier dependency now. The SINR takes on a vector dimension now and reads at layer l and subcarrier k

$$\left[\mathbf{SINR}^{\mathrm{OFDM},\,(l)}\right]_k = \frac{P_\mathrm{S}\left|\mathbf{s}^{(l)}\,\mathrm{diag}\left(\mathbf{G}_k\mathbf{H}_k\right)\right|^2}{P_\mathrm{S}\|\mathbf{s}^{(l)}\mathbf{G}_k\mathbf{H}_k\|_2^2 - P_\mathrm{S}\left|\mathbf{s}^{(l)}\,\mathrm{diag}\left(\mathbf{G}_k\mathbf{H}_k\right)\right|^2 + P_\mathrm{V}\|\mathbf{s}^{(l)}\mathbf{G}_k\|_2^2},$$

with the selection vector

$$\mathbf{s}^{(l)} = \left(0\ldots0,\ 1,\ 0\ldots0\right), \tag{8.8}$$

that contains an appropriate number of zeros and a single one at the lth entry.

8.2 SC-FDM Features

In this section, the specific impacts of the single carrier such as the modulation scheme in the LTE-A uplink will be analyzed. First, the main reason for SC-FDM employment, namely the PAPR reduction, is discussed. The transmit signal's PAPR is simulated for different bandwidths and compared to OFDM. Second, the SINR expressions are specialized to specific receiver types both, with DFT spreading and without DFT spreading. The impacts on the system performance in terms of BER are discussed.

8.2.1 Peak to Average Power Ratio

SC-FDM is employed as the physical layer modulation scheme for LTE uplink transmission, due to its lower PAPR compared to OFDM [6]. Lower PAPR, or similarly lower crest factor, leads to reduced linearity requirements for the power amplifiers and to relaxed resolution specifications for the digital-to-analog converters at the user equipments, entailing higher power efficiency.

The Vienna LTE-A uplink simulator calculates the discrete-time baseband PAPR with the default oversampling factor $N_o = 4$ [7]. The discrete time signal on each of the N_T transmit antennas is therefore calculated as

$$\left[\mathbf{s}^{(tx)}\right]_m = \frac{1}{\sqrt{N_{FFT}}} \sum_{k=0}^{N_{FFT}-1} \left[\mathbf{x}^{(pre)}\right]_k e^{j\frac{2\pi mk}{N_o N_{FFT}}}, \quad 0 \leq m \leq N_o N_{FFT} - 1,$$

where $\mathbf{x}^{(pre)}$ is the transmit vector right after precoding and before the IFFT at each transmit antenna. The PAPR of the stacked vector $\mathbf{s}^{(tx)}$ containing all N_T transmit vectors is calculated as

$$\mathrm{PAPR}\{\mathbf{s}_{tx}\} = \frac{\|\mathrm{diag}\left(\mathbf{s}^{(tx)}\mathbf{s}^{(tx)H}\right)\|_\infty}{\mathbb{E}\{|\left[\mathbf{s}^{(tx)}\right]_m|^2\}} \approx \frac{\|\mathbf{s}^{(tx)}\|_\infty^2}{\frac{1}{N_T N_{FFT}}\|\mathbf{s}^{(tx)}\|_2^2},$$

where the Euclidean norm in the denominator serves as an estimate for the ensemble average.

Figure 8.2 depicts the PAPR of OFDM and SC-FDM obtained for different system bandwidths. Already for a small bandwidth (1.4 MHz), there is a significant reduction for SC-FDM over OFDM transmissions. With increasing bandwidth OFDM's PAPR grows and the gains obtained by SC-FDM become more and more pronounced. The PAPR also depends on the modulation alphabet; the smaller the alphabet, the smaller the PAPR. This effect is illustrated in dotted lines in Fig. 8.2, where we depict the PAPR of 4-Quadrature Amplitude Modulation (QAM) as an example.

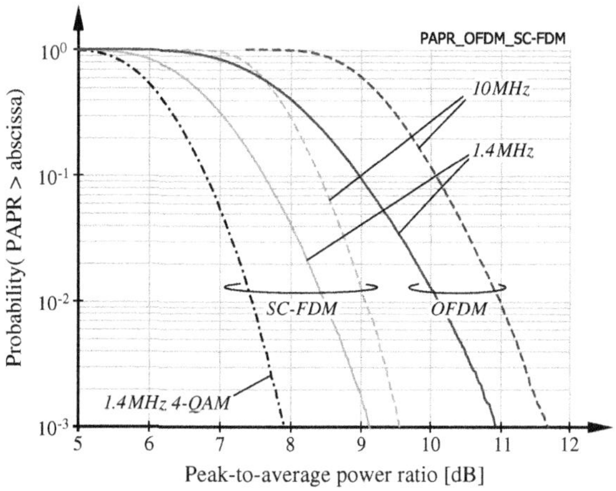

Fig. 8.2 PAPR for SC-FDM and OFDM for different bandwidths (1.4 and 10 MHz) and modulation alphabets (4/64 QAM)

8.2.2 BER Comparison Over Frequency Selective Channels

The additional spreading of SC-FDM leads to an SINR expression that is constant over all subcarriers, legitimating its name. The aim of this subsection is to analyze the SINR expression more in detail for the Single-Input Single-Output (SISO) case and draw conclusions on BER performance.

We focus on the two most prominent equalizer concepts and start with the ZF equalizer, for whom the SC-FDM Signal to Noise Ratio (SNR) expression (8.6) reduces to the harmonic mean

$$\mathbf{SNR}_{\mathrm{ZF}}^{\mathrm{SC-FDM}} = \frac{P_{\mathrm{S}}}{P_{\mathrm{V}}} \frac{1}{\frac{1}{N_{\mathrm{c}}} \sum_{k=1}^{N_{\mathrm{c}}} \frac{1}{|h_k|^2}}, \tag{8.9}$$

whereas the OFDM expression (8.8) is sub-carrier dependent and becomes proportional to the channel transfer function

$$\left[\mathbf{SNR}_{\mathrm{ZF}}^{\mathrm{OFDM}}\right]_k = \frac{P_{\mathrm{S}}}{P_{\mathrm{V}}} |h_k|^2. \tag{8.10}$$

The average OFDM SNR

$$\overline{\mathrm{SNR}_{\mathrm{ZF}}^{\mathrm{OFDM}}} = \frac{P_{\mathrm{S}}}{P_{\mathrm{V}}} \frac{1}{N_{\mathrm{c}}} \sum_{k=1}^{N_{\mathrm{c}}} |h_k|^2 \tag{8.11}$$

yields an upper bound on the Single-Carrier Frequency Division Multiple Access (SC-FDMA) SNR due to the harmonic mean—arithmetic mean inequality [8]

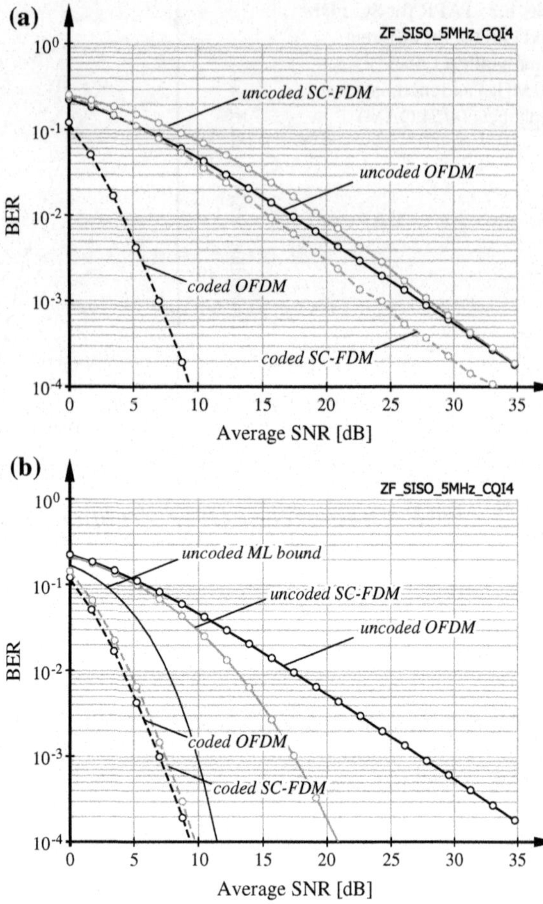

Fig. 8.3 BER comparison between OFDM and SC-FDM for a SISO PedB channel with 5 MHz bandwidth and fixed CQI = 4 transmission. **a** ZF receiver. **b** MMSE receiver

$$\mathrm{SNR}_{\mathrm{ZF}}^{\mathrm{SC-FDM}} \leq \overline{\mathrm{SNR}_{\mathrm{ZF}}^{\mathrm{OFDM}}}. \tag{8.12}$$

The equality in Eq. (8.12) holds if and only if the channel is frequency flat. The difference between the harmonic mean and the arithmetic mean becomes increasingly pronounced, the more selective the channel becomes. We, therefore, expect the (uncoded) BER of SC-FDM and ZF equalization to perform worse than OFDM, which is well reflected by simulation results. The BER simulations were carried out with CQI = 4 on a PedB channel [9]. This MCS employs 4-QAM and has an effective code-rate of 0.3008. As expected, the BER performance of SC-FDM is worse than in OFDM transmissions, both shown in Fig. 8.3a in solid lines. Due to the spreading SC-FDM already expends all channel diversity and coding does not increase the SNR slope of the BER curve. This manifests in an almost parallel shift of the BER curve for SC-FDM, as visual in Fig. 8.3a in dashed lines. While the SC-FDM in the uplink utilizes channel diversity by spreading, in OFDM transmissions of the downlink diversity supports coding across subcarriers which increases the BER slope considerably, cf. Fig. 8.3a dashed line. Closed form solutions for the BER achieved with ZF receivers in SC-FDM transmission are presented in [10].

To obtain the MMSE SINR expression a couple of mathematical transformations, as provided in the appendix of [5], are required to arrive at

$$\mathrm{SINR}_{\mathrm{MMSE}}^{\mathrm{SC-FDM}} = \frac{P_\mathrm{S}}{P_\mathrm{V}} \frac{1 - \frac{P_\mathrm{V}}{P_\mathrm{S}} \frac{1}{N_\mathrm{c}} \sum\limits_{k=1}^{N_\mathrm{c}} \frac{1}{\frac{P_\mathrm{V}}{P_\mathrm{S}} + |h_k|^2}}{\frac{1}{N_\mathrm{c}} \sum\limits_{k=1}^{N_\mathrm{c}} \frac{1}{\frac{P_\mathrm{V}}{P_\mathrm{S}} + |h_k|^2}} = \frac{P_\mathrm{S}}{P_\mathrm{V}} \left(\frac{1}{\frac{1}{N_\mathrm{c}} \sum\limits_{k=1}^{N_\mathrm{c}} \frac{1}{\frac{P_\mathrm{V}}{P_\mathrm{S}} + |h_k|^2}} - \frac{P_\mathrm{V}}{P_\mathrm{S}} \right). \tag{8.13}$$

The denominator of Eq. (8.13) is regularized and less sensitive to spectral notches.

An upper bound on the SINR can be obtained via the maximum of the transfer function h_k

$$\mathrm{SINR}_{\mathrm{MMSE}}^{\mathrm{SC-FDM}} \leq \frac{P_\mathrm{S}}{P_\mathrm{V}} \left(\frac{1}{\frac{1}{\frac{P_\mathrm{V}}{P_\mathrm{S}} + \max_k |h_k|^2}} - \frac{P_\mathrm{V}}{P_\mathrm{S}} \right) = \frac{P_\mathrm{S}}{P_\mathrm{V}} \max_k |h_k|^2. \tag{8.14}$$

In the low SNR regime $\frac{P_\mathrm{V}}{P_\mathrm{S}} \gg |h_k|^2$ this bound becomes tight. The higher the inverse SNR $\frac{P_\mathrm{V}}{P_\mathrm{S}}$ in relation to the maximum of the transfer function, the tighter the bound becomes. The average OFDM SNR can never be larger than its maximum entry and is only equal for frequency flat channels. The bound (8.14) is particularly tight at low SNR causing a lower BER when compared to OFDM transmissions. For the SISO case, ZF and MMSE equalizers perform equivalent for OFDM and 4-QAM. Again, this presumption is validated by our simulation, showing that the uncoded BER is lower for SC-FDM as for OFDM, cf. Fig. 8.3b in solid lines. There are also semi-closed form approximations for the BER of MMSE detection in SC-FDM transmission available [11]. Although the uncoded BER shows superior performance, the coded BER is lower for OFDM due to the coding gains stemming from channel diversity, cf., Fig. 8.3b dashed lines.

A bound for the Maximum Likelihood (ML) detection performance was derived in [12]. As bandwidth increases, the slope of the BER curve achieved with MMSE receivers reaches to the slope of ML detection, demonstrating the full exploitation of channel diversity by the MMSE equalizer, cf., Fig. 8.3b line.

8.3 Throughput Performance with Link Adaptation

In this section, we first investigate the throughput performance of LTE-A uplink transmissions, employing optimal rate adaptation and compare SC-FDM to OFDM transmissions with ZF and MMSE receivers. Then, we extend our single-user MIMO

Channel State Information (CSI) feedback algorithms proposed for LTE downlink transmissions in [13] to LTE uplink and evaluate their performance comparing to the throughput bounds developed in [14]. We also highlight some important basic differences between link adaptation in LTE up- and downlink transmissions.

8.3.1 Optimal Rate Adaptation

As demonstrated in the previous section, SC-FDM provides a significant advantage in terms of PAPR over OFDM, thus relaxing linearity requirements of radio frequency power amplifiers for user equipments. Yet, this comes at the cost of coded BER degradation since channel diversity is not taken advantage of and the performance is mostly dominated by the weakest subcarrier of a user, especially with ZF receivers; cf., (8.9). Such diversity loss cannot be recovered from by forward-error-correction channel coding since the DFT-spreading applied with SC-FDM effectively causes an averaging over SINR observed on all scheduled subcarriers according to (8.6). As a consequence, SC-FDM over frequency selective channels achieves worse throughput than OFDM transmissions. This is demonstrated in Fig. 8.4, where we cross-compare the achievable rate, as defined in Eqs. (8.15) and (8.16), and the actual throughput of SC-FDM and OFDM transmission as obtained by the Vienna LTE-A uplink simulator. We consider single-user transmission over 5 MHz bandwidth assuming $N_{\mathrm{T}} = N_{\mathrm{R}} = 2$ antennas at the user and the base station and $N_{\mathrm{L}} = 2$ spatial layers. The precoder is selected as a scaled identity matrix: $\mathbf{F} = 1/\sqrt{N_{\mathrm{L}}}\,\mathbf{I}_{N_{\mathrm{L}}}$. We consider transmission over independent and identically distributed frequency-selective Rayleigh fading channels, emphasizing the difference between OFDM and SC-FDM. The achievable rate in bits per OFDM/SC-FDM symbol with Gaussian signalling and equal power allocation over subcarriers and spatial layers is calculated as

$$R^{\mathrm{OFDM}} = \sum_{k=1}^{N_{\mathrm{c}}} \sum_{l=1}^{N_{\mathrm{L}}} \log_2\left(1 + \left[\mathbf{SINR}^{\mathrm{OFDM},\,(l)}\right]_k\right), \tag{8.15}$$

$$R^{\mathrm{SC-FDM}} = N_{\mathrm{c}} \sum_{l=1}^{N_{\mathrm{L}}} \log_2\left(1 + \mathbf{SINR}^{\mathrm{SC-FDM},\,(l)}\right), \tag{8.16}$$

with the receiver-specific post-de-spreading (post-equalization) SINRs from (8.6) and (8.8), respectively.

We observe a significant loss of achievable rate of SC-FDM transmission compared to OFDM in Fig. 8.4, which is especially pronounced with ZF receivers due to noise enhancement. In Fig. 8.4, we also show the actual rate achieved by LTE uplink SC-FDM transmission with optimal rate adaptation and compare to the performance obtained by OFDM transmission; the corresponding curves are denoted by *LTE rate*. We determine the performance of optimal rate adaptation by simulating

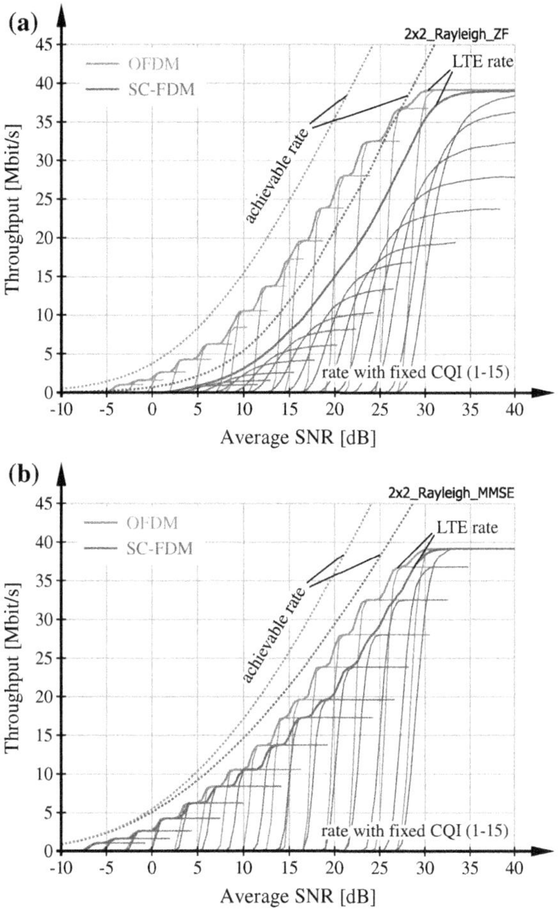

Fig. 8.4 Throughput comparison of OFDM and SC-FDM with rate adaptation and 2×2 Rayleigh fading channels of 5 MHz bandwidth. **a** ZF receivers. **b** MMSE receivers

all possible transmission rates, corresponding to CQI1–CQI15, and selecting at each subframe the largest rate that achieves error free transmission. The figure also shows the throughput of the individual CQIs. We observe a gap between the LTE throughput with OFDM and SC-FDM that is similar to the gap in terms of achievable rate. Notice that the performance loss with MMSE receivers is significantly smaller than with ZF detection, since MMSE avoids excessive noise enhancement.

We also observe in Fig. 8.4a that the gain achieved by instantaneous rate adaptation, as compared to rate adaptation based on the long-term average SNR, is much larger for ZF SC-FDM than for ZF OFDM; this is evident from the distance between the curves with rate adaptation (*LTE rate*) and the curves with *fixed CQI*. The reason for this behaviour is that the SNR of ZF SC-FDM shows strong variability around its means since it is dominated by the worst-case per-subcarrier SNR according to (8.9). The average SNR over subcarriers of ZF OFDM, however, approximately coincides with its mean value. This implies that the optimal CQI of ZF SC-FDM can vary significantly in-between subframes, as reflected by the large average SNR variation required to increase the rate with fixed CQI from zero to its respective maximum. Yet, for ZF OFDM the throughput of the individual CQIs follows almost a step function; hence, rate adaptation can be based on the long-term average SNR without substantial performance degradation. Notice, however, that instantaneous rate adaptation for ZF OFDM can be advantageous in case of frequency-correlated channels [15].

In case $N_{\mathrm{R}} > N_{\mathrm{L}}$, we can easily estimate the achievable rate of SC-FDM transmission: The per-layer SNR with ZF receivers is governed by the harmonic mean of the channel responses on the individual subcarriers, similar to (8.9)

$$\mathrm{SNR}_{\mathrm{ZF}}^{\mathrm{SC-FDM},\,(l)} = \frac{P_{\mathrm{S}}/P_{\mathrm{V}}}{\frac{1}{N_{\mathrm{c}}}\sum_{k=1}^{N_{\mathrm{c}}}\left[\left((\mathbf{H}_k\mathbf{F})^{\mathrm{H}}(\mathbf{H}_k\mathbf{F})\right)^{-1}\right]_{l,l}}, \tag{8.17}$$

where $\mathbf{H}_k \in \mathbb{C}^{N_{\mathrm{R}}\times N_{\mathrm{T}}}$ denotes the OFDM channel matrix at subcarrier k. Assuming constant precoding and semi-correlated Rayleigh fading

$$\mathbf{H}_k = \tilde{\mathbf{H}}_k\mathbf{C}_T^{\frac{1}{2}}, \quad \left[\tilde{\mathbf{H}}_k\right]_{i,j} \sim \mathcal{N}_{\mathbb{C}}(0,1), \tag{8.18}$$

with $\mathbf{C}_T \in \mathbb{C}^{N_{\mathrm{T}}\times N_{\mathrm{T}}}$ determining the spatial correlation at the user equipment side, the matrix in the denominator of (8.17) follows a complex-valued inverse Wishart distribution $\mathcal{W}_L^{\mathbb{C}}(N_{\mathrm{R}}, \mathbf{C})$ with N_{R} degrees of freedom and scale matrix $\mathbf{C} = \left(\mathbf{F}^{\mathrm{H}}\mathbf{C}_T\mathbf{F}\right)^{-1}$

$$\overline{\mathbf{H}} = \left((\mathbf{H}_k\mathbf{F})^{\mathrm{H}}(\mathbf{H}_k\mathbf{F})\right)^{-1} \sim \mathcal{W}_L^{\mathbb{C}}(N_{\mathrm{R}}, \mathbf{C}). \tag{8.19}$$

Letting $N_{\mathrm{c}} \to \infty$, we can replace the term in the denominator of (8.17) with its expected value

$$\frac{1}{N_{\mathrm{c}}}\sum_{k=1}^{N_{\mathrm{c}}}[\,\overline{\mathbf{H}}\,]_{l,l} \xrightarrow{N_{\mathrm{c}}\to\infty} \mathbb{E}\left([\,\overline{\mathbf{H}}\,]_{l,l}\right). \tag{8.20}$$

This expected value only exists in case $N_{\mathrm{R}} > N_{\mathrm{L}}$ [16]. For $N_{\mathrm{R}} = N_{\mathrm{L}}$, the diagonal elements of $\overline{\mathbf{H}}$ follow a heavy-tailed inverted Gamma distribution [17, 18] with a non-finite first moment. Yet, for $N_{\mathrm{R}} > N_{\mathrm{L}}$, which is a common situation in cellular networks since the base station is mostly equipped with far more antennas than the users, the expected value is finite and given by

$$\mathbb{E}\left([\,\overline{\mathbf{H}}\,]_{l,l}\right) = \frac{1}{N_{\mathrm{R}} - N_{\mathrm{L}}}[\mathbf{C}]_{l,l}. \tag{8.21}$$

Hence, we can estimate the achievable rate of SC-FDMA transmission over semi-correlated Rayleigh fading channels

$$R^{\text{SC-FDM}} \approx N_c \sum_{l=1}^{N_L} \log_2 \left(1 + \frac{P_S/P_V}{[\mathbf{C}]_{l,l}} (N_R - N_L)\right) \tag{8.22}$$

$$\approx N_c N_L \left[\log_2 \left(\frac{P_S/P_V}{\left(\prod_{l=1}^{N_L} [\mathbf{C}]_{l,l}\right)^{1/N_L}} \right) + \log_2 (N_R - N_L) \right]. \tag{8.23}$$

Here (8.23) resembles the high SNR approximation of the achievable rate of OFDM transmission with ZF detection as proposed in [19, Eq. (14)]; even more, for fixed N_L and letting N_R grow to infinity, (8.23) and [19, Eq. (14)] tend to the same limit, due to channel hardening on each subcarrier with a growing number of receive antennas.

In Fig. 8.5, we investigate the performance of the rate estimate (8.22) for $N_T = N_L = 4$ and varying number of receive antennas. We assume $\mathbf{F} = 1/\sqrt{N_L}\,\mathbf{I}_{N_L}$ and a correlation

$$\mathbf{C}_T = \begin{bmatrix} 1 & 0.9 & \dots & 0.9 \\ 0.9 & \ddots & & \vdots \\ \vdots & & & 0.9 \\ 0.9 & \dots & 0.9 & 1 \end{bmatrix},$$

and consider the smallest LTE bandwidth of $N_c = 72$ subcarriers. We observe that the proposed estimate performs very well even at this small bandwidth; notice, though, that a more realistic channel model with correlation over subcarriers may require larger bandwidth to validate the proposed estimate. Figure 8.5 also confirms the observation that single-user MIMO OFDM and SC-FDM with ZF detectors tend to the same limiting performance with increasing number of receive antennas.

This statement, however, will not hold true if the total number of layers grows proportionally with the number of receive antennas. For example, multi-user MIMO transmission with ZF equalization and single antenna users achieves only a diversity order of $N_R - N_L + 1$ [20], with N_L denoting the total number of layers being equal to the number of spatially multiplexed users. Hence, if N_L scales proportionally with N_R, channel hardening on each subcarrier will not occur and thus the performance of OFDM and SC-FDM will not coincide.

8.3.2 *Realistic Link Adaptation*

Instantaneous rate adaptation is an important tool for exploiting diversity of the wireless channel in LTE, by adjusting the transmission rate according to the current channel quality experienced by a user. LTE specifies a set of fifteen different MCSs; the selected MCS is signaled by the CQI.

LTE additionally supports spatial link adaptation by means of codebook based precoding with variable transmission rank. With this method, the precoding matrix

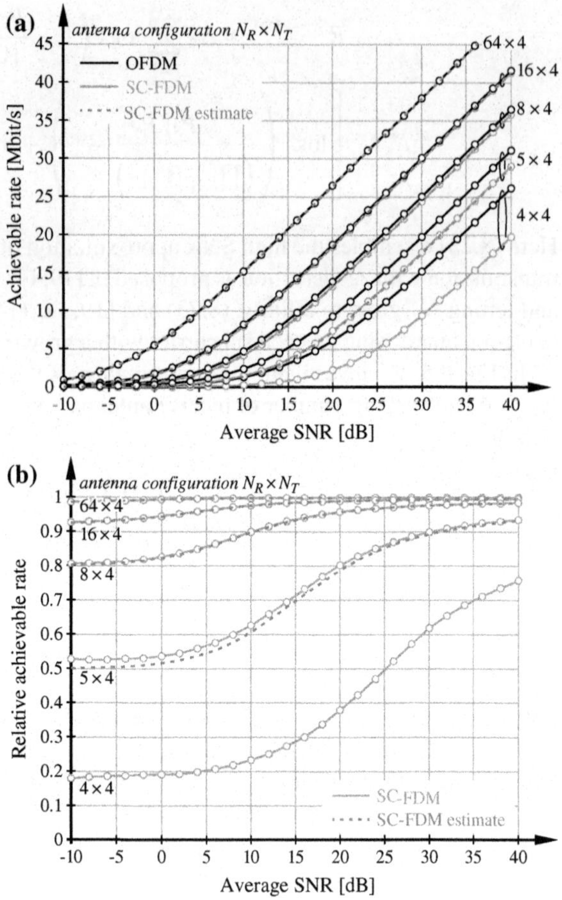

Fig. 8.5 Achievable rate of OFDM and SC-FDM with ZF equalizers and growing number of receive antennas at fixed number of streams $N_\mathrm{L} = 4$. **a** Comparison of achievable rate and the rate estimate of Eq. (8.22). **b** Relative achievable rate of SC-FDM with respect to OFDM

$\mathbf{F} \in \mathbb{C}^{N_\mathrm{T} \times N_\mathrm{L}}$ satisfying $\mathbf{F}^\mathrm{H}\mathbf{F} = 1/N_\mathrm{L}\,\mathbf{I}_{N_\mathrm{L}}$ is selected from a standard defined codebook $\mathcal{Q}_{N_\mathrm{L}}$ of scaled semi-unitary matrices; furthermore, the number of spatial layers N_L can be adjusted to achieve a favorable trade-off between beamforming and spatial multiplexing. The selected precoder and transmission rank are signaled, employing the PMI and the RI. In single-user MIMO LTE uplink transmission, the same precoder is applied on all RBs that are assigned to a specific user, whereas frequency-selective precoding is supported in LTE downlink.

There is a basic difference between the utilization of CQI, PMI and RI in up- and downlink directions of Frequency Division Duplex (FDD) systems. In downlink, the base station is reliant on CSI feedback from the users for link adaptation and multi-user scheduling [21], since channel reciprocity cannot be exploited in FDD. CQI, PMI and RI can be employed to convey such CSI from the users to the base station via dedicated feedback channels [15]. In the uplink, on the other hand, the base station can by itself determine the CSI, exploiting the Sounding Reference Signalss (SRSs)

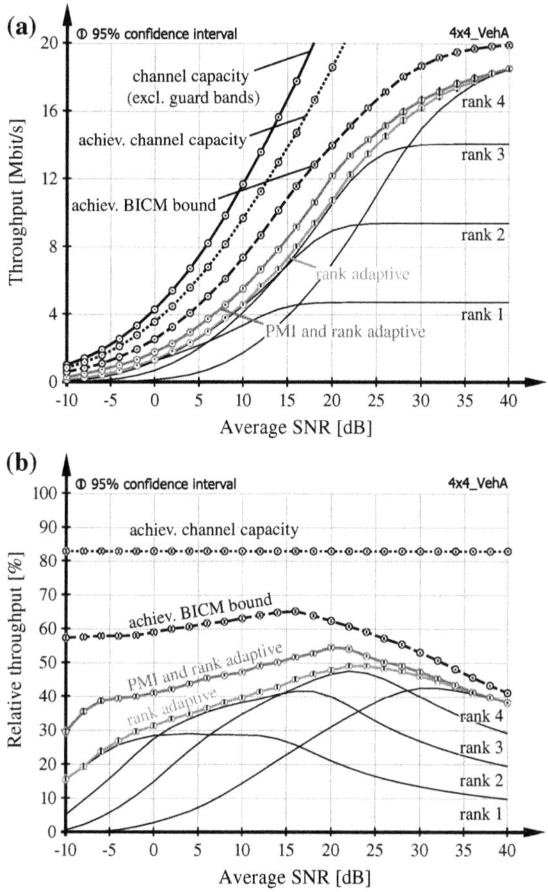

Fig. 8.6 Absolute and relative throughput of LTE uplink transmission over 4×4 VehA channels of 1.4 MHz bandwidth employing rate adaptation. We compare the performance of fixed rank, rank adaptive and PMI + rank adaptive transmission to the performance bounds proposed in [14]. **a** Absolute throughput. **b** Relative throughput

transmitted by the users. In this case, CQI, PMI and RI are employed by the base station to convey to the users its decision on link adaptation that has to be applied by the users during uplink transmission.

In Fig. 8.6, we evaluate the performance of single-user MIMO LTE uplink transmissions over $N_T = N_R = 4$ antennas with link adaptation, 1.4 MHz system bandwidth and ZF receiver. We do not consider signaling delays between the base station and the user. We employ the VehA channel model [9] and compare the absolute and relative (to channel capacity) throughput to the performance bounds proposed in [14]. Notice that the simulation setup is the same as employed in [14] for the investigation of LTE downlink transmission, thus, facilitating the comparison of up- and downlink performance. *Channel capacity* is obtained by applying Singular Value Decomposition (SVD)-based transceivers and water-filling power allocation over subcarriers and spatial streams. Notice that we do not account for guard band and CP overheads when calculating the channel capacity; that is, we only consider subcarriers that are available for data transmission. The *achievable channel capacity* takes overhead for pilot symbols (DMRS and SRS) into account, corresponding to a loss of 16.7 % in our simulation. The *achievable Bit Interleaved Coded Modulation (BICM) bound* additionally accounts for equal power allocation, codebook-based precoding, ZF detection as well as the applied BICM architecture as detailed in [14].

The performance of LTE uplink transmissions with full link adaptation (*PMI and rank adaptive*) is similar to the *achievable BICM bound* but shifted by approximately 3 dB. Notice that the saturation value is not the same because the highest CQI of LTE achieves 5.55 bit/channel use, whereas the BICM bound saturates at 6 bit/channel use. We also show the performance of LTE uplink when restricted to fixed precoding (*rank adaptive*) and fixed rank transmission (*rank 1, 2, 3, 4*). We observe that rank adaptive transmission even outperforms the envelope of the fixed rank transmission curves, since instantaneous rank adaptation selects the optimal rank in each subframe independently. In terms of relative throughput, we observe that LTE uplink with ZF receivers achieves around 40–50 % of channel capacity. Remember, though, that this does not include CP and guard band overheads.

8.4 Reference Symbols

In LTE uplink two types of reference signals are standardized. For CE and coherent detection, DMRS are exploited, while SRS are employed for channel sounding to enable frequency selective scheduling. For the purpose of CE we will consider DMRS only. The reference symbols are defined in [1] and are explained in more detail in [22, 23]. As shown in Fig. 8.7, DMRSs are multiplexed in the resource grid at OFDM symbol time $n = 3$ in every slot. In a Physical Uplink Shared Channel (PUSCH) transmission of the LTE-A uplink, an DMRS occupies all scheduled subcarriers. We assume that the user is assigned all N_c subcarriers starting at 0, i.e., $k \in \{0, 1, \ldots, N_c - 1\}$. We denote the Zadoff-Chu Zadoff-Chu (ZC) base sequence on N_c subcarriers for one slot by $\mathbf{x}_{\mathscr{Z}} \in \mathbb{C}^{N_c \times 1}$. The base sequences $\mathbf{x}_{\mathscr{Z}}$ are complex-valued exponential sequences lying on the unit circle fulfilling

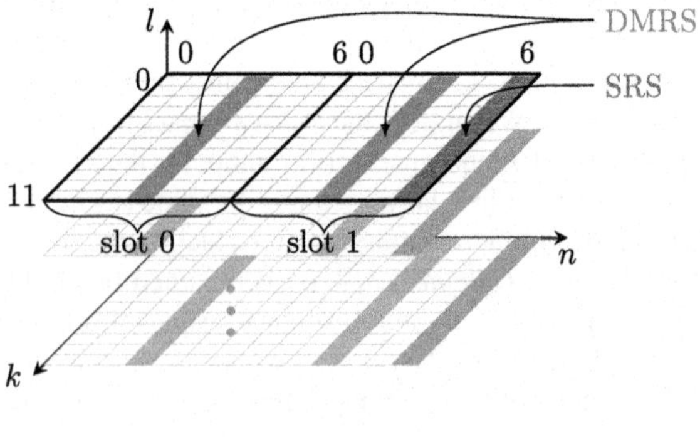

Fig. 8.7 The LTE-A uplink reference symbol allocation in two slots (one subframe)

$$|[\mathbf{x}_{\mathscr{X}}]_k| = 1. \tag{8.24}$$

In LTE-A the DMRS of different transmission layers in the same slot are orthogonal in terms of FD-CDM [22]. This is obtained by cyclically shifting the base sequence. Similar to [24], DMRS on layer l for one slot is given by

$$\mathbf{X}_{\mathscr{P}}^{(l)} = \text{Diag}\left(\mathbf{x}_{\mathscr{P}}^{(l)}\right) = \mathbf{T}^{(l)}\text{Diag}\left(\mathbf{x}_{\mathscr{X}}\right), \tag{8.25}$$

with the cyclic shift operator

$$\mathbf{T}^{(l)} = \text{Diag}\left(e^{j0}, \ldots, e^{j\alpha_l k}, \ldots, e^{j\alpha_l (N_c - 1)}\right), \tag{8.26}$$

and the layer dependent cyclic shift α_l. We further conclude from (8.24)–(8.26) that $(\mathbf{X}_{\mathscr{P}}^{(l)})^H = (\mathbf{X}_{\mathscr{P}}^{(l)})^{-1}$ which implies $\left(\mathbf{X}_{\mathscr{P}}^{(l)}\right)^H \mathbf{X}_{\mathscr{P}}^{(l)} = \mathbf{I}_{N_c}$. Exploiting (8.24), the product of two DMRSs from layers l and u with $l, u \in \{1, \ldots, N_L\}$, becomes

$$\begin{aligned}\left(\mathbf{X}_{\mathscr{P}}^{(l)}\right)^H \mathbf{X}_{\mathscr{P}}^{(u)} &= \left(\mathbf{T}^{(l)}\right)^H \mathbf{T}^{(u)}\, \text{Diag}\left(\mathbf{x}_{\mathscr{X}}\right)^H \text{Diag}\left(\mathbf{x}_{\mathscr{X}}\right) \\ &= \text{Diag}\left(e^{j0} \;\ldots\; e^{j\Delta\alpha_{u,l} k} \;\ldots\; e^{j\Delta\alpha_{u,l}(N_c - 1)}\right),\end{aligned} \tag{8.27}$$

with $\Delta\alpha_{u,l} = \alpha_u - \alpha_l$ being the cyclic phase shift between DMRS of two different spatial layers. The FD-CDM orthogonality can therefore be exploited as

$$\text{tr}\left(\left(\mathbf{X}_{\mathscr{P}}^{(u)}\right)^H \mathbf{X}_{\mathscr{P}}^{(l)}\right) = \left(\mathbf{x}_{\mathscr{P}}^{(u)}\right)^H \mathbf{x}_{\mathscr{P}}^{(l)} = \begin{cases} N_c & \text{for } u = l \\ 0 & \text{for } u \neq l. \end{cases} \tag{8.28}$$

After the transmission over a frequency selective channel, this orthogonality has to be exploited to separate all effective MIMO channels at the receiver.

8.5 Channel Estimation

For channel estimation we exploit the system model only at symbol times, where reference signals are allocated. For normal CP length this is the 4th symbol in each slot, i.e., $n = 3$ as shown in Fig. 8.7. Since we estimate the channel only at this single symbol time per slot, interpolation in time has to be carried out to obtain channel estimates for the whole resource grid. As illustrated in Fig. 8.1, the DMRS are added after DFT spreading, right before precoding. As the channel estimation takes place after the receiver's DFT, just before equalization, the system model for CE amounts to an OFDM system. The system model (8.2) therefore reads as

$$\mathbf{y} = \mathbf{H}^{(\text{eff})}\mathbf{x}_{\mathcal{P}} + \mathbf{v}', \tag{8.29}$$

with (pre-equalization) noise

$$\mathbf{v}' = \left(\mathbf{I}_{N_\text{R}} \otimes \mathbf{M}^H \mathbf{W}_{N_\text{FFT}} \mathbf{P}_\text{remCP}\right)\mathbf{v}, \tag{8.30}$$

and the stacked vector $\mathbf{x}_{\mathcal{P}}$ consisting of DMRS $\mathbf{x}_{\mathcal{P}}^{(l)} \in \mathbb{C}^{N_\text{c}\times 1}$ from all active spatial layers $l \in \{1, \ldots, N_\text{L}\}$, i.e., $\mathbf{x}_{\mathcal{P}} = \left(\left(\mathbf{x}_{\mathcal{P}}^{(1)}\right)^T, \ldots, \left(\mathbf{x}_{\mathcal{P}}^{(N_\text{L})}\right)^T\right)^T$. To consider the received signal separately for each receive antenna i, we can select the according part from $\mathbf{y}$ by left multiplying with the selector matrix $\mathbf{S}^{(i)}$ from (8.6). The received signal $\mathbf{y}^{(i)} = \mathbf{S}^{(i)}\mathbf{y}$ on antenna i is given by

$$\begin{aligned}\mathbf{y}^{(i)} &= \left(\mathbf{H}_{i,1}^{(\text{eff})}, \ldots, \mathbf{H}_{i,N_\text{L}}^{(\text{eff})}\right)\mathbf{x}_{\mathcal{P}} + \mathbf{v}'^{(i)} \\ &= \sum_{l=1}^{N_\text{L}} \mathbf{H}_{i,l}^{(\text{eff})}\mathbf{x}_{\mathcal{P}}^{(l)} + \mathbf{v}'^{(i)},\end{aligned} \tag{8.31}$$

with the pre-equalization noise $\mathbf{v}'^{(i)} = \mathbf{S}^{(i)}\mathbf{v}'$ on receive antenna i and $\mathbf{H}_{i,l}^{(\text{eff})} = \mathbf{S}^{(i)}\mathbf{H}^{(\text{eff})}\left(\mathbf{S}^{(l)}\right)^T$ being the (i, l)th block of $\mathbf{H}^{(\text{eff})}$. Since $\mathbf{H}_{i,l}^{(\text{eff})}$ is diagonal, we exploit the relations $\mathbf{X}_{\mathcal{P}}^{(l)} = \text{Diag}\left(\mathbf{x}_{\mathcal{P}}^{(l)}\right)$ and $\mathbf{H}_{i,l}^{(\text{eff})} = \text{diag}\left(\mathbf{H}_{i,l}^{(\text{eff})}\right)$ to estimate a channel vector rather than a matrix and rearrange terms in (8.31) leading to

$$\begin{aligned}\mathbf{y}^{(i)} &= \sum_{l=1}^{N_\text{L}} \mathbf{X}_{\mathcal{P}}^{(l)}\mathbf{H}_{i,l}^{(\text{eff})} + \mathbf{v}'^{(i)} \\ &= \underbrace{\left(\mathbf{X}_{\mathcal{P}}^{(1)}, \ldots, \mathbf{X}_{\mathcal{P}}^{(N_\text{L})}\right)}_{\mathbf{X}_{\mathcal{P}}}\mathbf{H}_i^{(\text{eff})} + \mathbf{v}'^{(i)},\end{aligned} \tag{8.32}$$

with the stacked vector $\mathbf{H}_i^{(\text{eff})} = \left(\left(\mathbf{H}_{i,1}^{(\text{eff})}\right)^T, \ldots, \left(\mathbf{H}_{i,N_\text{L}}^{(\text{eff})}\right)^T\right)^T$ of all effective channels from N_L active layers to receive antenna i for which we will drop the subscript in the following.

8.5.1 Minimum Mean Square Error Estimation

First we present a MMSE estimator where we exploit (8.32) and estimate the stacked vector $\mathbf{h}^{(i)}$ consisting of effective channels from all N_L active layers to receive antenna i. The MMSE CE for receive antenna i is given by

$$\hat{\mathbf{h}}_\text{MMSE}^{(i)} = \underset{\hat{\mathbf{h}}^{(i)}}{\text{argmin}}\ \mathbb{E}\left\{\left\|\hat{\mathbf{h}}^{(i)} - \mathbf{h}^{(i)}\right\|_2^2\right\}, \tag{8.33}$$

which leads to the well-known solution [25]

$$\hat{\mathbf{h}}^{(i)}_{\text{MMSE}} = \left(\sigma^2_{\mathbf{v}^{(i)}} \mathbf{C}^{-1}_{\mathbf{h}^{(i)}} + \mathbf{X}^H_{\mathscr{P}} \mathbf{X}_{\mathscr{P}}\right)^{-1} \mathbf{X}^H_{\mathscr{P}} \mathbf{y}^{(i)}, \tag{8.34}$$

with $\mathbf{C}_{\mathbf{h}^{(i)}} = \mathbb{E}\{\mathbf{h}^{(i)}\mathbf{h}^{(i)H}\}$.

8.5.2 *Correlation Based Estimation*

As a low complexity approach, we correlate (matched filter) the received signal with the reference symbol of layer l to obtain a channel estimate for the effective channel $\mathbf{h}^{(i,l)}$ from layer l to receive antenna i

$$\tilde{\mathbf{h}}^{(i,l)} = \left(\mathbf{X}^{(l)}_{\mathscr{P}}\right)^H \mathbf{y}^{(i)}. \tag{8.35}$$

Inserting our system model (8.32) and exploiting (8.27), we obtain

$$\begin{aligned} \tilde{\mathbf{h}}^{(i,l)} &= \left(\mathbf{X}^{(l)}_{\mathscr{P}}\right)^H \textstyle\sum_{u=1}^{N_\text{L}} \mathbf{X}^{(u)}_{\mathscr{P}} \mathbf{h}^{(i,u)} + \left(\mathbf{X}^{(l)}_{\mathscr{P}}\right)^H \mathbf{v}'^{(i)} \\ &= \mathbf{h}^{(i,l)} + \underbrace{\sum_{\substack{u=1\\u\neq l}}^{N_\text{L}} \left(\mathbf{T}^{(l)}\right)^H \mathbf{T}^{(u)} \mathbf{h}^{(i,u)}}_{\text{inter-layer interference}} + \tilde{\mathbf{v}}^{(i)}. \end{aligned} \tag{8.36}$$

Here $\tilde{\mathbf{v}}^{(i)}$ has the same distribution as $\mathbf{v}'^{(i)}$ since $(\mathbf{X}^{(l)}_{\mathscr{P}})^H$ is unitary and introduces phase changes only, cf. (8.25). Due to the allocation of DMRS on the same time and frequency resources on different spatial layers, the initial estimate $\tilde{\mathbf{h}}^{(i,l)}$ of one effective MIMO channel actually consists of a superposition of all N_L effective MIMO channels to receive antenna i. The unintentional contributions in (8.36), from layers $u \neq l$ are inter-layer interference, making it unsuited as initial estimate for coherent detection. Different methods to separate the different effective MIMO channels in (8.36) will be presented in the following.

8.5.2.1 DFT Based Channel Estimation

A well known approach for CE in LTE-A uplink is DFT based estimation [23], which aims to separate the MIMO channels contributing to (8.36) in time domain. For this the individual cyclic shift of each DMRS is exploited. Applying a DFT on the receive signal, the individual phase shifts will translate into a shifts in time domain. This makes a separation of Carrier to Interference Ratio (CIR)s from different MIMO channels possible by windowing. Applying an Inverse Discrete Fourier Transform (IDFT) on the initial estimate yields

$$\begin{aligned}\tilde{\mathbf{h}}_t^{(i,l)} &= \mathbf{W}_{N_c}^H \tilde{\mathbf{h}}^{(i,l)} \\ &= \mathbf{W}_{N_c}^H \sum_{u=1}^{N_L} \left(\mathbf{T}^{(l)}\right)^H \mathbf{T}^{(u)} \mathbf{W}_{N_c} \mathbf{h}_t^{(i,u)} + \mathbf{W}_{N_c}^H \tilde{\mathbf{v}}^{(i)} \\ &= \sum_{u=1}^{N_L} \boldsymbol{\Pi}^{\Delta N} \mathbf{h}_t^{(i,u)} + \mathbf{W}_{N_c}^H \tilde{\mathbf{v}}^{(i)},\end{aligned} \tag{8.37}$$

with permutation matrix $\boldsymbol{\Pi}$ which is obtained by cyclically left shifting the identity matrix as in [24]. The power of the permutation matrix is $\Delta N = \Delta\alpha_{u,l} N_c / 2\pi$ and depends on the difference in phase shift between layer u and l from Sect. 8.4. The time domain estimate $\tilde{\mathbf{h}}_t^{(i,l)}$ consists of N_L CIRs, from all N_L active layers to receive antenna i. An example of $\tilde{\mathbf{h}}_t^{(i,l)}$ for two active spatial layers is shown in Fig. 8.8a with the two contributing CIRs, $\mathbf{h}_t^{(i,1)}$ from layer one and $\mathbf{h}_t^{(i,2)}$ from layer two. Due to the cyclic shift of DMRS, these N_L contributions are shifted relative to each other such that the distance in time between them is maximized. The intended CIR, from layer one, is centered around the origin and can be separated by windowing. In case of a narrow transmission bandwidth or a long CIR, the time domain MIMO channels overlap, cf. Fig. 8.8b, which is referred to as CIR leakage. DFT based CE therefore leads to inferior performance for small scheduled user bandwidths or channels with high delay spread. To separate the different MIMO channels, windowing with window size 2β is carried out

$$\left[\check{\mathbf{h}}_t^{(i,l)}\right]_m = \begin{cases} \left[\tilde{\mathbf{h}}_t^{(i,l)}\right]_m & \text{for } m < \beta \text{ or } m > N_c - \beta \\ 0 & \text{otherwise,} \end{cases} \tag{8.38}$$

where β is chosen to be the CP length N_{cp}. To yield a frequency domain CE, a DFT is applied on the separated CIR

$$\hat{\mathbf{h}}_{\text{DFT}}^{(i,l)} = \mathbf{W}_{N_c} \check{\mathbf{h}}_t^{(i,l)}. \tag{8.39}$$

8.5.2.2 Averaging

In the initial estimate (8.36) obtained by correlation, we observe that the interference term is characterized by (8.27), which is a rotation on the unit circle in the complex plane. The phase shift between two adjacent points in frequency is determined by $\Delta\alpha_{u,l}$ and the number of distinct points on the unit circle is given by $\gamma = 2\pi/\Delta\alpha_{u,l}$. As the points are equally spaced on the unit circle, summing (8.27) over a complete turn will cancel the interference. For this we have to sum, or average, over γ consecutive elements of $\tilde{\mathbf{h}}^{(i,l)}$.

The phase shift $\Delta\alpha_{u,l} = \alpha_u - \alpha_l$ and therefore the periodicity γ of the interference is dependent on the two layers u and l. When estimating all MIMO channels, all

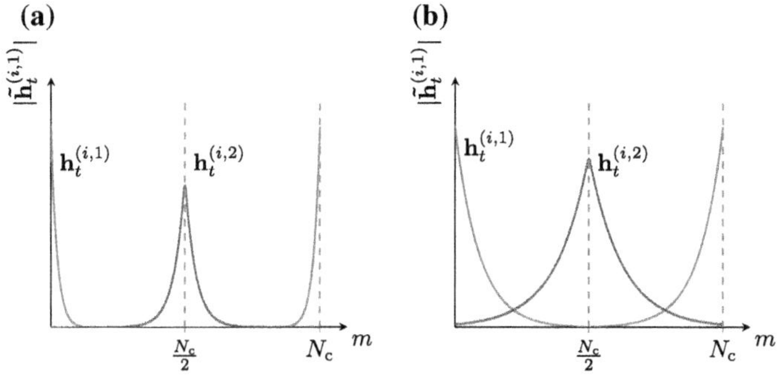

Fig. 8.8 Time domain separation of CIRs for DFT based CE with $N_L = 2$. **a** N_c large compared to CIR length. **b** N_c small compared to CIR length

possible combinations of $u, l \in \{1, \ldots, N_L\}$ appear in the interference term of (8.36). Therefore different values of $\Delta\alpha_{u,l} \in \{0, \pm\pi/2, \pm\pi\}$ will occur. According to [1], a relation between the possible occurring phase shifts and the number of active spatial layers can be established. In order to cancel all interference terms at once we have to sum over the largest possible number of points on the unit circle, i.e., the largest possibly occurring periodicity which we denote by $\bar{\gamma}$. For a given number of active streams $N_L = 1, 2, 3, 4$ this largest periodicity equals to $\bar{\gamma} = 1, 2, 4, 4$, respectively.

From this we conclude that the sum of $\bar{\gamma}$ consecutive elements in (8.27) will evaluate to zero, independent on the actual layers u and l. As a consequence for CE, when we sum (average) the initial estimate (8.36) over $\bar{\gamma}$ adjacent subcarriers k, all products of different ($u \neq l$) DMRS will vanish and the FD-CDM orthogonality will be preserved [26]. For this method however, we have to assume the channel to be piecewise flat in frequency domain, since only the average channel gain is estimated for $\bar{\gamma}$ consecutive subcarriers.

Within this context, the meaning of FD-CDM orthogonality can be further explained. DMRS are orthogonal in frequency domain, such that the inner product of two reference symbols from different layers is zero, as shown in (8.28). In terms of correlation based estimation, evaluating this inner product would also cancel all the inter-layer interference. However, with this method there would be only a single estimate of channel gain for all OFDM subcarriers. This estimate is only a useful one, if the channel is frequency flat. On a frequency selective channel, the inner product is split up into pieces of length of $\bar{\gamma}$. In this way, still the inter-layer interference is cancelled and a channel estimate for each part is obtained, representing the frequency selectivity of the channel.

Further, it suffices to sum over all points of interference on the unit circle while it actually does not matter where the summation starts and stops. Therefore we can augment this method to a sliding average. Here, the averaging window of size $\bar{\gamma}$ is shifted subcarrier by subcarrier over the scheduled bandwidth. As explained in [27], assuming the channel to be frequency flat on these $\bar{\gamma}$ consecutive subcarriers, the sliding average is given by

$$\left[\hat{\mathbf{h}}_{\mathrm{SAV}}^{(i,l)}\right]_k = \frac{1}{\bar{\gamma}^2} \sum_{t=k-\bar{\gamma}+1}^{k} \sum_{j=t}^{t+\bar{\gamma}-1} \left[\tilde{\mathbf{h}}^{(i,l)}\right]_j, \tag{8.40}$$

for $\bar{\gamma} \leq k \leq N_\mathrm{c} - \bar{\gamma} + 1$. The second sum describes the averaging of $\bar{\gamma}$ elements while the first sum describes the shift of this averaging window.

8.5.2.3 Quadratic Smoothing

The initial correlator estimate solves the decoupled least squares estimation problem for the channel $\mathbf{h}^{(i,l)}$ from layer l to receive antenna i, ignoring the other channels

$$\tilde{\mathbf{h}}^{(i,l)} = \underset{\mathbf{h}^{(i,l)}}{\mathrm{argmin}} \left\| \mathbf{y}^{(i)} - \mathbf{X}_{\mathscr{P}}^{(l)} \mathbf{h}^{(i,l)} \right\|_2^2. \tag{8.41}$$

The unitary structure of the DMRS yields

$$\begin{aligned} \tilde{\mathbf{h}}^{(i,l)} &= \left(\left(\mathbf{X}_{\mathscr{P}}^{(l)}\right)^H \mathbf{X}_{\mathscr{P}}^{(l)} \right)^{-1} \left(\mathbf{X}_{\mathscr{P}}^{(l)}\right)^H \mathbf{y}^{(i)} \\ &= \left(\mathbf{X}_{\mathscr{P}}^{(l)}\right)^H \mathbf{y}^{(i)}, \end{aligned} \tag{8.42}$$

which is identical to (8.35). This observation together with the previous assumption of correlated subcarriers leads to the idea of augmenting (8.41) by a frequency smoothness constraint. Quadratic smoothing [28] with a matrix $\mathbf{Q} \in \mathbb{R}^{(N_\mathrm{c}-1)\times N_\mathrm{c}}$

$$\mathbf{Q} = \begin{pmatrix} -1 & 1 & & & \\ & -1 & 1 & & \\ & & \ddots & \ddots & \\ & & & -1 & 1 \end{pmatrix} \tag{8.43}$$

can be formulated as a convex minimization problem

$$\hat{\mathbf{h}}_{\mathrm{QS}}^{(i,l)} = \underset{\mathbf{h}^{(i,l)}}{\mathrm{argmin}} \left\| \mathbf{y}^{(i)} - \mathbf{X}_{\mathscr{P}}^{(l)} \mathbf{h}^{(i,l)} \right\|_2^2 + \lambda \left\| \mathbf{Q}\mathbf{h}^{(i,l)} \right\|_2^2, \tag{8.44}$$

with the known result

$$\hat{\mathbf{h}}_{\mathrm{QS}}^{(i,l)} = \left(\mathbf{I}_B + \lambda \mathbf{Q}^H \mathbf{Q}\right)^{-1} \underbrace{\left(\mathbf{X}_{\mathscr{P}}^{(l)}\right)^H \mathbf{y}^{(i)}}_{\tilde{\mathbf{h}}^{(i,l)}}. \tag{8.45}$$

Similar to (8.40) this can be interpreted as another way to cope with the inter-layer interference in (8.36) by post processing. This method does not use the DMRS structure explicitly but suppresses the interference by smoothing. It is therefore not able to cancel the complete inter-layer interference.

The value of λ can be interpreted as the prior about the channel [29] as it reflects its frequency selectivity. The choice of λ poses another optimization problem where the optimal value depends on the number of used layers N_L. The minimum w.r.t. λ is rather broad and minima obtained for different channel models lie close to each other. This allows λ to be fixed for practical realizations such that the matrix inverse in (8.45) can be precomputed.

8.5.3 MSE and BER Comparison

We assume a single user 2×2 MIMO transmission with $N_c = 72$ subarriers, a fixed number of layers $N_L = 2$ and a TU channel model [9] at zero speed. We show the MSE curves of the proposed estimators in Fig. 8.9a. The DFT based CE (*D-bCE*)

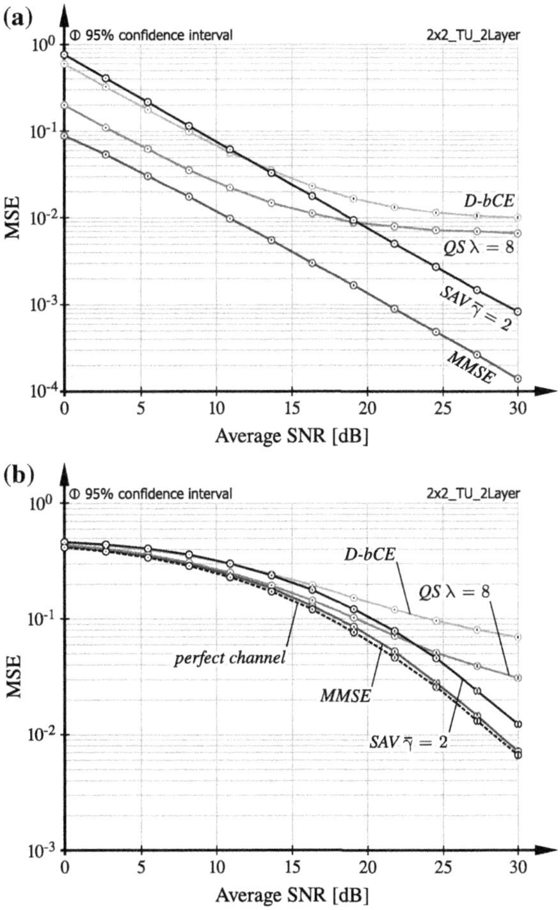

Fig. 8.9 Channel estimation performance comparison for block fading. **a** MSE curves of the proposed CE methods for a 2×2 transmission with $N_L = 2$ on a TU channel. **b** Resulting uncoded BER curves for the proposed CE schemes for 4-QAM and ZF equalization

shows the highest error flow of all estimators at high SNR while the *MMSE* estimator, of course shows best performance over the whole SNR range. Compared to these two methods, the Sliding-Averaging estimator (8.40), denoted by *SAV*, encounters an 8 dB SNR penalty when compared to *MMSE*, but comes closest to *MMSE* performance at high SNR. The quadratic smoothing estimation is denoted by *QS* and shows a significant improvement for low SNRs because it smooths over several observed channel coefficients. Quadratic smoothing performs uniformly better than *D-bC*E over the whole SNR range and comes close to 4 dB to *MMSE* at low SNR. The high error floor shows that *QS* is not able to cancel all the inter-layer interference.

In terms of BER performance, at high SNR, naturally the estimation method with lowest MSE leads to the smallest BER. At low SNR, the difference in CE MSE translates into very small differences in BER, meaning we cannot gain too much from a good low SNR MSE performance of *QS* or *MMSE* estimation. Considering estimation complexity and that *MMSE* as well as *QS* require prior channel knowledge, *SAV* estimation provides a good complexity performance trade-off.

References

1. 3rd Generation Partnership Project (3GPP), Evolved Universal Terrestrial Radio Access (E-UTRA) Physical Channels and Modulation, TS 36.211 (2015)
2. A. Wilzeck, Q. Cai, M. Schiewer, T. Kaiser, Effect of multiple carrier frequency offsets in MIMO SC-FDMA systems, in *Proceedings of the International ITG/IEEE Workshop on Smart Antennas* (2007)
3. E. Zöchmann, S. Pratschner, S. Schwarz, M. Rupp, MIMO transmission over high delay spread channels with reduced cyclic prefix length, in *19th International ITG Workshop on Smart Antennas (WSA)* (2015)
4. E. Zöchmann, S. Pratschner, S. Schwarz, M. Rupp, Limited feedback in OFDM systems for combating ISI/ICI caused by insufficient cyclic prefix length, in *Proceedings of the IEEE Asilomar Conference on Signals, Systems, and Computers* (2014)
5. E. Zöchmann, S. Schwarz, S. Pratschner, L. Nagel, M. Lerch, M. Rupp, Exploring the physical layer frontiers of cellular uplink - the Vienna LTE-A simulator, submitted to Eurasip Journal on Wireless Communications and Networking (2015). arXiv:1509.02351v1 [cs.IT]
6. H.G. Myung, J. Lim, D. Goodman, Single carrier FDMA for uplink wireless transmission. IEEE Veh. Technol. Mag. **1**(3), 30–38 (2006)
7. T. Jiang, Y. Wu, An overview: peak-to-average power ratio reduction techniques for OFDM signals. IEEE Trans. Broadcast **54**(2), 257 (2008)
8. P.S. Bullen, *Handbook of Means and Their Inequalities* (Springer Science & Business Media, Berlin, 2003)
9. Technical Specification Group Radio Access Network, Deployment aspects, 3rd Generation Partnership Project (3GPP). Technical report, TR 25.943 Version 12.0.0 (2014). http://www.3gpp.org/DynaReport/25943.htm
10. J.J. Sánchez-Sánchez, M. Aguayo-Torres, U. Fernández-Plazaola, BER analysis for zero-forcing SC-FDMA over Nakagami-m fading channels. IEEE Trans. Veh. Technol. **60**(8), 4077–4081 (2011)
11. J.J. Sánchez-Sánchez, U. Fernández-Plazaola, M.C. Aguayo-Torres, BER analysis for SC-FDMA over Rayleigh fading channels, in *6th IEEE International Conference on Broadband and Biomedical Communications (IB2Com)* (2011), pp. 43–47

12. M. Geles, A. Averbuch, O. Amrani, D. Ezri, Performance bounds for maximum likelihood detection of single carrier FDMA. IEEE Trans. Commun. **60**(7), 1945–1952 (2012)
13. S. Schwarz, M. Rupp, Throughput maximizing feedback for MIMO OFDM based wireless communication systems, in *IEEE International Workshop on Signal Processing Advances in Wireless Communications (SPAWC)* (San Francisco, 2011), pp. 316–320
14. S. Schwarz, M. Simko, M. Rupp, On performance bounds for MIMO OFDM based wireless communication systems, in *Signal Processing Advances in Wireless Communications SPAWC* (San Francisco, 2011), pp. 311–315
15. S. Schwarz, C. Mehlführer, M. Rupp, Calculation of the spatial preprocessing and link adaption feedback for 3GPP UMTS/LTE, in *6th Conference on Wireless Advanced (WiAD)* (London, 2010), pp. 1–6
16. D. Maiwald, D. Kraus, Calculation of moments of complex Wishart and complex inverse Wishart distributed matrices. IEE Proc. Radar Sonar Navig. **147**(4), 162–168 (2000)
17. A. Gupta, D. Nagar, *Matrix Variate Distributions*, vol. 104, Monographs and Surveys in Pure and Applied Mathematics (Chapman & Hall/CRC, Oregon, 2000)
18. R. Xu, Z. Zhong, J.-M. Chen, B. Ai, Bivariate gamma distribution from complex inverse Wishart matrix. IEEE Commun. Lett. **13**(2), 118–120 (2009)
19. D. Gore, R. Heath, A. Paulraj, Transmit selection in spatial multiplexing systems. IEEE Commun. Lett. **6**(11), 491–493 (2002)
20. A. Hedayat, A. Nosratinia, Outage and diversity of linear receivers in flat-fading MIMO channels. IEEE Trans. Signal Process. **55**(12), 5868–5873 (2007)
21. S. Schwarz, C. Mehlführer, M. Rupp, Low complexity approximate maximum throughput scheduling for LTE, in *Conference Record of the Forty Fourth Asilomar Conference on Signals, Systems, and Computers* (Pacific Grove, 2010), pp. 1563–1569
22. X. Hou, Z. Zhang, H. Kayama, DMRS design and channel estimation for LTE-advanced MIMO uplink, in *IEEE 70th Vehicular Technology Conference Fall (VTC 2009-Fall)* (2009), pp. 1–5
23. X. Zhang, Y. Li, Optimizing the MIMO channel estimation for LTE-advanced uplink, in *International Conference on Connected Vehicles and Expo (ICCVE)* (2012), pp. 71–76
24. C.-Y. Chen, D. Lin, Channel estimation for LTE and LTE-A MU-MIMO uplink with a narrow transmission band, in *IEEE International Conference on Acoustics, Speech and Signal Processing (ICASSP)* (2014), pp. 6484–6488
25. S.M. Kay, *Fundamentals of Statistical Signal Processing: Estimation Theory* (Prentice Hall, Upper Saddle River, 1993)
26. E. Dahlman, S. Parkvall, J. Skold, *4G: LTE/LTE-Advanced for Mobile Broadband* (Academic Press, Waltham, 2013)
27. S. Pratschner, E. Zöchmann, M. Rupp, Low complexity estimation of frequency selective channels for the LTE-A uplink. IEEE Wirel. Commun. Lett. **4**, 673–676 (2015)
28. S. Boyd, L. Vandenberghe, *Convex Optimization* (Cambridge University Press, Cambridge, 2009)
29. R. Gribonval, Should penalized least squares regression be interpreted as maximum a posteriori estimation? IEEE Trans. Signal Process. **59**(5), 2405–2410 (2011)

Chapter 9
Link Level Measurements

Martin Lerch

Simulators are a powerful tool to evaluate current mobile communications standards and to investigate new transmission and receiver techniques. The next step should then be to perform measurements in a real-world environment. Therefore, a very flexible wireless testbed, the Vienna MIMO Testbed was developed at TU Wien which allows for Long Term Evolution (LTE) measurements including the Vienna LTE-A Link Level Simulators as signal source and receiver. In Sect. 9.2 we provide an overview of the Vienna MIMO Testbed and the methodologies to perform LTE measurements. A measurement campaign that evaluated the impact of the transmit antenna configuration on the performance of the LTE Downlink (DL) is presented in Sect. 9.3. In order to perform reproducible and fully controllable measurements at velocities of up to 400 km/h our testbed was extended by an antenna on a rotary unit. In Sect. 9.4, this unit and the corresponding measurement methodology is described followed by a measurement campaign comparing different channel interpolation techniques for LTE Uplink (UL) transmissions.

9.1 Introduction

The decades after Marconi's invention were filled with wireless experiments. Although we understand many physical phenomena of wireless propagations today much better than in the past, our channel models still capture only a part of the complex physical propagation process. Nevertheless, in the last two decades, it has become a common method to entirely skip experimental validation and trust existing channel models when designing mobile communication systems. As the complexity of mobile communication standards also increases, simulation methods appear to be the Holy Grail to address open design questions. While these methods deliver

M. Lerch (✉)
Institute of Telecommunications, TU Wien, Vienna, Austria
e-mail: mlerch@nt.tuwien.ac.at

M. Rupp et al., *The Vienna LTE-Advanced Simulators*,
Signals and Communication Technology, DOI 10.1007/978-981-10-0617-3_9

quantitative results in acceptable time, many important issues are simplified or not modeled at all, trading off timely results against accuracy. Converting new algorithmic ideas into hardware on the other hand is quite time consuming and often lacks flexibility so that experimental evaluation remains no longer an attractive choice. With our testbed approach [1], we essentially combine the advantages of both worlds: design flexibility and timeliness under true physical conditions.

Although LTE cellular systems are already being rolled out and operated in many countries around the world, there are still unresolved issues in transmission technology. Focusing on point-to-point single-user LTE transmissions, there exist many open questions that can be best tackled by LTE measurements:

- Performance comparison of different kinds of receivers (receiver algorithms),
- Performance of novel and modified transmission schemes following the LTE standard,
- Performance measurements at extreme channels (for example, very high speed) for which channel models are very crude or even nonexistent,
- Comparison of different penetration scenarios or different antenna configurations.

9.2 The Vienna MIMO Testbed

Figure 9.1 exhibits the main hardware components of the Vienna MIMO Testbed required to convert a priori generated data into electromagnetic waves, transmit them over the air and finally to capture them before storing them in digital form for further evaluation. The major hardware components are:

- Three rooftop transmitters supporting four antennas each. The digital signal samples are converted with a precision of 16 bits and are transmitted with adjustable power within a continuous range of about −35–35 dBm per antenna.
- One indoor receiver with four channels that converts the received signals with a precision of 16 bits before the raw signal samples are saved to hard disk. The receive

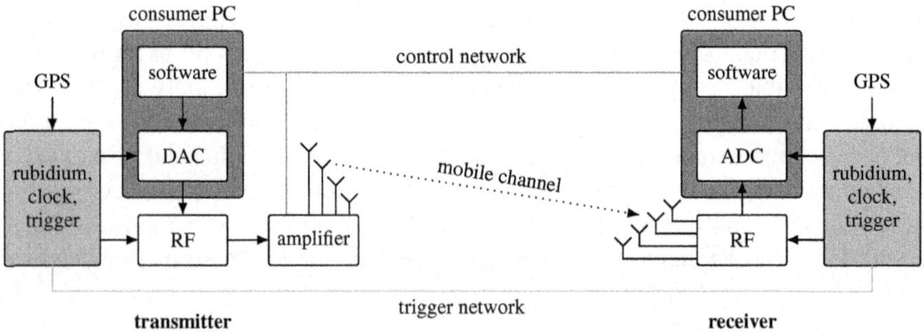

Fig. 9.1 Major components of the Vienna MIMO testbed

antennas are mounted on a positioning table, which allows for measurements at different positions within an area of about 1 m × 1 m, correspondingly 8λ × 8λ.
- The carrier frequency, the sample clock, and the trigger signals are generated separately at each station utilizing Global Positioning System (GPS) synchronized rubidium frequency standards. The synchronization of the triggers is based on exchanging timestamps in the form of User Datagram Protocol (UDP) packets over a trigger network [2]. The precision of this trigger mechanism does not require any further post-synchronization at the receiver. It is sufficient to measure the delay once and time-shift all signals according to the measured delay.
- A dedicated fiber-optic network is utilized to exchange synchronization commands as well as feedback information and general control commands.

The current setup supports a transmission bandwidth of up to 20 MHz at a center frequency of 2.503 GHz.

In typical measurements, the transmission of signals generated according to parameters of interest, is repeated with different values of transmit power in order to obtain results for a certain range of Signal to Noise Ratio (SNR). Furthermore, the transmission of such signals at all values of transmit power is repeated at different receive antenna positions in order to average over small-scale fading scenarios. As a rule of thumb, in a typical scenario approximately 30 measurements of different receive antenna positions are necessary to obtain sound results for an LTE signal with a 10 MHz bandwidth. In order to check whether we have measured sufficient channel realizations, we always include BCa bootstrap confidence intervals in our results [3]. While this process is usually the same for different kinds of measurements, they may differ in the way transmit signals are generated.

As illustrated in Fig. 9.2, we distinguish between two different methodologies as detailed in the following:

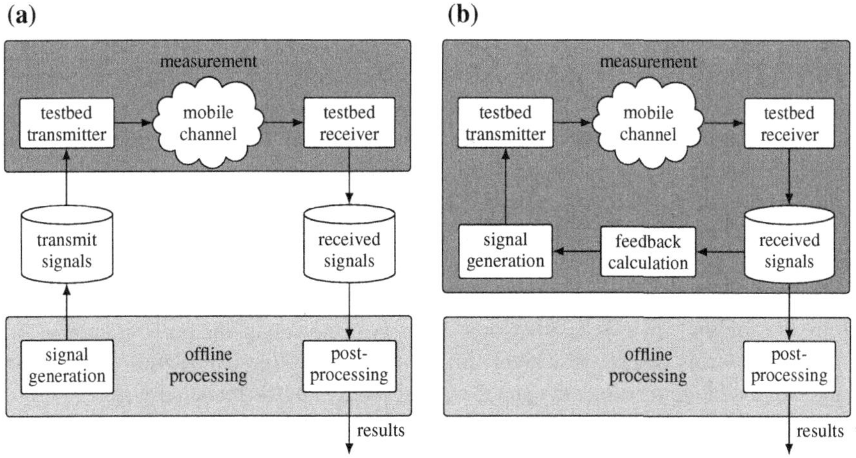

Fig. 9.2 Methodologies used in LTE measurements: **a** Brute-force approach. **b** Measurements with feedback

- *Brute force measurements*: All signals of interest are pre-generated, transmitted over the physical channel, and saved as raw signal samples to hard disk. The received signals are then evaluated offline. This approach is only feasible as long as the time duration of all the different transmit signals is small compared to the channel variations so that successively transmitted data sets appear to be transmitted over the same channel.
- *Measurements with feedback*: The transmit signals are generated on the fly utilizing channel state information obtained via a preceding transmission of training symbols. While the processing and evaluation of the actual data symbols can be computed offline, the demodulation of the training symbols, evaluation, and decision about the generation of the next transmit signal has to be performed in (quasi-) real time.

While brute force measurements typically take longer than the feedback approach and the number of different signals that need to be evaluated is much higher, results obtained by brute force measurements are typically more detailed and are certainly not contaminated by the quality of the feedback function. If the number of different transmit signals is not too large, a combination of both methodologies is possible. All signals of interest are pre-generated, but only those a feedback function decides for are transmitted. This approach reduces the number of signals that have to be evaluated and signals do not have to be generated during the measurement. Nevertheless, it should be noted that if the number of possible transmit signals is very large (for example, zero-forcing Multi-User MIMO or Interference Alignment (IA) [4]), only a feedback approach is feasible.

9.3 Evaluation of LTE MIMO Downlink Transmissions

In different measurement campaigns using Worldwide Inter-Operability for Microwave Access (WiMAX) [5], High-Speed Downlink Packet Access (HSDPA) [6] and LTE [7, 8] the impact of the transmit antenna configuration on the performance of Multiple-Input Multiple-Output (MIMO) transmission modes was investigated. Besides the transmit antenna configuration, the scenario the measurement was performed in has an impact on the performance. Furthermore, the impact of the antenna configuration depends on the receive SNR. In typical field tests, measurements are performed in different scenarios where the SNR is set by the actual scenario. Our approach [9] is different. We fix the scenario and measure over a wide range of transmit power allowing for deeper insights in the impact of the transmit antenna configuration on MIMO transmissions. Thereby, we evaluate the performance of the LTE DL in terms of physical layer throughput on the one hand, and on the other hand, we evaluate the channel capacity as theoretical performance metric.

Measurement Setup

The measurements were performed in an urban scenario at TU Wien in the city of Vienna, Austria using the MIMO testbed described in Sect. 9.2. The measurement

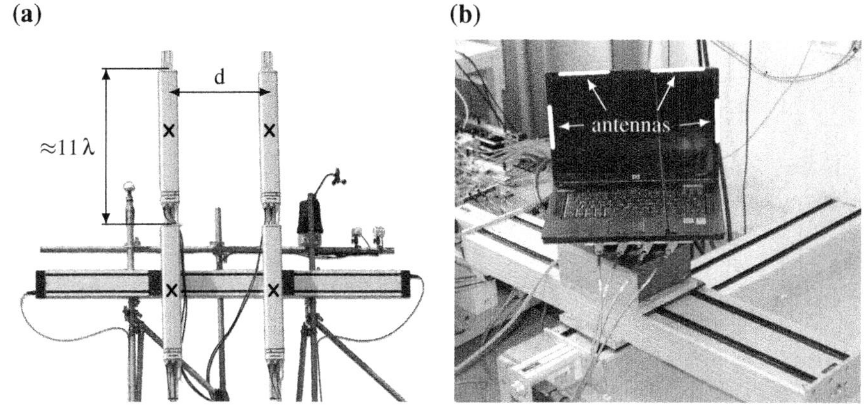

Fig. 9.3 Measurement setup in downtown Vienna, Austria: **a** Transmitter: Two separately shiftable pairs of vertically stacked antennas allow for measurements at vertical and horizontal transmit antenna configurations. **b** Receiver: A laptop carrying the receive antennas can be moved along the X, Y and Φ axis in order to measure different channel realizations

setup is shown in Fig. 9.3 where the transmitter is located outdoors on a rooftop and serves a single user located indoors in the opposite building. In order to implement the desired transmit antenna configurations, we use two pairs of vertically mounted commercial cross polarized antennas that can be moved separately along a linear guide. The four output channels of the transmitter are mapped to four out of the eight antenna elements to transmit with four channels over both, vertically stacked and horizontally spaced antennas. The receive antennas are two horizontally and two vertically polarized custom build patch-antennas mounted around the display of a laptop. This laptop is mounted on a XYΦ positioning table and can so be moved within an area of about $3\,\lambda \times 3\,\lambda$ as well as rotated within a range of about 210°.

Measurement Methodology

Measuring at just a single implementation of a transmit antenna configuration neither allows for a fair comparison of different antenna configurations nor does it lead to reproducible results. With our setup, different antenna elements are employed for different antenna configurations. For that reason, we repeat transmissions using all possible implementations of the antenna configurations under investigation. Then, as illustrated in Fig. 9.4, when averaging the results over all possible implementations, the same antenna elements are selected for every antenna configuration. Single antenna results are obtained in a similar way by averaging over the results for all single transmit antenna elements. Furthermore, the transmit antennas are at different positions for different antenna spacings. We, therefore, do not just measure at a single position of the transmit antennas. For every antenna spacing, we repeat the measurement at different random positions along the linear guide and average over the results. At the receiver site, results for different numbers of receive antennas are averaged in a similar way as it is done for the transmit antennas. For 2×2

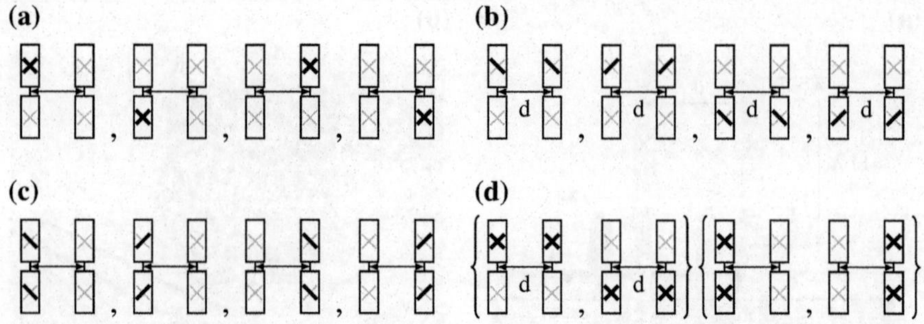

Fig. 9.4 Averaging over all possible implementations of an antenna configuration allows for a fair comparison with other antenna configurations. Every antenna element is used once for every configuration: **a** 2 × 2 cross polarized, **b** 2 × 2 horizontally spaced, **c** 2 × 2 vertically spaced and **d** 4 × 4 horizontally and vertically spaced

transmissions we average over the results obtained by the first two antennas and the results obtained by the second two antennas. Results for 1 × 1 are averaged over all four available receive antennas. Furthermore, measurements are repeated at different XYΦ positions of the receive antennas. Finally, the whole measurement procedure is repeated for different transmit powers. For the generation of the transmit signals and the evaluation after the transmission we modified the Vienna LTE-A Downlink Link Level Simulator [10] to work with the testbed. Thereby, to keep the number of different transmit signals low to use the brute-force approach described in Sect. 9.2, we use the open-loop transmit modes of LTE rather than the closed-loop modes. A summary of all measurement parameters is listed in Table 9.1.

Table 9.1 Measurement parameters

System bandwidth	10 MHz, scheduled for a single user
Transmission modes	Single antenna, Transmit Diversity (TxD),
	Open Loop Spatial Multiplexing (OLSM)
Center frequency	2.503 GHz ($\lambda \approx 12$ cm)
Transmit power	$-13 \ldots 35$ dBm
Transmit antennas	4 × Kathrein 80010541 cross polarized
Transmit antenna spacings	1.5 λ, 5.75 λ, 10 λ (horizontal)
	$\approx$11 λ (vertical)
Transmit antenna polarizations	2 × 2: cross polarized (X-pol), equally polarized
	4 × 4: double cross polarized

Evaluation

9.3.1 Physical Layer Throughput

By applying the brute-force approach, all possible combinations of Modulation and Coding Schemes (MCSs) and transmission rank $N_L = \{1, 2, 3, 4\}$ are transmitted and evaluated independently. Thereby, we obtain for every channel realization r, transmit power P_S and all combinations of the signal parameters MCSs and N_L a result in terms of physical layer throughput $D_r(P_S, \text{MCS}, N_L)$. The average throughput for a certain antenna configuration is then calculated as the average over all channel realizations r of the throughput of the respectively best performing combination of MCS and N_L to

$$D(P_S) = \frac{1}{R}\sum_{r=1}^{R} \max_{\text{MCS}, N_L} D_r(P_S, \text{MCS}, N_L). \tag{9.1}$$

In order to compare the impact of different antenna configurations in more detail, we furthermore evaluate the throughput for a fixed number of spatial streams N_L:

$$D_{N_L}(P_S) = \frac{1}{R}\sum_{r=1}^{R} \max_{\text{MCS}} D_r(P_S, \text{MCS}, N_L). \tag{9.2}$$

Figure 9.5 shows the measurement results for 2×2 transmissions when using equally polarized transmit antennas. If the transmission rank is fixed to $N_L = 1$, we do not observe a difference between different antenna spacings. That is different when transmitting two spatial streams ($N_L = 2$). The higher the spacing, the higher the measured throughput D_2. Furthermore, the average SNR, at which the two-stream transmission outperforms the single-stream transmission is shifted to

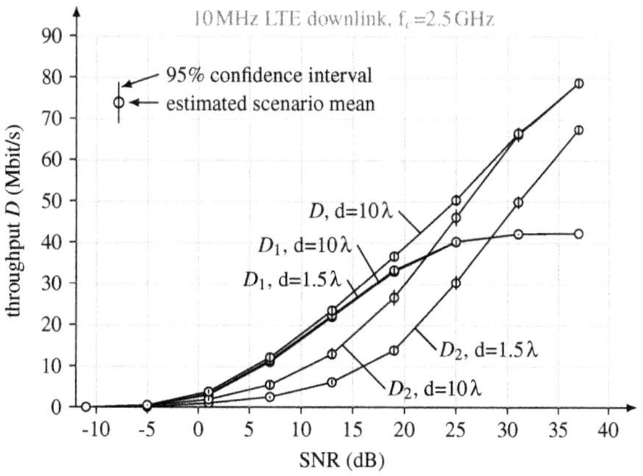

Fig. 9.5 Results for 2×2 LTE transmissions: While the performance of single stream transmissions is independent of the antenna spacing, the throughput when transmitting two streams increases with antenna spacing

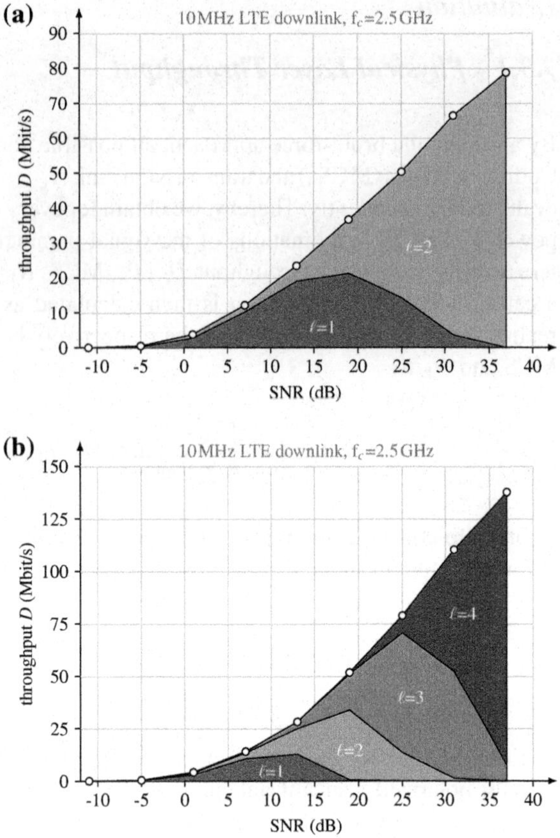

Fig. 9.6 Average contribution of rank N_L transmissions to the average throughput when using rank adaption: **a** 2×2 using equally polarized antennas with horizontal spacing of $d = 10\,\lambda$. **b** 4×4 double cross polarized antennas with horizontal spacing of $d = 10\,\lambda$

lower SNRs. The evaluation of D reflects the technique of rank adaption. In the lower SNR regime, the measured throughput is very close to the throughput for the single-stream transmission. With increasing SNR, the throughput gets close to the throughput of the two-stream transmission. In the region of moderate SNR, D is higher than the respective throughputs D_1 and D_2, as we evaluate average throughputs and the break-even point in terms of average SNR differs from channel realization to channel realization. This smooth transition is also shown in Fig. 9.6, where the average contribution of transmissions using N_L spatial streams to the average total throughput is illustrated. The results for all 2×2 antenna configurations when using rank adaption are given by Fig. 9.7. Large differences are only observed at moderate to high SNRs where transmissions with $N_L = 2$ outperform the single-stream transmissions. Thereby, the throughput increases with antenna spacing d when using equally polarized transmit antennas. The vertically stacked antennas perform about as good as the horizontally spaced antennas with spacing $d = 10\,\lambda$. Cross polarized transmit antennas outperform all other antenna configurations. Figure 9.8 shows the results of the 4×4 measurements using double cross polarized antennas. As for the

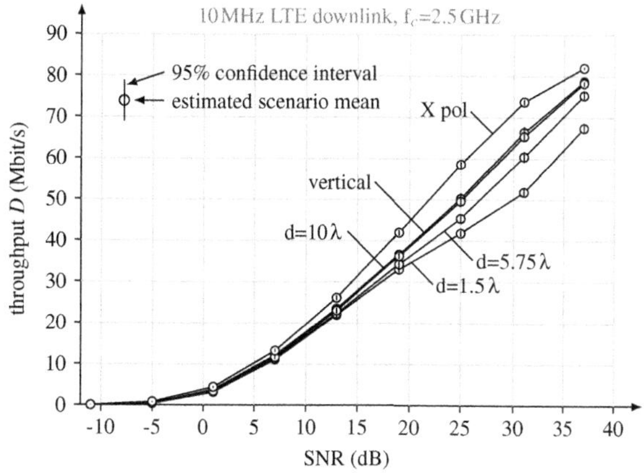

Fig. 9.7 Results for 2 × 2 LTE transmissions: The best performance for two antenna transmissions is observed when cross polarized antennas are employed

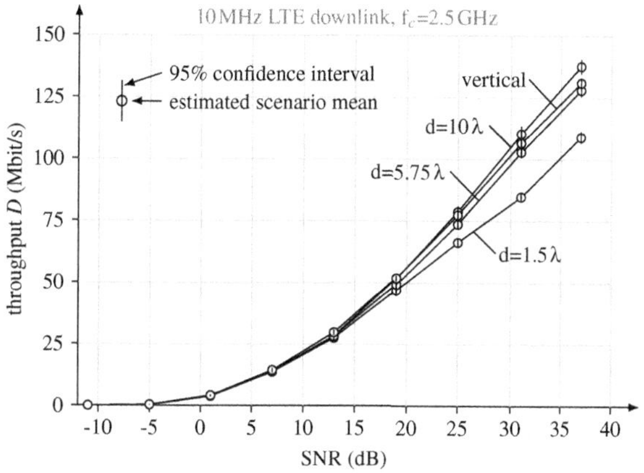

Fig. 9.8 Results for 4 × 4 LTE transmissions: The performance increases with antenna spacing

2 × 2 case, the performance increases with antenna spacing. The throughput with vertically stacked antennas is close to the throughput for horizontally spaced antennas with $d = 10\,\lambda$.

9.3.2 Channel Capacity

Besides results in terms of throughput, the receiver of the LTE simulator provides channel estimates for every subframe transmitted. We evaluate the channel capacity [11] as an upper bound for the data rate by these channel estimates $\mathbf{H}_k$ measured at the highest transmit power. For an Orthogonal Frequency Division Multiplexing (OFDM) transmission with N_c subcarriers, a total channel bandwidth B and N_T transmit antennas, the channel capacity $C(P_S)$ as a function of the measured channels $\mathbf{H}_k$, the measured noise power P_V and the transmit power P_S is given by

$$C\left(P_{\mathrm{S}}\right)=\frac{B}{N_{\mathrm{c}}}\sum_{k=1}^{N_{\mathrm{c}}}\log_2\det\left(\mathbf{I}+\frac{1}{N_{\mathrm{T}}}\frac{P_{\mathrm{S}}}{P_{\mathrm{V}}}\mathbf{H}_k\mathbf{F}_k\mathbf{F}_k^H\mathbf{H}_k^H\right). \tag{9.3}$$

Due to the guard band of 1 MHz specified in the 10 MHz LTE DL, $\mathbf{H}_k$ is estimated over a bandwidth of 9 MHz only. Therefore, we extrapolate the results to the full bandwidth of $B = 10$ MHz. Thereby, we assume full Channel State Information (CSI) at the transmitter and use the waterfilling algorithm to calculate the optimal precoder $\mathbf{F}_k$ for every channel realization. The waterfilling algorithm distributes a fixed value of total transmit power to the available layers according to the eigenvalues λ_l of $\mathbf{H}_k\mathbf{H}_k^H$. At low SNRs the maximum is achieved by assigning all power to the strongest eigenvalue. With increasing SNR the number of eigenvalues increases before the available transmit power is assigned equally to all eigenvalues. The eigenvalues obtained by the measurement are listed in Table 9.2 for 2 × 2 and in Table 9.3 for the 4 × 4 channels. A comparison of channel capacity and LTE physical layer throughput for three different transmit antenna configurations for 2 × 2 transmissions is depicted in Fig. 9.9. At low SNRs, both, the capacity and the LTE throughput is independent of the antenna configuration. With increasing SNR, the differences observed for the LTE throughput become visible for the channel capacity in a similar way. The same effects are observed when evaluating the eigenvalues in Table 9.2 where the measured eigenvalues are normalized to the strongest eigenvalue at $d = 1.5\,\lambda$. Thereby, the strongest eigenvalue λ_1 is independent of the transmit antenna configuration while the second eigenvalue λ_2 depends on the antenna configuration. The measured eigenvalues for 4 × 4 are listed in Table 9.3 where the two strongest eigenvalues are quite independent of the antenna configuration. The impact of the antenna configuration is visible for the two weakest eigenvalues only.

Finally, we were interested in how much of the channel capacity the LTE physical layer throughput could reach in our measurement. Therefore, we define the

Table 9.2 Normalized eigenvalues of measured 2 × 2 channels

	d=1.5 λ (dB)	**d=5.75 λ (dB)**	**d=10 λ (dB)**	**Vertical (dB)**	**X polarized (dB)**
λ_1	0.0	0.1	0.0	0.0	−0.27
λ_2	−20.2	−16.5	−14.4	−14.7	−10.9

Table 9.3 Normalized eigenvalues of measured 4 × 4 channels

	d=1.5 λ (dB)	**d=5.75 λ (dB)**	**d=10 λ (dB)**	**Vertical (dB)**
λ_1	0.0	0.2	0.1	−0.1
λ_2	−7.6	−7.7	−7.5	−7.5
λ_3	−21.0	−17.1	−15.0	−15.4
λ_4	−31.0	−26.1	−23.9	−25.6

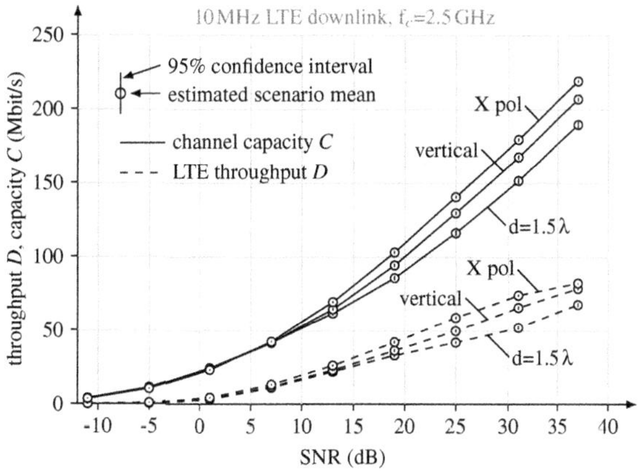

Fig. 9.9 Comparison of channel capacity and LTE physical layer throughput of 2 × 2 setups for three different transmit antenna configurations

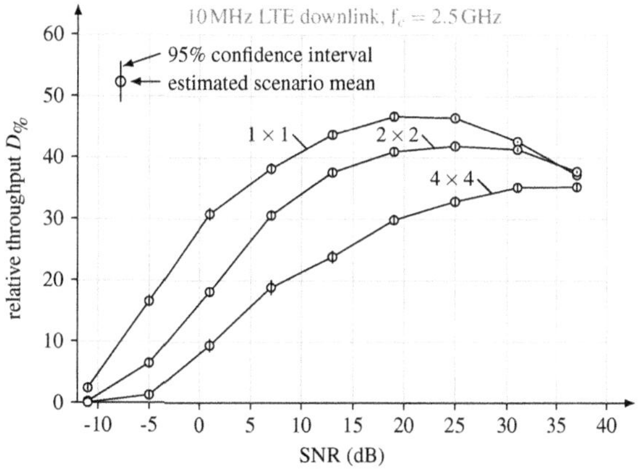

Fig. 9.10 Comparison of 1 × 1, 2 × 2 (cross polarized) and 4 × 4 ($d = 10\,\lambda$) in terms of relative throughput. The LTE throughput reaches nearly half of the channel capacity for 1 × 1 transmissions and decreases with increasing number of transmit antennas

relative throughput as the ratio of throughput and capacity: $D_{\%}\,(P_S) = 100 \cdot \frac{D(P_S)}{C(P_S)}$. In Fig. 9.10 we show a comparison of the respectively best performing antenna configurations in terms of relative throughput. For all different numbers of transmit antennas the relative throughput increases with SNR and reaches a maximum as the maximum data rates defined in the LTE standard are reached. This maximum decreases with increasing number of transmit antennas as the overhead increases.

9.4 Measurements at High Velocities

LTE is designed to support user velocities of up to 500 km/h whereas mobile communications experiments in high mobility environments such as high speed trains, motorways or airplanes are expensive and time-consuming. Although such experiments are feasible, they are not well suited to, for example, directly compare different transmission techniques or to measure at different velocities or SNRs. Such experiments require a fully controllable setup that allows for repeated transmissions under identical environmental conditions where all parameters are fixed except the one whose effect is being tested. In order to perform such experiments, the Vienna MIMO testbed was extended by antennas on the tip of a rotary unit [12, 13] that allows for fully controllable and repeatable measurements at velocities of up to 400 km/h. In this section, we give an overview of this measurement setup before the results of a measurement campaign comparing different channel interpolation methods for the LTE UL are presented.

9.4.1 Measurement Setup and Methodology

In our setup, as it is shown in Fig. 9.11, repeatable time-variant channels are generated by rotating the receive antennas around a central pivot. The received signals are then fed through the rod to rotary joints mounted inside the axis and are connected to the static receiver hardware of our testbed. A light barrier mounted on the axis captures the start of each turn of the rotating rod. This signal is connected to the trigger network of the testbed and triggers the signal transmission. Thereby, signal transmissions can be triggered at any desired angle of the rotating rod. The light barrier together with the trigger network allows for repeated transmissions over the same time-varying channel. Examples when multiple transmissions over the same channel are needed are:

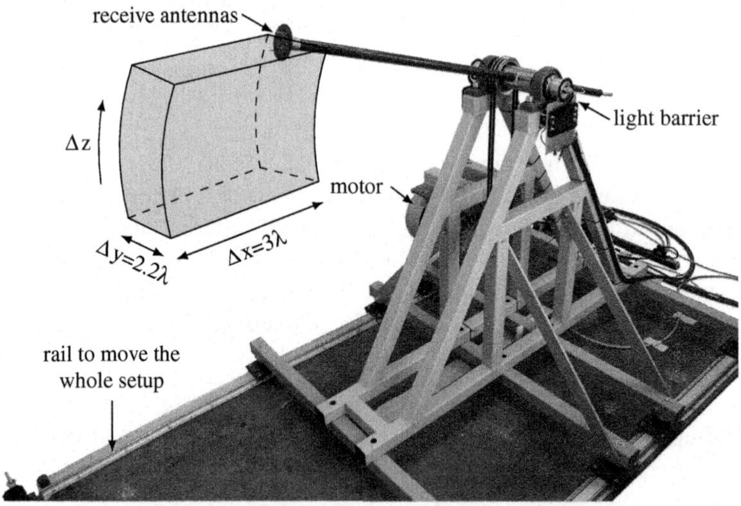

Fig. 9.11 Measurement setup to generate repeatable time-variant channels

- Measurements at different SNRs. The same signals need to be transmitted with different transmit powers over the same channel.
- When comparing different transmit signals or transmit modes.
- Brute-force LTE measurements where different signals need to be transmitted over the same channel.
- Measurements with feedback, if feedback information should be applied to transmissions over the same channels to obtain the CSI. For measurements with feedback delay the trigger can be delayed according to the desired delay.
- Measurements at different velocities. As the spatial length $\Delta z = T \cdot v$ of a signal with a certain temporal length T depends on the velocity v, it is not possible to transmit the same signal over the same channel at different velocities. Our approach for a fair comparison at different velocities is illustrated in Fig. 9.12. We transmit n realizations of the transmit signal of interest at the highest velocity. At half the maximum velocity, we transmit $2n$ realizations and so forth.

In order to measure different channel realizations within the same scenario, the whole setup can be moved along the x and y-axis. The area where typical measurements are performed is illustrated by the box in Fig. 9.11. While Δx and Δy are given by mechanical constraints of the setup, $\Delta z = T \cdot v$ depends on the length T of the transmit signals and the actual velocity v. Considering measurements at a velocity of $v = 400\,\text{km/h}$ and the transmission of single LTE subframes having a length of $T = 1\,\text{ms}$, the length of the path the receive antennas move during the transmission calculates to $\Delta z \approx 11\,\text{cm}$. For the rod having a length of 1 m, the length of the path corresponds to an angle of about 6°. Figure 9.12 illustrates the path of the receive antennas and the corresponding bending of the path over 1 ms when moving at 400 km/h.

9.4.2 LTE Uplink Fast Fading Channel Interpolation

In the measurement campaign reported in [14] we were interested in the performance of different channel interpolation techniques for the LTE UL. Compared to the LTE

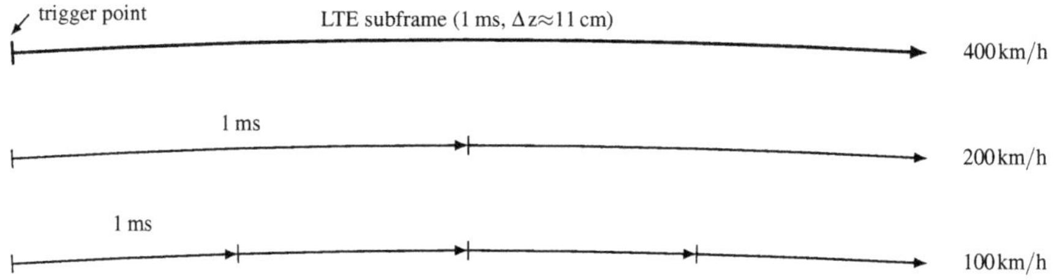

Fig. 9.12 Trajectories of the receive antennas when transmitting LTE subframes. Multiple transmissions of the transmit signals at lower velocities allows for transmissions over the same spatial channels at different velocities

DL, the temporal spacing of the Demodulation Reference Signals (DMRS) in the UL is about twice the spacing as in the DL. Furthermore, if frequency-hopping is performed the number of adjacent pilots transmitted in a certain subband is two for inter-subframe frequency hopping and one for intra-subframe frequency hopping where frequency hopping is performed on a per-slot basis. Due to this special structure of pilot symbols, channel estimation in the LTE UL is a challenging problem. The authors of [15] proposed an interpolation algorithm based on adaptive order polynomial fitting to mitigate Inter-Carrier Interference (ICI), in [16] the polynomial basis expansion model is employed and the estimation accuracy is improved by an autoregressive model. Our idea is to include channel estimates from the previous and from the subsequent subframe into the process of channel interpolation. The additional delay that is introduced by applying channel estimates from the subsequent subframe is not considered.

System Model

We consider continuous single antenna LTE UL transmissions with frequency hopping being disabled. Sounding Reference Signals (SRS) and the Physical Uplink Control Channel (PUCCH) are both not considered. Figure 9.13 illustrates the resulting time/frequency resource grid for three consecutive resource blocks that consist only of data symbols and pilot symbols. At the receiver side we perform a symbol-by-symbol Least Squares (LS) channel estimation in the frequency domain and calculate the Zero Forcing (ZF) equalizer by the different interpolation methods under investigation. Although we do not perform frequency hopping, we emulate it by considering the cases where only one or two pilot symbols are used.

- *Average*: Averaging the channel estimates from pilot positions p_0 and p_1. In the static case this method improves the channel estimation by 3 dB in terms of signal-to-noise ratio (SNR) but averages over temporal variations in the fast fading case.

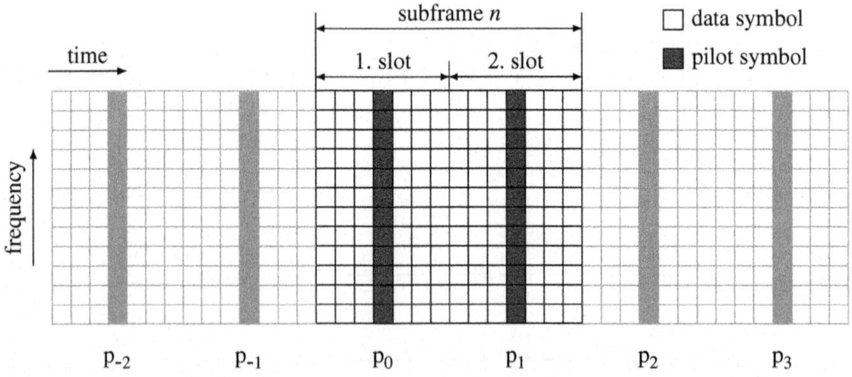

Fig. 9.13 Resource grid of the LTE uplink. Due to SC-FDMA modulation, symbols marked as data symbols are the DFT-precoded data symbols rather than the actual data symbols

- *1 point*: Data symbols of slot n are equalized by the channel estimates from pilot at position p_n. This method is applicable in the case of intra-subframe frequency hopping.
- *2 point linear*: Linear interpolation and extrapolation based on the estimates at pilot position p_0 and p_1. This method is applicable in the case inter-subframe frequency hopping is performed but intra-subframe frequency hopping is not activated.
- *4 point linear*: Linear interpolation based on the estimates at pilot position p_{-1}, p_0, p_1 and p_2.
- *4 point spline*: Spline interpolation based on the estimates at pilot position p_{-1}, p_0, p_1 and p_2.
- *6 point spline*: Spline interpolation using the estimates from pilot positions p_{-2} to p_3.

The resulting equalizers are then applied in the frequency domain on the DFT-precoded data symbols transmitted during subframe n in Fig. 9.13. The previous subframe and the subsequent subframe are only considered to obtain additional channel estimates.

Measurement

Both, the generation of transmit signals and the processing of the received signals is based on the Vienna LTE-A Uplink Link Level Simulator [17]. In order to measure the physical layer throughput by the brute-force approach described in Sect. 9.2, one subframe for each of the 15 different MCSs is pre-generated. Every subframe is repeated three times for transmissions at the maximum velocity of $v = 400\,\text{km/h}$ whereas the central subframe n is the subframe to be decoded and the neighboring subframes $n - 1$ and $n + 1$ are used to obtain the additional channel estimates. At half of the maximum velocity (200 km/h) two subframes of interest (n) are transmitted over the desired channel (Δz) and so forth. Figure 9.14 illustrates this idea of transmitting over the same spatial channel at different velocities whereas the number of subframes considered in the evaluation is given by $R(v) = \frac{400}{v}$ (Table 9.4).

Table 9.4 Measurement parameters.

Center frequency	2.506 GHz ($\lambda \approx 12\,\text{cm}$)
Velocities	50, 100, 200 and 400 km/h
Channel realizations	49 (within an area of about 3 $\lambda \times 2.2\lambda$)
Transmission mode	10 MHz LTE UL, single antenna transmission, normal cyclic prefix, all resources scheduled for a single user, no SRS, no PUCCH
Modulation and coding	15 different MCSs according to the 15 different Channel Quality Indicators (CQIs) defined in the standard
Receiver	Single antenna, LS channel estimation, ZF equalization

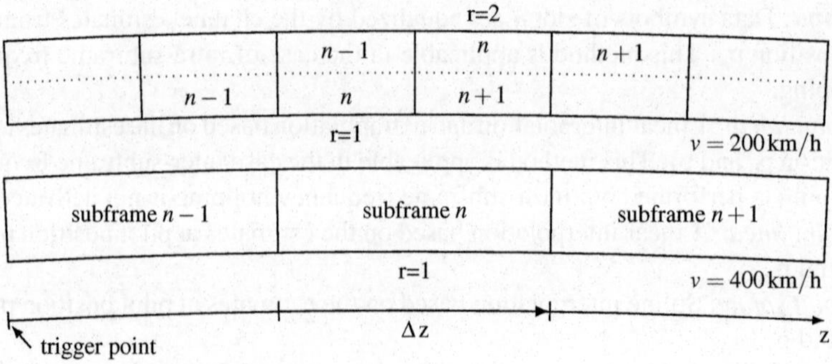

Fig. 9.14 Transmitting over the same spatial channels allows for a fair comparison at different velocities

Evaluation

As figure of merit for the comparison of different channel interpolation methods the physical layer throughput is considered. Furthermore, the SNR as well as the Signal-to-Interference Ratio (SIR) and the Signal to Interference and Noise Ratio (SINR) as measures for the amount of ICI are evaluated.

Physical Layer Throughput

By using the brute-force approach perfect knowledge of the best performing MCS is emulated for every channel realization and every value of transmit power by transmitting all different MCSs over the same channel. The independent evaluation of all received signals then yields a value of throughput D_m for every combination of measurement parameters whereas k denotes the channel realization, r being the temporal repetition, v the velocity, P_S the transmit power and I the channel interpolation method. The throughput D_m is maximized over the different MCSs by

$$\widehat{\mathrm{D}}_m\left(k, r, v, I, P_\mathrm{S}\right) = \max_{\mathrm{MCS}} \mathrm{D}_m\left(k, r, v, I, P_\mathrm{S}, \mathrm{MCS}\right) \tag{9.4}$$

before the average throughput

$$\mathrm{D}\left(v, I, P_\mathrm{S}\right) = \frac{1}{K}\frac{1}{R\left(v\right)}\sum_{k=1}^{K}\sum_{r=1}^{R(v)} \widehat{\mathrm{D}}_m\left(k, r, v, I, P_\mathrm{S}\right) \tag{9.5}$$

is obtained by averaging over all K different channel realizations and $R\left(v\right)$ temporal repetitions.

SIR, SINR and SNR

The power of each subcarrier is estimated in the frequency domain whereas we obtain the signal-plus-interference-plus-noise power P_SIN at data subcarrier positions, the interference-plus-noise power P_IN at the DC subcarrier where no data is transmitted and the noise power P_V by measuring at the same subcarrier positions during a noise gap when no signal is transmitted. These thereby obtained power estimates are averaged similar to Eq. (9.5) separately over all channel realizations and temporal repetitions. P_{SIN} and P_{IN} are furthermore averaged over all different MCSs. The estimated SIR then calculates to

$$\mathrm{SIR}\,(v,\,P_S) = \frac{\overline{P}_{SIN}\,(v,\,P_S) - \overline{P}_{IN}\,(v,\,P_S)}{\overline{P}_{IN}\,(v,\,P_S) - \overline{P}_{V}\,(v,\,P_S)}\,, \tag{9.6}$$

the SINR to

$$\mathrm{SINR}\,(v,\,P_S) = \frac{\overline{P}_{SIN}\,(v,\,P_S) - \overline{P}_{IN}\,(v,\,P_S)}{\overline{P}_{IN}\,(v,\,P_S)}\,, \tag{9.7}$$

and the SNR to

$$\mathrm{SNR}\,(v,\,P_S) = \frac{\overline{P}_{SIN}\,(v,\,P_S) - \overline{P}_{IN}\,(v,\,P_S)}{\overline{P}_{V}\,(v,\,P_S)}\,. \tag{9.8}$$

Results

The conditions in terms of SIR, SINR and SNR under which the measurement was performed are shown in Fig. 9.15. Due to the aforementioned methodology, the SNR is constant over the whole range of velocities. Comparing the SIR to analytical results [18], derived for two popular models, shows a higher SIR in our scenario. Both models, Jakes' spectrum and the uniform model are based on uniformly distributed scattering objects which is not the case in our scenario. The SINR is upper bounded by noise at low velocities and upper bounded by the ICI power at high velocities. While we observe a large decrease of SINR for increasing velocity at high SNR, the SINR curve flattens for lower SNRs. The impact of ICI on the throughput becomes nearly independent of the velocity and the performance is rather determined by noise and the quality of the channel interpolation method than by ICI.

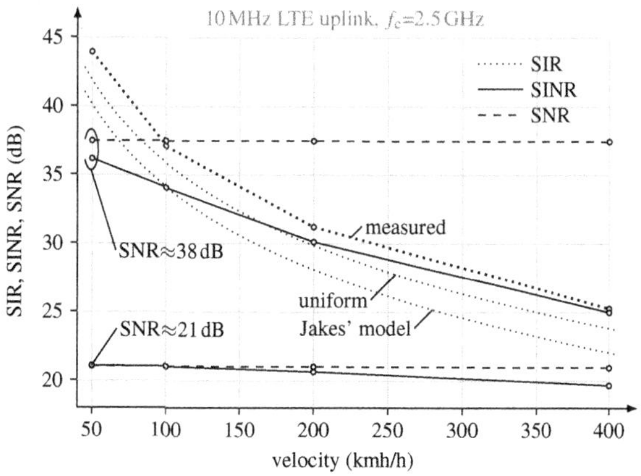

Fig. 9.15 Channel conditions under which the measurement was performed

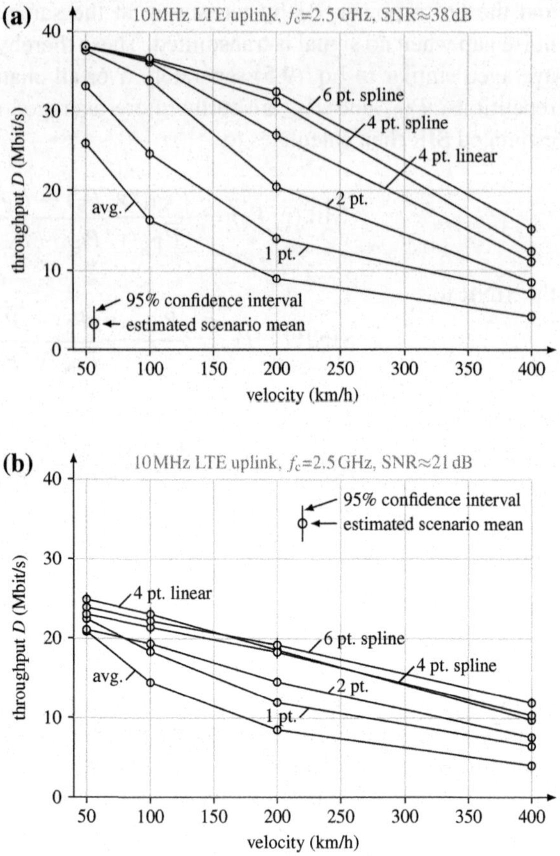

Fig. 9.16 Measurement results comparing different channel interpolation methods in terms of throughput for two different values of transmit power resulting in an average SNR of **a** ≈38 dB and **b** ≈21 dB

Figure 9.16 compares the considered channel interpolation methods in terms of physical layer throughput for two different values of SNR. As expected, the worst performance is observed when channel estimates from two pilots are averaged. The performance increases with the number of pilots in the channel interpolation. The highest gains at high SNR are observed between *1 point*, where no interpolation is performed and *2 point* interpolation and moreover when channel extrapolation in the *2 point* case is replaced by interpolation when performing *4 point linear* interpolation. Additional gains are observed when using spline interpolation, especially at high SNR and high velocities. At lower SNR, spline interpolation outperforms *4 point linear* interpolation only at moderate to high velocities. At low velocities, the gain of channel estimation SNR becomes visible for the *averaging* method as it performs as good as the *2 point* method. Furthermore, the throughput flattens at lower SNRs as the SINR flattens. At low velocities, the throughput is then determined by the SNR. The impact of the channel interpolation method is still visible at high velocities.

References

1. M. Lerch, S. Caban, M. Mayer, M. Rupp, The Vienna MIMO testbed: evaluation of future mobile communication techniques. Intel Technol. J. **18**, 58–69 (2014)
2. S. Caban, A. Disslbacher-Fink, J.A. García Naya, M. Rupp, Synchronization of wireless radio testbed measurements, in *Proceedings of IEEE International Instrumentation and Measurement Technology Conference (I2MTC2011)* (2011). doi:10.1109/IMTC.2011.5944089
3. S. Caban, J.A. García Naya, M. Rupp, Measuring the physical layer performance of wireless communication systems. IEEE Instrum. Meas. Mag. **14**(5), 8–17 (2011). doi:10.1109/MIM.2011.6041377
4. M. Mayer, G. Artner, G. Hannak, M. Lerch, M. Guillaud, Measurement based evaluation of interference alignment on the Vienna MIMO testbed, in *Proceedings of the Tenth International Symposium on Wireless Communication Systems (ISWCS'13)*, Ilmenau, 2013
5. S. Caban, J.A. García Naya, L. Castedo, C. Mehlführer, Measuring the influence of TX antenna spacing and transmit power on the closed-loop throughput of IEEE 802.16-2004 WiMAX, in *Proceedings of IEEE Instrumentation and Measurement Technology Conference (I2MTC2010)*, 2010
6. S. Caban, J.A. García Naya, C. Mehlführer, M. Rupp, Measuring the closed-loop throughput of $2x2$ HSDPA over TX power and TX antenna spacing, in *Proceedings of 2nd International Conference on Mobile Lightweight Wireless Systems (Mobilight-2010)*, 2010
7. K. Werner, J. Furuskog, M. Riback, B. Hagerman, Antenna configurations for $4x4$ MIMO in LTE—field measurements, in *71st IEEE Vehicular Technology Conference (VTC 2010-Spring)*, pp. 1–5, 2010. doi:10.1109/VETECS.2010.5493762
8. W. Xie, T. Yang, X. Zhu, F. Yang, Q. Bi, Measurement-based evaluation of vertical separation MIMO antennas for base station. IEEE Antennas Wirel. Propag. Lett. **11**, 415–418 (2012). doi:10.1109/LAWP.2012.2194688
9. M. Lerch, M. Rupp, Measurement-based evaluation of the lte mimo downlink at different antenna configurations, in *Proceedings of the 17th International ITG Workshop on Smart Antennas, WSA 2013*, Stuttgart, 2013
10. S. Schwarz, J.C. Ikuno, M. Simko, M. Taranetz, Q. Wang, M. Rupp, Pushing the limits of LTE: a survey on research enhancing the standard. IEEE Access **1**, 51–62 (2013). doi:10.1109/ACCESS.2013.2260371
11. G. Foschini, M. Gans, On limits of wireless communications in a fading environment when using multiple antennas. Wirel. Pers. Commun. **6**, 311–335 (1998). doi:10.1023/A:1008889222784
12. S. Caban, J. Rodas, J.A. García-Naya, A methodology for repeatable, off-line, closed-loop wireless communication system measurements at very high velocities of up to 560 km/h, in *Proceedings of International Instrumentation and Measurement Technology Conference (I2MTC 2011)*, Binjiang, 2011. doi:10.1109/IMTC.2011.5944019
13. S. Caban, R. Nissel, M. Lerch, M. Rupp, Controlled OFDM measurements at extreme velocities, in *Proceedings of ExtremeCom'2014, Galapagos Islands, Ecuador*, 2014
14. M. Lerch, Experimental comparison of fast-fading channel interpolation methods for the LTE uplink, in *Proceedings of the 57th International Symposium ELMAR-2015, Zadar*, 2015. doi:10.1109/ELMAR.2015.7334482
15. B. Karakaya, H. Arslan, H. Cirpan, Channel estimation for LTE Uplink in High Doppler Spread, in *Wireless Communications and Networking Conference (WCNC 2008)*, pp. 1126–1130, 2008. doi:10.1109/WCNC.2008.203
16. L. Yang, G. Ren, B. Yang, Z. Qiu, Fast time-varying channel estimation technique for LTE uplink in HST environment. IEEE Trans. Vehic. Technol. **61**(9), 4009–4019 (2012). doi:10.1109/TVT.2012.2214409
17. J. Blumenstein, J.C. Ikuno, J. Prokopec, M. Rupp, Simulating the long term evolution uplink physical layer, in *Proceedings of the 53rd International Symposium ELMAR-2011, Zadar*, 2011
18. P. Robertson, S. Kaiser, The effects of Doppler spreads in OFDM (A) mobile radio systems. IEEE Vehic. Technol. Conf. Fall **1**, 329–333 (1999)

Part II
System Level Simulations

In the second part of the book, the focus is on *system level*, where performance evaluation requires to encompass a large number of network elements and upscales the number of interconnecting links. These have become an indispensable tool for predicting the behavior of wireless cellular systems [1–3]. The main interest lies in network-related issues such as resource allocation, mobility management, and network planning [4–8]. Hence, computational complexity needs to be decreased substantially in order to make the problem feasible. A widely accepted solution is the application of *link abstraction models* that specify the interaction between link- and system level simulators [2]. This approach is expected to persist in simulation tools for the fifth generation of wireless cellular networks (5G) [9].

Exhaustive simulations enable to study any scenario at any desired depth of detail. On the other hand, this requires to run a separate simulation for every setup and each choice of parameters. As an example, in a heterogeneous cellular network, the amount of combinations of deployment parameters rises exponentially with the number of network tiers. Hence, it is expedient to use insights from *analytical models* to shrink the search space, and to run targeted simulations. Results from theory provide overall insights without misinterpreting or being biased by any specific scenario.

Current analytical models for system level evaluation of mobile cellular networks focus on an accurate statistical description of the *aggregate interference*, i.e., the cumulative impact from all co-channel interferers, as it is one of the main performance limiting factors [10]. The two key interference-shaping factors are the spatial distribution of concurrently transmitting BS and the path losses, which encompass signal attenuation by distance and fading [10–15].

Tractable interference statistics have mainly been reported in the field of stochastic geometry. This powerful mathematical framework recently gained momentum as the only available tool that provides a rigorous approach to modeling, analysis, and design of networks with a substantial amount of nodes per unit area [10–12, 15–26]. When closed-form expressions are desired, it imposes its own particular limitations, typically including spatial stationarity and isotropy of the scenario [10, 15, 27]. Hence, the potential to consider an asymmetric interference impact is very limited and notions such as *cell-center* and *cell-edge* are, in general, not accessible.

The second part of this book places particular emphasis on system models of low complexity that enable to analytically evaluate interference statistics of arbitrary interferer topologies at any desired user location. The second major focus is on a systematic and reproducible simulation methodology.

This part is organized as follows. First, basic concepts of system level simulations are explained in Chap. 10. Moreover, advanced features such as runtime-precoding and 3D channel modeling are introduced. In Chaps. 11−13, two circular as well as a stochastic interference model are presented that enable to analytically predict system level performance. Their applicability against LTEA system level simulations is validated. Chapter 14 provides a systematic and reproducible simulation methodology for evaluating two-tier heterogeneous cellular networks. Advanced system level applications are presented in Chap. 15.

References

[1] L. Chen, W. Chen, B. Wang, X. Zhang, H. Chen, D. Yang, System-level simulation methodology and platform for mobile cellular systems. IEEE Commun. Mag. **49**(7), 148–155 (2011). doi:10.1109/MCOM.2011.5936168

[2] S. Ahmadi, *LTE-Advanced: A Practical Systems Approach to Understanding 3GPP LTE Releases 10 and 11 Radio Access Technologies*, ser. ITPro collection. (Elsevier Science, 2013)

[3] J.C. Ikuno, M. Wrulich, M. Rupp, System level simulation of LTE networks, in *IEEE Vehicular Technology Conference (VTC Spring)*. Taipei, Taiwan, (2010)

[4] Y. Gao, Y. Li, H. Yu, S. Gao, Performance of dynamic comp cell selection in 3gpp lte system level simulation, in *IEEE International Conference Communications Software and Networks (ICCSN)*, pp. 210–213. Xi'an, China, (2011). doi:10.1109/ICCSN.2011.6013577

[5] D. Jiang, H. Wang, E. Malkamaki, E. Tuomaala, Principle and performance of semi-persistent scheduling for VoIP in LTE system, in *International Conference Wireless Communication Networking and Mobile Computing* (*WiCom*), pp. 2861–2864. Shanghai, China, (2007). doi:10.1109/WICOM.2007.710

[6] O. Yilmaz, S. Hamalainen, J. Hamalainen, System level analysis of vertical sectorization for 3GPP LTE, in *International Symposium Wireless Communication System* (*ISWCS*), pp. 453–457. Tuscany, Italy, (2009). doi:10.1109/ISWCS.2009.5285317

[7] M. Sawahashi, Y. Kishiyama, A. Morimoto, D. Nishikawa, M. Tanno, Coordinated multipoint transmission/reception techniques for LTE-advanced [coordinated and distributed MIMO]. IEEE Wirel. Commun. **17**(3), 26–34, (2010). doi:10.1109/MWC.2010.5490976

[8] K. Kusume, G. Dietl, T. Abe, H. Taoka, S. Nagata, System level performance of downlink MU-MIMO transmission for 3GPP LTE-Advanced, in *IEEE Vehicular Technology Conference* (*VTC*), pp. 1–5. Taipei, Taiwan, (2010). doi:10.1109/VETECS.2010.5493979

[9] Y. Wang, J. Xu, L. Jiang, Challenges of system-level simulations and performance evaluation for 5G wireless networks, IEEE Access, **2**, 1553–1561 (2014). doi:10.1109/ACCESS.2014.2383833

[10] M. Haenggi, R.K. Ganti, *Interference in Large Wireless Networks*, ser. Foundations and Trends in Networking. 1em plus 0.5em minus 0.4em NoW Publishers, vol. 3, (2009)

[11] J. Andrews, R. Ganti, M. Haenggi, N. Jindal, S. Weber, A primer on spatial modeling and analysis in wireless networks, IEEE Commun. Mag. **48**(11), 156–163, (2010)
[12] M. Win, P. Pinto, L. Shepp, A mathematical theory of network interference and its applications. Proc. **97**(2), 205–230, (2009)
[13] K. Gulati, A. Chopra, B.L. Evans, K.R. Tinsley, Statistical modeling of co-channel interference, in *IEEE Global Telecommunications Conference* (*GLOBECOM*), Honolulu, HI, USA, (2009)
[14] M. Kountouris, N. Pappas, Approximating the interference distribution in large wireless networks, in *International Symposium Wireless Communication System* (*ISWCS*), Barcelona, Spain, (2014)
[15] M. Haenggi, J. Andrews, F. Baccelli, O. Dousse, M. Franceschetti, Stochastic geometry and random graphs for the analysis and design of wireless networks, IEEE J. Sel. Areas Commun. **27**(7), 1029–1046, (2009). doi:10.1109/JSAC.2009.090902
[16] F. Baccelli, S. Zuyev, Stochastic geometry models of mobile communication networks, in *Frontiers in queueing: Models And Applications in Science and Engineering*, pp. 227–243, (1996)
[17] T. Brown, Cellular performance bounds via shotgun cellular systems, IEEE J. Sel. Areas Commun. **18**(11), 2443–2455, (2000), doi:10.1109/49.895048
[18] F. Baccelli, M. Klein, M. Lebourges, S. Zuyev, Stochastic geometry and architecture of communication networks. *Telecommun. Syst.*, **7**, 209–227, (1997). doi:10.1023/A: 1019172312328
[19] F. Baccelli, B. Blaszczyszyn, *Stochastic Geometry and Wireless Networks: Volume I Theory*, ser. Foundation and Trends in Networking. Now Publishers, March 2009. doi:10.1561/ 1300000006
[20] F. Baccelli, B. Blaszczyszyn, *Stochastic Geometry and Wireless Networks, Volume II—Applications*, ser. Foundations and Trends in Networking, F. Baccelli and B. Blaszczyszyn, Eds. vol 2 (Now Publishers, 2009). doi:10.1561/1300000026
[21] C. Ren, J. Zhang, W. Xie, D. Zhang, Performance analysis for heterogeneous cellular networks based on Matern-like point process model, in *International Conference Information Science and Technology* (*ICIST*), pp. 1507–1511, Yangzhou, China, (2013). doi:10.1109/ ICIST.2013.6747823
[22] A. Guo, M. Haenggi, Spatial stochastic models and metrics for the structure of base stations in cellular networks. IEEE Trans. Wireless Commun. **12**(11), 5800–5812, (2013)
[23] M. Di Renzo, A. Guidotti, G. Corazza, Average rate of downlink heterogeneous cellular networks over generalized fading channels: A stochastic geometry approach, IEEE Trans. Commun., **61**(7), 3050–3071, (2013)
[24] M. Haenggi, *Stochastic Geometry for Wireless Networks*, ser. Stochastic Geometry for Wireless Networks. (Cambridge University Press, 2012)
[25] S. Weber, J.G. Andrews, Transmission capacity of wireless networks. Found. Trends Netw. **5** (2–3), pp. 109–281, 2012, doi:10.1561/1300000032
[26] T. Bai, R. Vaze, R.W. Heath Jr., Analysis of blockage effects on urban cellular networks. IEEE Trans. Wirel. Commun. **13**(9), 5070–5083, (2014). doi:10.1109/TWC.2014.2331971
[27] H. ElSawy, E. Hossain, M. Haenggi, Stochastic geometry for modeling, analysis, and design of multi-tier and cognitive cellular wireless networks: A survey, IEEE Commun. Surv. Tutorials, **15**(3), 996–1019, (2013)

Notation and Common Simulation Parameters

The subsequent table summarizes commonly employed parameters.

Table 1 Frequently used parameters

Symbol	Annotation
(r, ϕ)	User location in polar coordinates
(R, Ψ)	Transmitter-site location in polar coordinates
P_M	Macro base station transmit power in [W]
μ_M	Macro base station density in [m^{-2}]
P_M	Small cell transmit power in [W]
μ_S	Small cell density in [m^{-2}]
η	Small cell occupation probability/-ratio, $0 \leq \eta \leq 1$
$\ell(\cdot)$	Distance-dependent path loss law, $0 \leq \ell(\cdot) \leq 1$
R_I	Radius of indoor area or building
L_W	Wall penetration loss, $0 \leq L_\mathrm{W} \leq 1$
S, I	Aggregate signal and interference powers in [W]
γ	Signal-to-Interference Ratio $\gamma = S/I$
τ	Normalized rate (spectral efficiency) in [bit/s/HZ]

Common simulation parameters for Chaps. 11–14 are summarized in Table 2.

Table 2 Common simulation parameters

Parameter	Value
Carrier frequency	$f_c = 2.14$ GHz
LTEA bandwidth	20 MHz
Macro site deployment	Hexagonal grid, one ring
Inter-macro site distance	500 m
eNodeB transmit power	$P_\mathrm{M} = 46$ dBm
eNodeB minimum coupling loss	$c_B = -70$ dB
Shadow fading	Spatially correlated log-normal
Fast fading	Time-correlated Rayleigh
Receiver type	Zero forcing
Noise power density	−174 dBm/Hz
Traffic model	Full buffer
Channel knowledge	Perfect
Simulation length	100 Transmission Time Interval (TTI)
Number of simulation runs	100

Chapter 10
Basic System Level Simulations and Advanced Features

Markus Rupp, Stefan Schwarz and Martin Taranetz

10.1 Survey on Existing Simulation Tools

This chapter gives an overview on existing simulation tools, outlines basic concepts of system level simulations and introduces the advanced features of runtime-precoding and 3D channel modeling. There are several ways to categorize existing system-level simulation tools. First, we may distinguish between simulators, which are implemented as 'modules' of a larger suite, and those, which are specifically designed for LTE-A [1]. Examples for the former include the *Riverbed SteelCentral NetModeler* (formerly *OPNET Modeler Suite*) [2], *OMNeT++* [3], *IT++* [4], *ns-2* [5, 6], *GNS3* [7], *openWNS* [8] and *Hurricane II* [9]. The main drawback of these solutions is their low level of detail, hence leaving most implementation work to the user. Consequently, results often lack accuracy and verification. On the other hand, technology-specific simulators are mainly developed by network operators and vendors, and are typically not intended for commercial distribution [1, 10–12]. Such tools yield a broad range of possibilities for parameter calibration and statistical evaluation. Thus, they are key instruments for the standardization process and the development of new technologies.

While these two classes of simulators largely vary in complexity, scalability and usability, probably the most relevant difference for scientific research is their accessibility. The authors strongly believe that *open access* is a key prerequisite for reproducible simulation studies. The short list of openly available, technology-specific approaches includes *LTE-Sim* [13], the tool presented in [12] and the *Vienna LTE-A*

M. Rupp (✉) · S. Schwarz · M. Taranetz
Institute of Telecommunications, TU Wien, Vienna, Austria
e-mail: mrupp@nt.tuwien.ac.at

S. Schwarz
e-mail: sschwarz@nt.tuwien.ac.at

M. Taranetz
e-mail: martin.taranetz@tuwien.ac.at

M. Rupp et al., *The Vienna LTE-Advanced Simulators*,
Signals and Communication Technology, DOI 10.1007/978-981-10-0617-3_10

downlink system level simulator. While the first lacks of detailed Multiple-Input Multiple-Output (MIMO) modeling, the second provides a rather limited set of features. In this book, we present the *Vienna LTE-A downlink system level simulator* [14].

10.2 The Vienna LTE-A Downlink System Level Simulator

The Vienna LTE-A system level simulator is implemented in object-oriented MATLAB. For further information, see, e.g., [15]. It is made openly available for download under an academic, non-commercial use license. Its rich set of features and easy adaptability has led to numerous publications from researchers all over the globe, including studies on energy-efficient cell-coordination schemes [16], handover algorithms in self-optimizing networks [17], and resource allocation techniques for femtocell networks [18] as well as for machine-to-machine communication [19]. On top of that, the *open accessibility* warrants the reproducibility of these contributions. Today (December 2015), the simulator counts more than 30,000 downloads and undergoes permanent peer-review from a substantially large online community. With some 100,000 lines of code, employing a large forum with active users is the only method to guarantee its quality.

10.3 Physical Layer Modeling

This section provides a brief introduction to modeling concepts of the physical layer of LTE-A on system level. The LTE-A PHY procedures can conceptually be described as a Bit Interleaved Coded Modulation (BICM)-system [20], as shown in Fig. 10.1. It comprises a transmitter including channel coder, bit interleaver and modulator ($\mathcal{M}$). In LTE-A, coding and interleaving is achieved by a turbo-coder in combination with rate matching. The symbol mapping employs 4-, 16- and 64-QAM with Gray mapping, respectively. Signal propagation over an $N_R \times N_T$ MIMO channel is commonly modeled by slowly-varying, position-dependent macro-scale fading L_0, small-scale

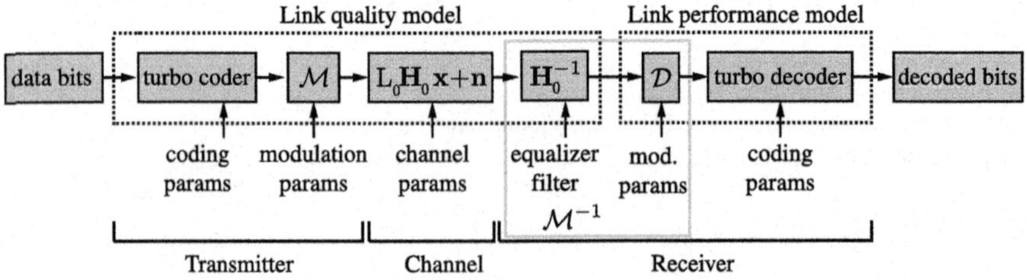

Fig. 10.1 Separation of an LTE link into link quality- and link performance model. The link can equivalently be described as an LTE BICM transmitter–receiver chain [20]

fading $\mathbf{H}_0$ and Additive White Gaussian Noise (AWGN). The matrix representation follows from the assumption that the cyclic prefix exceeds the channel length, hence omitting inter-symbol interference. The channel coefficients are typically calculated from a power-delay profile or a ray-based spatial channel model, such as the Wireless World Initiative New Radio (WINNER) model [21] or 3GPP's 3D model [22]. The receiver encompasses an equalizer filter and a demodulator ($\mathscr{M}^{-1}$) as well as a turbo decoder, which provides de-interleaving and channel decoding. In the current version (v1.8r1375) of the Vienna LTE-A system level simulator, low complexity models for Zero Forcing (ZF)- and Minimum Mean Square Error (MMSE) receivers are available. The former approaches the average performance of an optimal receiver by exploiting Multi-User (MU) diversity, which is typically present in system level scenarios [23].

10.4 Link-to-System Mapping

The objective of the *link abstraction model* or *link-to-system mapping* is to predict the performance of the presented LTE-A link, given a parameterization of the inputs. For simplification, the model can be divided into a *link quality*- and a *link performance model*, as indicated in Fig. 10.1. The link quality model measures the quality of the received signal after equalization. Since the metric has to represent the quality of the input to the turbo decoder, the post-equalization Signal to Interference and Noise Ratio (SINR) is a straightforward choice [20]. The link performance model translates this measure into Block Error Ratio (BLER) and further into (area) spectral efficiency and effective throughput, based on the employed Modulation and Coding Scheme (MCS). The translation is carried out by means of an AWGN BLER curve of the corresponding MCS. The curves are obtained from LTE link level simulations, thus forming the *only* computationally costly physical layer evaluation, which is required for the link abstraction model.

In order to accurately map the system level SINR to the link level curves, an effective SINR is calculated. This mapping is commonly denoted as Effective Signal to Interference and Noise Ratio Mapping (ESM) in literature [24]. It compresses the vector of post-equalization SINR values of the allocated Orthogonal Frequency Division Multiplexing (OFDM) subcarriers to a single effective SINR value that can be further mapped to a BLER value. The general process is illustrated in Fig. 10.2. Several approaches of an ESM have been reported in literature, including Exponential Effective Signal to Interference and Noise Ratio Mapping (EESM) and Mutual Information Effective Signal to Interference and Noise Ratio Mapping (MIESM) [25, 26], respectively. The Vienna LTE-A simulator employs a MIESM, which already proved beneficial in Release 5 of UMTS [27], and was shown to outperform all other approaches (e.g., EESM [28]) in both complexity and performance. In comparison to other methods, it does not require an empirical calibration step as long as the employed codes perform close to capacity [20].

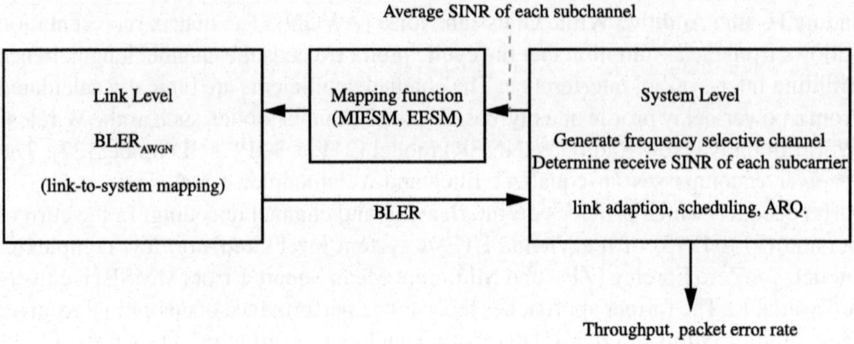

Fig. 10.2 PHY link-to-system mapping procedure [29]

The MIESM method compresses the SINR values of the assigned Resource Blocks (RBs) for each User Equipment (UE) and TTI into an *effective SINR*, yielding an AWGN-equivalent representation in terms of mutual information. The non-linear ESM mapping is expressed as

$$\gamma_{\text{eff}} = I_k^{-1}\left(\frac{1}{N}\sum_{n=1}^{N} I_k(\gamma_n)\right), \tag{10.1}$$

where N is the length of the SINR vector and I_k denotes the BICM capacity for the chosen modulation at the given SINR value γ_n. Thus, MIESM effectively averages the subcarriers in the Mutual Information (MI) domain and then remaps the average MI value to Signal to Noise Ratio (SNR) [20]. For a modulation encoding k bits per symbol, the BICM capacity is expressed as [30]:

$$I_k(\gamma) = k - \mathbb{E}\left\{\frac{1}{k}\sum_{i=1}^{k}\sum_{b=0}^{1}\sum_{z\in\mathscr{X}_b^i}\log\frac{\sum_{\hat{x}\in\mathscr{X}}\exp(-|Y-\sqrt{\gamma}(\hat{x}-z)|^2)}{\sum_{\bar{x}\in\mathscr{X}_b^i}\exp(-|Y-\sqrt{\gamma}(\bar{x}-z)|^2)}\right\}, \tag{10.2}$$

where $\mathscr{X}$ is the set of 2^k constellation symbols, $\mathscr{X}_b^i$ denotes the set of symbols for which bit i equals b and Y is complex normal Random Variable (RV) with zero mean and unit variance.

The model in Fig. 10.1 is a simplification of the actual link abstraction model, as it does not account for interference from other base stations. Its expansion to the whole network is illustrated in Fig. 10.3. The figure identifies the main components of the model as network layout, time-variant fading and scheduling. It also illustrates the corresponding input–output relations to the link quality- and link performance model, respectively.

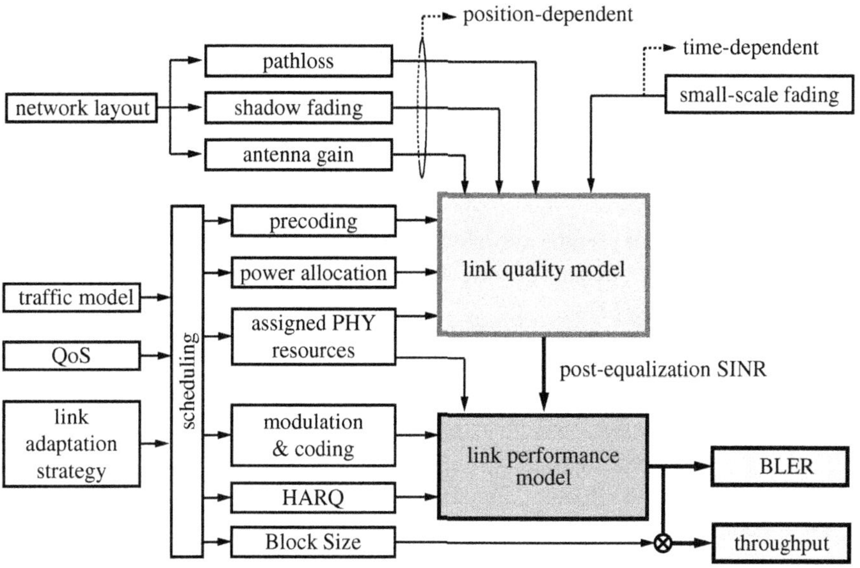

Fig. 10.3 LTE link abstraction model as employed in the Vienna LTE-A downlink system level simulator

10.5 Advanced Features: Runtime Precoding

10.5.1 Relevance

In existing open-source system level simulation tools, UE association is limited to a single eNodeB with all antennas being mounted at the same site. In such scenarios, the fading as experienced over a MIMO link can be decomposed into a slowly varying, position-dependent macro-scale component and a faster changing small-scale component, as shown in Fig. 10.3. Macro-scale fading is determined by the network layout and comprises antenna directivity, path loss and shadowing. Small-scale fading represents fast, frequency-selective channel variations over time. As explained in Sect. 10.3, it is commonly modeled by a normalized $N_\mathrm{R} \times N_\mathrm{T}$ channel matrix $\mathbf{H}_0$, where N_T and N_R denote the number of transmit- and receive antennas, respectively. All entries of H_i have unit mean power in ensemble average.

In the single-eNodeB-single-site case, the macro-scale parameter L_0 is a scalar, which is applied on all entries of $\mathbf{H}_0$. Thus, both L_0 and $\mathbf{H}_0$ can be computed *off-line* and *independently* from each other. Such separation further enables to determine the optimal precoder for each transmission rank a-priori with minimum loss of accuracy [20]. The *effective channel* $\mathbf{H} = \mathbf{G}\mathbf{H}_0\mathbf{F}$, which encompasses precoder $\mathbf{F}$ and receive filter $\mathbf{G}$, can be stored in *channel traces* and may be reused in all simulations with the *same MIMO setting*.

If the desired signal is received from *multiple* eNodeBs (e.g., in certain Coordinated Multi-Point (CoMP) schemes, as indicated in Sect. 10.5.2a) or from a single eNodeB with geographically separated antennas (e.g., in Distributed Antennas System (DAS)- and Remote Radio Head (RRH) deployments, as shown in Sect. 10.5.2b), a-priori computation of the optimal precoders and the corresponding receive filters is no longer possible. In this case, only the normalized small-scale fading matrices $\mathbf{H}_i$ for each transmitter site can be pre-generated. The *composite channel* of the desired signal does not become available until *runtime*. It is obtained by stacking the matrices $\mathbf{H}_i$ from each transmitter-site and weighting them with the corresponding macro-scale losses, L_i, i.e.,

$$\mathbf{H}'_0 = [L_1\mathbf{H}_1\ L_2\mathbf{H}_2\ L_3\mathbf{H}_3\ \ldots]\,. \tag{10.3}$$

Then, the *effective channel* is calculated as $\mathbf{H} = \mathbf{G}\mathbf{H}'_0\mathbf{F}$. In this case, the optimal precoder, $\mathbf{F}$, and the corresponding receive filter $\mathbf{G}$ are determined *at runtime*. The challenge is to enable coherent signal reception from spatially distributed transmission points while keeping additional computational complexity at a minimum. Subsequently, we introduce the so called *runtime-precoding* method.

10.5.2 Implementation

The functionality of *runtime-precoding* is implemented in the UE's link quality model, as indicated in Fig. 10.3. Its major building blocks are outlined in Algorithm 1. The model enables coherent signal reception from spatially distributed sources, *which can be selected at runtime*. At first, it collects the normalized small-scale fading channels and macro-scale losses for both desired and interfering signals. Then, the channels are stacked according to Eq. 10.3.

In the next step, the corresponding precoders are determined. By default, they may be chosen from a 3GPP standard-compliant codebook [31]. Note there is no

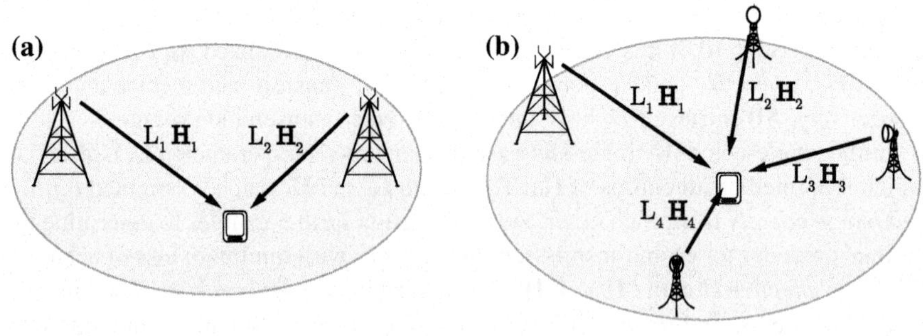

Fig. 10.4 Spatially distributed transmission points. **a** CoMP scenario. **b** eNodeB with RRHs

Algorithm 1: Proposed UE link quality model with runtime-precoding functionality.

Result: post equalization SINR
collect macroscopic path losses and normalized channel matrices from all transmitting sites;
calculate *composite desired channel* by stacking channel matrices of desired signal;
determine precoder or beamformer;
if *there are interferers* **then**
  calculate *composite interfering channels* by stacking channel matrices of interferers;
  determine precoders or beamformers for interfering channels;
else
  noise power only;
end
calculate receive filters and effective channel matrices;
determine post equalization SINR and store for link performance model;
calculate feedback based on actual channel;

standardized method to determine the Precoding Matrix Indicator (PMI). The Vienna LTE-A downlink system level simulator employs a scheme that maximizes the mutual information between transmitted and received symbols [32]. Nonetheless, the availability of the *full channel at runtime* allows researchers to apply arbitrary precoders and beamformers, yielding a profound novelty in open-source LTE-A system level simulation tools. Note that in general, beamforming strategies are found by solving multi-objective optimization problems [33]. The implementation of 3GPP's 3D channel model enables the investigation of *elevation beamforming* and full-dimension MIMO [22].

10.5.3 Performance Evaluation

In this section, we evaluate the price to pay for enabling coherent multi-point transmission in system level simulations. For this purpose, we extend the UE link quality model of the Vienna LTE-A simulator by the runtime-precoding functionality according to Algorithm 1. Then we measure simulation run times with the new- and the legacy model. The results are compared with run times as obtained with the Vienna LTE-A Downlink Link Level simulator [34]. For a meaningful comparison, all simulations were carried out on the same hardware, an Intel(R) Core(TM) i7-3930K CPU @ 3.20 GHz, equipped with 32 GB of DDR3 1333 quad-channel RAM.

The common setup, which is employed in both link- and system level simulations, is summarized in Table 10.1. We carry out simulations with the LTE bandwidths $B = \{1.4, 3, 5, 10, 20\}$ MHz and the $N_\mathrm{T} \times N_\mathrm{R}$ antenna configurations $\{2 \times 2, 4 \times 2, 4 \times 1\}$ for various simulation lengths (measured in multiples of 1 TTI), in particular $N_\mathrm{TTI} = \{100, 500, 1000\}$ on system level and $N_\mathrm{TTI} = \{10, 100, 500\}$ on link level, respectively. Moreover, we perform system level simulations with $K = \{1, 10, 100\}$ UEs at a simulation length of 100 TTI. A round robin scheduler is employed. Note that

Table 10.1 Simulation parameters as employed for the simulation run time evaluation

Parameter	Value
Frequency	2.14 GHz
Number of eNodeBs	1
Transmit power	5 W
eNodeB antenna gain pattern	Omni-directional
LTE transmission mode	CLSM
Path loss in dB	$\max\left(10\log_{10}\left(\frac{4\pi d f}{c_0}\right)^2, 0\right)$
Shadow fading	None
Channel model	ITU-R Ped-A, block fading
Receiver type	Zero forcing
Noise power spectral density	−160 dBm/Hz
UE position	Random in circle with $R = 250\,\text{m}$
UE antenna gain pattern	Omni-directional
Traffic model	Full buffer
Channel knowledge	Perfect
Feeback	AMC: CQI, MIMO: PMI and RI
Feedback delay	3 TTI

the round robin scheduler does not increase in complexity with the number of physical RBs (i.e., increasing B). Other scheduling algorithms may have a considerable impact on the simulation run time in multi-user scenarios [20].

Figure 10.5 shows the obtained simulation run times. Each point was computed by averaging over ten simulation runs. It is observed that on system level the results scale approximately linearly with the simulation length N_{TTI}, the bandwidth B and the number of UEs K. Compared to this, the link level results exhibit a slightly non-linear scaling with B (note that on link level, only a single link is evaluated, i.e., $K = 1$). Both link- and system level run times show a non-linear dependence on the number of transmit- and receive antennas, N_{T} and N_{R}, respectively. From these observations, we can derive the following generic run time estimator (for a better understanding of the scaling with N_{T} and N_{R}, additional simulations with the MIMO configurations $\{2 \times 1, 4 \times 1\}$ were carried out).

$$\begin{aligned} &T_{[\mathrm{s}]}(N_{\mathrm{TTI}}, B, N_{\mathrm{T}}, N_{\mathrm{R}}, K) \\ &\quad = (c_0 + N_{\mathrm{TTI}} \cdot K \cdot B_{[\mathrm{MHz}]} \cdot (c_1 N_{\mathrm{T}} + c_2 N_{\mathrm{R}} + c_3 N_{\mathrm{T}} N_{\mathrm{R}}) \\ &\qquad + \cdots + N_{\mathrm{TTI}} \cdot K \cdot B^2_{[\mathrm{MHz}]} \cdot (c_4 N_{\mathrm{T}} + c_5 N_{\mathrm{R}} + c_6 N_{\mathrm{T}} N_{\mathrm{R}})) \\ &\qquad \times \cdots \times 1\,\mathrm{s}, \end{aligned} \tag{10.4}$$

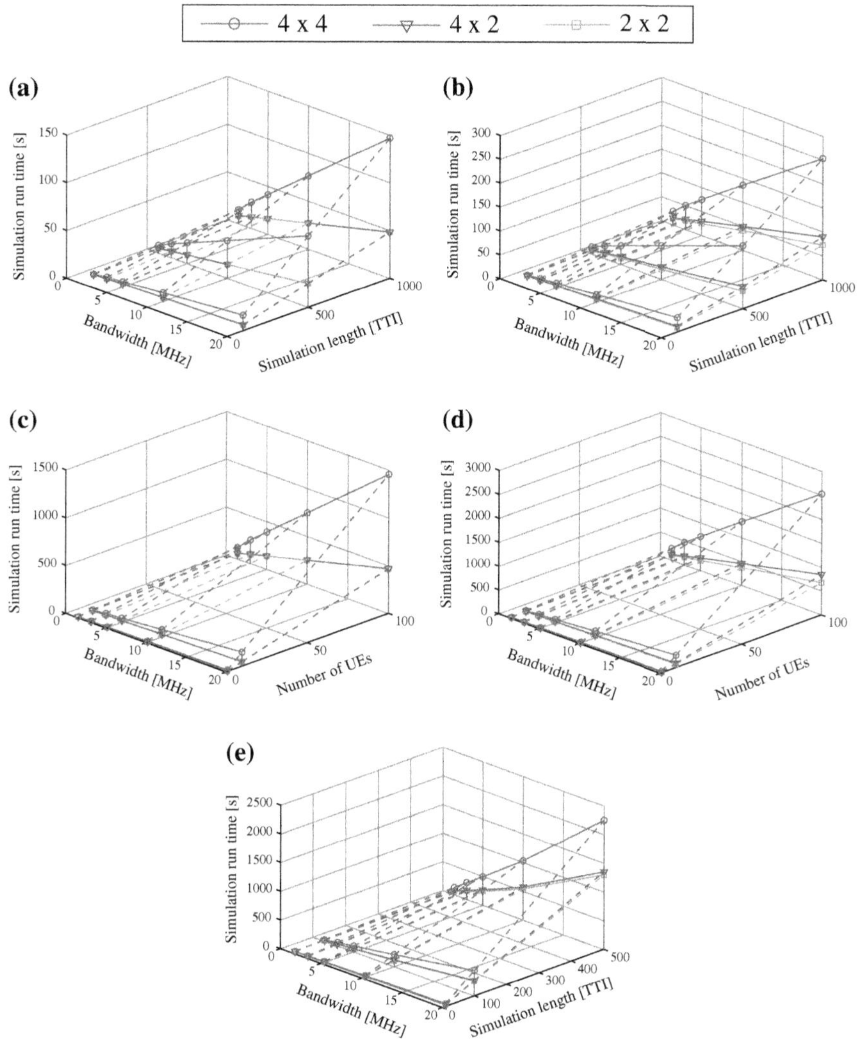

Fig. 10.5 Simulation run times (s) for various LTE bandwidths (MHz), antenna configurations $\{N_T \times N_R\}$ and number of UEs, **a** system level, *old* model, 1 UE, **b** system level, *new* model, 1 UE, **c** system level, *old* model, 100 TTIs, **d** system level, *new* model, 100 TTIs, **e** link level, 1 UE

where $B_{[\text{MHz}]} = B/10^6$ Hz. Next, we compute the coefficients $c_0, \ldots, c_6$ by linear least squares. The results are summarized in Table 10.2. Compared to the legacy link quality model, simulations with the new model require 1.25× longer for initialization (represented by the coefficient c_0), which is still 52.5× faster than link level simulations. It slightly decreases the scaling with $B_{[\text{MHz}]} \cdot N_R$ (referring to c_2) by 1.4× while increasing c_3 (according to the scaling with $B_{[\text{MHz}]} \cdot N_T \cdot N_R$) by 2.7×. The latter term is of particular relevance for investigating massive MIMO scenarios

Table 10.2 Coefficients for the runtime estimator in Eq. 10.4 as obtained by linear least squares

Simulation type	c_0	c_1	c_2	c_3	c_4	c_5	c_6
Legacy link quality model	0.8	0	8.2×10^{-4}	2.7×10^{-4}	0	0	0
New link quality model	1.0	0	5.7×10^{-4}	7.4×10^{-4}	0	0	0
Link level simulations	52.5	0	235.0×10^{-4}	0	2.3×10^{-4}	11.0×10^{-4}	0

Values are provided for system level simulations with the legacy- and the new link quality model as well as for link level simulations

with a large number of transmit antennas. On link level, c_2 (referring to a scaling proportional to $B_{[\mathrm{MHz}]} \cdot N_\mathrm{R}$) is $41.2\times$ larger than on system level, while $c_3 = 0$ (corresponding to the scaling proportional to $B_{[\mathrm{MHz}]} \cdot N_\mathrm{R} \cdot N_\mathrm{T}$). On the other hand, the link level simulation run times scale with $N_\mathrm{T} \cdot B^2_{[\mathrm{MHz}]}$ and $N_\mathrm{R} \cdot B^2_{[\mathrm{MHz}]}$ (referring to c_4 and c_5), while system level simulations with both legacy- and new link quality model exhibit no dependency on $B^2_{[\mathrm{MHz}]}$, i.e., $c_4 = 0$, $c_5 = 0$ and $c_6 = 0$, respectively. A more detailed analysis of the link level simulation times shows that the non-linearity mainly arises from the symbol demapping and decoding, as described in Sect. 10.3.

10.6 Advanced Features: 3D Channel Modeling

Fjolla Ademaj

10.6.1 Introduction

Developing realistic channel models is one of the greatest challenges in describing wireless communications. Their quality is crucial for accurately predicting the performance of a wireless cellular system. Broadly speaking, channel models can be divided into two categories, *deterministic* and *stochastic*. Deterministic models describe the channel for a *specific* propagation environment between transmitter and receiver. This method can be tedious to evaluate and does not allow for general statements in an ensemble of environments. In stochastic models, the channel characteristics are often condensed to a statistical description, e.g., the *typical* Power Delay Profile (PDP).

In order to close the gap between the two approaches, 3rd Generation Partnership Project (3GPP) has introduced the Spatial Channel Model (SCM) [35]. Unlike traditional channel models, it incorporates not only a random PDP but also a random angular profile Access Point (AP). The model represents scatterers through statistical parameters without being physically positioned. The SCM belongs to the class of *geometric stochastic* model and separately defines *large scale parameters* (e.g., shadow fading, delay spread and angular spreads) as well as *small scale parameters* (e.g., delays, cluster powers, and arrival- and departure angles). Both parameter sets are randomly drawn from tabulated distributions. The large scale parameters encompass the geometric positions of the BSs and the UEs, respectively. Moreover, they are used to parameterize the statistics of the small scale parameters.

The SCM model in [35] includes six different scenarios, each of them representing a unique environment. Initially it was targeted for a bandwidth of only 5 MHz, and a carrier frequency of 2 GHz. Later, it was extended to the Spatial Channel Model Extension (SCME). The SCME follows the same procedure as the SCM, but supports bandwidths of up to 100 MHz and a frequency range of 2–6 GHz. In the course of the WINNER projects, the model was extended for 15 different scenarios, including urban-, rural- and moving environments [21, 36]. The WINNER model is recommended as a baseline for evaluating radio interface technologies in the International Telecommunication Union-Radiocommunication Sector (ITU-R) [37].

Recently there has been a significant interest in enhancing system performance through the use of antenna systems having a two-dimensional array structure that provides adaptive control over both the elevation- and the azimuth dimension. The additional control over the elevation dimension enables a variety of strategies such as sector-specific elevation beamforming (e.g., adaptive control over the vertical pattern beamwidth and/or downtilt), advanced sectorization in the vertical domain, and UE-specific elevation beamforming. Vertical sectorization can improve average system performance through the higher gain of the vertical sector patterns, but vertical sectorization generally does not need additional standardization support. UE-specific elevation beamforming promises to increase the SINR statistics seen by the UEs by pointing the vertical antenna pattern in the direction of the UE while spraying less interference to adjacent sectors by virtue of being able to steer the transmitted energy in elevation. In order to enable evaluation of the possible specification method to enhance the performance with 3D beamforming or with Full Dimensional (FD)-MIMO, a new channel model is needed which enables modelling in both vertical- and horizontal dimension of the environment. Recently, 3GPP introduced a new 3D SCM for LTE-A [22].

As of this writing, only few simulation studies, including reports from the 3GPP TSG RAN WG1 meetings, have been published that claim the practical implementation of the model [38, 39]. However, the employed tools are mainly developed by network operators and vendors, and thus typically intended for commercial use. We provides a guideline for the practical implementation of the model. The MATLAB source code is openly available for download on our webpage www.nt.tuwien.ac.at/vienna-lte-a-simulators under an academic, non-commerical use license. It is provided as a stand-alone package that is directly applicable for system level simulation tools and can straightforwardly be ported to link level.

Based on [40], the contribution of this section outlines as follows. The guideline for a computationally efficient implementation of the 3GPP 3D channel model is presented in Sect. 10.6.3. In Sect. 10.6.4, the implementation is validated against results from the 3GPP standard with the Vienna LTE-A Downlink System Level Simulator. An outline of new opportunities and challenges is given in Sect. 10.6.5.

10.6.2 3GPP 3D Channel Model

The 3GPP 3D channel model characterizes wireless communication channels of typical European cities. It is a 3D geometric stochastic model, describing the scattering environment between BS and UE in both azimuth and elevation dimensions. The scatterers are represented by statistical parameters without having a real physical location. In 3GPP TR 36.873 [22], three scenarios, Urban Macro cell (UMa), Urban Micro cell (UMi) and UMa-high rise (UMa-H) are specified. They represent typical urban macro-cell and micro-cell environments. Both UMa and UMa-H scenarios, consider a BS height of 25 m, thus surpassing the surrounding buildings. UMa-H also specifies such environments with one high-rise building per Evolved Base Station (eNodeB) sector. UMi, considers a BS height of 10 m, lying below the rooftop level. All three environments are assumed to be densely populated with buildings and take into account both indoor- and outdoor UEs.

The 3GPP 3D channel model specifies three propagation conditions, Line-Of-Sight (LOS), Non Line of Sight (NLOS) and Outdoor-to-Indoor (O-to-I). For each of these conditions it defines different parameters for mean propagation path loss, macroscopic fading and microscopic fading. All three scenarios in [22], UMa, UMi and UMa-H, consider 80 % of the UEs to be located indoors. The probability of being in LOS is determined separately for indoor and outdoor UEs and depends on the height of the UE as well as the break point distance. The break point distance characterizes the gap between transmitter and receiver at which the Fresnel zone is barely broken for the first time [41]. For an indoor UE, LOS refers to the signal propagation outside the building in which the UE is located. For each UE location, large scale parameters are generated according to its geographic position as well as the propagation conditions at this location. The large scale parameters incorporate shadow fading, the Ricean K-factor (only in the LOS case), delay spread, azimuth angle spread of departure- and arrival, as well as zenith angle spread of departure- and arrival.

The small scale parameters incorporate delays, cluster powers as well as angles of departure and -arrival in azimuth and elevation direction, respectively. The model considers N clusters of scatterers, where each cluster is resolvable to M paths. A simplified sketch of the model is given in Fig. 10.6. The channel coefficients are defined per cluster n, BS antenna element s and UE antenna element u as

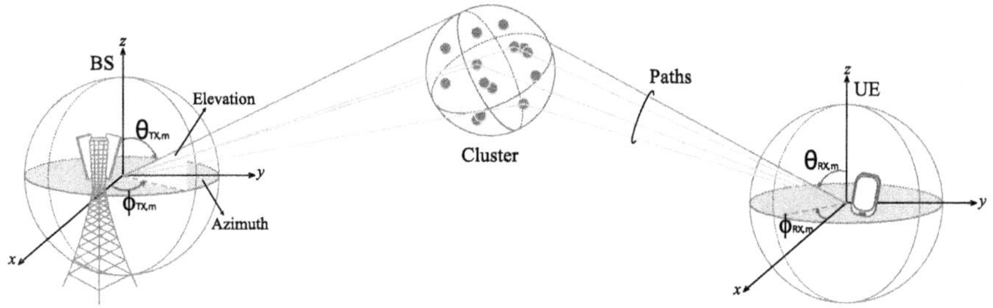

Fig. 10.6 Scattering concept in the 3D model. Figure demonstrates a link that is resolvable to M paths. Elevation- and azimuth angles at BS and UE are denoted as θ and ϕ, respectively

$$
\begin{aligned}
H_{u,s,n}(t) = \sqrt{\frac{P_n}{M}} \sum_{m=1}^{M} & \begin{bmatrix} F_{rx,u,\theta}\left(\theta_{n,m,ZOA}, \phi_{n,m,AOA}\right) \\ F_{rx,u,\phi}\left(\theta_{n,m,ZOA}, \phi_{n,m,AOA}\right) \end{bmatrix}^T \\
& \times \begin{bmatrix} \exp\left(j\Psi_{n,m}^{\theta\theta}\right) & \sqrt{K_{n,m}^{-1}} \exp\left(j\Psi_{n,m}^{\theta\phi}\right) \\ \sqrt{K_{n,m}^{-1}} \exp\left(j\Psi_{n,m}^{\phi\theta}\right) & \exp\left(j\Psi_{n,m}^{\phi\phi}\right) \end{bmatrix} \\
& \times \begin{bmatrix} F_{tx,s,\theta}\left(\theta_{n,m,ZOD}, \phi_{n,m,AOD}\right) \\ F_{tx,s,\phi}\left(\theta_{n,m,ZOD}, \phi_{n,m,AOD}\right) \end{bmatrix} \\
& \times \exp\left(j2\pi\lambda_0^{-1}\left(\hat{r}_{rx,n,m}^T \bar{d}_{rx,u}\right)\right) \\
& \times \exp\left(j2\pi\lambda_0^{-1}\left(\hat{r}_{tx,n,m}^T \bar{d}_{tx,s}\right)\right) \\
& \times \exp\left(j2\pi\, v_{n,m} t\right),
\end{aligned}
\tag{10.5}
$$

where P_n is the power of path n, $F_{rx,u,\theta}$ and $F_{rx,u,\phi}$ are field patterns of the receive antenna element u in the direction of the spherical basis vectors, $\hat{\theta}$ in the zenith direction, $\hat{\phi}$ in the azimuth direction. The expressions $F_{tx,s,\theta}$ and $F_{tx,s,\phi}$ are field patterns of the transmit antenna element s in the direction of $\hat{\theta}$ and $\hat{\phi}$, respectively. The departure- and arrival angels in zenith and azimuth direction are denoted with θ and ϕ, respectively. The term $K_{n,m}$ represents cross polarization power ratios for each path m and cluster n, and $\Psi_{n,m}^{AB}$ are random initial phases for four different polarization combinations $AB = \{\theta\theta, \theta\phi, \phi\theta, \phi\phi\}$. The terms $\hat{r}_{rx,n,m}$ and $\hat{r}_{tx,n,m}$ are the receiver and transmitter spherical unit vectors expressed in Cartesian coordinates. They are defined as

$$
\hat{r}_{n,m} = \begin{pmatrix} \sin\theta\cos\phi \\ \sin\theta\sin\phi \\ \cos\theta \end{pmatrix}. \tag{10.6}
$$

The parameters $\bar{d}_{rx,u}$ and $\bar{d}_{tx,s}$ are the location vectors of receive and transmit antenna elements, respectively. Considering a BS with coordinates $\left(s_x, s_y, s_z\right)$, and a planar antenna array, the location vector per antenna element is

$$\bar{d}_{tx,s} = \begin{pmatrix} s_x \\ s_y \\ s_z \end{pmatrix} + \begin{pmatrix} 0 \\ (p-1)d_H \\ (q-1)d_V \end{pmatrix} \tag{10.7}$$

where $p \in P$ and $q \in Q$. The terms P and d_H denote the number of antenna elements and the element spacing in the horizontal direction, while Q and d_V are the number of antenna elements and the element spacing in the vertical direction, respectively. The last component of (10.5), $v_{n,m}$, represents the Doppler frequency component of the UE moving at velocity $\bar{v}$. Further details on the calculation of the variables in (10.5) can be referred from [22].

Antenna Modeling

The 3GPP 3D channel model enables to scrutinize 2-dimensional (2D) planar antenna arrays, also known as rectangular arrays. The antenna elements can either be linearly polarized (*co-pol*) or cross prolarized (*cross-pol*), as shown in Fig. 10.7. In this regard, the model represents a compromise between practicality and precision as it does not include the mutual coupling effect as well as different propagation effects of horizontally and vertically polarized waves. Our well-structured implementation will substantially facilitate the implementation of further techniques for modeling different polarization modes such as the one proposed in [42].

The antenna elements are equidistantly spaced in the y- and the z-direction. For static electrical beam steering, also known as *electrical tilting*, a complex weight is applied to each antenna element in the vertical direction. For an antenna element in the qth row, it is given as

$$w_q = \frac{1}{\sqrt{Q}} \exp\left(-j\frac{2\pi}{\lambda}(q-1)d_V \cos\theta_{\text{etilt}}\right), \tag{10.8}$$

where Q represents the total number of antenna elements in the vertical direction and θ_{etilt} is the steering angle in the vertical plane. Unlike the conventional approach of applying an array factor to the field pattern of a single element of a uniform

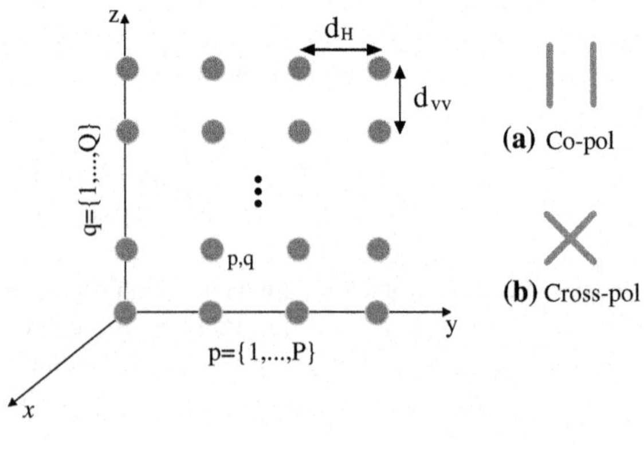

Fig. 10.7 Geometry and polarization modes of a planar antenna array. The antenna elements in horizontal and vertical direction are indexed by p and q, respectively

antenna array, in the 3D model the beamforming weights are applied to the channel coefficients for each antenna element

$$[\mathbf{H}^{\mathrm{c}}_{i,n}(t)]_{a,b} = \sum_{u \varepsilon P_a} w_u \sum_{s \varepsilon P_b} w_s H_{i,u,s,n}(t), \qquad (10.9)$$

where $[\mathbf{H}^{\mathrm{c}}_n(t)]_{a,b}$ represents the weighted and combined channel coefficients. The terms $\mathscr{P}_a$ and $\mathscr{P}_b$ denote the sets of antenna elements that belong to receive antenna port a with $a \in \{1, \ldots, N_{\mathrm{Rx}}\}$ and transmit antenna port b with $b \in \{1, \ldots, N_{\mathrm{Tx}}\}$, respectively. The terms ω_u and ω_s are complex weights that account for phase shifts as applied for static beamforming (e.g., electrical downtilting), respectively. The relative position of each element in the array is incorporated in the channel coefficients $H_{u,s,n}(t)$, where n denotes the cluster index, s and u are the BS and UE antenna elements, respectively.

In the following, a detailed procedure on the implementation of the model for simulations is provided.

10.6.3 Implementation on System-Level

In this section, we describe the necessary steps to integrate the 3D channel model into an existing simulation tool. The target is to compute a $N_{\mathrm{Rx}} \times N_{\mathrm{Tx}}$ MIMO-channel matrix $\mathbf{H}(t,f)$ for each sampling point on the time-frequency grid, where N_{Tx} and N_{Rx} refer to the number of transmit- and receive antenna *ports*, respectively. On link level, channel realizations are typically calculated per OFDM symbol and LTE-A subcarrier [34]. On system level, they are commonly generated per physical RB and TTI [11] (Fig. 10.8).

In the 3GPP 3D channel model, the channel coefficients depend on the UE location in the 3D space and, thus, have to be calculated *at runtime*. Hence, in order to reduce complexity, the challenge is to perform computationally intensive tasks *off-line* or *on demand*, whenever possible. We will follow the stepwise procedure as specified in [22, Sect. 7.3] and illustrated in Fig. 10.9, and explain its expedient partition for implementation. The steps are denoted as '*Step N*' with $N \in \{1, \ldots, 12\}$.

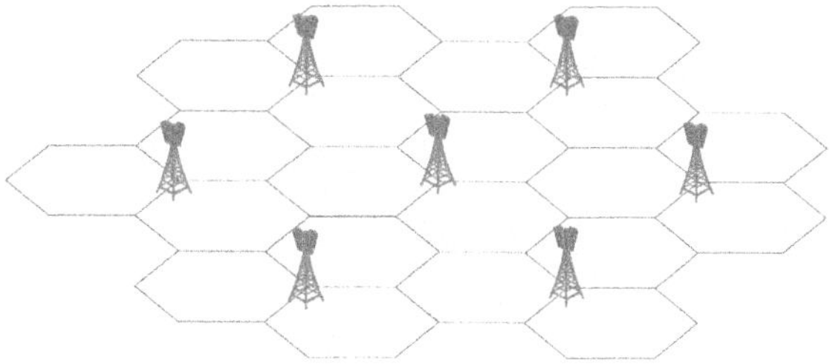

Fig. 10.8 Hexagonal-grid macro-cell scenario with seven BSs and 21 sectors. This scenario is used in simulations throughout the paper

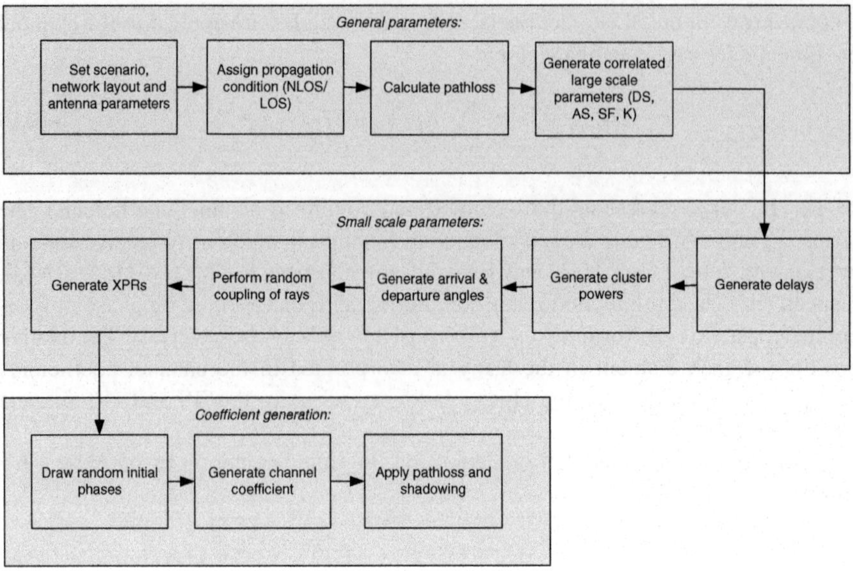

Fig. 10.9 Procedure for generating channel coefficients in 3GPP 3D channel model

[GP] The first step is to generate the *general parameters*. It starts with specifying the network layout, the scenario environment and the antenna array parameters (*Step 1*). Currently, the standard specifies three scenarios, 3D-UMa, 3D-UMi and 3D-UMa-H, and various planar antenna array structures, defining the location and polarization of each antenna element, as well as the element-to-port mapping. *Step 2* is to assign the propagation condition, i.e., either LOS or NLOS, separately for indoor and outdoor UEs. The decision is based on the LOS probability as specified in [22, Table 7.2-2], which is dependent on the UE's height and its distance to the macro-site. Different path loss models are applied for LOS, NLOS and O-to-I, as defined in [22, Table 7.2-1]. The experienced path loss is calculated in *Step 3*. In *Step 4*, the large scale parameters are generated. The detailed procedure is described in [21, Sect. 3.3.1]. For each UE, a vector of large scale parameters is generated as

$$LSP = \left(\delta_{\mathrm{SK}}\ \delta_{\mathrm{K}}\ \delta_{\mathrm{DS}}\ \delta_{\mathrm{ASD}}\ \delta_{\mathrm{ASA}}\ \delta_{\mathrm{ZSD}}\ \cdots\ \delta_{\mathrm{ZSA}}\right). \tag{10.10}$$

In case the UE is not in LOS of BS, $\delta_{\mathrm{K}} = 0$. These tasks can be performed offline, i.e., before entering the actual simulation loop. Similar to the generation of the shadow fading, they have to be performed only *once per site*.

[SSP] The next step is to generate *small scale parameters*. In the 3GPP 3D channel model, channel coefficients $H_{u,s,n}(t)$ are determined individually for each cluster n and each receiver- and transmitter *antenna element* pair $\{u, s\}$, respectively. The calculation of $H_{u,s,n}(t)$ requires to generate delays (*Step 5*), cluster powers (*Step 6*) as well as arrival- and departure angles for both azimuth and elevation (*Step 7*). After coupling the rays within a cluster (*Step 8*), Cross Polarization Power Ratios (XPRs) and random initial phases are drawn (*Step 9 and 10*). Together with the calculation of the spherical unit vectors and the Doppler frequency component (both *Step 11*), all parameters mentioned above are commonly applied to each antenna element pair $\{u, s\}$ and thus *have to be determined only once per antenna array and eNodeB*. The Doppler component accounts for the time variance of the channel. The frequency selectivity is determined by the channel impulse response $H_{i,u,s,n}(t)$ and the sampling frequency, which is directly related to the system bandwidth.

[CG] After generating the channel coefficients for each antenna element pair $\{u, s\}$, the channel coefficients for an antenna array are combined according to the antenna element-to-port mapping given in Sect. 10.6.2. Then, the combined channel $\mathbf{H}_n^{\mathrm{c}}(t)$, for each cluster n is sampled based on the delay taps m defined as

$$m = \left\lfloor \frac{\tau_n}{T_s} \right\rfloor, \tag{10.11}$$

where $m \in \{1, \ldots, \mathrm{N}\}$, T_s is the sample interval in the time domain and τ_n denotes the actual delay of the nth cluster. The sampled $N_{\mathrm{Rx}} \times N_{\mathrm{Tx}}$ channel matrix is denoted as $\hat{\mathbf{H}}_m(t)$, with an element $[\hat{\mathbf{H}}_m(t)]_{a,b}$, referring to the sampled channel coefficient for receive antenna port a and transmit antenna port b, respectively. It is important to note that the model is designed such that the channel impulse response before sampling has unit sum power on average over t, i.e., $\mathbb{E}_t\left[\sum_n |[\mathbf{H}_n^{\mathrm{c}}(t)]_{a,b}|^2\right] = 1$, when assuming antenna elements with *omni-directional* gain pattern and 0 dB gain, as well as an XPR of one. In order not to change the sum power after the sampling, we multiply the sampled channel coefficients $[\hat{\mathbf{H}}_m(t)]_{a,b}$ by a factor of $\sqrt{\sum_n |[\mathbf{H}_n^{\mathrm{c}}(t)]_{a,b}|^2 / \sum_m |[\hat{\mathbf{H}}_m(t)]_{a,b}|^2}$. The final channel impulse response in the frequency domain is obtained by performing a Fast Fourier Transform (FFT) over the sampled and normalized channel impulse response

$$\tilde{\mathbf{H}}[k, t] = \sum_{m=0}^{\mathrm{N}-1} \hat{\mathbf{H}}_m(t) e^{-j2\pi k \frac{m}{\mathrm{N}}}, \tag{10.12}$$

where, $k = 0, 1, \ldots, \mathrm{N} - 1$. The term N represents the number of FFT samples which is the maximum number of delay taps m. For example, assuming a transmission bandwidth of 10 MHz, according to [43], the sampling interval is $T_s = 65$ ns and the number of FFT samples is $\mathrm{N} = 1024$.

Considering a UE with a fixed location in the 3D space, ***[SSP]*** and ***[CG]*** have to be carried out only in the first time instant of the simulation. Afterwards, the channel will remain static over time (no Doppler effect). If the UE moves at a certain speed, represented by the vector $\mathbf{v} \in \mathbb{R}^3$, in principle, ***[SSP]*** would have to be performed at runtime in each time instant of the simulation. This also implies the generation

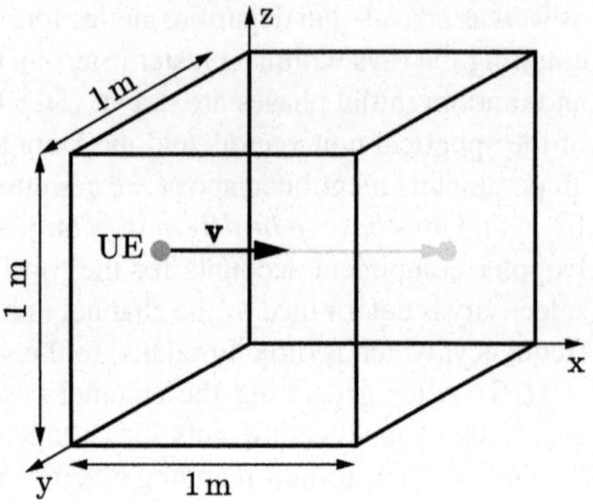

Fig. 10.10 UE travels through cube with an edge length of 1 m

of new clusters and random initial phases, i.e., a complete change of the multi-path propagation environment. Thus, it is considered reasonable from a physical perspective (see, e.g., [21]) as well as in view of computational complexity to partition the scenario into equally sized cubes. As long as the UE resides within the same cube, it is assumed to experience the same path loss, shadow fading, propagation conditions (LOS/NLOS/O-to-I) and large scale parameters, as generated in ***[GP]***. Then, ***[SSP]*** has to be carried out only once at the beginning of the simulation and each time the UE transfers to another cube. Assuming a spatial resolution of 1 m and a temporal resolution of 1 ms, referring to the length of one LTE sub-frame, also denoted as *TTI*, a UE moving at $\mathbf{v} = [27.78, 0, 0]\,\mathrm{m/s}$ requires 36 ms to travel from one face of the cube to the other, as indicated in Fig. 10.10. In this case, ***[SSP]*** is performed every 36 sub-frames. Within a cube, channel variations are caused by the slightly changing angles of arrival and departure (and thus the antenna element field patterns) as well as the phase shift due to the Doppler effect. They can be incorporated into ***[CG]*** thus yielding the only variable components that have to be recalculated in each time instant of the simulation. If the UE trace is known, e.g., in train and car scenarios, the simulation complexity can be reduced even further. In such scenarios, the Doppler frequency component in (10.5) has to be calculated only in the first time instant of the simulation, and can be reused in subsequent time instances as long as the user stays within the same cube.

10.6.4 Calibration

Following the steps in Sect. 10.6.3, we incorporated the 3D channel model in the Vienna LTE-A Downlink System Level Simulator (current version v1.8r1375) [11]. The simulator is implemented in object-oriented MATLAB and is made openly available for download under an academic, non-commercial use license. It is built

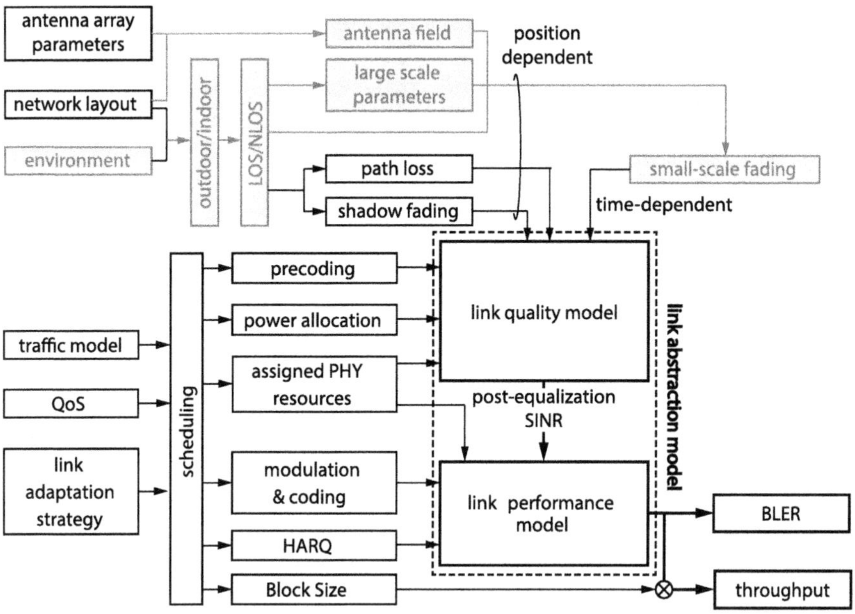

Fig. 10.11 Enhanced link abstraction model for enabling 3D channel modeling

according to the commonly employed structure for system level simulation tools (see, e.g., in [24, 44]), as illustrated in Fig. 10.11, thus serving as a representative example. Its centerpiece is the *link abstraction model* that specifies the interaction between link- and system level simulations [11, 24]. This structure is expected to persist in simulation tools for the fifth generation of mobile cellular networks (5G) [44]. The enhancements that were necessary to enable the 3D channel model are depicted by the boxes shaded in gray at the top of the figure.

For calibration purposes, we carry out simulations with the setup as specified in [22, Table 8.2-2] and summarized in Table 10.3. Two scenarios, 3D-UMa and 3D-UMi are investigated. We provide the calibration results for large scale parameter statistics, using the *circular angle spread* method, as recommended in [45]. This method is used to generate the angular statistics a posteriori from the angular parameters generated in *Step 11*. We thus consider this procedure to provide a more reliable verification of our implementation than the previously used method in [40]. Figure 10.12 depicts the obtained statistics for zenith spread of departure- and arrival. In accordance with the results in [46], the distributions show similar characteristics for 3D-UMa and 3D-UMi scenarios. Furthermore, they exhibit a good agreement with results from [22] (dash-dotted curves), which were obtained by averaging over 21 sources as reported in [47]. In Fig. 10.13 we provide the calibration results for largest and smallest singular values as referred from [22, Table 8.2-2]. The singular values are generated on a RB basis at $t = 0$ by considering channel matrices where

Table 10.3 Simulation parameters for calibration as referred from [22]

Parameter	Value
Carrier frequency	2 GHz
LTE bandwidth	10 MHz
Macro-site deployment	Hexagonal grid
Scenarios	3D-UMa, 3D-UMi
BS antenna height (UMa)	25 m
BS antenna height (UMi)	10 m
BS antenna configuration	$N_{\text{Tx}} = 4$
UE antenna configuration	$N_{\text{Rx}} = 2$
Polarized antenna modeling	Model 2 [22]
BS antenna polarization	X-pol ($\pm 45°$)
UE antenna polarization	X-pol ($0/+90°$)
Antenna elements per port	$M = 10$
Vertical antenna element spacing	0.5λ
Horizontal antenna element spacing	0.5λ
Maximum antenna element gain	8 dBi
UE antenna pattern	Isotropic antenna gain
Electrical downtilt	12°
UE distribution	Uniform in cell [22, Table 6-1]

path loss and shadowing are not yet applied to the channel coefficients. The results show a good agreement with the results from [22] (dash-dotted curves), which were obtained by averaging over 21 sources as reported in [47].

10.6.5 Opportunities

The integration of the 3D channel model into existing link- and system level simulation tools paves the way for more advanced studies on the performance of a mobile cellular system in realistic environments. Existing channel models only support linear antenna arrays in the azimuth. With the introduction of the third dimension, not only higher-order MIMO schemes but also a higher number of antenna elements per antenna array can be investigated. Currently, the 3GPP LTE-A standard supports up to eight antenna ports. However, recent trends aim at 100 and more antenna ports per eNodeB [48]. A main enabler for this so called *massive MIMO* approach will be the adoption of higher carrier frequencies, also termed *millimeter-wave communication*, as it enables to considerably decrease the size of the antenna arrays. On the one hand, this may lead to higher complexity of the hardware, larger energy consumption and a greater demand for signal processing capabilities. On the other

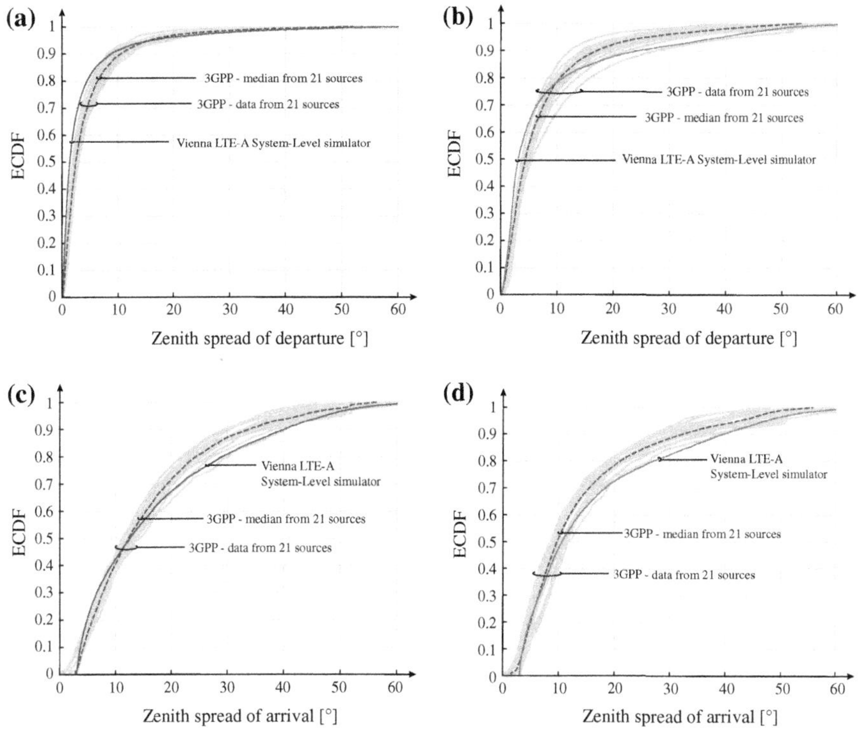

Fig. 10.12 Large scale parameter statistics. *Gray lines* refer to results reported by 21 sources from [47]. *Dashed curves* denote the 3GPP reference results from [22, Figs. 8.2-11, 8.2-13]. **a** UMa-Zenith spread of departure. **b** UMi-Zenith spread of departure. **c** UMa-Zenith spread of arrival. **d** UMi-Zenith spread of arrival

hand, it will enable a much more accurate bundling of energy towards the intended receiver, which is a key prerequisite for aggressive frequency reuse. In dense urban environments, where UEs move in three dimension (consider, e.g., shopping malls, skyscrapers, and more) it is conceivable that the spectral efficiency per unit sphere might replace the area spectral efficiency as a figure of merit. Other important use cases are scenarios with high user mobility, as the number of commuters is expected to increase substantially. People have become used to services following them wherever they travel. Mobile cellular access has even become a key argument to choose the means of transportation. Sharp, steerable beams might be an expedient solution to this issue, as they could follow a vehicle along its path.

Improvements targeting planar antenna arrays are to be further investigated. New virtualization models of antenna arrays, considering a full-connection between antenna elements, weighted in both horizontal-and vertical direction will lead to a better understanding of the 3D beamforming. Moreover, new two-dimensional codebook designs are necessary for the evaluation of FD-MIMO.

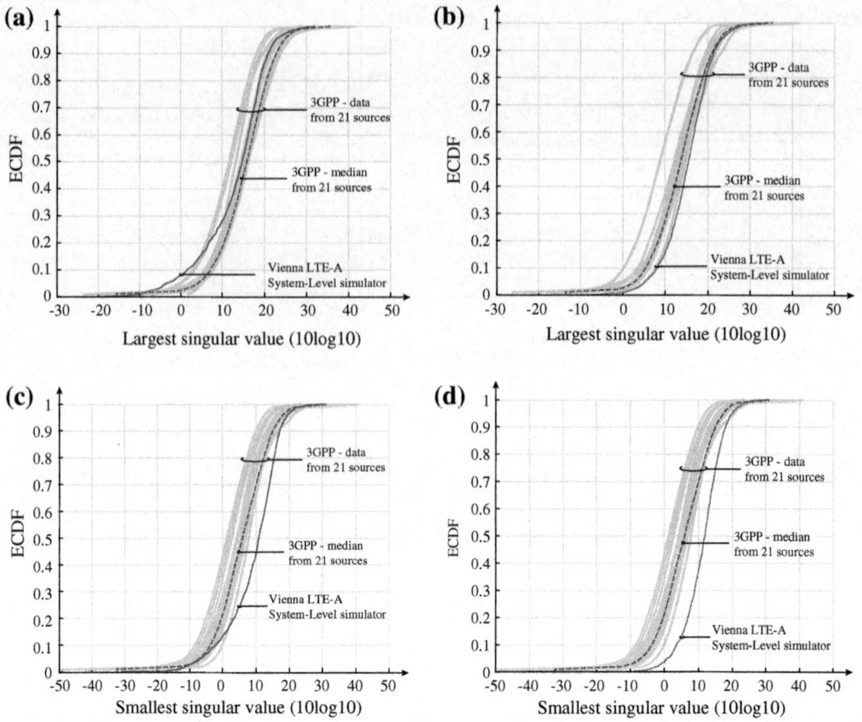

Fig. 10.13 Largest and smallest singular value ECDF in logarithmic scale. *Gray lines* refer to results reported by 21 sources from [47]. *Dashed curves* denote the 3GPP reference results from [22, Figs. 8.2-17, 8.2-19]. **a** UMa-Largest (1st) singular value. **b** UMi-Largest (1st) singular value. **c** UMa-Smallest (2nd) singular value. **d** UMi-Smallest (2nd) singular value

References

1. M. Gerasimenko, S. Andreev, Y. Koucheryavy, A. Trushanin, V. Shumilov, M. Shashanov, S. Sosnin, Performance comparison of system level simulators for 3GPP LTE Uplink, in *Internet of Things, Smart Spaces, and Next Generation Networking*. Lecture Notes in Computer Science, vol. 7469, ed. by S. Andreev, S. Balandin, Y. Koucheryavy (Springer, Berlin, 2012), pp. 186–197
2. Riverbed (2015), http://www.riverbed.com/products/performance-management-control/opnet.html?redirect=opnet
3. OMNeT++ (2015), http://www.omnetpp.org
4. IT++ (2015), http://itpp.sourceforge.net/4.3.1/
5. ns-2 (2015), http://www.isi.edu/nsnam/ns
6. ns-3 (2015), http://www.nsnam.org/
7. GNS3 (2015), http://www.gns3.com/
8. S. Max, D. Bltmann, R. Jennen, M. Schinnenburg, Evaluation of IMT-advanced scenarios using the open wireless network simulator, in *International ICST Conference on Simulation Tools and Techniques* (Torremolinos, Malaga, Spain, 2010), p. 26
9. Hurricane II WAN Emulation and Network Simulation (2015), http://packetstorm.com/packetstorm-products/hurricane-ii-software

10. Nomor System Level Simulation (2015), http://www.nomor.de/home/solutions-and-products/system-level-simulation
11. M. Taranetz, T. Blazek, T. Kropfreiter, M.K. Müller, S. Schwarz, M. Rupp, Runtime precoding: enabling multipoint transmission in LTE-advanced system level simulations. IEEE Access **3**, 725–736 (2015). Jun
12. Y. Li, F. Yu, S. lei Zheng, C. lei Yang, LTE system level simulation with MATLAB, in *International Conference Internet Technology and Applications (iTAP)*, pp. 1–4 (2011). doi:10.1109/ITAP.2011.6006434
13. G. Piro, L. Grieco, G. Boggia, F. Capozzi, P. Camarda, Simulating LTE cellular systems: an open-source framework. IEEE Trans. Veh. Technol. **60**(2), 498–513 (2011). Feb
14. The Vienna LTE Simulators (2015), http://www.nt.tuwien.ac.at/ltesimulator
15. Mathworks, MATLAB Documentation (2015), http://www.mathworks.com/help/matlab/index.html
16. K. Abdallah, I. Cerutti, P. Castoldi, Energy-efficient coordinated sleep of LTE cells, in *IEEE International Conference on Communication (ICC)* (Ottawa, ON, Canada, 2012), pp. 5238–5242. doi:10.1109/ICC.2012.6363707
17. M. Carvalho, P. Vieira, An enhanced handover oscillation control algorithm in LTE self-optimizing networks, in *IEEE International Symposium Wireless Personal Multimedia Communication (WPMC)* (Brest, France, 2011)
18. M.M. Selim, M. El-Khamy, M. El-Sharkawy, Enhanced frequency reuse schemes for interference management in LTE femtocell networks, in *International Symposium on Wireless Communication System (ISWCS)* (Paris, France, 2012), pp. 326–330
19. S.Y. Shin, D. Triwicaksono, Radio resource control scheme for machine-to-machine communication in lte infrastructure, in *International Conference ICT Convergence (ICTC)* (Jeju Island, South Korea, 2012)
20. J.C. Ikuno, System level modeling and optimization of the LTE downlink, Ph.D. dissertation, E389, Technische Universität Wien (2013), http://theses.eurasip.org/theses/479/system-level-modeling-and-optimization-of-the-lte/
21. WINNER II WP1, WINNER II channel models, IST-4-027756 WINNER II Deliverable D1.1.2 (2007)
22. 3rd Generation Partnership Project (3GPP). Study on 3D channel model for LTE, 3rd Generation Partnership Project (3GPP), TR 36.873 (2014)
23. R.W. Heath Jr., M. Airy, A. Paulraj, Multiuser diversity for MIMO wireless systems with linear receivers, in *Asilomar Conference on Signals, Systems and Computers (Asilomar)*, vol. 2 (Pacific Grove, CA, USA, 2001), pp. 1194–1199. doi:10.1109/ACSSC.2001.987680
24. S. Ahmadi, *LTE-Advanced: A Practical Systems Approach to Understanding 3GPP LTE Releases 10 and 11 Radio Access Technologies*. ITPro collection (Elsevier Science, 2013)
25. I. Latif, F. Kaltenberger, R. Knopp, Link abstraction for multi-user MIMO in LTE using interference-aware receiver, in *IEEE Wireless Communication and Networking Conference (WCNC)* (Shanghai, China, 2012), pp. 842–846. doi:10.1109/WCNC.2012.6214489
26. M. Döttling et al., Assessment of Advanced Beamforming and MIMO Technologies, IST2003-507581 WINNER, Technical Report, D2.7 ver 1.1 (2005)
27. S. Caban, M. Rupp, C. Mehlführer, M. Wrulich, Evaluation of HSDPA and LTE: From Testbed Measurements to System Level Performance. (Wiley, 2011)
28. Nortel Networks, OFDM Exponential Effective SIR Mapping Validation, EESM Simulation Results, 3rd Generation Partnership Project (3GPP). TR **R1–040089** (2004)
29. IEEE 802.16 Broadband Wireless Access Working Group, IEEE 802.16m Evaluation Methodology Document (EMD), IEEE, Technical Report (2008)
30. G. Caire, G. Taricco, E. Biglieri, Capacity of bit-interleaved channels. Electron. Lett. **32**, 1060–1061 (1996)
31. 3rd Generation Partnership Project (3GPP), Evolved Universal Terrestrial Radio Access (E-UTRA); Further advancements for E-UTRA physical layer aspects, 3rd Generation Partnership Project (3GPP), TR 36.814 (2010)

32. S. Schwarz, M. Wrulich, M. Rupp, Mutual information based calculation of the precoding matrix indicator for 3GPP UMTS/LTE, in *International ITG Workshop Smart Antennas (WSA)* (Bremen, Germany, 2010), pp. 52–58. doi:10.1109/WSA.2010.5456388
33. E. Björnson, E. Jorswieck, Optimal resource allocation in coordinated multi-cell systems. Found. Trends Commun. Inform. Theory **9**(2–3), 381-p (2013)
34. S. Schwarz, J.C. Ikuno, M. Simko, M. Taranetz, Q. Wang, M. Rupp, Pushing the limits of LTE: A survey on research enhancing the standard. IEEE Access **1**, 51–62 (2013)
35. 3rd Generation Partnership Project (3GPP), Spatial channel model for Multiple Input Multiple Output (MIMO) simulations, 3rd Generation Partnership Project (3GPP), TR 36.996 (2003)
36. WINNER I WP5, Final report on link level and system level channel models, IST-2003-507581 WINNER I Deliverable D5.4 (2005)
37. M. ITU-R, Guidelines for evaluation of radio interface technologies for IMT-Advanced, Report (2009)
38. A. Kammoun, H. Khanfir, Z. Altman, M. Debbah, M. Kamoun, Preliminary results on 3D channel modeling: From theory to standardization. IEEE J. Sel. Areas Commun. **32**(6), 1219–1229 (2014). doi:10.1109/JSAC.2014.2328152
39. Z. Hu, R. Liu, S. Kang, X. Su, J. Xu, Work in progress: 3D beamforming methods with user-specific elevation beamfoming, in *International Conference on Communications and Networking in China (CHINACOM)* (2014), pp. 383–386. doi:10.1109/CHINACOM.2014.7054323
40. F. Ademaj, M. Taranetz, M. Rupp, Implementation, validation and application of the 3GPP 3D MIMO channel model in open source simulation tools, in *IEEE International Symposium Wireless Commununication System (ISWCS)* (2015)
41. H. Masui, T. Kobayashi, M. Akaike, Microwave path-loss modeling in urban line-of-sight environments. IEEE J. Sel. Areas Commun. **20**(6), 1151–1155 (2002). doi:10.1109/JSAC.2002.801215
42. A. Weber, A. Bestard, Modeling of X-pol antennas for LTE system simulation, in *8th International Symposium on Wireless Communication Systems (ISWCS)* (2011), pp. 221–225. doi:10.1109/ISWCS.2011.6125342
43. 3rd Generation Partnership Project (3GPP), LTE Physical Layer - General Description, 3rd Generation Partnership Project (3GPP), TS 36.201 (2009)
44. Y. Wang, J. Xu, L. Jiang, Challenges of system-level simulations and performance evaluation for 5G wireless networks. IEEE Access **2**, 1553–1561 (2014). doi:10.1109/ACCESS.2014.2383833
45. 3rd Generation Partnership Project (3GPP), Spatial channel model for Multiple Input Multiple Output (MIMO) simulations, 3rd Generation Partnership Project (3GPP), TR 325.996 (2012)
46. 3GPP TSG RAN WG-1, R1-140048: Phase 2 calibration results for 3D channel model, 3rd Generation Partnership Project (3GPP), Technical Report (2014)
47. 3GPP TSG RAN WG-1, R1-143469: Summary of 3D-channel model calibration results, 3rd Generation Partnership Project (3GPP), Technical Report (2014)
48. F. Rusek, D. Persson, B.K. Lau, E. Larsson, T. Marzetta, O. Edfors, F. Tufvesson, Scaling up MIMO: opportunities and challenges with very large arrays. IEEE Signal Process. Mag. **30**(1), 40–60 (2013). doi:10.1109/MSP.2011.2178495

Chapter 11
Modeling Regular Aggregate Interference by Symmetric Structures

Markus Rupp, Stefan Schwarz and Martin Taranetz

In this chapter, downlink co-channel interference statistics in wireless cellular networks with hexagonal grid layout are investigated. The main target is to facilitate the analysis at user locations outside the center of the cell of interest.

The proposal of a cellular structure for mobile networks dates back to 1947. Two Bell Labs engineers, Douglas H. Ring and W. Rae Young were the first to mention the idea in an internal memorandum [1]. Almost two decades later, in 1966, Richard H. Frenkiel and Philip T. Porter, shaped a *hexagonal cellular array of areas* to propose the first mobile phone system. Although never proposed as innovative research solution, the hexagon model gained high popularity within the research community and is still extensively utilized nowadays [2–7]. It serves either as the system model itself, or as a reference system for more involved simulation scenarios. On the other hand, its geometric structure renders closed-form analysis of *aggregate interference statistics* difficult [8]. Hence, simulation results often lack a mathematical back up.

Recently, closed-form results have been reported with system models based on stochastic geometry [9–11]. The stochastic approach is based on an ensemble of network realizations and is therefore not applicable when a fixed structure of the network is given. Since the *well-planned* deployment of macro-sites is not expected to vanish in the medium term, it is thus considered imperative to make interference analysis in the hexagonal grid model more convenient.

Current work on regular grid models has mainly focused on link-distance statistics [12, 13]. The authors also account for fading and provide closed-form approximations for the co-channel interference of a *single link*. However, convenient expressions

M. Rupp (✉) · S. Schwarz · M. Taranetz
Institute of Telecommunications, TU Wien, Vienna, Austria
e-mail: mrupp@nt.tuwien.ac.at

S. Schwarz
e-mail: sschwarz@nt.tuwien.ac.at

M. Taranetz
e-mail: martin.taranetz@tuwien.ac.at

M. Rupp et al., *The Vienna LTE-Advanced Simulators*,
Signals and Communication Technology, DOI 10.1007/978-981-10-0617-3_11

for the moments and the distribution of *aggregate co-channel interference* are not available yet.

Based on the work in [14, 15], the contributions of this chapter are:

- A *circular interference model* to facilitate interference analysis in cellular networks with regular grid layout is introduced. Particular focus is placed on the hexagonal grid due to its ubiquity in wireless communication engineering [2–7]. The key idea is to consider the power of the interfering BSs as being uniformly spread along the perimeter of the hexagon.
- It is proposed to model interference statistics in a hexagonal scenario by a single Gamma Random Variable RV. Its shape- and scale parameters are determined in closed form by employing the *circular model.* The analysis yields key insights on the formative components of the interference distribution. A scenario with regularly arranged macro-sites and randomly distributed small cells demonstrates the model's expedient application in heterogeneous cellular networks.

The chapter forgoes hexagonal grid setups with more than one ring of interferers as well as further performance analysis, which is enabled by the Gamma approximation. Both are considered straightforward and of no particular relevance for this thesis.

The remainder of this chapter is structured as follows: Sect. 11.1 provides preliminaries on the Gamma distribution. Section 11.2 specifies the hexagonal reference-system model. In Sect. 11.3 the circular interference model and its dual pendant are introduced. Section 11.4 investigates Gamma-distributed interference and its parametrization by the proposed circular interference model. In Sect. 11.5, the accuracy of the Gamma approximation is verified. In Sect. 11.6, the circular model is applied for modeling the interference from the macro BSs in a two-tier heterogeneous cellular network. Section 11.7 provides a comparison of the circular model against Long Term Evolution-Advanced LTE-A system level simulations. Section 11.8 concludes the chapter.

11.1 Preliminaries on the Gamma Distribution

In the current- as well as the subsequent chapter, particular focus is placed upon the Gamma distribution due to its wide range of useful properties for wireless communication engineering, some of which are outlined in this section.

The Probability Density Function (PDF) of a Gamma distributed RV X with *shape parameter* k and *scale parameter* θ, i.e., $G \sim \Gamma[k, \theta]$, is defined as

$$f_G(x) = \frac{1}{\theta^k \Gamma(k)} x^{k-1} e^{-x/\theta}. \tag{11.1}$$

Its mean and variance are given by $\mathbb{E}[G] = k\theta$ and $\mathrm{Var}[G] = k\theta^2$.

The Gamma distribution exhibits the *scaling property*, i.e., if $G \sim \Gamma[k, \theta]$, then $a\,G \sim [k, a\,\theta]$, $\forall a > 0$, and the *summation property*, i.e., if $G_i \sim \Gamma[k_i, \theta]$ with $i = 1, 2, \ldots, N$, then $\sum_{i=1}^{N} G_i \sim \Gamma[\sum_{i=1}^{N} k_i, \theta]$.

Consider an arbitrary distribution with mean ν and variance σ^2. Then, the distribution $\Gamma[k, \theta]$ with the same first- and second order moments has the parameters

$$k = \frac{\nu^2}{\sigma^2}, \quad \theta = \frac{\sigma^2}{\nu}. \tag{11.2}$$

These simple moment-matching identities can be exploited for accurately approximating fading distributions [16], such as generalized-K [17, 18] and log-normal [16–19], as well as aggregate interference statistics [11, 20, 21].

Additionally, the Gamma distribution covers the power fading distribution of various single- and multi-antenna schemes under the Rayleigh fading assumption. Conventional Single-Input Single-Output (SISO) yields an exponential distribution $\text{Exp}[1/\theta]$, which is equivalent to $\Gamma[1, \theta]$. The power fading of Maximum Ratio Transmission (MRT) with N_{Tx} transmit antennas and one receive antenna can be modeled by $\Gamma[N_{\text{Tx}}, \theta]$, Maximum Ratio Combining (MRC) with one transmit antenna and N_{Rx} receive antennas is characterized by $\Gamma[N_{\text{Rx}}, \theta]$. Furthermore, MRC is often studied in the presence of Nakagami-m fading. Let $Y \sim \text{Nakagami}[m, \Omega]$ and $G = Y^2$. Then, $G \sim \Gamma[m, \Omega/m]$.

According to [22], the quotient $\gamma = S/I$ of two RVs $S \sim \Gamma[k_S, \theta_S]$ and $I \sim \Gamma[k_I, \theta_I]$ is distributed as

$$f_\gamma(x) = \frac{(\theta_I/\theta_S)^{k_S}}{B(k_S, k_I)} \left(1 + \frac{\theta_I}{\theta_S} x\right)^{-k_S - k_I} x^{k_S - 1}, \; x > 0 \tag{11.3}$$

with $B(\cdot, \cdot)$ denoting the Beta function. Interpreting γ as a Signal-to-Interference Ratio (SIR) allows to determine the success probability $\mathbb{P}[\gamma > \delta]$ for a given threshold δ as

$$\mathbb{P}[\gamma > \delta] = \frac{\Gamma(k_S + k_I)}{\Gamma(k_S)} \left(\frac{\theta_S}{\delta\, \theta_I}\right)^{k_I} {}_2\bar{F}_1\left(k_I, k_S + k_I, 1 + k_I, -\frac{\theta_S}{\delta\, \theta_I}\right), \tag{11.4}$$

where ${}_2\bar{F}_1(\cdot, \cdot, \cdot, \cdot)$ is a regularized hypergeometric function [11].

These observations motivate the application of the Gamma RV as a sensible compromise between accuracy and tractability. Further properties of Gamma RVs will be discussed as needed.

11.2 Hexagonal Reference Model

The reference hexagonal setup is composed of a central cell and six interfering BSs, as shown in Fig. 11.1. The interferers are equipped with omnidirectional antennas and are located at the edges of a hexagon with radius R (marked as '+' in Fig. 11.1). All BSs are assumed to transmit with the same power. The signal from the ith interfering BS with polar coordinates (R, Ψ_i) to a user with polar coordinates (r, ϕ) experiences macroscopic path loss and fading. It is assumed that $0 < r \leq R/2$, so as to assure

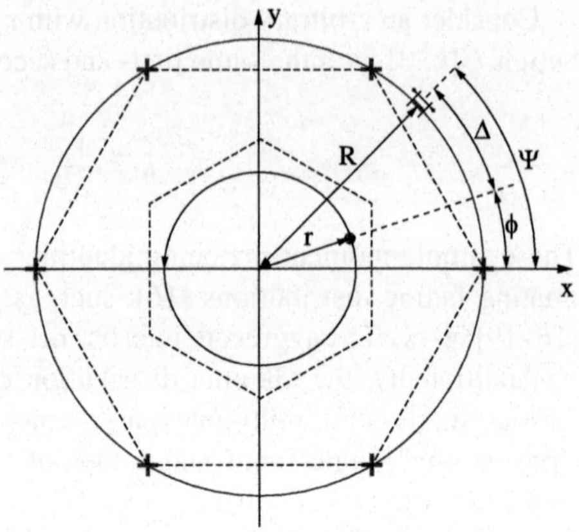

Fig. 11.1 System model. Center cell with user at (r, ϕ). Interfering BSs are located at (R, Ψ_i), where $\Psi_i = 2\pi\, i/M,\ i = 1, \ldots, M$

that the user is associated with the central BS. The path loss is modeled by the exponential law

$$\ell\left(d_{r,\Delta_i}^{(M)}\right) = \min\left(b_{\mathrm{P}}, \frac{1}{c_{\mathrm{P}}}\left(d_{r,\Delta_i}^{(M)}\right)^{-\alpha}\right), \tag{11.5}$$

where b_{P} denotes the intercept, c_{P} is a constant, α refers to the path loss exponent and

$$d_{r,\Delta_i}^{(M)} = \sqrt{R^2 + r^2 - 2Rr\cos(\Delta_i)}, \tag{11.6}$$

with $\Delta_i = \phi - \Psi_i$ and $\Psi_i = 2\pi\, i/M,\ i = 1, \ldots, M$. In the remainder of this chapter, it is assumed that $d_{r,\Delta_i}^{(M)} > (b_{\mathrm{P}} c_{\mathrm{P}})^{-1/\alpha}$. Exemplifying from [23], a minimum coupling loss of 70 dB and free space propagation at an LTE-A frequency of $f_{\mathrm{c}} = 2.14$ GHz yield $b_{\mathrm{P}} = 10^{-7}$, $c_{\mathrm{P}} = 8.05 \times 10^{-3}$ and $(b_{\mathrm{P}} c_{\mathrm{P}})^{-1/\alpha} = 0.028$ m, hence justifying this assumption.

In the hexagonal scenario, $M = 6$. The terms r and Δ_i denote the user's distance to the center and its angle-difference to the ith interfering BS, respectively. Motivated by Sect. 11.1, fading is modeled by an independent and Identically Distributed (I.I.D.) Gamma RV $G_i \sim \Gamma[k_0, \theta_0]$, where k_0 and θ_0 refer to shape- and scale parameter, respectively.

11.3 Circular Interference Model

In a one-tier hexagonal grid scenario, as presented in Sect. 11.2, the experienced aggregate interference power at position (r, ϕ) can be expressed as

$$I_6(r, \phi) = \sum_{i=1}^{6} P_{\mathrm{M}}\, G_i\, \ell\left(d_{r,\Delta_i}^{(6)}\right), \tag{11.7}$$

where P_{M} denotes the transmit power, G_i is the fading and $\ell(d^{(6)}_{r,\Delta_i})$ refers to the path loss at distance $d^{(6)}_{r,\Delta_i}$, as specified in Eqs. (11.5) and (11.6), respectively. Each sum term can be regarded as a RV G_i, which is weighted by the factor $P_{\mathrm{M}}\,\ell(d^{(6)}_{r,\Delta_i})$. Hence, the statistics of $I_6(r, \phi)$ outside the cell-center, i.e., $r > 0$, are accessible via a sum of differently weighted RVs. Since, in general, this does not lead to closed-form results, as detailed in Chap. 12, in this chapter a circular interference model is proposed in order to facilitate the statistical analysis.

11.3.1 Proposed Model

In the proposed circular interference model, the power of the six reference BSs is spread uniformly along a circle of radius R. This is achieved by evenly distributing the total transmit power $6\,P_{\mathrm{M}}$ among M equally spaced BSs and considering the limiting case $M \to \infty$. By generalizing Eq. (11.7), this is expressed as

$$I_{\mathrm{C}}(r) = \lim_{M\to\infty} \frac{6\,P_{\mathrm{M}}}{M} \sum_{i=1}^{M} G_i\,\ell\left(d^{(M)}_{r,\Delta_i}\right) = \frac{6\,P_{\mathrm{M}}}{2\pi} \mathbb{E}\left[G_i\right] \int_{-\pi}^{\pi} \ell\left(d_{r,\Delta}\right) d\Delta, \tag{11.8}$$

with $\ell(\cdot)$ from Eq. (11.5) and $d^{(M)}_{r,\Delta_i}$ from Eq. (11.6). The terms $d_{r,\Delta}$ and Δ denote distance and angle-difference between the user and an infinitesimal interfering circular segment, as illustrated in Fig. 11.1.

Assuming a path loss exponent $\alpha = 2$, i.e., free space propagation, Eq. (11.8) can explicitly be evaluated as

$$I_{\mathrm{C}}(r) = 6\,P_{\mathrm{M}}\,\mathbb{E}\left[G_i\right] \frac{1}{c_{\mathrm{P}}} \frac{1}{R^2 - r^2}. \tag{11.9}$$

An intuitive interpretation of this result is provided in the next section by the model's pendant.

In the remainder of this chapter, $\alpha = 2$ is employed. It represents the worst case of low interference attenuation. However, previously- as well as all subsequently presented analysis can be carried out in closed-form for $\alpha = 2n$ with $n \in \mathbb{N}$. Values α other than these require the evaluation of elliptic integrals (see, e.g., [24]). Thus, a practical first order estimate for arbitrary values of α is achieved by evaluating the performance with $2n$ and $2(n+1)$, where $2n < \alpha < 2(n+1)$.

11.3.2 The Dual Model

Consider a user in a hexagonal scenario, which is moved along a circle of radius r from $-\pi$ to π, as indicated in Fig. 11.1. The *average expected* interference along the circle can be calulated as

$$I'_C(r) = \frac{1}{2\pi} \int_{-\pi}^{\pi} \mathbb{E}\left[I_6(r,\phi')\right] d\phi' \tag{11.10}$$

$$= \sum_{i=1}^{6} P_M \, \mathbb{E}\left[G_i\right] \frac{1}{2\pi} \int_{-\pi}^{\pi} \ell(d_{r,\Delta}) d\phi'. \tag{11.11}$$

The result is obtained by plugging Eq. (11.7) into Eq. (11.10), exchanging sum and integral, and exploiting the linearity of the expectation. It is equivalent to $I_C(r)$ in Eq. (11.8) and, consequently, also yields Eq. (11.9). Thus, the result is *independent of the user's angle-position*. It can be interpreted as the *average expected* interference, i.e., the interference experienced by a *typical* user in a hexagonal scenario at distance r.

From Eq. (11.9) it is observed that the average expected interference increases by either increasing the transmit power P_M, decreasing the distance of the interferers R, or moving the user further away from the origin, which is reflected by the parameter r. The fading enters the equation only via the expectation, i.e., Eqs. (11.8) and (11.11) hold for arbitrary fading distributions with finite mean. Finally, note that the circular interference model is not restricted to hexagons. By replacing '6' by 'N' in Eqs. (11.7)–(11.11), it can generally be applied for substituting any convex regular N-polygonal model, as validated in Sect. 11.5.1.

11.4 Statistics of Aggregate Interference

In this section, aggregate interference in a hexagonal scenario with I.I.D. Gamma fading is investigated. Motivated by the findings in Sect. 11.1, it is proposed to approximate its statistics by a *single Gamma RV*. The corresponding shape- and scale parameters are dependent on the distance and can be determined by applying the previously presented circular model.

11.4.1 Interference Statistics at the Center

Assume I.I.D. Gamma fading with $G_i \sim \Gamma[k_0, \theta_0]$. Then, according to Sect. 11.3, interference can be considered as a sum of Gamma RVs, which are weighted by the factors $P_M \, \ell(d^{(6)}_{r,\Delta_i})$, i.e., the received power without fading.

At the center of a hexagonal scenario (i.e., at $r = 0$), all weighting factors are equal, i.e., $P_M \, \ell(d^{(6)}_{r,\Delta_i}) = P_M \, \ell(R)$. By virtue of the scaling- and summation property of a Gamma RV (cf. Sect. 11.1), the resulting interference is distributed as

$$I_6(0,\phi) \sim \Gamma\left[6\,k_0, \theta_0 P_M \ell(R)\right]. \tag{11.12}$$

11.4.2 Interference Statistics Outside the Center

Outside the center (i.e., at $r > 0$), the distances $d^{(6)}_{r,\Delta_i}$ and, thus, also the weighting factors $P_\mathrm{M}\,\ell(d^{(6)}_{r,\Delta_i})$ generally differ from each other. Hence, a non-uniform impact of the interferers is observed. Then, the interference statistics are only accessible via evaluating the distribution of a sum of Gamma RVs with varying scale parameter. This method is particularized in Chap. 12.

The current chapter resorts to the following first order estimate. It is proposed to approximate the typically experienced interference distribution at distance r, $0 < r \leq R/2$, by

$$\hat{I}(r) \sim \Gamma[\hat{k}(r), \hat{\theta}(r)]. \tag{11.13}$$

The rationale for this model are findings in prior work, where out-of-cell interference in stochastic networks is appropriately assessed by a Gamma distribution [11]. If it can be proven as accurate, it considerably facilitates further performance analysis by applying the methods in Sect. 11.1.

The distribution in Eq. (11.13) is fully determined by the distance-dependent shape- and scale parameters $\hat{k}(r)$ and $\hat{\theta}(r)$, respectively. In order to evaluate the two parameters, firstly the proposed circular interference model is employed to determine expectation and variance of $\hat{I}(r)$. Then, it is exploited that $\mathbb{E}[\hat{I}(r)] = \hat{k}(r)\,\hat{\theta}(r)$ and $\mathrm{Var}[\hat{I}(r)] = \hat{k}(r)\,\hat{\theta}^2(r)$.

As discussed in Sect. 11.3.1, the distinct received powers from the interfering BSs can be *averaged* along a circle of radius r. Thus, the typical impact of one interferer is calculated as

$$P_\mathrm{M}\frac{1}{2\pi}\int_{-\pi}^{\pi} \ell\left(d_{r,\Delta}\right) d\Delta = \frac{P_\mathrm{M}}{c_\mathrm{P}}\frac{1}{R^2 - r^2}, \tag{11.14}$$

and yields the average expected interference at distance r as

$$\mathbb{E}\left[\hat{I}(r)\right] = 6\,k_0\theta_0\frac{P_\mathrm{M}}{c_\mathrm{P}}\frac{1}{R^2 - r^2}. \tag{11.15}$$

The variance of the aggregate interference comprises two components:

1. The variance of the fading, which calculates as

$$\mathrm{Var}_f\left[\hat{I}(r)\right] = 6\,k_0\left(\theta_0\frac{P_\mathrm{M}}{c_\mathrm{P}}\frac{1}{R^2 - r^2}\right)^2. \tag{11.16}$$

2. The variance of the received power without fading, which is caused by the unequal distances d_{r,Δ_i}. With

$$\frac{1}{2\pi}\int_{-\pi}^{\pi}\left(P_{\mathrm{M}}\ell\left(d_{r,\Delta}\right)-P_{\mathrm{M}}\ell(R)\right)^{2}d\Delta=\left(\frac{P_{\mathrm{M}}}{c_{\mathrm{P}}R^{2}}\right)^{2}\frac{2r^{2}R^{4}+r^{4}R^{2}-r^{6}}{\left(R^{2}-r^{2}\right)^{3}}, \tag{11.17}$$

the second variance component is obtained as

$$\mathrm{Var}_{d}\left[\hat{I}(r)\right]=6k_{0}\left(\theta_{0}\frac{P_{\mathrm{M}}}{c_{\mathrm{P}}R^{2}}\right)^{2}\frac{2r^{2}R^{4}+r^{4}R^{2}-r^{6}}{\left(R^{2}-r^{2}\right)^{3}}. \tag{11.18}$$

Since the two components are statistically independent, the overall variance is calculated as

$$\begin{aligned}\mathrm{Var}\left[\hat{I}(r)\right]&=\mathrm{Var}_{f}\left[\hat{I}(r)\right]+\mathrm{Var}_{d}\left[\hat{I}(r)\right]\\&=6\,k_{0}\left(\theta_{0}\frac{P_{\mathrm{M}}}{c_{\mathrm{P}}}\frac{1}{R^{2}-r^{2}}\right)^{2}\left(1+\frac{2r^{2}R^{4}+r^{4}R^{2}-r^{6}}{R^{6}-r^{2}R^{4}}\right)\end{aligned} \tag{11.19}$$

where $\mathrm{Var}_{f}[\hat{I}(r)]$ and $\mathrm{Var}_{d}[\hat{I}(r)]$ refer to Eqs. (11.16) and (11.18), respectively.

Finally, the distance-dependent shape- and scale parameter are derived from Eqs. (11.15) and (11.19) as

$$\hat{k}(r)=6\,k_{0}\frac{R^{4}(R^{2}-r^{2})}{R^{6}+r^{2}R^{4}+r^{4}R^{2}-r^{6}}, \tag{11.20}$$

$$\hat{\theta}(r)=\theta_{0}\frac{P_{\mathrm{M}}}{c_{\mathrm{P}}}\frac{1}{R^{2}-r^{2}}\left(1+\frac{2r^{2}R^{4}+r^{4}R^{2}-r^{6}}{R^{6}-r^{2}R^{4}}\right). \tag{11.21}$$

11.5 Numerical Results and Discussion

In this section, the accuracy of the circular model and the proposed Gamma approximation are verified by numerical evaluation.

11.5.1 Validation of Expected Aggregate Interference

First, the expected interference powers in the hexagonal reference scenario and the proposed circular interference setup are compared to each other. The transmit power and inter-site distance are specified as $P_{\mathrm{M}} = 40$ W and $R = 500$ m, based on the standard 3rd Generation Partnership Project (3GPP) macro cell scenario from [23]. Intercept and constant of the path loss $\ell(\cdot)$ are set $b_{\mathrm{P}} = 1$ and $c_{\mathrm{P}} = 1$ for simplicity.

Table 11.1 System parameters for validation. Transmit power and circle radius are referred from [23]

Parameter	Value
Transmit power	$P_M = 40\,W$
Circle radius	$R = 500\,m$
Path loss intercept	$b_P = 1$
Path loss constant	$c_P = 1$
Path loss exponent	$\alpha = 2$
Fading distribution	$G_i \sim \Gamma[1, 1]$

Fading is assumed to be distributed as $G_i \sim \Gamma[1, 1]$. The parameters are summarized in Table 11.1.

Consider a user which is moved along a semi circle $\{(r, \phi) | \phi \in [0, \pi]\}$, as indicated in Fig. 11.1. The expected interference in the hexagonal scenario is calculated as

$$\mathbb{E}\left[I_6(r, \phi)\right] = \sum_{i=1}^{6} P_M \, \mathbb{E}[G_i] \, \ell\left(d_{r,\Delta_i}^{(6)}\right), \tag{11.22}$$

with $I_6(r, \phi)$ from Eq. (11.7) and $\mathbb{E}[G_i] = 1$. For the circular model, $\mathbb{E}\left[I_C(r)\right] = I_C(r)$, with $I_C(r)$ from Eq. (11.9).

Figure 11.2 depicts the evaluated results of Eqs. (11.9) and (11.22) for various distances r. It is observed that

- At cell-center, i.e., at $r = 0$ m, the expected interference powers in the hexagonal- and circular scenario ($\mathbb{E}[I_6(0, \phi)]$ and $I_C(0)$) are equal.
- Outside the center, i.e., at $r > 0$ m, $\mathbb{E}[I_6(r, \phi)]$ fluctuates around $I_C(r)$. The deviation is weak in the middle of the cell ($r = 125$ m), and strong at cell-edge

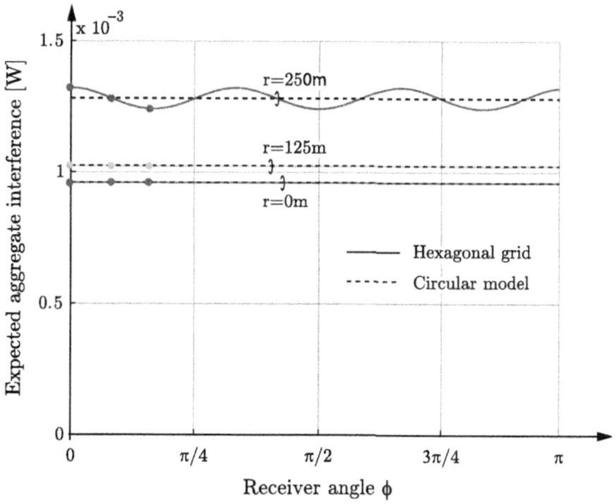

Fig. 11.2 Expected aggregate interference experienced at position (r, ϕ) in circular- ($I_C(r)$) and hexagonal model ($\mathbb{E}[I_6(r, \phi)]$), respectively. Receiver distances $r = \{0, 125, 250\}$ m refer to cell-center, middle of cell and cell-edge, respectively

($r = 250$ m). Note that $\mathbb{E}[I_6(r, \phi)]$ is not symmetric about $I_C(r)$ due to the concavity of the path loss model.

The relative error of the circular interference model is calculated as

$$\varepsilon(r, \phi) = \left| \frac{\mathbb{E}[I_6(r, \phi)] - I_C(r)}{\mathbb{E}[I_6(r, \phi)]} \right|, \tag{11.23}$$

with $I_C(0)$ and $\mathbb{E}[I_6(0, \phi)]$ from Eqs. (11.9) to (11.22), respectively. The largest error occurs at cell-edge, i.e.,

$$\max_{r,\phi} \epsilon(r, \phi) = \max_{\phi} \epsilon(250, \phi). \tag{11.24}$$

In the specified scenario, $\max_\phi \epsilon(250, \phi) = 3.2\,\%$, as shown in Fig. 11.3.

11.5.2 Validation of Gamma Approximation

In this section, the accuracy of the Gamma approximation in Eq. (11.13) and its parameterization by Eqs. (11.20) and (11.21) are verified. The exact position-dependent distributions of $I_6(r, \phi)$ are obtained by evaluating Theorem 12.1.

In order to capture a *representative* profile of distributions, three user distances $r = \{0, 125, 250\}$ m and three angle-positions $\phi = \{0, \frac{\pi}{12}, \frac{\pi}{6}\}$ are considered, as illustrated by bold dots in Fig. 11.4. The distances correspond to cell-center, middle of cell and cell-edge, respectively. The angle $\phi = 0$ represents a user, which is moved

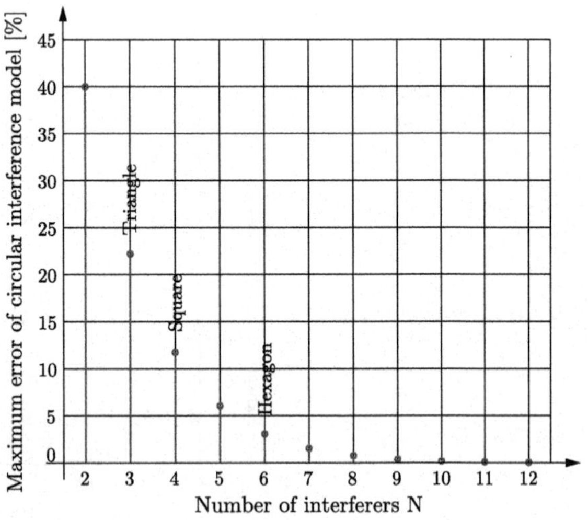

Fig. 11.3 Maximum deviation of circular interference model from expected interference in *convex regular N-polygonal* models. The labeled cell-shapes can be arranged in a grid without overlapping areas

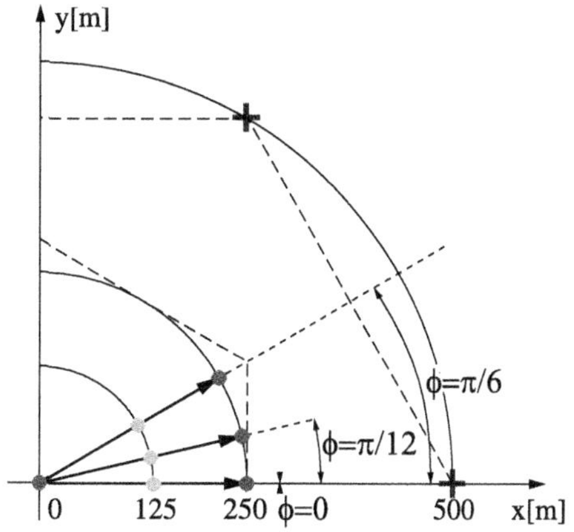

Fig. 11.4 Setup for evaluation. Cutout of Fig. 11.1 (*upper right* quadrant)

directly towards its strongest interferer, $\phi = \frac{\pi}{6}$ refers to the path with two equidistant dominant interferers, and $\phi = \frac{\pi}{12}$ is a variation thereof.

Fading is specified as $G_i \sim \Gamma[2, 1]$. This corresponds to a 1×2 Single-Input Multiple-Output SIMO system with Rayleigh-fading and Multiple-Input Single-Output (MISO) at the user, or, equivalently, a 2×1 MISO system with MRT at the BS.

The Cumulative Distribution Function (CDF) of the Gamma approximation, $F_{\hat{I}}(x; \hat{k}(r), \hat{\theta}(r))$ and the CDF $F_6(x; r, \phi)$ of $I_6(r, \theta)$ are evaluated at each distance r and angle ϕ, respectively. The accuracy of the Gamma approximation is *quantified* by Kolmogorov–Smirnov KS statistics, which formulate as

$$D_{\mathrm{KS}}(r, \phi_m) = \sup_x \left| F_{\hat{I}}\left(x; \hat{k}(r), \hat{\theta}(r)\right) - F_6\left(x; r, \phi\right) \right| . \tag{11.25}$$

Results are depicted in Fig. 11.5. The Gamma approximation most closely resembles the experienced interference distributions at $\phi = \frac{\pi}{12}$. In this case, the difference between exact- and approximated CDFs is less than 1 % for $r < 159$ m and 2.75 % at cell-edge ($r = 250$ m). The largest deviation occurs at $\phi = \frac{\pi}{6}$, where the user is moved centrally in between its two dominant interferers (upper curve). Then, the distributions differ by less than 1 % for $r < 155$ m and by 3.7 % at cell-edge.

For *qualitative* evaluation, Fig. 11.6 depicts the exact CDFs and the corresponding Gamma approximations at the specified representative user positions, which are denoted by bold dots in Figs. 11.4 and 11.5, respectively.

The Gamma CDFs perfectly fit at cell center ($r = 0$ m) and in the middle of the cell ($r = 125$ m). At cell-edge ($r = 250$ m), the Gamma approximation closely resembles the experienced interference of a user at $\phi = \frac{\pi}{12}$. The probability of high interference values at $\phi = \frac{\pi}{6}$ is slightly underestimated by at most 3.7 % (cf. Fig. 11.5).

Fig. 11.5 KS statistics at position (r, ϕ_m), comparing Gamma approximation and exact distribution. Receiver distances $r = \{0, 125, 250\}$ m refer to cell-center, middle of cell and cell-edge, respectively

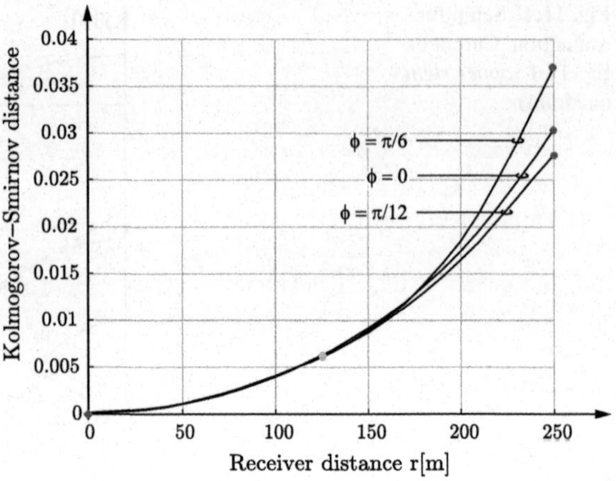

Fig. 11.6 Aggregate interference at particular user positions (see *bold dots* in Fig. 11.4): Exact CDFs, as obtained by numerically evaluating [25] for a hexagon scenario *dashed lines* and corresponding Gamma approximations *solid lines*

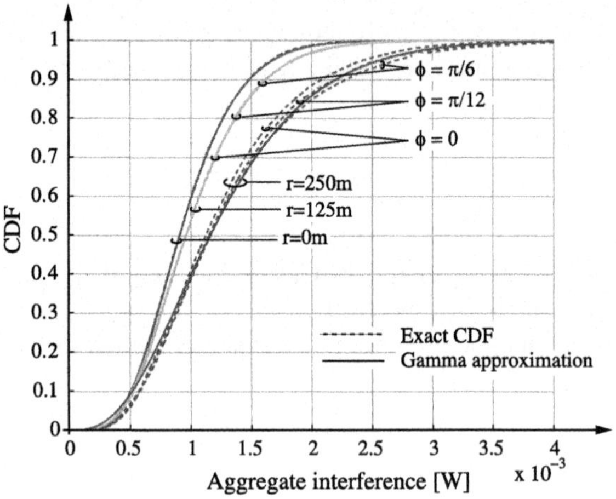

11.6 Application in Heterogeneous Networks

In this section, aggregate interference statistics in a two-tier heterogeneous cellular network with regularly placed macro-BSs and randomly distributed small cell BSs are investigated. The interference contribution from each tier is approximated by a single Gamma RV and the total interference is calculated as the sum of the two. The accuracy of the approximations is verified by extensive Monte Carlo simulations.

The macro-tier comprises six hexagonally arranged BSs at distance $R = 500$ m, each transmitting with $P_M = 40$ W. Small cell BSs are distributed according to a Poisson Point Process PPP of density $\mu_S = 10^{-4}$ m^{-2} and transmit with a power of $P_S = 0.4$ W. As indicated in Fig. 11.7, they are excluded from a ball (in fact, it is a disc, but *ball* is the more common term in literature, e.g., in [11]) of radius $R_{Ex} = (P_M/P_S)^{-1/\alpha} R/2$ around the user so as to ensure user association to the central

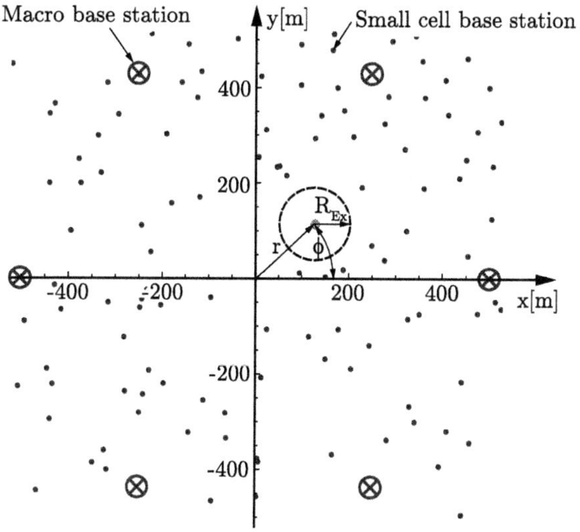

Fig. 11.7 Snapshot of a heterogeneous network deployment. Macro-BSs are arranged on a hexagon. Small cell BSs are randomly distributed around a user at (r, ϕ) and excluded from a ball of radius R_{Ex}

Table 11.2 Parameters for numerical evaluation of heterogeneous scenario

Parameter	Value
Macro-BS transmit power	$P_{\mathrm{M}} = 40\,\mathrm{W}$
Inter macro-site distance	$R = 500\,\mathrm{m}$
Small cell BS transmit power	$P_{\mathrm{S}} = 0.4\,\mathrm{W}$
Small cell density	$\mu_{\mathrm{S}} = 10^{-4}\mathrm{m}^{-2}$
Path loss intercept	$b_{\mathrm{P}} = 1$
Path loss constant	$c_{\mathrm{P}} = 1$
Path loss exponent	$\alpha = 4$
Fading distribution	$G_i \sim \Gamma[1, 1]$

macro-BS at cell-edge. In both tiers, the path loss $\ell(\cdot)$ is modeled according to Eq. (11.5), with intercept $b_{\mathrm{P}} = 1$, constant $c_{\mathrm{P}} = 1$ and exponent $\alpha = 4$. Fading is assumed to be distributed as $G_i \sim \Gamma[1, 1]$. The parameters are summarized in Table 11.2.

In the first step, the interference contribution from the macro-tier is approximated by a Gamma RV $\hat{I}_{\mathrm{M}}(r) \sim \Gamma[\hat{k}_{\mathrm{M}}(r), \hat{\theta}_{\mathrm{M}}(r)]$. According to Sect. 11.4, it can be parameterized by the circular interference model. Recalculating Eqs. (11.15), (11.16), (11.18) and (11.19) for $\alpha = 4$ yields

$$\hat{k}_{\mathrm{M}}(r) = \frac{6\,k_0 R^8 \left(r^2 - R^2\right)\left(r^2 + R^2\right)^2}{r^{14} - 7r^{12}R^2 + 23r^{10}R^4 - 41r^8R^6 + 39r^6R^8 - 25r^4R^{10} - 9r^2R^{12} - R^{14}}, \tag{11.26}$$

$$\hat{\theta}_{\mathrm{M}}(r) = \frac{\theta_0 P_{\mathrm{M}} \left(-r^{14} + 7r^{12}R^2 - 23r^{10}R^4 + 41r^8R^6 - 39r^6R^8 + 25r^4R^{10} + 9r^2R^{12} + R^{14}\right)}{c_{\mathrm{P}} \left(r^2 - R^2\right)^4 \left(r^2 + R^2\right) R^8}. \tag{11.27}$$

Secondly, the contribution of the small cell tier is also approximated by a Gamma RV $\hat{I}_S \sim \Gamma[\hat{k}_S, \hat{\theta}_S]$. Along the lines of [26, Eqs. (2.19) and (2.21)], mean and variance of the actual interference $I_{A,S}$ from the PPP model are determined as

$$\begin{aligned} \mathbb{E}[I_{A,S}] &= P_S\, \mathbb{E}[G_i] 2\pi \mu_S \int_{R_{Ex}}^{\infty} r\, \ell(r) dr \\ &= P_S\, k_0\, \theta_0 \frac{1}{c_P} \pi \mu_S\, R_{Ex}^{-2}, \end{aligned} \tag{11.28}$$

$$\begin{aligned} \mathrm{Var}[I_{A,S}] &= \mathbb{E}[G_i^2] 2\pi \mu_S \int_{R_{Ex}}^{\infty} r\, \ell(r)^2 dr \\ &= (1 + k_0) k_0\, \theta_0^2\, P_S^2 \frac{1}{c_P^2} \frac{\mu_S \pi}{3} R_{Ex}^{-6}. \end{aligned} \tag{11.29}$$

Then, exploiting the identities $\mathbb{E}[\hat{I}_S] = \hat{k}_S\, \hat{\theta}_S$ and $\mathrm{Var}[\hat{I}_S] = \hat{k}_S\, \hat{\theta}_S^2$ yields

$$\hat{k}_S = \frac{3 R_{Ex}^2\, k_0\, \theta_0\, \mu_S}{(1 + k_0)}, \tag{11.30}$$

$$\hat{\theta}_S = \frac{P_S (1 + k_0)}{3\, c_P\, R_{Ex}^{\alpha}}. \tag{11.31}$$

Finally, the PDF of the total aggregate interference, $\hat{I}_A(r) = \hat{I}_M(r) + \hat{I}_S$, at user distance r is calculated as

$$\begin{aligned} f_{\hat{I}_A}(x; r) &= \hat{\theta}_M(r)^{-\hat{k}_M(r)} \hat{\theta}_S^{-\hat{k}_S} e^{-\frac{x}{\hat{\theta}_S}} x^{\hat{k}_M(r) + \hat{k}_S - 1} \\ &\times {}_1\tilde{F}_1 \left(\hat{k}_M(r); \hat{k}_M(r) + \hat{k}_S; \left(\frac{1}{\hat{\theta}_S} - \frac{1}{\hat{\theta}_M(r)} \right) x \right), \end{aligned} \tag{11.32}$$

where ${}_1\tilde{F}_1(\cdot;\cdot;\cdot)$ denotes the regularized confluent hypergeometric function.

In order to verify the accuracy of this approximation, Monte Carlo simulations are carried out. The results for a *typical* user at distance r are obtained by averaging over 10^6 uniformly distributed angle-positions on $[0, 2\pi]$. For each position, 10^5 fading- and 10^4 spatial realizations of the small cell deployment are generated. The small cell BSs are distributed over a circular area of radius 10R.

Figure 11.8a depicts the individual interference contributions from the macro- and the small cell tier at various user distances r. It is observed that the CDFs for the macro tier, which correspond to the approximation in Eqs. (11.26), and (11.27), show an accurate fit with the Monte Carlo simulations. This corroborates the claim in Sect. 11.3.1 that the circular model is also applicable for path loss exponents other than $\alpha = 2$. The interference CDF of the small cell tier, which refers to the approximation in Eqs. (11.30) and (11.31), is independent of the user distance r due to the fixed exclusion radius R_{Ex}. It is also in close agreement with the simulations. Figure 11.8b shows the CDFs of the aggregate interference from both macro- and

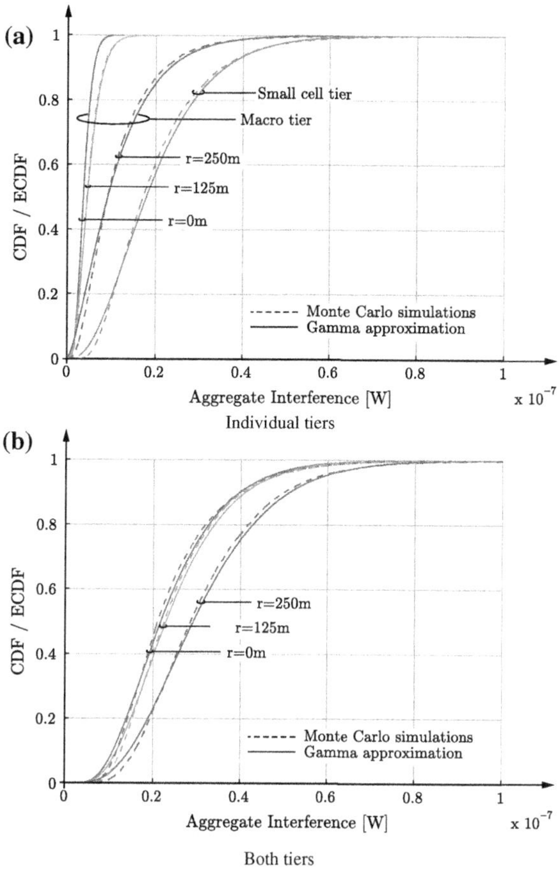

Fig. 11.8 CDFs of interference from macro- and small cell tier **a** and aggregate interference from both tiers **b**. *Solid lines* indicate results as obtained by approximating the contribution of each tier by a Gamma RV. *Dashed lines* show results from Monte Carlo simulations. User distances $r = \{0, 125, 250\}$ m refer to cell-center, middle of cell and cell-edge, respectively

small cell tier. It is found that the approximation by a sum of two independently parameterized Gamma RVs almost perfectly captures the actual interference characteristics at the cell center ($r = 0$ m) and in the middle of the cell ($r = 125$ m). It even provides an accurate fit at cell-edge ($r = 250$ m).

11.7 LTE-Advanced System Level Simulations

In this section, the validity of the Gamma distribution for approximating aggregate interference in symmetric interference scenarios is evaluated. In the first part, the corresponding system model is introduced.

11.7.1 System Model

The system model is composed of a central macro site and six neighboring nodes, which are arranged according to a hexagonal grid, as illustrated in Fig. 11.9. Each site employs a single eNodeB, which is equipped with an omni-directional antenna. For systematic investigations, the UEs are equidistantly distributed along concentric circles of radius $r = \{50, 120, 210\}$ m, referring to cell-center, middle of cell and cell-edge, respectively. Each circle encompasses 24 UEs, which are uniquely identified by the tuple (r, ϕ), where ϕ denotes the angle position. The signal experiences free-space path loss, fast fading according to a time-correlated Rayleigh channel, and spatially-correlated log-normal shadowing with 8 dB standard deviation. Hereinafter, the combination of these three mechanisms is termed *composite fading*. The free-space path loss law is defined as $\min(b_P, 1/c_P d^{-2})$. In this section, $b_P = 10^{-7}$ and $f_c = 2.14$ GHz, yielding $c_P = (4\pi\ f_c/c_0)^2 = 8.0465 \times 10^3$, where c_0 is the speed of light. The shadow fading maps are computed by applying the method in [27]. The results in this section are obtained by averaging over 100 channel realizations and 100 TTIs. The simulation parameters are summarized in Tables 9.6 and 11.3, respectively.

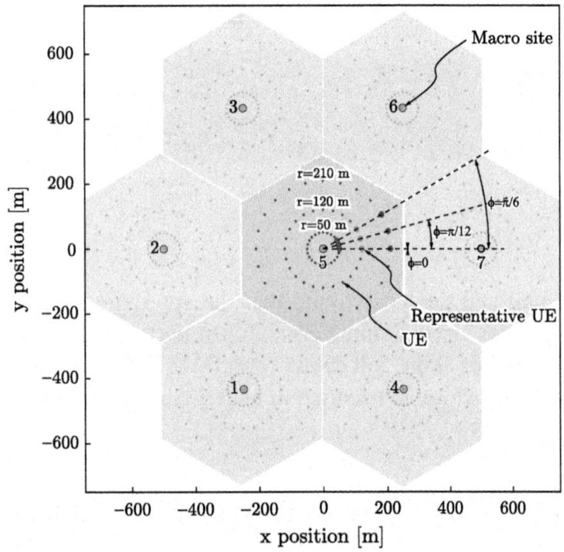

Fig. 11.9 *Hexagonal grid* setup with central cell and six interfering eNodeBs. UEs are equidistantly distributed along circles of radius r = {50, 120, 210} m. *Bold dots* indicate representative UE positions. The corresponding angles are given by $\phi = \{0, \frac{\pi}{12}, \frac{\pi}{6}\}$, respectively. In the case of BS collaboration, eNodeB 7 does not contribute to the aggregate interference

Table 11.3 Specific parameters for simulations of homogeneous macro cellular network

Parameter	Value
Antenna configuration	$N_{Tx} \times N_{Rx} = 1 \times 1$
eNodeB antenna gain in dB	$A(\theta) = 0\,\text{dB}$
Path loss	$\ell(d) = \min(b_P,\ 1/c_P\, d^{-2})$
Scheduler type	Round robin

11.7.2 Validation of Gamma Approximation

In this section, the introduced circular interference model is validated. The particular focus lies on the accuracy of the Gamma distribution as an approximation for both composite fading and aggregate interference. The system model is referred from Sect. 11.7.1.

Firstly, the average aggregate interference is measured along each of the three UE circles. The results are depicted as solid lines in Fig. 11.10. In accordance with Sect. 11.5.1, it is observed that average aggregate interference is almost constant at the cell-center and in the middle of the cell. At cell-edge, the curves exhibit fluctuations due to the vicinity of the dominant interferers. Results from the circular interference model accurately assess the average behavior, as shown by the dashed lines.

In the next step, the Empirical Cumulative Distribution Function (ECDF) of the aggregate interference is computed at nine representative UE locations, which are marked by bold dots in Fig. 11.9. Similar to Sect. 11.5.2, the angle positions $\phi = \{0, \frac{\pi}{6}, \frac{\pi}{12}\}$ refer to UEs with one dominant interferer (eNodeB 7), two equidistant dominant interferers (eNodeBs 6 and 7) and a variation thereof. Solid lines in Fig. 11.11 depict the results. In accordance with Sect. 11.5.2, the interference distributions are dominated by the UEs' distances to the origin while their angle positions have only minor impact. The latter is illustrated by the enlarged section in Fig. 11.11.

Finally, the introduced circular model is applied to approximate the aggregate interference distribution at a certain distance r by a Gamma RV. The first step consists in determining the parameters k_0 and θ_0 of the Gamma distribution $\Gamma[k_0, \theta_0]$ that models the composite fading (cf. Sect. 11.4). This is achieved by applying Algorithm 2. The intital values k_0' and θ_0' are obtained from Maximum Likelihood Estimation (MLE). MLE maximizes the *likelihood* $\mathcal{L}(k_0', \theta_0'|x) = f(x|k_0', \theta_0')$, where $f(\cdot)$ denotes a Gamma PDF with shape k_0' and scale θ_0', and x are the given outcomes.

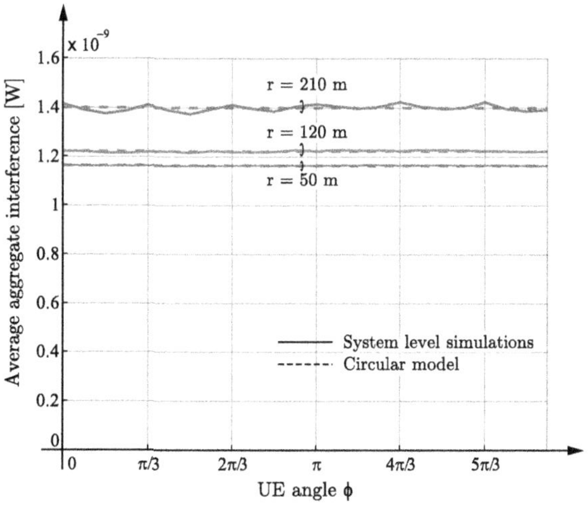

Fig. 11.10 Average aggregate interference power along the three UE circles in Fig. 11.9. *Solid curves* refer to simulation results, *dashed curves* denote results from circular interference model

Using a step size of $\Delta = 0.001$ and $N_{\text{iterations}} = 100$ yields a KS distance of 0.0512 between simulated- and approximated composite fading distribution. For comparison, employing only MLE achieves a KS distance of 0.0917.

Algorithm 2: Iterative algorithm for improving KS distance between empirical composite fading distribution and Gamma approximation. The term $F(x; k, \theta)$ denotes the CDF of a Gamma distribution with shape k and scale θ, respectively.

Data: empirical CDF of composite fading from simulations: $F_{\text{fading}}(x)$;
initial shape- and scale parameter: k'_0, θ'_0;
stepsize: Δ;
number of iterations: $N_{\text{iterations}}$;
Result: shape- and scale parameter: k_0, θ_0;
set $k_0 = k'_0$ and $\theta_0 = \theta'_0$;
for i = *1 to* $N_{\text{iterations}}$ **do**
    compute $\{k^*, t^*\} = \arg\min_{\{k,t\}} \sup_x |F_{\text{fading}}(x) - F(x; k, \theta)|$, with $k \in [k_0 - \Delta, k_0, k_0 + \Delta]$ and $\theta \in [\theta_0 - \Delta, \theta_0, \theta_0 + \Delta]$;
    if k^* *equals* k_0 ***and*** θ^* *equals* θ_0 **then**
        break;
    else
        set $k_0 = k^*$ and $\theta_0 = \theta^*$;
    end
end

Then, for each UE distance, the parameters of the aggregate interference distribution, $\hat{k}(r)$ and $\hat{\theta}(r)$, are calculated with Eqs. (11.20) and (11.21), respectively. The corresponding CDF curves are depicted as dashed lines in Fig. 11.11. It is observed that the approximated distributions slightly underestimate the occurrence of high interference values. In order to quantify the deviation from the simulated curves,

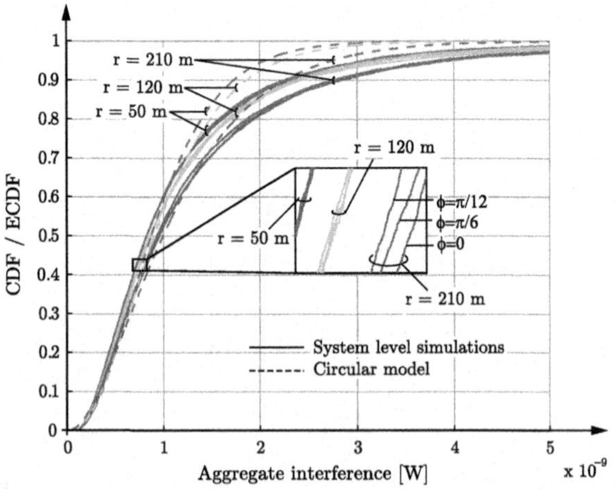

Fig. 11.11 Aggregate interference distributions at representative UE locations $r = \{50, 120, 210\}$ m and $\phi = \{0, \frac{\pi}{12}, \frac{\pi}{6}\}$, as marked by bold dots in Fig. 11.9. *Solid lines* refer to ECDF curves from simulations, *dashed lines* denote Gamma approximations as obtained with circular model

Table 11.4 KS distances between Gamma approximations and simulated ECDF curves at representative UE locations

	$\phi = 0$	$\phi = \frac{\pi}{12}$	$\phi = \frac{\pi}{6}$
r = 50 m	0.0713	0.0768	0.0762
	0.1391	0.1491	0.1454
	0.0720	0.0797	0.0773
r = 120 m	0.0697	0.0659	0.0698
	0.1347	0.1393	0.1300
	0.0815	0.0708	0.0701
r = 210 m	0.0565	0.0496	0.0466
	0.1183	0.1289	0.1274
	0.0823	0.0828	0.0840

For each r, the first two rows correspond to the Gamma approximation as obtained with the circular model. In the first row, composite fading is estimated with Algorithm 2, while in the second row it is assessed with MLE, respectively. The third row refers to the direct application of MLE on the distribution of the aggregate interference

the first row in Table 11.4 provides the KS distance for each UE location (r, ϕ). The values range from 0.05 at $r = 210$ m to 0.08 at $r = 50$ m.

For comparison, each simulated curve is also approximated by two further Gamma distributions. The first distribution adapts the circular model and estimates the composite fading by MLE, i.e., it employs the parameters k_0 and θ_0 that were used above to initialize Algorithm 2. The second distribution is computed by applying MLE directly to the simulated aggregate interference. The corresponding KS distances are likewise listed in the second- and third row of Table 11.4 for each UE location (r, ϕ). The first observation is that Algorithm 2 considerably improves the performance of the circular model, such that it even exceeds pure MLE of the aggregate interference. Hence, the accuracy of the circular model crucially depends on the precision of the composite fading approximation. Secondly, the results of the MLE range from 0.07 to 0.08, indicating that the assumption of Gamma-distributed interference itself induces a systematic error.

In summary, the circular model achieves a remarkable accuracy of fit despite its simplicity, thus corroborating its applicability.

11.8 Conclusion

In this chapter, a circular interference model for aggregate interference analysis in regular grid deployments is introduced. Particular focus is placed on characterizing a user at an eccentric location. The expected interference from the circular model is identified as the interference that is experienced by a typical user in a hexagonal grid at a certain distance from the origin. At cell-edge, it deviates by at most 3.2 % from the actual values.

In a second step, the corresponding interference statistics are approximated by a single Gamma RV. By means of the circular model, the distance-dependent shape- and scale parameters are determined in closed form and unveil the two key formative components of the distribution as the variance of the fading and the variance of the path loss due to the eccentric user location, respectively. Qualitative- and quantitative comparisons with the exact distributions confirm the accuracy of the Gamma approximation, yielding KS statistics no higher than 3.7 %.

The circular model's expedient adaption for representing the *well-planned part* of a two-tier heterogeneous cellular network is demonstrated. The example merges a fully regular macro-deployment with completely randomly distributed small cells and models the interference contribution from each tier by a single Gamma RV. The resulting aggregate interference distribution shows a remarkably good fit with Monte Carlo simulations. Hence, the model enables to accurately capture the impact of *both* user eccentricity *and* heterogeneity of the network with only few key parameters.

In the last part of the chapter, it is shown that the circular model enables an accurate prediction of the interference statistics in an LTE-A hexagonal grid scenario. Deviations from the simulation results mainly stem from the inaccurate approximation of the composite fading. The remainder of the approximation error is caused by the assumption of Gamma distributed aggregate interference itself.

The presented circular model does not allow to account for power control and coordination among BSs. This is a major motivation for the next chapter, which extends the model by non-uniform power profiles.

References

1. J. Gertner, *The Idea Factory: Bell Labs and the Great Age of American Innovation* (Penguin Group, New York, 2012)
2. P. Marsch, G. Fettweis, Static clustering for cooperative multi-point (CoMP) in mobile communications, in *IEEE International Conference on Communications* (*ICC*) Kyoto, June (2011). doi:10.1109/icc.2011.5963458
3. C. Ball, R. Mullner, J. Lienhart, H. Winkler, Performance analysis of closed and open loop MIMO in LTE, in *European Wireless Conference* (*EW*), pp. 260–265, Aalborg, May (2009). doi:10.1109/EW.2009.5358012
4. A. Farajidana, W. Chen, A. Damnjanovic, T. Yoo, D. Malladi, C. Lott, 3GPP LTE downlink system performance, in *IEEE Global Telecommunication Conference* (*GLOBECOM*) Honolulu, December (2009). doi:10.1109/GLOCOM.2009.5425327
5. J. Giese, M. Amin, S. Brueck, Application of coordinated beam selection in heterogeneous LTE-advanced networks, in *International Symposium on Wireless Communications Systems* (*ISWCS*), pp. 730–734, Aachen, November (2011)
6. Y. Liang, A. Goldsmith, G. Foschini, R. Valenzuela, D. Chizhik, Evolution of base stations in cellular networks: denser deployment versus coordination, in *IEEE International Conference on Communications* (*ICC*), pp. 4128–4132, Beijing, May (2008). doi:10.1109/ICC.2008.775
7. T. Novlan, R. Ganti, J. Andrews, Coverage in two-tier cellular networks with fractional frequency reuse, in *IEEE Global Telecommunications Conference* (*GLOBECOM*), Houston, December (2011). doi:10.1109/GLOCOM.2011.6133767

8. F. Di Salvo, A characterization of the distribution of a weighted sum of Gamma variables through multiple hypergeometric functions. Integral Transforms Special Funct. **19**(8), 563–575 (2008). doi:10.1080/10652460802045258
9. M. Haenggi, J. Andrews, F. Baccelli, O. Dousse, M. Franceschetti, Stochastic geometry and random graphs for the analysis and design of wireless networks. IEEE J. Sel. Areas Commun. **27**(7), 1029–1046 (2009). doi:10.1109/JSAC.2009.090902
10. M. Haenggi, *Stochastic Geometry for Wireless Networks* (Cambridge University Press, Cambridge, 2012)
11. R.W. Heath Jr., M. Kountouris, T. Bai, Modeling heterogeneous network interference using Poisson point processes. IEEE Trans. Signal Process. **61**(16), 4114–4126 (2013). doi:10.1109/TSP.2013.2262679
12. Y. Zhuang, Y. Luo, L. Cai, J. Pan, A geometric probability model for capacity analysis and interference estimation in wireless mobile cellular systems, in *IEEE Global Telecommunications Conference (GLOBECOM)* Houston, December (2011). doi:10.1109/GLOCOM.2011.6134503
13. K.B. Baltzis, Hexagonal vs Circular Cell Shape: A Comparative Analysis and Evaluation of the Two Popular Modeling Approximations, *Cellular Networks-Positioning, Performance Analysis, Reliability* (InTech, Rijeka, 2011). doi:10.5772/626
14. M. Taranetz, M. Rupp, A circular interference model for wireless cellular networks, in *Proceedings of the International Wireless Communications & Mobile Computing Conference*, Nicosia (2014)
15. M. Taranetz, System level modeling and evaluation of heterogeneous cellular networks, Ph.D. dissertation, Institute of Telecommunications, TU Wien (2015). http://theses.eurasip.org/theses/611/system-level-modeling-and-evaluation-of/
16. I. Kostic, Analytical approach to performance analysis for channel subject to shadowing and fading. IEEE Proc. Commun. **152**(6), 821–827 (2005)
17. S. Al-Ahmadi, H. Yanikomeroglu, On the approximation of the generalized-K PDF by a gamma PDF using the moment matching method, in *IEEE Wireless Communications and Networking Conference (WCNC)* Budapest, April (2009). doi:10.1109/WCNC.2009.4917849
18. S. Al-Ahmadi, H. Yanikomeroglu, On the use of high-order moment matching to approximate the generalized-K distribution by a gamma distribution, in *IEEE Global Telecommunication Conference (GLOBECOM)* Honolulu, November (2009). doi:10.1109/GLOCOM.2009.5425259
19. J. Zhang, M. Matthaiou, Z. Tan, H. Wang, Performance analysis of digital communication systems over composite $\eta - \mu$ gamma fading channels. IEEE Trans. Veh. Technol. **61**(7), 3114–3124 (2012). doi:10.1109/TVT.2012.2199344
20. R.W. Heath Jr., T. Wu, Y.H. Kwon, A. Soong, Multiuser MIMO in distributed antenna systems with out-of-cell interference. IEEE Trans. Signal Process. **59**(10), 4885–4899 (2011). doi:10.1109/TSP.2011.2161985
21. S. Kusaladharma, C. Tellambura, Aggregate interference analysis for underlay cognitive radio networks. IEEE Wireless Commun. Lett. **1**(6), 641–644 (2012). doi:10.1109/WCL.2012.091312.120600
22. C.A. Coelho, J.T. Mexia, On the distribution of the product and ratio of independent generalized gamma-ratio random variables. Sankhyā: Indian J. Statist. **69**(12), 221–255 (2007)
23. 3rd Generation Partnership Project (3GPP), Evolved universal terrestrial radio access (E-UTRA); radio frequency (RF) system scenarios, in *3rd Generation Partnership Project (3GPP)*, TR 36.942, October (2014)
24. I.S. Gradshteyn, I.M. Ryzhik, *Table of Integrals, Series, and Products*, 7th edn. (Elsevier/Academic Press, Amsterdam, 2007)

25. P. Moschopoulos, The distribution of the sum of independent gamma random variables. Ann. Inst. Statist. Math. **37**(1), 541–544 (1985)
26. F. Baccelli, B. Blaszczyszyn, *Stochastic Geometry and Wireless Networks: Volume I Theory*, Foundation and Trends in Networking (Now Publishers, Hanover, 2009). doi:10.1561/1300000006
27. H. Claussen, Efficient modelling of channel maps with correlated shadow fading in mobile radio systems, in *IEEE International Symposium on Personal, Indoor and Mobile Radio Communications* (*PIMRC*), vol. 1, pp. 512–516, Berlin, September (2005)

Chapter 12
Modeling Asymmetric Aggregate Interference by Symmetric Structures

Markus Rupp, Stefan Schwarz and Martin Taranetz

In this chapter, the circular model from Chap. 11 is extended by non-uniform power profiles along the circles. The enhanced model enables to aggregate given interferer deployments such that the original interference statistics are accurately preserved while the amount of relevant interferers is reduced considerably.

Scaling up the number of base stations per unit area is one of the major trends in mobile cellular systems of the fourth (4G)- and fifth generation (5G) [1], making it increasingly difficult to characterize aggregate interference statistics with system models of low complexity. Tractable interference statistics have mainly been reported in the field of stochastic geometry. When closed-form expressions are desired, this mathematical framework imposes its own particular limitations, typically including spatial stationarity and isotropy of the scenario [2–4]. Hence, the potential to consider an asymmetric interference impact is very limited and notions such as *cell-center* and *cell-edge* are, in general, not accessible. Based on [5, 6], the contributions of this chapter outline as follows:

- A new *circular interference model* is introduced. The key idea is to map arbitrary *out-of-cell* interferer deployments onto circles of uniformly spaced nodes such that the original aggregate interference statistics can accurately be reproduced. The model greatly reduces complexity as the number of participating interferers is significantly reduced.
- A *mapping scheme* that specifies a procedure for determining the power profiles of arbitrary interferer deployments is proposed. Its performance is evaluated by means of Kolmogorov–Smirnov (KS) statistics. The test scenarios are modeled by

M. Rupp (✉) · S. Schwarz · M. Taranetz
Institute of Telecommunications, TU Wien, Vienna, Austria
e-mail: mrupp@nt.tuwien.ac.at

S. Schwarz
e-mail: sschwarz@nt.tuwien.ac.at

M. Taranetz
e-mail: martin.taranetz@tuwien.ac.at

M. Rupp et al., *The Vienna LTE-Advanced Simulators*,
Signals and Communication Technology, DOI 10.1007/978-981-10-0617-3_12

Poisson Point Processes (PPPs) so as to confront the regular circular structure with complete spatial randomness. It is shown that the individual spatial realizations exhibit largely diverging power profiles.

- A new finite sum representation for the Probability Density Function (PDF) of the *sum of Gamma Random Variables (RVs) with integer-valued shape parameter* is introduced to further enhance and validate interference analysis with the circular model. Its restriction to integer-valued shape parameters is driven by relevant use cases for wireless communication engineering and the availability of *exact* solutions. The key strength of the proposed approach lies in the ability to decompose the interference distribution into the contributions of the individual interferers.
- Statistics of aggregate interference with *asymmetric interference impact* are investigated. The asymmetry is induced by eccentrically placing a user in a generic, isotropic scenario. This setup is achieved by applying the introduced circular model with uniform power profiles. The model enables to employ the proposed finite sum representation. It is shown that the partition of the interference distribution is particularly useful to identify candidate Base Stations (BSs) for user-centric BS collaboration schemes. Moreover, the framework allows to predict the corresponding Signal-to-Interference Ratio (SIR)- and rate statistics.

The main focus of this chapter is on downlink transmission in cellular networks. A comparable framework for the uplink is found in [7].

This chapter is organized as follows. Sections 12.1 and 12.2 introduce the circular interference model and the new finite sum representation for the sum of Gamma RVs with integer-valued shape parameter, respectively. Section 12.3 presents a mapping scheme and validates the applicability of the circular model. Section 12.4 investigates aggregate interference statistics and the performance of BS collaboration schemes at eccentric user locations. In Sect. 12.5, the rate performance as predicted by the circular model is compared against Long Term Evolution-Advanced (LTE-A) system level simulations.

12.1 Circular Interference Model

Consider the serving BS to be located at the origin. The proposed circular interference model is composed of C concentric circles of interferers, as shown in Fig. 12.1. On circle $c \in \{1, \ldots, C\}$ of radius R_c, N_c interfering nodes are spread out equidistantly. The interferer locations are expressed in terms of polar coordinates as $(R_c, \Psi_{c,n})$, where $\Psi_{c,n} = 2\pi n/N_c - \phi_c$, with $n \in \{1, \ldots, N_c\}$ and $\phi_c \in [0, 2\pi)$. Each node is unambiguously assigned to a tuple (c, n) and labeled as $\mathcal{T}_{c,n}$. The central BS is denoted as $\mathcal{T}_{0,0}$. Some of the interferers on the circles may also become serving nodes when BS collaboration schemes are applied, as will be shown later in Sect. 12.4.3.

The interferers on the circles do not necessarily represent real physical sources. As illustrated in Fig. 12.2, they rather correspond to the N_c mapping points of an angle-dependent *power profile* $p_c[n]$, with $\sum_{n=1}^{N_c} p_c[n] = 1$. Exemplary profiles of a single

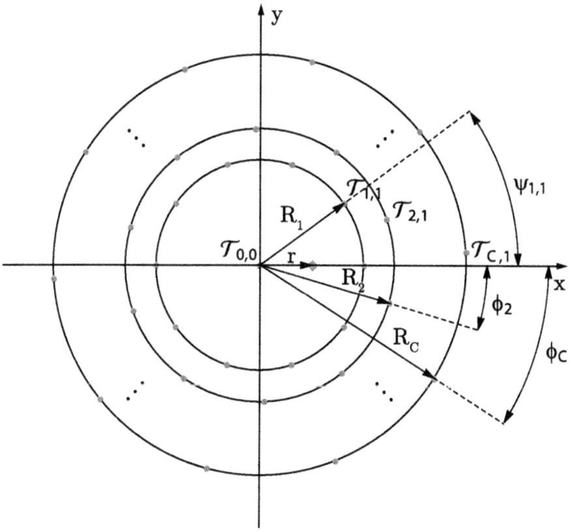

Fig. 12.1 Circular interference model with C *circles* of radius R_c and phase ϕ_c, $c \in \{1, \ldots, C\}$, and user at $(r, 0)$. $\mathcal{T}_{c,n}$ denotes the nodes of the model

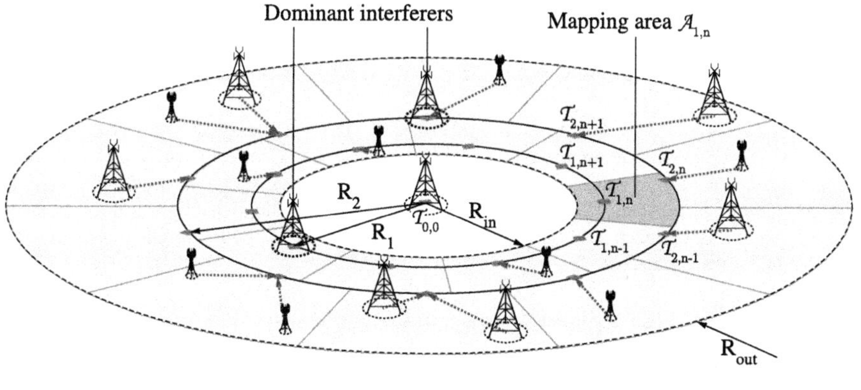

Fig. 12.2 Circular interference model with two *circles*, i.e., $C = 2$. Characteristics of an arbitrary heterogeneous interferer deployment are condensed to *circles* of equidistantly spaced nodes $\mathcal{T}_{c,n}$ such that the original interference statistics can accurately be reproduced. A mapping scheme is presented in Sect. 12.3. The original BSs are distributed within an annulus of inner radius R_{in} and outer radius R_{out}

circle are shown in Fig. 12.3. Intuitively, $p_c[n]$ condenses the interferer characteristics of an annulus with inner radius R_c (R_{in} in case of $c = 1$) and possibly outer radius R_{c+1} (R_{out} in case of $c = C$) such that the circular model equivalently reproduces the original BS deployment in terms of interference statistics. This technique enables to represent substantially large networks by a *finite- and well-defined* constellation of nodes. By reducing the number of relevant interferers, it greatly reduces complexity and thus allows to apply finite sum-representations as those introduced in Sect. 12.2.

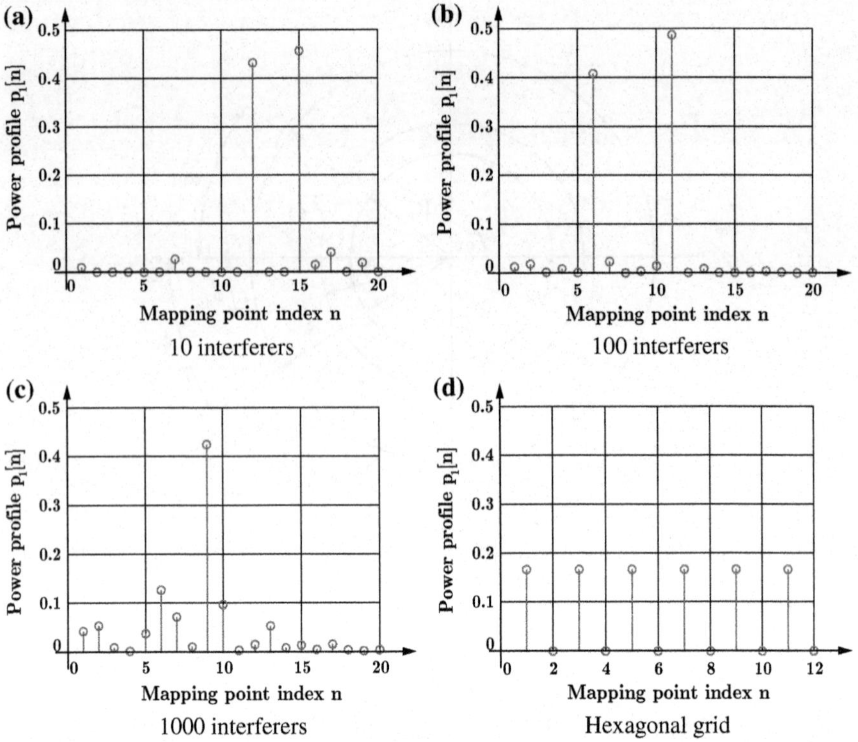

Fig. 12.3 Power profiles of circular models with one *circle*, i.e., $C = 1$, for three stochastic interference scenarios (**a–c**) with $N_1 = 20$ mapping points, and for a hexagonal grid with $N_1 = 12$ mapping points, respectively. The stochastic BS distributions are modeled by a PPP of intensity $\lambda = 10^{-6}\,\mathrm{m}^{-2}$. The expected number of interferers as denoted by the figure labels, is varied by altering the scenario size

Table 12.1 Parameters of the circular interference model

Symbol	Annotation
R_{in}	Inner radius of mapping region, $R_{\mathrm{in}} \geq 0$
R_{out}	Outer radius of mapping region, $R_{\mathrm{out}} > R_{\mathrm{in}}$
C	Number of interferer circles, $C \in \mathbb{N}^+$
R_c	Radius of circle c, $c \in \{1, \ldots, C\}$, $R_c > 0$
ϕ_c	Phase of circle c, $c \in \{1, \ldots, C\}$ $\phi_c \in [-\frac{\pi}{N_c}, \frac{\pi}{N_c}]$
N_c	Number of mapping points, $c \in \{1, \ldots, C\}$, $N_c \in \mathbb{N}^+$
P_c	Total transmit power of circle c, $c \in \{1, \ldots, C\}$, $P_c > 0$
$p_c[n]$	Power profile of circle c, $c \in \{1, \ldots, C\}$, $n \in \{1, \ldots, N_c\}$, $p_c[n] \in [0, 1]$

Table 12.1 summarizes the parameters of the model. Typically, the size of the mapping region, as specified by R_{in} and R_{out}, is predetermined by the scenario. The freely selectable variables are the amount of circles C and, for each circle, the phase

ϕ_c, the radius R_c and the number of mapping points N_c, respectively. Section 12.3 presents systematic experiments to provide a reference for the parameter setting and proposes a mapping scheme to determine power profiles $p_c[n]$ and transmit powers P_c, respectively.

A signal from node $\mathscr{T}_{c,n}$, located at $(R_c, \Psi_{c,n})$, to a user at $(r, 0)$ experiences *path loss* $\ell(d_{c,n}(r))$, where $d_{c,n}(r) = \sqrt{R_c^2 + r^2 - 2R_c r \cos(\Psi_{c,n})}$ (cf. Fig. 12.1) and $\ell(\cdot)$ is an arbitrary distance-dependent path loss law, as well as *fading*, which is modeled by statistically independent RVs $G_{c,n}$. The received power from node $\mathscr{T}_{c,n}$ at position $(r, 0)$ is determined as

$$P_{\mathrm{Rx},c,n}(r) = P_c\, p_c[n]\, \ell(d_{c,n}(r))\, G_{c,n}, \tag{12.1}$$

where P_c denotes the total transmit power of circle c. It is important to note that the term $P_{\mathrm{Rx},c,n}(r)$ can be interpreted as a RVs $G_{c,n}$, which is scaled by a factor of $P_c\, p_c[n]\, \ell(d_{c,n}(r))$.

The nodes employ omnidirectional antennas with unit antenna gain. Characteristics of antenna directivity are incorporated into the power profile. In general, the central cell will have an irregular shape that can be determined by Voronoi tessellation [8]. For simplicity, the *small ball* approximation from [8] is applied. A user is considered as *cell-edge user*, if it is located at the edge of the central Voronoi cell's inscribing ball. This approximation misses some poorly covered areas at the actual cell-edge with marginal loss of accuracy [8].

Let $\mathscr{S}$ and $\mathscr{I}$ denote the sets of nodes $\mathscr{T}_{c,n}$ corresponding to desired signal and interference, respectively. Then, the aggregate signal- and interference powers are calculated as

$$S(r) = \sum_{\{(c,n)|\mathscr{T}_{c,n}\in\mathscr{S}\}} P_{\mathrm{Rx},c,n}(r), \tag{12.2}$$

$$I(r) = \sum_{\{(c,n)|\mathscr{T}_{c,n}\in\mathscr{I}\}} P_{\mathrm{Rx},c,n}(r), \tag{12.3}$$

with $P_{\mathrm{Rx},c,n}(r)$ from Eq. 12.1. The set $\mathscr{S}$ may include the central node $\mathscr{T}_{0,0}$ as well as nodes on the circles, if collaboration among the BSs is employed. The incoherence assumption is exploited for a more realistic assessment of the co-channel interference [9]. Following the interpretation of Eqs. 12.1, 15.6 and 15.7 can be viewed as sums of scaled RVs, which are supported by a vast amount of literature for certain fading distributions such as Rayleigh, log-normal and Nakagami-m [10–29].

This chapter places particular focus upon the Gamma distribution due to its wide range of useful features for wireless communication engineering. Preliminary information is provided in Sect. 11.1. The next section introduces a new theorem on the sum of Gamma RVs. The theorem is presented before validating the accuracy of the circular model as it is later exploited for this purpose.

12.2 Distribution of the Sum of Gamma Random Variables

As explained in Sect. 11.1, the Gamma distribution exhibits the *summation property*, i.e., if $G_i \sim \Gamma[k_i, \theta]$ with $i = 1, 2, \ldots, N$, then $\sum_{i=1}^{N} G_i \sim \Gamma[\sum_{i=1}^{N} k_i, \theta]$. While this feature is convenient to apply, it is the *sum of Gamma RVs* with *distinct* scale parameters that has attracted a lot of attention in describing wireless communications though. Most commonly, it emerged in the performance analysis of diversity combining receivers and the study of *aggregate co-channel interference under Rayleigh fading* [10–20]. Therefore, communication engineers have considerably pushed the search for closed form statistics.

Representatively, Moschopoulos' much-cited series expansion in [21] was extended for correlated Gamma RVs in [10]. Other approaches based on the inverse Mellin transform (e.g., [30, 31]) paved the way for representations with a single integral as shown, e.g., in [12] or a Lauricella hypergeometric series as employed, e.g., in [11, 16].

All the aforementioned contributions focus on the sum of Gamma RVs with *real-valued* shape parameter. The resulting integrals and *infinite* series, despite being composed of elementary functions, typically yield a slow rate of convergence. Therefore, an accurate approximation by a truncated series requires to keep a high amount of terms and complicates further analysis.

The sum of Gamma RVs with *integer* shape parameter has mainly been reported in statistical literature. Initial approaches focused on the moment generating function and results were obtained in the form of series expansions [22]. Based on the work of [23, 24] was among the first to formulate a convenient closed form solution. Soon after, the Generalized Integer Gamma (GIG) distribution was published in [25]. This approach was also adopted in wireless communication engineering [13, 15]. In comparison to RVs with *real-valued* shape parameter, the PDF of the sum of RVs with *integer* shape parameter allows an *exact* representation by a *finite* series.

12.2.1 Proposed Finite Sum Representation

In the analysis of aggregate interference statistics, it is particularly desirable to identify the main distribution-shaping factors, i.e., the interfering sources with the highest impact. However, the expressions in [13, 15] are not suitable for this task due to multiple nested sums and recursions. The proposed finite-sum representation in this chapter avoids recursive functions and enables to *straightforwardly trace the main determinants* of the distribution characteristics.

Theorem 12.1 *Let $G_l \sim \Gamma[k_l, \theta_l]$ be L independent Gamma RVs with $k_l \in \mathbb{N}^+$ and all θ_l different. Then, the PDF of $Y = G_1 + \cdots + G_L$ can be expressed as*

$$f_Y(y) = \sum_{l=1}^{L} \frac{\Lambda_l}{\theta_l^{k_l}} h_{k_l-1,l}(0) e^{-y/\theta_l} \qquad (12.4)$$

with

$$\Lambda_l = \frac{(-1)^{k_l+1}}{(k_l-1)!} \prod_{i=1,i\neq l}^{L} \left(1 - \frac{\theta_i}{\theta_l}\right)^{-k_i}, \qquad l = 1, \ldots, L \tag{12.5}$$

$$h_{\delta+1,l}(\zeta) = h_{1,l}(\zeta) h_{\delta,l}(\zeta) + \frac{d}{d\zeta} h_{\delta,l}(\zeta), \qquad \delta = 0, \ldots, k_l - 1 \tag{12.6}$$

and

$$h_{1,l}(0) = -y + \sum_{i=1,i\neq l}^{L} k_i \left(\frac{1}{\theta_i} - \frac{1}{\theta_l}\right)^{-1}, \qquad l = 1, \ldots, L \tag{12.7}$$

$$h_{1,l}^{(m)}(0) = m! \sum_{i=1,i\neq l}^{L} k_i \left(\frac{1}{\theta_i} - \frac{1}{\theta_l}\right)^{-m-1}, \qquad m = 1, \ldots, k_l - 1 \tag{12.8}$$

Proof The proof is provided in [6, Appendix C].

The uniqueness of θ_l can be assumed without loss of generality. In case of some θ_l being equal, the corresponding RVs are added up by virtue of the *summation property of Gamma RVs* (cf. Sect. 11.1).

Superscript (m) of $h_{1,l}^{(m)}(\zeta)$ denotes the mth derivative of $h_{1,l}(\zeta)$. The recursive determination of $h_{\delta,l}(\zeta)$ in Eq. 12.6 seemingly interrupts the straightforward calculation of $f_Y(y)$. However, $h_{\delta,l}(\zeta)$ is a function of only $h_{1,l}(\zeta)$ and its higher order derivatives. Therefore, the function series in Eq. 12.6 can be evaluated *in advance* up to the highest required degree $\delta_{\max} = \max_l k_l - 1$.

Thus, the proposed approach enables the *exact* calculation of $f_Y(y)$ in a *component-wise* manner. In the next step, it is shown how to apply Theorem 12.1 in the proposed circular model.

12.2.2 Application in Circular Interference Model

Assume that $G_{c,n} \sim \Gamma[k_{c,n}, \theta_{c,n}]$ in Eq. 12.1, with $k_{c,n} \in \mathbb{N}^+$ and $\theta_{c,n} > 0$. Then, Eqs. 15.6 and 15.7 represent sums of scaled Gamma RVs $P_{\text{Rx},c,n}(r) \sim \Gamma[k_{c,n}, \theta'_{c,n}(r)]$, where $\theta'_{c,n}(r) = P_c\, p_c[n] \cdot \ell(d_{c,n}(r))\, \theta_{c,n}$. Therefore, their PDFs can be determined by applying Theorem 12.1.

The theorem requires all scale parameters to be different. Thus, let $\boldsymbol{\theta}_{\mathcal{I}}(r)$ denote the vector of unique scale parameters $\theta'_{c,n}(r)$ with (c, n) from the set $\{(c, n)|\mathcal{T}_{c,n} \in \mathcal{I}\}$. A second vector $\mathbf{k}_{\mathcal{I}}$ contains the corresponding shape parameters. By virtue of the *summation property*, if $\theta'_{c,n}(r)$ occurs multiple times in the set, the respective shape parameter in $\mathbf{k}_{\mathcal{I}}$ is calculated as the sum of shape parameters $k_{c,n}$ of the according entries. The vectors $\boldsymbol{\theta}_{\mathcal{S}}(r)$ and $\mathbf{k}_{\mathcal{S}}$ are obtained equivalently. Then, the PDFs of $S(r)$ and $I(r)$ are expressed as

$$f_S(\gamma; r) = \sum_{l=1}^{L_{\mathcal{S}}} \frac{\Lambda_l}{\theta_l(r)^{k_l}} h_{k_l-1,l}(0) e^{-\gamma/\theta_l(r)}, \tag{12.9}$$

$$f_I(\gamma; r) = \sum_{l=1}^{L_{\mathcal{I}}} \frac{\Lambda_l}{\theta_l(r)^{k_l}} h_{k_l-1,l}(0) e^{-\gamma/\theta_l(r)}, \tag{12.10}$$

with Λ_l and $h_{\delta,l}(\cdot)$ as defined in Eqs. (12.5) and (12.6). Subscript l indicates the lth components of the vectors $\mathbf{k}_{\mathcal{S}}$ $(\boldsymbol{\theta}_{\mathcal{S}}(r))$ and $\mathbf{k}_{\mathcal{I}}$ $(\boldsymbol{\theta}_{\mathcal{I}}(r))$ and $L_{\mathcal{S}}$ and $L_{\mathcal{I}}$ are their corresponding lengths, respectively.

Hence, employing Theorem 12.1 allows to evaluate the *exact* distributions of the aggregate signal- and interference from the circular model by *finite* sums. In the following section, this fact is exploited to verify the accuracy of the model.

12.3 Mapping Scheme for Stochastic Network Deployments

This section presents a procedure to determine the power profiles $p_c[n]$ and the corresponding powers P_c of the circular model for completely random interferer distributions. Then, systematic experiments are carried out to provide a reference for selecting the free variables C and N_c, respectively. The parameters R_c and ϕ_c are also specified by the procedure. The accuracy of the approximation is measured by means of the KS distance. It is defined as

$$D_{\text{KS}}(r) = \sup_x \left| F_{I,\text{original}}(x; r) - F_{I,\text{circular}}(x; r) \right|, \tag{12.11}$$

where r refers to the user's eccentricity and $F_{I,\text{original}}(x; r)$ and $F_{I,\text{circular}}(x; r)$ denote the aggregate-interference Cumulative Distribution Functions (CDFs) of the original deployment and the circular model, respectively. The corresponding PDFs are obtained by applying Theorem 12.1.

12.3.1 Mapping Procedure

Let $\mathcal{N}$ denote a (possibly heterogeneous) set of BSs that are arbitrarily distributed within an annulus $\mathcal{A}$ of inner radius R_{in} and outer radius R_{out}, as shown in Fig. 12.2. Radius R_{out} as well as the number of nodes in $\mathcal{N}$ could be substantially large. Given a circular model with C circles and N_c nodes per circle, the parameters P_c, R_c and ϕ_c as well as the power profile $p_c[n]$ can be determined by Algorithm 3.

Algorithm 3: Mapping procedure for circular model.

Data: number of circles C; nodes per circle N_c;
original base station deployment $\mathscr{N}$;
inner- and outer radius of mapping region $\mathscr{A}$: R_{in} and R_{out};
Result: P_c, $p_c[n]$, R_c and ϕ_c for all $c \leq C$;
for $c = 1$ to C **do**
  determine R_c and ϕ_c based on the strongest interferer that has not yet been mapped;
end
for $c = 1$ to C **do**
  specify mapping region $\mathscr{A}_c$ with inner radius R_c and outer radius R_{c+1};
  if $c = 1$ **then** set inner radius of $\mathscr{A}_c$ to R_{in}; **end**
  if $c = C$ **then** set outer radius of $\mathscr{A}_c$ to R_{out}; **end**
  compute P_c and $p_c[n]$ for $\mathscr{A}_c$;
end

The presented procedure employs the origin as a reference point and therefore does *not* depend on the user location. The computation of P_c and $p_c[n]$ outlines as follows. Let $\mathscr{T}_{c,n}$ denote node n on circle c. Assume that its associated mapping area $\mathscr{A}_{c,n}$ is bounded by the circles of radius R_c and R_{c+1} (in the case of $c = 1$, the inner radius is set to R_{in}; for $c = C$ the outer radius is set to R_{out}) as well as the perpendicular bisectors of the two line segments $\overline{\mathscr{T}_{c,n}\mathscr{T}_{c,n-1}}$, and $\overline{\mathscr{T}_{c,n}\mathscr{T}_{c,n+1}}$, as illustrated in Fig. 12.2. This yields an even division of circle c's mapping area $\mathscr{A}_c$, which can be formulated as $\mathscr{A}_c = \bigcup_{n\in\{1,\ldots,N_c\}} \mathscr{A}_{c,n}$. The average received power at the origin from all considered BSs in $\mathscr{A}_c$ is calculated as

$$P_{\text{Rx},\mathscr{A}_c} = \sum_{i\,\in\,\mathscr{N}\cap\mathscr{A}_c} P_{\text{Tx},i}\;\ell(d_i)\mathbb{E}[G_i], \tag{12.12}$$

where $P_{\text{Tx},i}$, d_i and G_i correspond to transmit power, distance and experienced fading of interferer i, respectively. Then, the total transmit power P_c is obtained by mapping $P_{\text{Rx},\mathscr{A}_c}$ back on the circle, which formulates as $P_c = P_{\text{Rx},\mathscr{A}_c}\,\ell(R_c)^{-1}$. Hence, in this scheme the average received powers from the original deployment and the circular model are equivalent at the origin. The segmentation of $\mathscr{A}_c$ into areas $\mathscr{A}_{c,n}$ yields the corresponding power profile

$$p_c[n] = \frac{1}{P_{\text{Rx},\mathscr{A}_c}}\left(\sum_{i\,\in\,\mathscr{N}\cap\mathscr{A}_{c,n}} P_{\text{Tx},i}\;\ell(d_i)\right), \tag{12.13}$$

with $P_{\text{Rx},\mathscr{A}_c}$ from Eq. 12.12.

In the presented procedure, the parameters R_c and ϕ_c are set such that the cth dominant interferer coincides with a node on circle c, as illustrated in Fig. 12.2. This ensures that $R_1 \geq R_{\text{in}}$ (in a heterogeneous network, as investigated in Sect. 12.3.3, non-dominant interferers between R_{in} and R_1 are mapped “back” on circle 1 by the receive-power dependent weighting in Eq. 12.13) and $R_C \leq R_{\text{out}}$, and is especially suitable for completely random interferer distributions, as demonstrated in the next section. In fully regular scenarios, on the other hand, a circle comprises multiple, equally dominant nodes, making it expedient to specify R_c and ϕ_c according to the structure of the grid. For example, the circular model allows to *perfectly* represent a hexagonal grid setup, when the number of mapping points is set as a multiple of six. Then, the nodes on the circle coincide with the actual interferer locations. An exemplary power profile for $N_\mathrm{I} = 12$ is shown in Fig. 12.3d.

Algorithm 3 is one of many possible mapping approaches. It is a heuristic, based on my experience and observations and is thus *not claimed* to be optimal. Its refinement yields an interesting topic for further work. In the next two sections, systematic experiments in *completely random* scenarios are performed to provide a reference for setting C and N_c. For reasons of clarity, Sect. 12.3.2 is limited to *homogeneous* BS deployments. Heterogeneous setups are then evaluated in Sect. 12.3.3. It is refrained from stochastic scenarios with a certain degree of regularity, since measuring spatial inhomogeneity is itself an ongoing topic of research [32]. Completely random- and fully regular scenarios are considered as limiting cases, encompassing every conceivable practical deployment in between.

12.3.2 Performance Evaluation of Homogeneous Base Station Deployments

The *original* interferer deployment $\mathcal{N}$ is modeled by a PPP of intensity λ. Such process is considered *most challenging* for the regularly structured circular model, as it represents complete spatial randomness. Signal attenuation is modeled by a log-distance dependent path loss law $\ell(x) = \min(b_\mathrm{P}, 1/c_\mathrm{P}\, x^{-4})$, and Gamma fading with $k = 2$ and $\theta = 1$, referring to a 2×1 Multiple-Input Single-Output (MISO) setup and Maximum Ratio Transmission (MRT). In this chapter, $b_\mathrm{P} = 1$ and $c_\mathrm{P} = 1$ for simplicity. The BSs transmit with power $P_{T1} = 40\,\mathrm{W}$ and are distributed within an annular regions of inner radius $R_\mathrm{in} = 500\,\mathrm{m}$ and $R_\mathrm{out} = \sqrt{N_\mathrm{I}/(\pi\lambda) + R_\mathrm{in}^2}$. Radius R_in ensures that the inscribing ball of the central cell has a minimum radius of 250 m, assuming that the central BS also transmits with P_{T1}. The outer radii R_out are chosen such that, on average, N_I BSs locations are generated within the corresponding annulus. Consider a PPP of intensity λ within an annulus of inner radius R_in and outer radius R_out. The expected number of generated nodes is calculated as $N_\mathrm{I} = \lambda(R_\mathrm{out}^2 - R_\mathrm{in}^2)\pi$. In order to cover a wide range of scenarios, $N_\mathrm{I} = \{100, 1000\}$ and $\lambda = \{0.5 \times 10^{-6}, 10^{-6}\}\,\mathrm{m}^{-2}$ are studied. The parameter settings are summarized in Table 12.2.

For each scenario snapshot, eight circular models with $C = \{1, 2, 3, 4\}$ and two distinct values of N_c are set up according to Sect. 12.3.1. In the case of $\lambda = 0.5 \times 10^{-6}\,\mathrm{m}^{-2}$, $N_c = \{10, 20\}$ and, for $\lambda = 10^{-6}\,\mathrm{m}^{-2}$, $N_c = \{20, 40\}$, respectively. Then, the *aggregate interference distributions* are determined. The distributions for the original interferer deployment are only obtained via simulations (by averaging over

Table 12.2 System setup for evaluation

Parameter	Value
Transmit power	$P_{T1} = 40\,\text{W}$ ($P_{T2} = 4\,\text{W}$)
Node density	$\lambda = \{0.5 \times 10^{-6}, 10^{-6}\}\,\text{m}^{-2}$ ($\lambda_2 = 0.5 \times 10^{-5}\,\text{m}^{-2}$)
Expected number of interferers	$N_\text{I} = \{100, 1000\}$
Path loss	$\ell(x) = \min(b_\text{P}, 1/c_\text{P}\, x^{-4})$, $b_\text{P} = 1$, $c_\text{P} = 1$, $x > 0$
Fading	$G_{c,n} \sim \Gamma[2, 1]$

1000 spatial realizations and 10,000 fading realizations), since the vast amount of nodes hampers the application of Theorem 12.1 due to complexity issues. On the other hand, the circular models comprise at most 43 *active* nodes and therefore enable to utilize the theorem. This number is obtained for $C = 4$ and $N_c = 40$, and stems from the fact that in a homogeneous BSs deployment, the dominant interferers are also the closest ones. Therefore, the presented scheme only maps a single BSs on each circle $c < C$, i.e., except for $c = C$ there is only one active node per circle.

Figure 12.4 depicts KS distances over the user eccentricity r. The first important observation is that the accuracy considerably improves with an increasing number of circles C. This mainly results from accurately capturing the first few dominant BSs that have the largest impact on the aggregate interference distribution, as later shown in Sect. 12.4. A second remarkable observation is that doubling the amount of nodes per circle from $N_c = 10$ to $N_c = 20$ for $\lambda = 0.5 \times 10^{-6}\,\text{m}^2$ (cf. Fig. 12.4a, b), and from $N_c = 20$ to $N_c = 40$ for $\lambda = 10^{-6}\,\text{m}^2$ (cf. Fig. 12.4c, d) does not achieve smaller KS distances, respectively. This result indicates that it is rather the number of circles C than the number of nodes per circle N_c that impacts the accuracy. As shown in the examples, good operating points for homogeneous macro-BS deployments are $N_c = 20$ and $C = 4$, independent of the deployment parameters. Lastly, it should be noted that the circular model allows to represent 1000 and more interferers by some 10 nodes with KS distances at the cell-edge not exceeding 0.05.

12.3.3 Performance Evaluation of Heterogeneous Base Station Deployments

In this section, a second independent PPP of intensity $\lambda_2 = 0.5 \times 10^{-5}\,\text{m}^{-2}$ is added on top of the PPP scenarios with $\lambda = 0.5 \times 10^{-6}\,\text{m}^{-2}$ in Sect. 12.3.2. The corresponding nodes transmit with normalized power $P_{T2} = 4\,\text{W}$, thus representing a dense overlay of low power BSs. For simplicity, they are distributed within annuli of inner radius R_in and outer radii R_out as specified above. Note that in order to ensure that the inscribing ball of the central cell has a radius of 250 m, an inner radius of $1 + (P_{T1}/P_{T2})^{-1/\alpha}$ would be sufficient. This yields a total number of {1100, 11 000} expected interferers, respectively. For each snapshot, Algorithm 3 is applied with

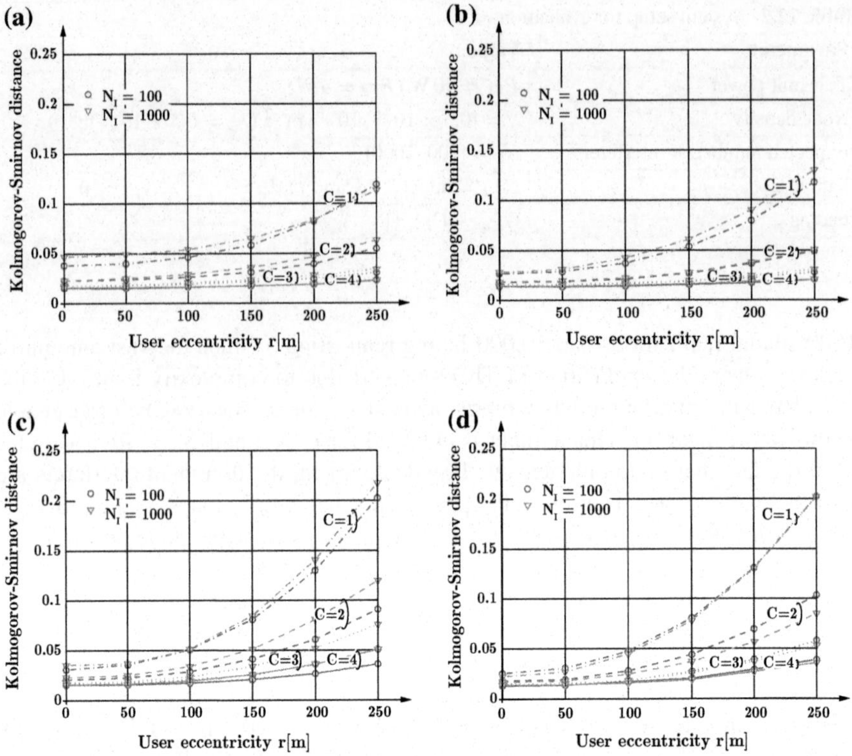

Fig. 12.4 KS distance over user eccentricity r. Plot markers {"o", "▽"} refer to various scenario sizes with $N_I = \{100, 1000\}$ expected interferers, respectively. *Different line* styles denote *circular* models with $C = \{1, 2, 3, 4\}$. Figure labels refer to the corresponding number of nodes per *circle*, N_c, and the spatial density λ of the original interferer deployment. *Black* bars depict 95 % confidence intervals **a** $N_c = 10, \lambda = 0.5 \times 10^{-6}\,\text{m}^{-2}$, **b** $N_c = 20, \lambda = 0.5 \times 10^{-6}\,\text{m}^{-2}$, **c** $N_c = 20, \lambda = 0.5 \times 10^{-6}\,\text{m}^{-2}$, **d** $N_c = 40, \lambda = 0.5 \times 10^{-6}\,\text{m}^{-2}$

$C = \{2, 4, 6, 8, 10, 12\}$ and $N_c = \{10, 20\}$. The performance evaluation is carried out along the lines of Sect. 12.3.2 and the parameters are summarized in Table 12.2.

Figure 12.5 depicts the results in terms of KS distances. It is observed that, in accordance with Sect. 12.3.2, accuracy is rather improved by increasing the number of circles C than by employing more nodes per circle (i.e., increasing N_c). In the heterogeneous scenarios the number of circles has to be roughly tripled in order to achieve a performance similar to the homogeneous cases (cf. Fig. 12.5), although mapping 11 times as many interferers.

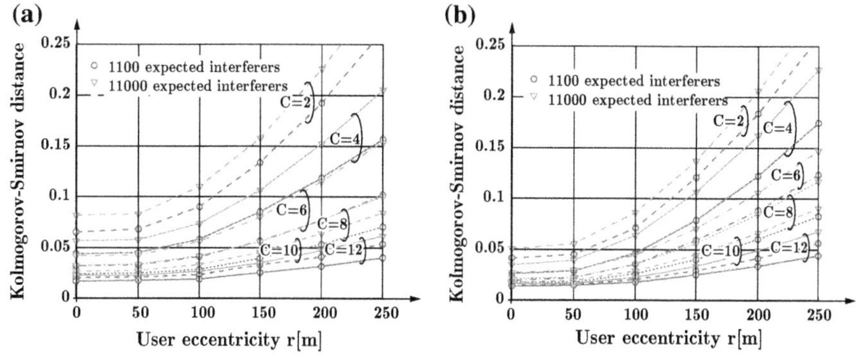

Fig. 12.5 KS distance over user eccentricity r for *heterogeneous* PPP scenarios with $\lambda = 0.5 \times 10{-}6\,\mathrm{m}^{-2}$ ($P_{T1} = 40\,\mathrm{W}$) and $\lambda_2 = 0.5 \times 10^{-5}\,\mathrm{m}^{-2}$ ($P_{T2} = 4\,\mathrm{W}$). Plot markers {"$\circ$", "∇"} refer to various scenario sizes with {1100, 11,000} expected interferers, respectively. *Different line* styles denote circular models with $C = \{2, 4, 6, 8, 10, 12\}$. Figure labels refer to the corresponding number of nodes per circle N_c. *Black* bars depict 95 % confidence intervals. **a** $N_c = 10$, **b** $N_c = 20$

12.4 Interference and Rate at Eccentric User Locations

In this section, user-centric BS collaboration schemes in scenarios with *asymmetric interferer impact* are investigated. The asymmetry can either arise from *non-uniform* power profiles or user locations outside the center of an otherwise isotropic scenario. The particular emphasis of this section is on the latter, since it is found less frequently in literature. In order to generate a generic, circularly symmetric scenario, the introduced circular model is applied, which enables to employ Theorem 12.1 for the analysis of the interference statistics. Note that, in fact, the circular model generates a *rotationally symmetric* scenario due to the finite number of nodes. However, by setting N_c sufficiently large, the scenario can be considered as *quasi-circularly symmetric*.

12.4.1 Generic Circularly Symmetric Scenario

The network is composed of a central BS and two circles of interferers with $R_1 = 500\,\mathrm{m}$ and $R_2 = 1000\,\mathrm{m}$, as depicted in Fig. 12.6. Each circle employs 10 interferers and a uniform power profile, i.e., $p_c[n] = 1/10$. The respective total transmit powers are specified as $P_1 = 400\,\mathrm{W}$ and $P_2 = 800\,\mathrm{W}$, respectively. The interferer locations are assumed to be rotated by $\phi_1 = -\pi/10$ and $\phi_2 = 0$, respectively. BS $\mathcal{T}_{0,0}$ is located at the origin and $P_0 = 40\,\mathrm{W}$.

The parameters of the circular model are summarized in Table 12.3 and the modeling of the signal propagation is referred from Table 12.2, respectively. The first goal is to identify the nodes, which dominate the interference statistics at eccentric user locations. Then, these insights are applied for user-centric BS coordination and -cooperation.

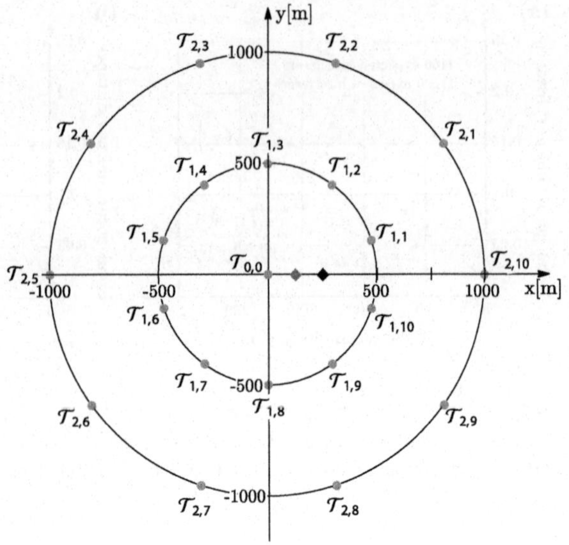

Fig. 12.6 Circular model with *two circles* of radius $R_1 = 500\,\text{m}$ and $R_2 = 1000\,\text{m}$, respectively. Each *circle* employs 10 BSs. The BS positions are rotated by $\phi_1 = -\pi/10$ and $\phi_2 = 0$. Users at (125 m, 0) and (250 m, 0) are denoted as *bold dots* and refer to *middle of cell* and *cell-edge*, respectively

Table 12.3 Parameters of circular model for numerical evaluation

Circle	Parameters					
1	$R_1 = 500\,\text{m}$	$N_1 = 10$	$P_1 = 400\,\text{W}$	$\phi_1 = -\frac{\pi}{10}$	$p_1[n] = \frac{1}{10}$	$n \in \{1, \ldots, 10\}$
2	$R_2 = 1000\,\text{m}$	$N_2 = 10$	$P_2 = 800\,\text{W}$	$\phi_2 = 0$	$p_2[n] = \frac{1}{10}$	$n \in \{1, \ldots, 10\}$

12.4.2 Components of Asymmetric Interference

In the first step, only the inner circle of interferers is assumed to be present, i.e., the set $\mathscr{I}$ comprises the 10 nodes $\mathcal{T}_{1,n}$, $n = 1, \ldots, 10$, of circle 1. The target is to determine the impact of the closest nodes on the aggregate interference statistics. For this purpose, two representative user locations at $r = R_1/4$ and $r = R_1/2$ are investigated, referring to *middle of cell* and *cell-edge*, respectively.

The PDF of the aggregate interference is obtained by Theorem 12.1. Its evaluation is simplified by the scenario's symmetry about the x-axis: (i) equal node-to-user distances from upper- and lower semicircle, i.e., $d_{1,n} = d_{1,10-n+1}$, (ii) uniform power profile $p_1(n) = 1/10$, and (iii) equal scale parameters $\theta_{1,n} = 1$. Thus, $\theta'_{1,n}(r) = \theta'_{1,10-n+1}(r)$, with $\theta'_{1n}(r) = P_1/10\,\ell(d_{1,n}(r))$. The vectors $\boldsymbol{\theta}_{\mathscr{I}}(r)$ and $\mathbf{k}_{\mathscr{I}}$ are of length $L^{\mathscr{I}} = 5$, with $[\boldsymbol{\theta}_{\mathscr{I}}(r)]_l = \theta'_{1,l}(r)$ and $[\mathbf{k}_{\mathscr{I}}]_l = 4$, respectively. Hence, the distribution of aggregate interference at distance r from the center formulates as

$$f_I(x; r) = \sum_{l=1}^{5} \frac{\Lambda_l}{\theta'_{1,l}(r)^4} h_{3,l}(0) e^{-x/\theta'_{1,l}(r)}, \tag{12.14}$$

where

$$\Lambda_l = -\frac{1}{6} \prod_{i=1, i \neq l}^{5} \left(1 - \frac{\theta_i}{\theta_l}\right)^{-4}, \qquad l = 1, \ldots, 5, \tag{12.15}$$

$$h_{3,l} = \left(h_{1,l}(0)\right)^3 + 3h_{1,l}(0)h_{1,l}^{(1)}(0) + h_{1,l}^{(2)}(0), \tag{12.16}$$

with

$$h_{1,l}(0) = -y + 4 \sum_{i=1, i \neq l}^{5} \left(\frac{1}{\theta_i} - \frac{1}{\theta_l}\right)^{-1}, \tag{12.17}$$

$$h_{1,l}^{(1)}(0) = 4 \sum_{i=1, i \neq l}^{5} \left(\frac{1}{\theta_i} - \frac{1}{\theta_l}\right)^{-2}, \tag{12.18}$$

$$h_{1,l}^{(2)}(0) = 8 \sum_{i=1, i \neq l}^{5} \left(\frac{1}{\theta_i} - \frac{1}{\theta_l}\right)^{-3}. \tag{12.19}$$

Figure 12.7 shows $f_I(x; r)$ for $r = 125$ m (narrow solid curve) and $r = 250$ m (wide solid curve), referring to middle of cell and cell-edge, respectively. The dots denote results as obtained with the approach in [12], which requires numerical evaluation of a line-integral and confirms the accuracy of the proposed finite-sum representation.

In Eq. 12.14, each sum term refers to a pair of transmitters $\{\mathcal{T}_{1,l}, \mathcal{T}_{1,10-l+1}\}$. The contribution of each pair to the final PDF is rendered visible by truncating the sum in Eq. 12.14 at L' with $L' \in \{1, \ldots, 5\}$, i.e., only the first L' sum terms are taken into account. Dashed curves in Fig. 12.7 depict results for $L' = 1$ and $L' = 2$.

It is observed that (i) in the *middle of the cell*, body and tail of the PDF are mainly shaped by the *four closest* interferers while (ii) at *cell-edge* the distribution is largely dominated by the *two closest* interferers, and (iii) interference at $r = 250$ m yields a larger variance than at $r = 125$ m due to higher diversity of the transmitter-to-user distances. The results verify link-level simulations in [33]. They emphasize the strong impact of interference asymmetry due to an eccentric user location, which is commonly overlooked in stochastic geometry analysis. In the next section, the above findings are exploited for BS coordination and -cooperation and the resulting SIR- and rate statistics are investigated.

12.4.3 Transmitter Collaboration Schemes

In this section, SIR- and rate statistics in the full two-circle scenario, as shown in Fig. 12.6, are studied. Motivated by the observations in Sect. 12.4.2, three schemes of *BS collaboration* are discussed:

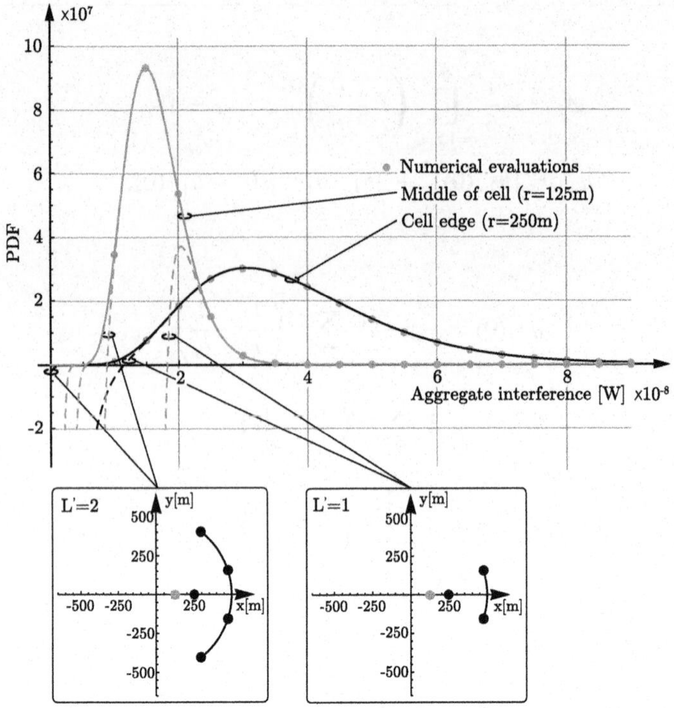

Fig. 12.7 Distribution of aggregate interference at user distances $r = 125$ m and $r = 250$ m, respectively. *Dots* refer to results as obtained with the approach in [12]. *Dashed curves* show contribution from dominant interferers

1. *No collaboration among nodes*: This scenario represents the *baseline*, where $\mathcal{S} = \{\mathcal{T}_{0,0}\}$ and $\mathcal{I}$ comprises all nodes on the circle, i.e., $\mathcal{I} = \{\mathcal{T}_{c,n}\}$ with $c \in \{1, 2\}$ and $n \in \{1, \ldots, 10\}$.
2. *Interference coordination* (cf., e.g., Enhanced Intercell Interference Coordination (eICIC) in the 3GPP LTE-A standard [34]): The nodes coordinate such that co-channel interference from the two strongest interferers of the inner circle, $\mathcal{T}_{1,1}$ and $\mathcal{T}_{1,10}$, is eliminated. This could be achieved, e.g., by joint scheduling. Then, $\mathcal{S} = \{\mathcal{T}_{0,0}\}$ and $\mathcal{I}$ is composed of $\{\mathcal{T}_{1,n}\}$ with $n \in \{2, \ldots, 9\}$ and $\{\mathcal{T}_{2,n}\}$ with $n \in \{1, \ldots, 10\}$.
3. *Transmitter cooperation* (cf., e.g., Coordinated Multi-Point (CoMP) in the 3GPP LTE-A standard [35]): The signals from the two closest nodes of the inner circle, $\mathcal{T}_{1,1}$ and $\mathcal{T}_{1,10}$, can be exploited as useful signals and are incoherently combined with the signal from $\mathcal{T}_{0,0}$. Then, $\mathcal{S} = \{\mathcal{T}_{0,0}, \mathcal{T}_{1,1}, \mathcal{T}_{1,10}\}$ and, as above, $\mathcal{I}$ comprises $\{\mathcal{T}_{1,n}\}$ with $n \in \{2, \ldots, 9\}$ and $\{\mathcal{T}_{2,n}\}$ with $n \in \{1, \ldots, 10\}$.

For each *collaboration* scheme, the PDFs of aggregate signal and -interference, $f_S(x; r)$ and $f_I(x; r)$, are calculated using Theorem 12.1. The *SIR* at user location $(r, 0)$ is defined as $\gamma(r) = S(r)/I(r)$. According to [36], the PDFs of $\gamma(r)$ is calculated as

$$f_\gamma(\gamma; r) = \int_0^\infty z f_S(z\gamma; r) f_I(z; r) dz, \tag{12.20}$$

where z is an auxiliary variable, $f_S(\cdot; r)$ and $f_I(\cdot; r)$ refer to Eqs. (12.9) and (12.10), and the integration bounds are obtained by exploiting the fact that $f_S(\gamma; r) = 0$ and $f_I(\gamma; r) = 0$ for $x < 0$, respectively.

Evaluating Eqs. (12.9) and (12.10) yields sums of elementary functions of the form $a\gamma^b e^{-c\gamma}$, with the auxiliary parameters $a \in \mathbb{R}$, $b \in \mathbb{N}^+$ and $c > 0$. Therefore, $f_S(\gamma; r)$ and $f_I(\gamma; r)$ can generically be written as

$$f_S(\gamma; r) = \sum_s a_s \gamma^{b_s} e^{-c_s \gamma}, \tag{12.21}$$

$$f_I(\gamma; r) = \sum_i a_i \gamma^{b_i} e^{-c_i \gamma}, \tag{12.22}$$

and allow to straightforwardly evaluate Eq. (12.20) as

$$\begin{aligned} f_\gamma(\gamma; r) &= \sum_s \sum_i \int_0^\infty z\, a_s (z\gamma)^{b_s} e^{-c_s(\gamma z)}\, a_i z^{b_i} e^{-c_i z} dz \\ &= \sum_s \sum_i a_s a_i \gamma^{b_s} (c_i + c_s \gamma)^{-i-b_s-b_i} \Gamma(i + b_s + b_i). \end{aligned} \tag{12.23}$$

The *normalized ergodic rate* τ as a function of the SIR $\gamma(r)$ is calculated by the modified Shannon capacity formula $\tau(\gamma(r)) = \alpha_\text{B} \log_2 (1 + \alpha_\text{SIR} \gamma(r))$, where α_B and α_SIR are coefficients for the calibration against link level simulations with $0 < \alpha_\text{B} \leq 1$ and $0 < \alpha_\text{SIR} \leq 1$. Since $\tau(\cdot)$ is a function of the RV $\gamma(r)$, its distribution is obtained by a transformation as

$$f_\tau(x; r) = \frac{1}{\alpha_\text{B}\alpha_\text{SIR}} 2^{x/\alpha_\text{B}} f_\gamma\left(\frac{1}{\alpha_\text{SIR}}\left(2^{x/\alpha_\text{B}} - 1\right); r\right) \log_e(2), \tag{12.24}$$

with $f_\gamma(\cdot; \cdot)$ from Eq. (12.23).

The distributions $f_\gamma(\gamma; r)$ and $f_\tau(\tau; r)$ are analyzed at $r = 125$ m and $r = 250$ m referring to *middle of the cell*, and *cell-edge*, respectively. In this chapter, $\alpha_\text{B} = 1$ and $\alpha_\text{SIR} = 1$. For reasons of clarity, CDF curves are presented. In order to verify the analysis, Monte Carlo simulations are carried out, employing the system model from Sect. 12.4.1 and the signal propagation model from Table 12.2. The results are computed by averaging over 10^7 channel realizations for each BSs collaboration scheme and each user location, and are denoted as bold dots in Figs. 12.8 and 12.9, respectively.

Figure 12.8 shows the obtained SIR distributions. It is observed that

- In the case of *no collaboration* (solid lines in Fig. 12.8), the curves have almost equal shape in the middle of the cell and at cell-edge. The distribution in the middle

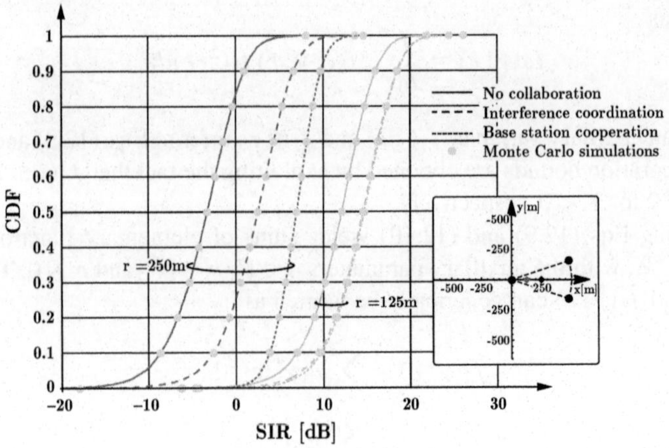

Fig. 12.8 SIR CDF curves for user locations in the *middle of the cell* ($r = 125$ m) and at *cell-edge* ($r = 250$ m), respectively. Three cases are depicted: (i) no collaboration among BSs (*solid*), (ii) interference coordination (*dashed*), (iii) cooperation among BSs (*dotted*)

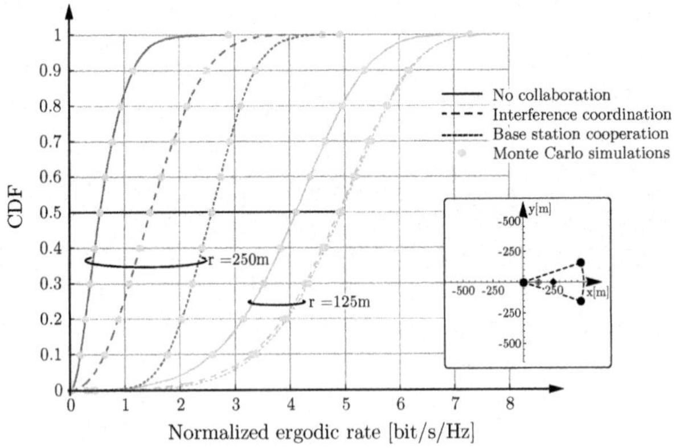

Fig. 12.9 Normalized-rate CDF curves for user locations in the *middle of the cell* ($r = 125$ m) and at *cell-edge* ($r = 250$ m), respectively. Three cases are depicted: (i) no collaboration among BSs (*solid*), (ii) interference coordination (*dashed*), (iii) cooperation among BSs (*dotted*)

of the cell is slightly steeper due to the lower variance of the interferer impact. Their medians, hereafter used to represent the distributions' position, differ by 15.5 dB.

- When the central node $\mathcal{T}_{0,0}$ *coordinates* its channel access with the user's two dominant interferers, $\mathcal{T}_{1,1}$ and $\mathcal{T}_{1,10}$, the SIR improves by 2.4 dB in the middle of the cell and 5.9 dB at cell-edge (dashed curves in Fig. 12.8), compared to *no collaboration*.

- *BS cooperation* enhances the SIR by 10.2 dB at cell-edge in comparison to *no collaboration* (left dotted curve in Fig. 12.8). Note that the CDF curve also has a steeper slope than without coordination, indicating lower variance of the SIR.
- In the middle of the cell, *cooperation* achieves hardly any additional improvement, as recognized from the overlapping rightmost curves in Fig. 12.8. This remarkable result states that *interference coordination* already performs close to optimal at this user location. Note that in realistic networks *coordination* is typically far less complex than *cooperation*.

The curves reflect findings from [37], stating that even in the best case, gains of transmitter cooperation are much smaller than largely envisioned. Figure 12.9 depicts the corresponding rate distributions. The results show that

- Notably, the rate statistics of all three collaboration schemes indicate lower variance at cell-edge than in the middle of the cell.
- In terms of median value, *BS coordination* shows rate improvements by 18.7 % in the middle of the cell and by 167 % at cell-edge.
- *Cooperation* between the central node $\mathscr{T}_{0,0}$ and the user's two closest interferers, $\mathscr{T}_{1,1}$ and $\mathscr{T}_{1,10}$, achieves a rate enhancement of 19.8 % in the middle of the cell and 355.7 % at cell-edge. Similar to the SIR, it is observed that in the middle of the cell, interference coordination already performs close to optimal.

In summary, collaboration among the BSs that were identified as main contributors to the shape of the interference distribution by Theorem 12.1, achieved large performance enhancements in terms of SIR and rate. It was further shown that the efficiency of such schemes considerably depends on the user eccentricity, or equivalently, the asymmetry of the interference impact.

12.5 LTE-Advanced System Level Simulations

In this section, SINR and spectral efficiency at eccentric UE locations are evaluated by LTE-A system level simulations, validating results from theory. The system model is directly adopted from Sect. 11.7.1. The corresponding circular model from Sect. 12.1 encompasses one circle ($C = 1$) with radius $R_1 = 500$ m and six transmitters. According to Sect. 12.3, such model allows to *exactly* reproduce a regular grid model in terms of aggregate interference characteristics, provided that the composite fading follows a Gamma distribution. In order to omit the error, which is induced by the Gamma approximation as detailed in Sect. 11.7.2, *composite fading* is modeled by free space path loss and Rayleigh fading in both simulations and analysis (i.e., $G_{c,n} \sim \Gamma[1, 1]$ in Eq. (12.1)).

The particular aim of this section is to verify results from Sect. 12.4. Accordingly, two scenarios, namely *no collaboration among transmitters* and *interference coordination* are defined. In the latter case, eNodeB 7 in Fig. 11.9 does not contribute to the co-channel interference. From the representative UE locations, as specified

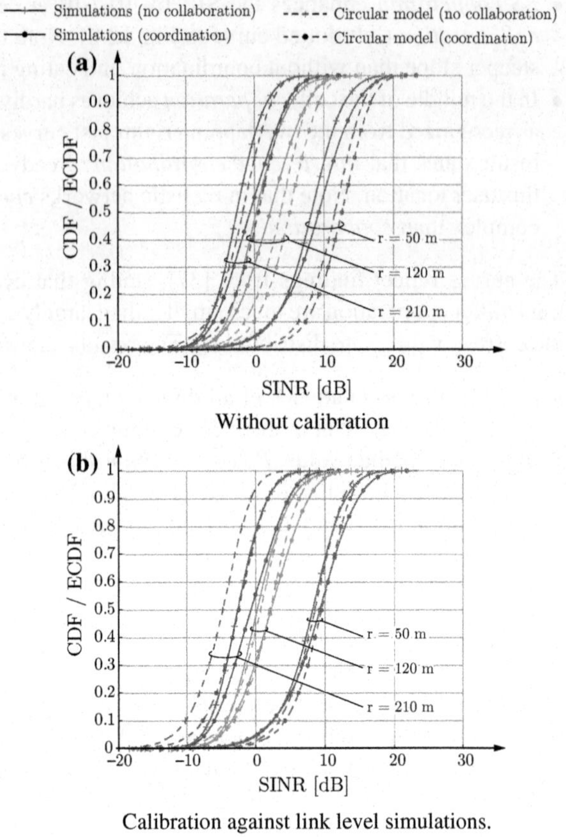

Fig. 12.10 SINR at various UE distances $r = \{50, 120, 210\}$ m and angle position $\phi = 0$, considering *no collaboration*- and *interference coordination* among the eNodeBs. The *curves* show results as obtained by system level simulations and the circular model from Sect. 12.4 with-(**b**) and without calibration (**a**), respectively

in Sect. 11.7.1 and marked by bold dots in Fig. 11.9, the particular interest of this section is on the angle position $\phi = 0$.

Figure 12.10a depicts SINR distributions for both collaboration schemes, comparing results from system level simulations and the circular model. It is observed that the SINR consistently deteriorates for a UE moving from cell-center to cell-edge. In accordance with Sect. 12.4, it is improved by interference coordination, with the largest gains being achieved at cell-edge. The simulated median values increase by 1.2, 1.8 and 1.9 dB at the cell-center, middle of the cell and cell-edge, respectively. Furthermore, it is seen that the curves from the simulations are steeper, i.e., have a smaller variance than those obtained with the circular model. This is mainly caused by the fact that the simulator employs a Zero Forcing (ZF) receiver and measures the *post-equalization SINR*. In order to more accurately capture the receiver characteristics, the first adaption of the circular model concerns the shape of the fading distribution. It is set to $k = 2$, i.e., $G_{c,n} \sim \Gamma[2, 1]$.

In a practical system, performance is decreased by a variety of design constraints. Hence, the circular model has to be further calibrated against simulations

[38]. This chapter employs the modified channel capacity formula $\tau(\gamma(r)) = \alpha_{\rm B}\log_2(1+\alpha_{\rm SIR}\gamma(r))$, as introduced in Sect. 12.4.3. The term $\gamma(r)$ denotes the SIR at distance r (note that in this section the UEs of interest have angle position $\phi = 0$), and $\alpha_{\rm B}$ and $\alpha_{\rm SIR}$ are freely adjustable calibration parameters, with $0 < \alpha_{\rm B} \leq 1$ and $0 < \alpha_{\rm SIR} \leq 1$.

Figure 12.10 depicts the simulation results from both baseline- and coordination scheme in terms of spectral efficiency versus SINR. The dots refer to the simulation results at various user distances $r = \{50, 120, 210\}$ m. For comparison, the figure also shows the *channel capacity* and the performance of a SISO LTE-A system over an Additive White Gaussian Noise (AWGN) channel. The latter is obtained from link level simulations and can expediently be used to predict the optimal performance of the system. Its discontinuous behavior stems from LTE-A's Adaptive Modulation and Coding (AMC) scheme [39]. It is observed that the results from the system level simulations lie below this curve, since they encompass the actual channel code performance [40]. In order to achieve reliable upper performance bounds, the link level results are employed as a reference for the calibration of the circular model, yielding $\alpha_{\rm B} = 0.85$ and $\alpha_{\rm SIR} = 0.5$, respectively. The term $\alpha_{\rm B}$ shifts the capacity curve in Fig. 12.11, while $\alpha_{\rm SIR}$ determines its scale. The calibration is carried out such that the curve is tangent to the link level results and achieves the maximum spectral efficiency in the scenario without collaboration, as denoted by '×' in Fig. 12.11a. Intuitively, the first term accounts for overhead, such as pilot symbols, while the second term represents the limits of the AMC in LTE-A [41, 42].

The dashed curves in Fig. 12.11a show the adapted SIR distributions. Except for $r = 210$ m, they exhibit a better fit than the uncalibrated curves in both shape and scale. The deviation at the cell-edge ($r = 210$ m) is partly compensated by

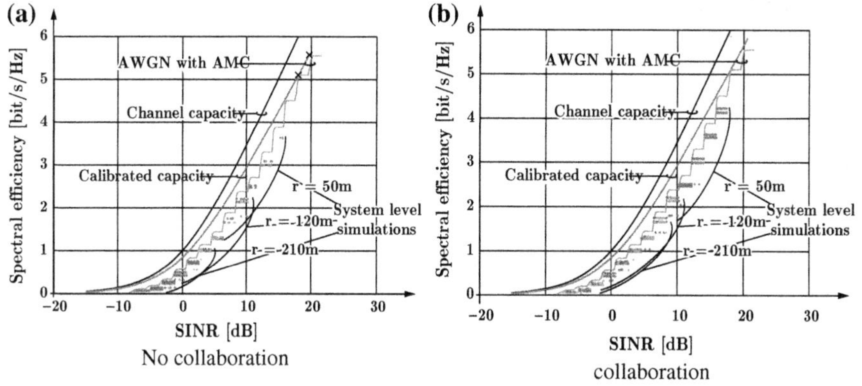

Fig. 12.11 Spectral efficiency [bit/s/Hz] versus SINR [dB] for baseline- (**a**) and coordination scheme (**b**). *Dots* refer to results from system level simulations at various user distance $r = \{50, 120, 210\}$ m. The *curves* refer to the Shannon channel capacity, the *calibrated* capacity (with $\alpha_{\rm B} = 0.85$ and $\alpha_{\rm SIR} = 0.5$) and the performance of a SISO LTE-A system over an AWGN channel employing AMC, as obtained from link level simulations

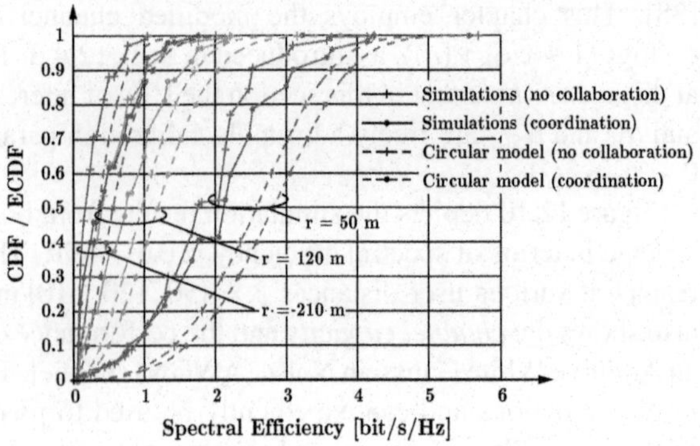

Fig. 12.12 Spectral efficiency [bit/s/Hz] at various UE distances $r = \{50, 120, 210\}$ and angle position $\phi = 0$, considering *no collaboration*- and *interference coordination* among the eNodeBs. The curves show results as obtained by system level simulations and the circular model with $\alpha_B = 0.85$ and $\alpha_{SIR} = 0.5$, respectively

overestimating the link level performance in the low-SINR regime (cf. Figure 12.11). The corresponding spectral efficiency distributions are obtained by applying Eq. 12.24 from Sect. 12.4.3 Note that the term *spectral efficiency* is employed instead of *normalized rate* in this chapter. They are depicted in (Fig. 12.12) together with the simulation results. It is observed that, despite the large number of simulated TTIs, the curves from the simulations exhibit a discontinuous behavior due to the AMC. As expected, the circular model provides reliable upper performance bounds that are tightest at $r = 210$ m.

In conclusion, the circular model from Sect. 12.1 is well suited to make a first-order prediction of the SINR- and spectral efficiency performance. The presented calibration against link level simulations has to be carried out only once for each MISO- or Single-Input Multiple-Output (SIMO) transmission scheme. This method forfeits a certain amount of accuracy while avoiding tedious ad-hoc calibration against each system level simulation run.

References

1. N. Bhushan, J. Li, D. Malladi, R. Gilmore, D. Brenner, A. Damnjanovic, R. Sukhavasi, C. Patel, S. Geirhofer, Network densification: the dominant theme for wireless evolution into 5G. IEEE Commun. Mag. **52**(2), 82–89 (2014). doi:10.1109/MCOM.2014.6736747
2. H. ElSawy, E. Hossain, M. Haenggi, Stochastic geometry for modeling, analysis, and design of multi-tier and cognitive cellular wireless networks: A survey. IEEE Commun. Surveys Tutor. **15**(3), 996–1019 (2013)
3. M. Haenggi, R.K. ganti, Interference in large wireless networks, in, *Foundations and Trends in Networking* (Now Publishers, The Netherlands, 2009), pp. 127–248
4. M. Haenggi, J. Andrews, F. Baccelli, O. Dousse, M. Franceschetti, Stochastic geometry and random graphs for the analysis and design of wireless networks. IEEE J. Sel. Areas Commun. **27**(7), 1029–1046 (2009). doi:10.1109/JSAC.2009.090902
5. M. Taranetz, M. Rupp, A circular interference model for heterogeneous cellular networks. IEEE Trans. Wirel. Commun. **15**(2), 1432–1444 (2015). doi:10.1109/TWC.2015.2490068

6. M. Taranetz, System level modeling and evaluation of heterogeneous cellular networks, Ph.D. dissertation, Institute of Telecommunications, TU Wien, 2015, http://theses.eurasip.org/theses/611/system-level-modeling-and-evaluation-of/
7. H. Tabassum, F. Yilmaz, Z. Dawy, M.-S. Alouini, A framework for uplink intercell interference modeling with channel-based scheduling. IEEE Trans. Wirel. Commun. **12**(1), 206–217 (2013). doi:10.1109/TWC.2012.120412.112244
8. R.W. Heath Jr., M. Kountouris, T. Bai, Modeling heterogeneous network interference using poisson point processes. IEEE Trans. Signal Process. **61**(16), 4114–4126 (2013). doi:10.1109/TSP.2013.2262679
9. R. Prasad, A. Kegel, Improved assessment of interference limits in cellular radio performance. IEEE Trans. Veh. Technol. **40**(2), 412–419 (1991)
10. M.-S. Alouini, A. Abdi, M. Kaveh, Sum of gamma variates and performance of wireless communication systems over Nakagami-fading channels. IEEE Trans. Veh. Technol. **50**(6), 1471–1480 (2001). doi:10.1109/25.966578
11. V. Aalo, T. Piboongungon, G. Efthymoglou, Another look at the performance of MRC schemes in Nakagami-m fading channels with arbitrary parameters. IEEE Trans. Commun. **53**(12), 2002–2005 (2005). doi:10.1109/TCOMM.2005.860089
12. I. Ansari, F. Yilmaz, M.-S. Alouini, O. Kucur, On the sum of gamma random variates with application to the performance of maximal ratio combining over Nakagami-m fading channels, in, *IEEE International Workshop Signal Processing Advances in Wireless Commununication (SPAWC)*, Cesme, June 2012, pp. 394–398
13. T.A. Tsiftsis, G.K. Karagiannidis, S.A. Kotsopoulos, N.C. Sagias, Performance of MRC diversity receivers over correlated Nakagami-*m* fading channels, in, *International Symposium on Communication Systems, Networks and Digital Signal Processing (CSNDSP)*, Patras, July 2006
14. E.K. Al-Hussaini, A. Al-Bassiouni, Performance of MRC diversity systems for the detection of signals with Nakagami fading. IEEE Trans. Commun. **33**(12), 1315–1319 (1985). doi:10.1109/TCOM.1985.1096243
15. G. Karagiannidis, N. Sagias, T. Tsiftsis, Closed-form statistics for the sum of squared Nakagami-m variates and its applications. IEEE Trans. Commun. **54**(8), 1353–1359 (2006). doi:10.1109/TCOMM.2006.878812
16. G. Efthymoglou, T. Piboongungon, V. Aalo, Performance of DS-CDMA receivers with MRC in Nakagami-m fading channels with arbitrary fading parameters. IEEE Trans. Veh. Technol. **55**(1), 104–114 (2006). doi:10.1109/TVT.2005.861204
17. P. Lombardo, G. Fedele, M. Rao, MRC performance for binary signals in Nakagami fading with general branch correlation. IEEE Trans. Commun. **47**(1), 44–52 (1999). doi:10.1109/26.747812
18. Q. Zhang, Exact analysis of postdetection combining for DPSK and NFSK systems over arbitrarily correlated Nakagami channels. IEEE Trans. Commun. **46**(11), 1459–1467 (1998). doi:10.1109/26.729390
19. L.-L. Yang, H.-H. Chen, Error probability of digital communications using relay diversity over Nakagami-m fading channels. IEEE Trans. Wirel. Commun. **7**(5), 1806–1811 (2008)
20. D.J. Torrieri, M.C. Valenti, The outage probability of a finite ad hoc network in Nakagami fading. IEEE Trans. Commun. **60**(11), 3509–3518 (2012)
21. P. Moschopoulos, The distribution of the sum of independent gamma random variables. Ann. Inst. Stat. Math. **37**(1), 541–544 (1985)
22. D.G. Kabe, On the exact distribution of a class of multivariate test criteria. Ann. Math. Stat. **33**(3), 1197–1200 (1962)
23. E. Scheuer, Reliability of an m-out-of-n system when component failure induces higher failure rates in survivors. IEEE Trans. Rel. **37**(1), 73–74 (1988). doi:10.1109/24.3717
24. S. Amari, R. Misra, Closed-form expressions for distribution of sum of exponential random variables. IEEE Trans. Rel. **46**(4), 519–522 (1997). doi:10.1109/24.693785
25. C.A. Coelho, The generalized integer gamma distribution—a basis for distributions in multivariate statistics. J. Multivar. Anal. **64**(1), 86–102 (1998). doi:10.1006/jmva.1997.1710

26. A.A. Abu-Dayya, N.C. Beaulieu, Outage probabilities in the presence of correlated lognormal interferers. IEEE Trans. Veh. Technol. **43**(1), 164–173 (1994)
27. N.C. Beaulieu, A.A. Abu-Dayya, P.J. McLane, Estimating the distribution of a sum of independent lognormal random variables. IEEE Trans. Commun. **43**(12), 2869–2873 (1995)
28. J. Hu, N.C. Beaulieu, Accurate simple closed-form approximations to rayleigh sum distributions and densities. IEEE Commun. Lett. **9**(2), 109–111 (2005)
29. N.B. Mehta, J. Wu, A.F. Molisch, J. Zhang, Approximating a sum of random variables with a lognormal. IEEE Trans. Wirel. Commun. **6**(7), 2690–2699 (2007)
30. S.B.P. Iman Mabrouk, The exact density function of a sum of independent gamma random variables as an inverse Mellin transform. Int. J. Appl. Math. Stat. **41**(11) (2013)
31. S.B. Provost, On sums of independent gamma random variates. Stat. J. Theor. Appl. Stat. **20**(4) (1989)
32. J. Andrews, R. Ganti, M. Haenggi, N. Jindal, S. Weber, A primer on spatial modeling and analysis in wireless networks. IEEE Commun. Mag. **48**(11), 156–163 (2010)
33. S. Plass, X.G. Doukopoulos, R. Legouable, Investigations on link-level inter-cell interference in OFDMA systems, in, *Symposium on Communications and Vehicular Technology (SCVT)*, Liege, November 2006, pp. 49–52
34. 3rd Generation Partnership Project (3GPP), Evolved universal terrestrial radio access (E-UTRA); mobility enhancements in heterogeneous networks, in *3rd Generation Partnership Project (3GPP)*, TR 36.839, January 2013
35. 3rd Generation Partnership Project (3GPP), Coordinated multi-point operation for LTE physical layer aspects, in *3rd Generation Partnership Project (3GPP)*, TR 36.819, September 2013
36. J.H. Curtiss, On the distribution of the quotient of two chance variables. Ann Math. Stat. **12**(4), 409–421 (1941)
37. A. Lozano, R.W. Heath Jr., J. Andrews, Fundamental limits of cooperation. IEEE Trans. Inf. Theory **59**(9), 5213–5226 (2013). doi:10.1109/TIT.2013.2253153
38. J.C. Ikuno, System level modeling and optimization of the LTE downlink, Ph.D. dissertation, E389, Technische Universität Wien, 2013, http://theses.eurasip.org/theses/479/system-level-modeling-and-optimization-of-the-lte/
39. 3rd Generation Partnership Project (3GPP), Evolved universal terrestrial radio access (E-UTRA); physical layer procedures, in *3rd Generation Partnership Project (3GPP)*, TS 36.213, January 2015
40. S. Schwarz, M. Simko, M. Rupp, On performance bounds for MIMO OFDM based wireless communication systems, in, *IEEE International Workshop Signal Processing Advances in Wireless Commununication (SPAWC)*, San Francisco, June 2011, pp. 311–315. doi:10.1109/SPAWC.2011.5990419
41. A. Burr, A. Papadogiannis, T. Jiang, MIMO truncated Shannon bound for system level capacity evaluation of wireless networks, in, *IEEE Wireless Commununication and Networking Conference Workshops (WCNCW)*, Paris, April 2012
42. 3rd Generation Partnership Project (3GPP), Evolved universal terrestrial radio access (E-UTRA); radio frequency (RF) system scenarios, *3rd Generation Partnership Project (3GPP)*, TR 36.942, October 2014

Chapter 13
Analysis of Urban Two-Tier Heterogeneous Cellular Networks

Markus Rupp, Stefan Schwarz and Martin Taranetz

This chapter presents a system model that is based on techniques from stochastic geometry and enables the analysis of *indoor* downlink performance in urban two-tier heterogeneous cellular networks. Chapters 11 and 12 mainly deal with eccentric receiver locations, which, in general, strongly restrict stochastic models from yielding convenient expressions. In this chapter, the issue is resolved by proposing a *virtual building approximation*. Moreover, two other important limitations are addressed, namely *shadowing* and the *separation between indoor- and outdoor environments*.

In the analysis on stochastic geometry, *shadowing* is typically incorporated by log-normally distributed Random Variables (RVs) [1–3] or neglected at all [4–12]. A recent study on blockage effects in urban environments indicates its dependency on the link length [13]. It follows the intuition that a longer link increases the likelihood of buildings to intersect with it. Such propagation characteristics have been discussed by the 3rd Generation Partnership Project (3GPP) only recently in a technical report on 3-dimensional channel modeling [14]. Secondly, scenarios comprising *both indoor- and outdoor environments* have not received much attention in analytical studies due to the imposed inhomogeneities on signal propagation. The designated area of operation for small cell Base Stations (BSs) is indoors. Existing approaches either neglect the wall partitioning [2, 15], as indicated in Fig. 13.1a, oversimplify the macro-tier topology [7–9] or omit cross-tier interference [1].

In this chapter, a two-tier cellular network with outdoor macro- and indoor-deployed small cell BSs is considered. Referring to the work in [17, 18], the contributions are:

M. Rupp (✉) · S. Schwarz · M. Taranetz
Institute of Telecommunications, TU Wien, Vienna, Austria
e-mail: mrupp@nt.tuwien.ac.at

S. Schwarz
e-mail: sschwarz@nt.tuwien.ac.at

M. Taranetz
e-mail: martin.taranetz@tuwien.ac.at

M. Rupp et al., *The Vienna LTE-Advanced Simulators*,
Signals and Communication Technology, DOI 10.1007/978-981-10-0617-3_13

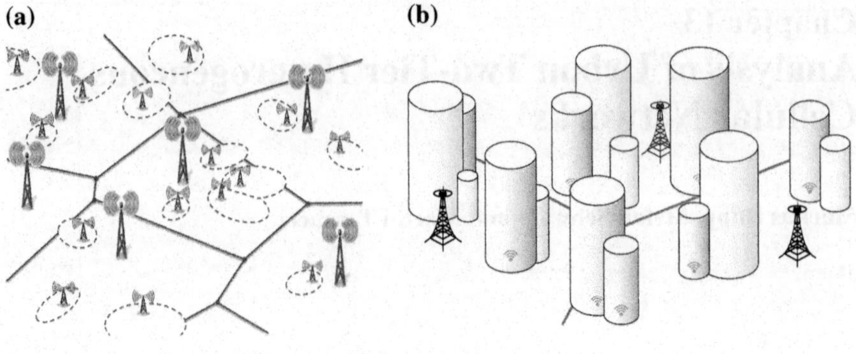

Fig. 13.1 Model environments of two-tier heterogeneous cellular networks. **a** Flat plane [16]. **b** Urban

- A tractable model for urban environment topologies is introduced. It comprises an outdoor environment, which is partly covered by circular building objects with a certain density. A method to extract its parameters from real-world data is provided. Based on concepts from random shape theory, the model is applied to characterize both *signal propagation* and *network deployment*.
- A novel *virtual building approximation* to simplify aggregate interference analysis is presented. The key idea is to establish a user-centric interference environment by shifting the centers of the typical building and its exclusion regions to the user location.
- Assuming that a building is served by a small cell BS with a certain *occupation probability*, the normalized ergodic rate of a *typical* indoor user is evaluated with respect to building density and wall penetration loss. Based on these results, the impact of distinguishing Line-Of-Sight (LOS)- and Non Line of Sight (NLOS) macro BSs is identified.

The chapter is organized as follows. Section 13.1 provides preliminaries on random shape theory and indoor area coverage. Section 13.2 presents the system model. In Sect. 13.3 analytical expressions for the coverage probability of indoor users are derived. Section 13.4 studies the accuracy of the model by numerical evaluations. In Sect. 13.5, the model is extended by the consideration of LOS macro BSs.

13.1 Preliminaries

13.1.1 Random Shape Theory

In this section, concepts from random shape theory are reviewed (see, e.g., [19, 20]), so as to make the model formulation in this chapter more accessible.

Let $\mathcal{O}$ denote a set of objects on $\mathbb{R}^n$, which are closed and bounded, i.e., have finite area and perimeter. For instance, $\mathcal{O}$ could be a collection of circles or rectangles on $\mathbb{R}^2$, or a combination of cubes in $\mathbb{R}^3$. For each object in $\mathcal{O}$, a *center point* is determined, which has to be well-defined but does not necessarily relate to the object's center of gravity. Non-symmetric objects additionally require to specify the orientation in space by a directional unit vector.

A *Random Object Process* (*ROP*) is constructed by randomly sampling objects from $\mathscr{O}$ and placing their corresponding center points at the points of some Point Process (PP). The orientation of each object is independently determined according to some probability distribution.

In general, a ROP is difficult to analyze, particularly when there are correlations between sampling, location and orientation of the objects. For the sake of tractability, this chapter employs a *Boolean scheme*, which satisfies the following properties: (i) the center points form a Poisson Point Process (PPP), (ii) the attributes of the objects such as orientation, shape and size are mutually independent, and (iii) for each object, sampling, location and orientation are also independent.

The scheme is used to model an urban environment, where the objects of the process refer to buildings. In the next section, a method to parameterize this model with real-world data is presented.

13.1.2 Indoor Coverage Ratio

Define the *indoor coverage ratio* as the fraction of the total area on $\mathbb{R}^2$ that is covered by buildings. Then, consider a Boolean model with the center points being distributed according to a stationary PPP on $\mathbb{R}^2$ with intensity λ_B, i.e., $\Lambda(dx) = \lambda_\mathrm{B} dx$. Further, let $\mathscr{C}$ denote a *random closed set* on $\mathbb{R}^2$, representing a *generic building*. According to [21, Definition 3.1.8], the resulting indoor coverage ratio is determined as

$$p_\mathrm{I} = 1 - e^{-\lambda_\mathrm{B}\,\mathbb{E}[|\mathscr{C}|]}, \tag{13.1}$$

where $|\cdot|$ denotes the Lebesgue measure on $\mathbb{R}^2$.

As an example, assume that $\mathscr{C}$ is a random closed ball that is centered at the origin and has random radius $R \in \mathbb{R}^+$, i.e., $\mathscr{C} = \mathscr{B}(0, R)$. Then,

$$p_\mathrm{I} = 1 - e^{-\lambda_\mathrm{B}\,\mathbb{E}[R^2]\pi}. \tag{13.2}$$

Expediently, the indoor coverage ratio can also be extracted from real-world data. For example, the OpenStreetMap project provides open access to shape files, as illustrated in Fig. 13.2. In order to determine the degree of coverage by buildings, these files can be processed by a simple MATLAB script. For the University of Texas at Austin and Vienna's inner district (see Fig. 13.2a, b) ratios of $p_\mathrm{I} = 0.25$ and $p_\mathrm{I} = 0.6$ were measured. Related work in [22] evaluated the indoor coverage of various Turkish cities. The ratios ranged from 0.13 to 0.39.

In the following, the Boolean model is applied to model the deployment of a heterogeneous cellular network in accordance with the characteristics of an urban environment.

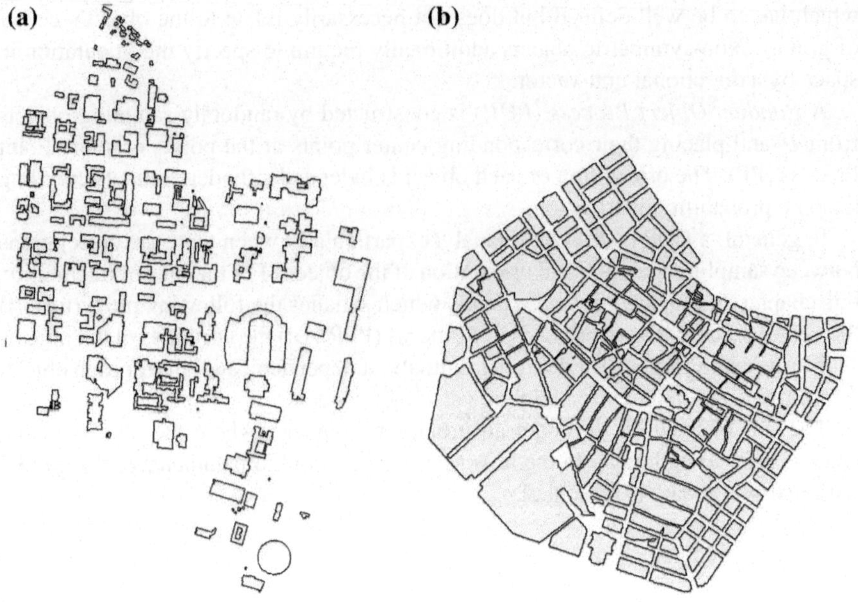

Fig. 13.2 Building footprints as extracted from OpenStreetMap data. **a** University of Texas at Austin. **b** Downtown Vienna

13.2 System Model

13.2.1 Topology Model for Urban Environments

Consider a two-tier cellular network comprising outdoor macro- and indoor small cell BSs, as shown in Fig. 13.3. Buildings are modeled by a Boolean scheme of circles on the $\mathbb{R}^2$ plane. The centers of the circles form a PPP Φ_B of intensity λ_B [21]. For simplicity, it is assumed that all circles have a fixed radius R_I. A point on the plane is said to be *indoors*, if it is covered by a building, and *outdoors* otherwise. Indoor- and outdoor environment are partitioned by a wall penetration loss, which is hereafter assumed constant for all buildings and denoted as L_W unless specified otherwise.

13.2.2 Network Deployment

Macro BSs are distributed according to a PPP Φ_M of intensity μ_M. Note that these BSs are required to be located outdoors. Thus, the macro BS process can equivalently be constructed by independently thinning (see, e.g., [21]) an initial PPP of density $\mu'_M = \mu_M/p_O$, where p_O equals the probability that a point on $\mathbb{R}^2$ is not covered

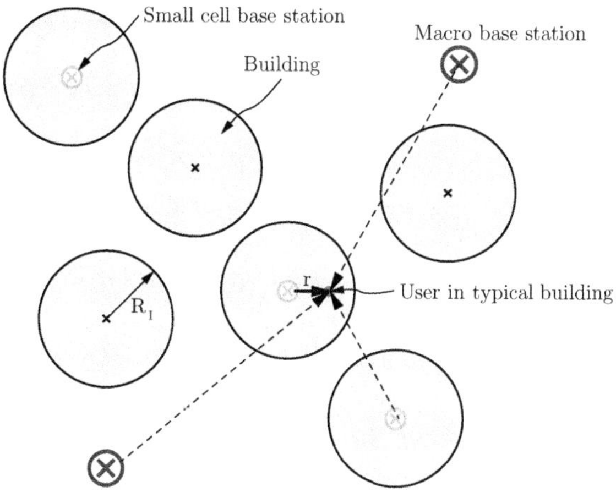

Fig. 13.3 Urban two-tier heterogeneous cellular network. Macro BSs are deployed in an outdoor environment. Buildings are modelled as a random process of *circles* and are assumed to have a fixed radius R_I. Only a fraction of buildings is occupied by small cell BSs. The figure depicts a typical indoor user with macro BSs and neighboring small cell BSs (*dashed lines*)

by a building. According to Sect. 13.1.2, the thinning probability is determined as $p_\mathrm{O} = 1 - p_\mathrm{I} = \exp(-\lambda_\mathrm{B} R_\mathrm{I}^2 \pi)$.

A building will deploy an indoor small cell BS with a certain *occupation probability* η. Assume the indoor small cell BSs to be located at the center points of the occupied buildings. Then, their spatial distribution can be modeled by a PPP Φ_S of intensity $\lambda_\mathrm{B}\,\eta$, which results from independently thinning the object center PPP Φ_B.

13.2.3 User Association

The aim of this chapter is to characterize the coverage and rate performance of indoor users. Noting that the buildings are assumed to form a Boolean scheme, the centers of the buildings form a PPP on the plane [13]. Therefore, by Slivnyak's theorem [21], when fixing a typical building at the origin, the centers of the other buildings still form a PPP. The performance of users will be investigated inside the typical building. Separate association rules are defined, depending on whether or not this building is occupied by a small cell BS.

Case 1 *[Typical Building with Small Cell BS]*: Consider a typical building at the origin, which is equipped with an indoor small cell BS. For simplicity, it is assumed that all users inside this building are associated with the small cell at the origin. The cases in which indoor users at the edge of the typical building may receive stronger signals from a close-by outdoor macro BS are omitted, thus underestimating the coverage probability. Similar to the analysis in [23], *exclusion guard regions* are imposed on both macro- and small cell tier, where no BSs from the corresponding tier are allowed to distribute. For simplicity, it is assumed that the exclusion region for macro BSs is a ball of radius R_I centered at the origin, ensuring that no macro BSs are located inside the typical building. The exclusion region of the small cell tier is defined as a ball of radius $2\,R_\mathrm{I}$ in order to prevent overlapping association regions of two small cells.

Case 2 *[Typical Building without Small Cell BS]:* When the typical building is not occupied by a small cell BS, the user is either associated to the dominant macro BS or a small cell BS in the immediate vicinity. The former is regarded as being of greater relevance and the latter is omitted, which leads to a lower bound on coverage probability. In this case, the indoor user will be served by the nearest BS of the macro-tier. The same exclusion regions as defined in Case 1 are employed for macro- and small cell BSs.

13.2.4 Virtual Building Approximation

Without loss of generality, a *typical indoor user* is assumed to be located at $(r, 0)$. Note that the exclusion regions as defined in Sect. 13.2.3 are centered at the origin rather than at the user. Consequently, the interference field as observed by the user is asymmetric and renders analysis difficult in general. Thus, the following approximation is proposed.

Let (R, θ) denote the position of an interference. Its distance to a user located at $(r, 0)$ is determined as

$$d(r) = \sqrt{R^2 + r^2 - 2Rr\cos\theta}. \tag{13.3}$$

Since typically $R \gg r$, $d(r)$ is approximated as

$$d(r) \approx R, \tag{13.4}$$

which is independent of the angle θ. As shown in Fig. 13.4, the approximation in Eq. (13.4) is equivalent to shifting all the BSs along with the exclusion regions by a vector $(r, 0)$, as if the typical building was centered at the user location. Thus, this

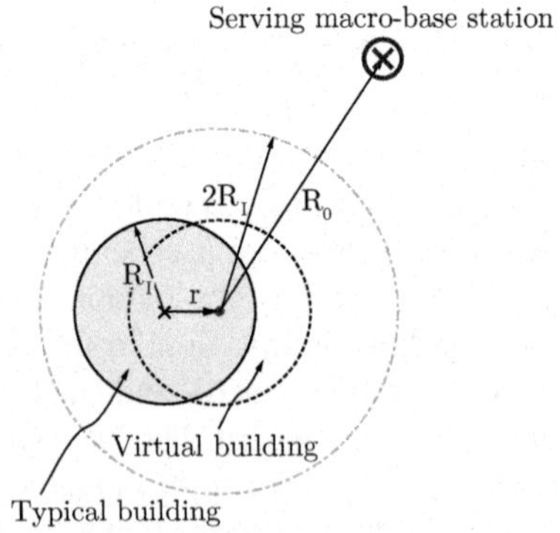

Fig. 13.4 Target area without small cell BS (*gray shaded*) and user-centric *virtual building* (*dashed*). *Dashed-dotted circles* denotes the shifted small cell exclusion region. The indoor user is assumed to be served by the nearest macro BS at distance R_0

approach is referred to as *virtual building approximation*, and is applied to simplify further analysis.

13.2.5 Signal Propagation

13.2.5.1 Macro BS to Indoor User

A signal originating from a macro BS experiences small scale fading, log-distance dependent path loss, attenuation due to building blockage and wall penetration, L_W. Small scale fading is modeled by a Gamma RV G_i, with $G_i \sim \Gamma[1, 1]$, i.e., Rayleigh fading. Along the lines of [13, Theorem 1], the number of obstructing blockages along a link of length R is a Poisson RV with parameter $\beta_B R$, where $\beta_B = 2\lambda_B R_I$ in the introduced topology model. For analytical tractability the *expected blockage attenuation* as referred from [13, Theorem 6] is employed. Combining blockage- and log-distance path loss along a link of length R yields

$$\ell(R) = e^{-\beta_B R(1-L_B)} \ell_O(R), \tag{13.5}$$

where L_B refers to the attenuation of a single blockage, also denoted as *building penetration loss* and $\ell_O(R) = \min(b_O, 1/c_O\, R^{-\alpha_O})$, with intercept b_O, constant c_O and outdoor path loss exponent α_O. In this chapter, $b_O = 1$ and $c_O = 1$ for simplicity. Note that the exponential term in Eq. (13.5) incorporates the condition that the macro BS is deployed outdoors.

Equation 13.5 reveals a major strength of the model: *Shadowing* is entirely characterized by the parameters of the underlying environment topology, which can, e.g., be extracted from openly-available online data, as demonstrated in Sect. 13.1.2. For comparison, the variance of log-normally distributed shadowing is typically obtained from measurements, which substantially exacerbates the finding of topologies with similar characteristics.

13.2.5.2 Small Cell BS to Indoor User

When user and small cell BS are situated in the same building, the signal experiences small scale fading and path loss $\ell_I(r) = \min(b_I, 1/c_I\, r^{-\alpha_I})$, with intercept b_I, constant c_I and indoor path loss exponent α_I. For simplicity, $b_I = 1$ and $c_I = 1$ in this chapter.

The signals from all other small cell BSs are subject to small scale fading, log-distance path loss $\ell_O(\cdot)$ as defined above, and attenuation by a factor L_W^2, as caused by the indoor-to-outdoor and outdoor-to-indoor wall penetration. Since the transmit power of a small cell BS is typically low, only small cell interferers from *neighboring* buildings are taken into account. Two buildings are defined as being *neighbors* to each other, if the segment connecting their centers is not intersected by any other building.

13.3 Performance Analysis

In this section, analytical expressions for the coverage probability of an indoor user at position $(r, 0)$ are derived, regarding both buildings with- and without small cell deployment. The network is assumed to be interference limited, as is typically the case in urban areas [24]. Thus, thermal noise is neglected in the analysis.

13.3.1 Typical Building with Small Cell Base Station

Let $\Phi_M = \{X_i\}$ and $\Phi_S = \{X_j\}$ denote the point processes of macro- and small cell BSs, respectively. Further, define $R_i = |OX_i|$ and $R_j = |OX_j|$ as the distances of macro BS i and small cell BS j to the origin O. Assume the typical building to be occupied by a small cell BS. Then, the Signal-to-Interference Ratio (SIR) at distance $r, 0 < r \leq R_I$, is determined as

$$\gamma_S(r) = \frac{P_S G_0 \ell_I(r)}{\sum_{i:R_i \in \Phi_M \setminus \mathscr{B}(0,R_I)} P_M G_i L_W \ell(R_i) + \sum_{j:X_j \in \Phi_S \setminus \mathscr{B}(0,2R_I)} S_j P_S G_j L_W^2 \ell_O(R_j)} \tag{13.6}$$

where the terms P_M and P_S denote macro- and small cell BS transmit powers, $\ell_I(\cdot)$ and $\ell_O(\cdot)$ are indoor- and outdoor path loss laws as specified in Sect. 13.2.5, $\ell(\cdot)$ corresponds to the combined blockage- and path loss attenuation, as defined in Eq. (13.5), and $\mathscr{B}(0, R)$ refers to a ball of radius R, which is centered at the origin. The RVs S_j are Bernoulli distributed and, by [13, Theorem 1], have parameters $\exp(-\beta_B R_j - p_B)$, where $p_B = \lambda_B R_I^2 \pi$. They indicate whether or not an interfering small cell BS is in a neighboring building of the typical user.

Theorem 13.1 *Consider a user at distance r, $0 < r \leq R_I$, away from the center of a small cell BS-occupied building. Then, its coverage probability is determined as*

$$P_{c,S}(\delta|r) = \mathbb{P}\left[\gamma_S(r) > \delta|r\right] = e^{-2\pi(\mu_S I_S + \mu_M I_M)}, \tag{13.7}$$

where

$$I_S = \int_{2R_I}^{\infty} \left(\frac{\delta L_W^2 \ell_O(t) e^{-(\beta_B t + p_B)}}{\ell_I(r) + \delta L_W^2 \ell_O(t)} \right) t \, dt, \tag{13.8}$$

$$I_M = \int_{R_I}^{\infty} \left(1 - \frac{\frac{P_S}{P_M} \ell_I(r)}{\frac{P_S}{P_M} \ell_I(r) + \delta L_W \ell(t)} \right) t \, dt. \tag{13.9}$$

Proof The proof is provided in [18, Appendix E].

Note that Eqs. (13.8) and (13.9) correspond to the two interference contributions in Eq. (13.6).

13.3.2 Typical Building Without Small Cell Base Station

Assume a dominant macro BS to be located at distance R_O, with $R_O > R_I$, away from the center of the typical building and consider that this building is not occupied by a small cell BS. Then, the SIR at distance r, $0 < r \leq R_I$, calculates as

$$\gamma_M(R_O) = \frac{P_M G_O \ell(R_O)}{\sum_{i:R_i \in \Phi_M \setminus \mathscr{B}(0,R_O)} P_M G_i \ell(R_i) + \sum_{j:X_j \in \Phi_S \setminus \mathscr{B}(0,2R_I)} S_j P_S G_j L_W \ell_O(R_j)} \tag{13.10}$$

Note that (i) the expression is independent of r and (ii) the factor L_W is omitted, since attenuation due to wall penetration is experienced by all signals and therefore cancels out in the SIR term.

Theorem 13.2 *Consider a user at distance r, $0 < r \leq R_I$, away from the center of a typical building without small cell BS and assume that it is associated with its dominant macro BS. Then, its coverage probability is determined as*

$$P_{c,M}(\delta) = \mathbb{P}\left[\mathbb{E}_{R_O}\left[\gamma_M(R_O) > \delta\right]\right] = \int_{R_I}^{\infty} P_{c,M}(\delta|R) f_{R_O}(R) dR, \tag{13.11}$$

where

$$P_{c,M}(\delta|R_O) = e^{-2\pi(\mu_S I_S + \mu_M I_M)}, \tag{13.12}$$

with

$$I_S = \int_{2R_I}^{\infty} \left(\frac{\delta L_W \frac{P_S}{P_M} \ell_O(t) e^{-(\beta_B t + p_B)}}{\ell(R_O) + \delta L_W \frac{P_S}{P_M} \ell_O(t)} \right) t\, dt, \tag{13.13}$$

$$I_M = \int_{R_O}^{\infty} \left(1 - \frac{\ell(R_O)}{\ell(R_O) + \delta\, \ell(t)} \right) t\, dt, \tag{13.14}$$

and

$$f_{R_O}(R) = \begin{cases} 2\pi \mu_M R\, e^{-\pi \mu_M (R^2 - R_I^2)}, & R \geq R_I \\ 0, & \text{otherwise} \end{cases}. \tag{13.15}$$

Proof The conditional coverage probability $P_{c,M}(\delta|R)$ in Eq. (13.12) is derived along the lines of Eq. (13.7). Averaging over the dominant macro BS distance leads to Eq. (13.11). According to [21, Example 1.4.7], the term $f_{R_O}(R)$ in Eq. (13.15) is the nearest neighbor distance distribution of a homogeneous PPP outside a ball of radius R_I.

13.3.3 *Typical Indoor User*

The coverage probability of a *typical* indoor user at distance $r, 0 < r \leq R_{\rm I}$, is obtained by linearly combining $P_{c,{\rm S}}(\delta|r)$ from Eq. (13.7) and $P_{c,{\rm M}}(\delta)$ from Eq. (13.11) according to the small cell occupation probability η. Then,

$$P_c(\delta|r) = \eta\, P_{c,{\rm S}}(\delta|r) + (1-\eta)\, P_{c,{\rm M}}(\delta). \tag{13.16}$$

13.4 Numerical Evaluation

In this section, the performance of a typical user at the edge of a building, i.e., $r = R_{\rm I}$ is evaluated numerically. At this location, the proposed *virtual building approximation* is expected to perform worst. The normalized ergodic rate is employed as a metric. Along the lines of Sect. 12.4.3, it is defined as $\tau(r) = \mathbb{E}_{\gamma(r)}[\alpha_{\rm B}\log_2(1 + \alpha_{\rm SIR}\min(\delta_{\max}, \gamma(r)))]$ and can be reformulated in terms of coverage probability as

$$\tau(r) = \frac{\alpha_{\rm B}}{\log(2)} \int_0^{\delta_{\max}} \frac{\alpha_{\rm SIR}}{1+\alpha_{\rm SIR}\delta} P_c(\delta|r)d\delta, \tag{13.17}$$

with $P_{\rm c}(\delta|r)$ from Eq. (13.16) and $\delta_{\max} = 2^6 - 1$, referring to 64-Quadrature Amplitude Modulation (QAM), which is the highest modulation order in the current LTE-A standard [25]. The terms $\alpha_{\rm B}$ and $\alpha_{\rm SIR}$ denote calibration parameters, with $0 < \alpha_{\rm B} \leq 1$ and $0 < \alpha_{\rm SIR} \leq 1$. In this chapter, $\alpha_{\rm B} = 1$ and $\alpha_{\rm SIR} = 1$.

Parameters for numerical evaluation are listed in Table 13.1. To verify the accuracy of the *virtual building approximation*, Monte Carlo simulations are carried out, using the system model as introduced in Sect. 13.2. The density of the macro BSs is chosen such that the inscribing ball of the typical cell has $R_{\rm C} = 250$ m and the BSs are distributed over a field of $15\,R_{\rm C} \times 15\,R_{\rm C}$. The results are estimated from averaging over 500 fading- and 500 spatial realizations.

Figure 13.5 depicts the normalized ergodic rate $\tau(r)$ over the indoor coverage ratio $p_{\rm I}$, as defined in Sect. 13.1.2. Note that when fixing the average building *size*, $p_{\rm I}$

Table 13.1 Parameters for numerical evaluation

Parameter	Value
Macro-to-small cell power ratio	$P_{\rm S}/P_{\rm M} = 10^{-2}$
Macro BS density	$\mu_{\rm M} = 4.61 \times 10^{-6}\,{\rm m}^{-2}$
Outdoor path loss exponent	$\alpha_{\rm O} = 4$
Indoor path loss exponent	$\alpha_{\rm I} = 2$
Radius of building area	$R_{\rm I} = 25$ m

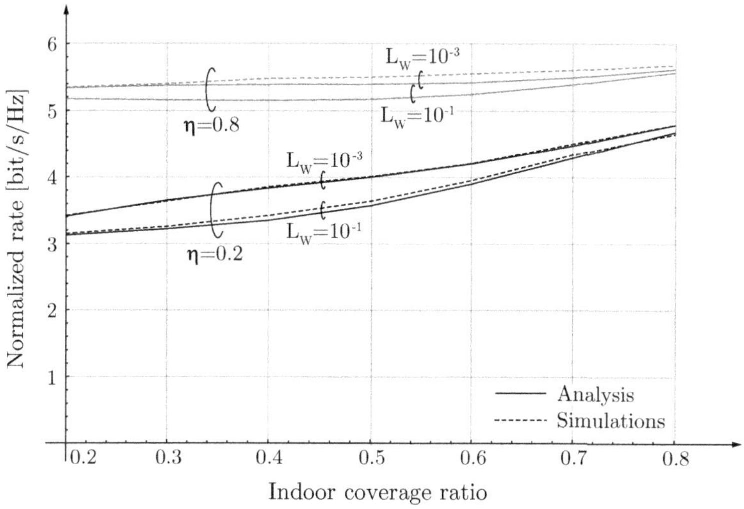

Fig. 13.5 Normalized rate [bit/s/Hz] over area-ratio, which is covered by buildings. *Solid-* and *dashed lines* denote results from analysis and simulations, respectively. *Curves* are shown for varying small cell occupation probability η and wall penetration loss L_W

scales with the *density* of the buildings. Solid- and dashed lines correspond to analysis and simulations, respectively. Results are shown for a sparse- and a dense small cell deployment, as quantified by the occupation probability η. For both scenarios, weak- and strong wall partitioning are investigated. The wall penetration loss is correlated to the building penetration loss L_B, as introduced in Eq. (13.5). This work employs the conservative setting $L_B = L_W$, which can be replaced by more elaborated models in future work. It is observed that

- The achievable normalized rate improves with increasing building density. This result follows the intuition that obstructions due to large objects establish a safeguard against interference [13]. Note that for constant occupation probability, the small cell density grows in proportion to the building density. Therefore, the results render the existence of a *hotspot limited regime* in urban environments questionable, supporting simulation results in [8, 9, 26, 27] and Chap. 14.
- Low isolation by wall penetration deteriorates performance in both deployment scenarios. Intuitively, the isolation of the interfering small cell BSs is decreased when the wall penetrations become weaker. The impact of penetration loss on coverage probability, however, becomes minor especially when the building density is high. Intuitively, this indicates that *the number of penetrations rather than the loss per penetration dominates the effect of partitioning between indoor and outdoor environment.*
- Even though a user at the edge of a typical building is evaluated, the analytical results closely resemble the simulations. This confirms the accuracy of the *virtual building approximation* as well as the inclusion of macro-interferers in the immediate vicinity of the typical building, as claimed in Sect. 13.2.3.

13.5 LOS- and NLOS Macro Base Stations

In this section, previous analyses are refined by taking into account whether the path between a macro BS and the indoor user is obstructed by any other building. The distinction between LOS- and NLOS links recently gained momentum with the study of *millimeter* wave communication [28, 29]. In contrast, it has commonly been neglected in the analysis of conventional cellular *microwave* networks. Typically, signal propagation is modeled by a single combination of shadow fading distribution, path loss law and wall penetration. Nonetheless, numerous measurement campaigns clearly indicate the differences between LOS- and NLOS microwave link characteristics [30–36].

In this section, a framework is established which enables to account for both link states. The idea is to partition the macro BSs process Φ_M into two independent non-homogeneous PPPs Φ_L and Φ_N, referring to LOS- and NLOS nodes such that $\Phi_M = \Phi_L \cup \Phi_N$. Each process employs its own model for signal propagation, incorporating log-distance dependent path loss, blockage and wall penetration. LOS- and NLOS model are hereinafter referred to as $\ell_L(\cdot)$ and $\ell_N(\cdot)$, respectively.

Consider a typical building at the origin and an indoor user at distance r, $0 < r \leq R_I$. Further, let a macro BS be located at distance R away from the origin. Then, using the Boolean scheme as specified in Sect. 13.2.1 and the *virtual building approximation* as introduced in Sect. 13.2.4, the probability that the link between user and BS is obstructed by any other building is obtained as $v(R) = \exp(-\beta_B(R - R_I))$. This term can be interpreted as *LOS probability*, following the intuition that it becomes increasingly unlikely to experience a connection with a distant BS.

According to [13], the shadowing of different links can be considered as uncorrelated with minor loss of accuracy. This allows to apply the thinning property of PPPs and yields the intensities of the independent BS processes Φ_L and Φ_N as $\mu_M\, v(R)$ and $\mu_M(1 - v(R))$, respectively.

13.5.1 Distance Distributions of Associated Macro Base Stations

Consider a typical building without small cell BS. The following lemmas provide Probability Density Functions (PDFs) for the distance between an indoor user and its associated macro BS, given that the BS is either in LOS or NLOS. The expressions extend results in [28] by conditioning on deploying a *virtual building* around the user, as illustrated in Fig. 13.6.

Lemma 13.1 *Consider a typical building without a small cell BS and an indoor user at distance r, $0 < r \leq R_I$, away from its center, which is associated with the closest LOS macro BS. Then, applying the virtual building approximation, its distance to the serving BS is distributed as*

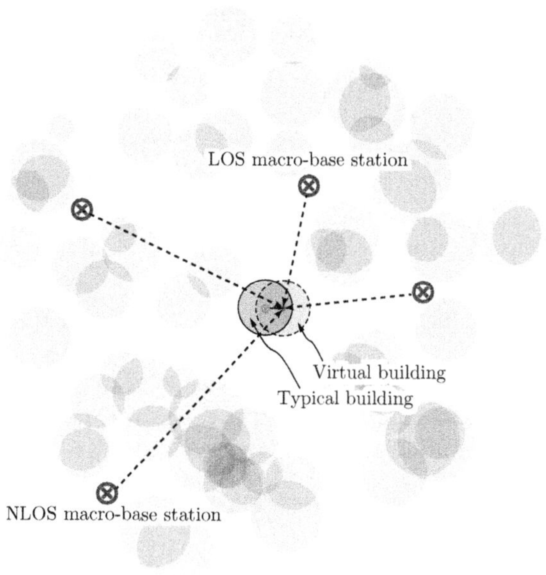

Fig. 13.6 Snapshot of an urban environment as obtained by a Boolean scheme with circles of random radius. The figure depicts an indoor user in a typical building, which is occupied by a small cell BS. The scenario encompasses LOS- and NLOS macro BSs. The *dashed circle* indicates the virtual building

$$f_{\mathrm{L}}(x) = \begin{cases} \frac{B_{\mathrm{L}}\hat{f}_{\mathrm{L}}(x)}{A_{\mathrm{L}}} e^{-2\pi\mu_{\mathrm{M}} \int_{R_{\mathrm{I}}}^{\Psi_{\mathrm{L}}(x)} (1-v(t))tdt}, x \geq R_{\mathrm{I}} \\ 0, \quad \text{otherwise} \end{cases}, \tag{13.18}$$

where

$$B_{\mathrm{L}} = 1 - e^{-2\pi\mu_{\mathrm{M}} \int_{R_{\mathrm{I}}}^{\infty} v(r)rdr} \tag{13.19}$$

$$\hat{f}_{\mathrm{L}}(x) = \begin{cases} \frac{1}{B_{\mathrm{L}}} 2\pi\mu_{\mathrm{M}} x v(x) e^{-2\pi\mu_{\mathrm{M}} \int_{R_{\mathrm{I}}}^{x} v(r)rdr}, & x \geq R_{\mathrm{I}} \\ 0, & \text{otherwise} \end{cases} \tag{13.20}$$

$$A_{\mathrm{L}} = B_{\mathrm{L}} \int_{R_{\mathrm{I}}}^{\infty} e^{-2\pi\mu_{\mathrm{M}} \int_{R_{\mathrm{I}}}^{\Psi_{\mathrm{L}}(x)} (1-v(t))tdt} \hat{f}_{\mathrm{L}}(x)dx, \tag{13.21}$$

and $\Psi_{\mathrm{L}}(x) = \ell_{\mathrm{L}}^{-1}(\ell_{\mathrm{N}}(x))$. *The term* B_{L} *denotes the probability that the user receives at least one LOS BS and* $\hat{f}_{\mathrm{L}}(x)$ *is the corresponding conditional distance distribution function of the closest node. The quantity* A_{L} *captures the likelihood to be associated with the closest LOS BS.*

Proof The proof is derived along the lines of [13, Theorem 10] by excluding BS from a ball of radius R_{I} around the user.

Lemma 13.2 *Consider an indoor user at distance* r, $0 < r \leq R_{\mathrm{I}}$ *away from the center of a typical building without a small cell BS. Let the user be attached to the closest NLOS BS. Then, employing the virtual building approximation, the PDF of its distance to the serving BS is expressed as*

$$f_{\mathrm{N}}(x) = \begin{cases} \frac{B_{\mathrm{N}}\hat{f}_{\mathrm{N}}(x)}{A_{\mathrm{N}}} e^{-2\pi\mu_{\mathrm{M}} \int_{R_{\mathrm{I}}}^{\Psi_{\mathrm{N}}(x)} v(t)t dt} & , x \geq R_{\mathrm{I}} \\ 0 & , \text{otherwise} \end{cases}, \tag{13.22}$$

where

$$B_{\mathrm{N}} = 1 - e^{-2\pi\mu_{\mathrm{M}} \int_{R_{\mathrm{I}}}^{\infty} (1-v(r)) r dr} \tag{13.23}$$

$$\hat{f}_{\mathrm{N}}(x) = \begin{cases} \frac{1}{B_{\mathrm{N}}} 2\pi\mu_{\mathrm{M}} x (1 - v(x)) e^{-2\pi\mu_{\mathrm{M}} \int_{R_{\mathrm{I}}}^{x} (1-v(r)) r dr} & , x \geq R_{\mathrm{I}} \\ 0 & , \text{otherwise} \end{cases} \tag{13.24}$$

$$A_{\mathrm{N}} = 1 - A_{\mathrm{L}}, \tag{13.25}$$

with A_{L} *from Eq.* (13.21) *and* $\Psi_{\mathrm{N}}(x) = \ell_{\mathrm{N}}^{-1}(\ell_{\mathrm{L}}(x))$. *The term* B_{N} *refers to the probability that the user receives at least one NLOS BS and* $\hat{f}_{\mathrm{N}}$ *is the according conditional pdf of the distance to the closest node.*

Proof As above, the proof follows [13, Theorem 10] and is therefore omitted.

13.5.2 SINR and Coverage Analysis

Given an indoor user at distance r, $0 < r \leq R_{\mathrm{I}}$ away from the center of a typical building with small cell BS, its SIR is determined as

$$\gamma_{\mathrm{S}}(r) = \frac{P_{\mathrm{S}} G_0 \ell_{\mathrm{I}}(r)}{\sum_{\substack{i:X_i \in \Phi_{\mathrm{L}} \\ \setminus \mathcal{B}(0,R_{\mathrm{I}})}} P_{\mathrm{M}} G_i \ell_{\mathrm{L}}(R_i) + \sum_{\substack{j:X_j \in \Phi_{\mathrm{N}} \\ \setminus \mathcal{B}(0,R_{\mathrm{I}})}} P_{\mathrm{M}} G_j \ell_{\mathrm{N}}(R_j) + \sum_{\substack{k:X_k \in \Phi_{\mathrm{S}} \\ \setminus \mathcal{B}(0,2R_{\mathrm{I}})}} S_k P_{\mathrm{S}} G_k L_W \ell_{\mathrm{L}}(R_k)}, \tag{13.26}$$

where the first- and second sum in the denominator denote the aggregate interference from the LOS- and NLOS macro BSs, respectively. The third sum refers to the contribution from the small cell tier.

When the building is not occupied by a small cell BS, the user associates with the dominant macro BS at distance R_0, with $R_0 > R_{\mathrm{I}}$. Depending on whether the serving BS is in LOS or NLOS, Eq. (13.10) reformulates as

$$\gamma_{\mathrm{L}}(R_0) = \frac{P_{\mathrm{M}} G_0 \ell_{\mathrm{L}}(R_0)}{\sum_{\substack{i:X_i \in \Phi_{\mathrm{L}} \\ \setminus \mathcal{B}(0,R_0)}} P_{\mathrm{M}} G_i \ell_{\mathrm{L}}(R_i) + \sum_{\substack{j:X_j \in \Phi_{\mathrm{N}} \\ \setminus \mathcal{B}(0,\Psi_{\mathrm{L}}(R_0))}} P_{\mathrm{M}} G_j \ell_{\mathrm{N}}(R_j) + \sum_{\substack{k:X_k \in \Phi_{\mathrm{S}} \\ \setminus \mathcal{B}(0,2R_{\mathrm{I}})}} S_k P_{\mathrm{S}} G_k L_W \ell_{\mathrm{L}}(R_k)}, \tag{13.27}$$

or

$$\gamma_{\mathrm{N}}(R_0) = \frac{P_{\mathrm{M}} G_0 \ell_{\mathrm{N}}(R_0)}{\sum_{\substack{i:X_i \in \Phi_{\mathrm{L}} \\ \setminus \mathcal{B}(0,\Psi_{\mathrm{N}}(R_0))}} P_{\mathrm{M}} G_i \ell_{\mathrm{L}}(R_i) + \sum_{\substack{j:X_j \in \Phi_{\mathrm{N}} \\ \setminus \mathcal{B}(0,R_0)}} P_{\mathrm{M}} G_j \ell_{\mathrm{N}}(R_j) + \sum_{\substack{k:X_k \in \Phi_{\mathrm{S}} \\ \setminus \mathcal{B}(0,2R_{\mathrm{I}})}} S_k P_{\mathrm{S}} G_k L_W \ell_{\mathrm{L}}(R_k)}. \tag{13.28}$$

Note that by virtue of the virtual building approximation from Sect. 13.2.4 both $\gamma_{\rm L}(\cdot)$ and $\gamma_{\rm N}(\cdot)$ are independent of the user's location within the building. The following theorems extend Theorems 13.1 and 13.2 with respect to NLOS- and NLOS macro BSs.

Theorem 13.3 *Consider an indoor user at distance r, $0 < r \leq R_{\rm I}$ away from the center of a typical building with a small cell BS. Then, its coverage probability is calculated as*

$$P_{c,{\rm S}}(\delta|r) = \mathbb{P}\left[\gamma_{\rm S}(r) > \delta|r\right] = e^{-2\pi(\mu_{\rm S}I_{\rm S}+\mu_{\rm M}(I_{\rm L}+I_{\rm N}))}, \tag{13.29}$$

with $\gamma_{\rm S}(\cdot)$ from Eq. (13.26) *and*

$$I_{\rm S} = \int_{2R_{\rm I}}^{\infty} \frac{\delta L_{\rm W}\ell_{\rm L}(t)e^{-(\beta_{\rm B}t+p_{\rm B})}}{\ell_{\rm I}(r)+\delta L_{\rm W}\ell_{\rm L}(t)} t dt, \tag{13.30}$$

$$I_{\rm L} = \int_{R_{\rm I}}^{\infty} \left(1 - \frac{\frac{P_{\rm S}}{P_{\rm M}}\ell_{\rm I}(r)}{\frac{P_{\rm S}}{P_{\rm M}}\ell_{\rm I}(r)+\delta\ell_{\rm L}(t)}\right) tv(t)dt, \tag{13.31}$$

$$I_{\rm N} = \int_{R_{\rm I}}^{\infty} \left(1 - \frac{\frac{P_{\rm S}}{P_{\rm M}}\ell_{\rm I}(r)}{\frac{P_{\rm S}}{P_{\rm M}}\ell_{\rm I}(r)+\delta\,\ell_{\rm N}(t)}\right) t(1-v(t))dt. \tag{13.32}$$

Proof The proof is provided in [18, Appendix F].

Theorem 13.4 *Consider a typical building without a small cell BS and an indoor user at distance r, $0 < r \leq R_{\rm I}$ away from its center. Given that the user is associated with the closest LOS macro BS, its coverage probability is determined as*

$$P_{c,{\rm L}}(\delta) = \mathbb{P}\left[\mathbb{E}_{R_0}[\gamma_{\rm L}(R_0) > \delta]\right] = \int_{R_{\rm I}}^{\infty} e^{-2\pi(\mu_{\rm S}I_{\rm S}+\mu_{\rm M}(I_{\rm L}+I_{\rm N}))} f_{\rm L}(R)dR, \tag{13.33}$$

with $\gamma_{\rm L}(\cdot)$ from Sect. 13.5.2, $f_{\rm L}(\cdot)$ from Eq. (13.18) and

$$I_{\rm S} = \int_{2R_{\rm I}}^{\infty} \frac{\delta P_{\rm S}L_{\rm W}\ell_{\rm L}(t)e^{-(\beta_{\rm B}t+p_{\rm B})}}{P_{\rm M}\ell_{\rm L}(R)+\delta P_{\rm S}L_{\rm W}\ell_{\rm L}(t)} t\, dt \tag{13.34}$$

$$I_{\rm L} = \int_{R}^{\infty} \left(1 - \frac{\ell_{\rm L}(R)}{\ell_{\rm L}(R)+\delta\ell_{\rm L}(t)}\right) t\, v(t)dt \tag{13.35}$$

$$I_{\rm N} = \int_{\Psi_{\rm L}(R)}^{\infty} \left(1 - \frac{\ell_{\rm L}(R)}{\ell_{\rm L}(R)+\delta\ell_{\rm N}(t)}\right) t\,(1-v(t))dt. \tag{13.36}$$

When the user is served by the closest NLOS macro BS, its coverage probability is calculated as

$$P_{\rm c,N}(\delta) = \mathbb{P}\left[\mathbb{E}_{R_0}[\gamma_{\rm N}(R_0) > \delta]\right] = \int_{R_{\rm I}}^{\infty} e^{-2\pi(\mu_{\rm S}I_{\rm S}+\mu_{\rm M}(I_{\rm L}+I_{\rm N}))} f_{\rm N}(R)dR \tag{13.37}$$

where $\gamma_{\mathrm{N}}(\cdot)$ *and* $f_{\mathrm{N}}(\cdot)$ *are obtained from Eq. (13.22) and Sect. 13.5.2, and*

$$I_{\mathrm{S}} = \int_{2R_{\mathrm{I}}}^{\infty} \frac{\delta P_{\mathrm{S}} L_{\mathrm{W}} \ell_{\mathrm{L}}(t) e^{-(\beta_{\mathrm{B}} t + p_{\mathrm{B}})}}{P_{\mathrm{M}} \ell_{\mathrm{N}}(R) + \delta P_{\mathrm{S}} L_{\mathrm{W}} \ell_{\mathrm{L}}(t)} t\, dt \tag{13.38}$$

$$I_{\mathrm{L}} = \int_{\psi_{\mathrm{N}}(R)}^{\infty} \left(1 - \frac{\ell_{\mathrm{N}}(R)}{\ell_{\mathrm{N}}(R) + \delta \ell_{\mathrm{L}}(t)}\right) t\, v(t) dt \tag{13.39}$$

$$I_{\mathrm{N}} = \int_{R}^{\infty} \left(1 - \frac{\ell_{\mathrm{N}}(R)}{\ell_{\mathrm{N}}(R) + \delta \ell_{\mathrm{N}}(t)}\right) t\, (1 - v(t)) dt. \tag{13.40}$$

Proof For a given BS distance R_0, the proofs for $\mathbb{P}[\gamma_{\mathrm{N}}(R_0) > \delta]$ and $\mathbb{P}[\gamma_{\mathrm{L}}(R_0) > \delta]$ are carried out along the lines of [18, Appendix F]. Averaging over R_0 yields Eqs. (13.33) and (13.37), respectively.

Finally, Theorems 13.3 and 13.4 enable to extend Eq. (13.16). The coverage probability of a *typical* indoor user at distance r, $0 < r \leq R_{\mathrm{I}}$, which experiences LOS- and NLOS macro BSs, is expressed as

$$P_c(\delta|r) = \eta P_{c,\mathrm{S}}(\delta|r) + (1-\eta)(A_{\mathrm{L}} P_{c,\mathrm{L}}(\delta) + (1 - A_{\mathrm{L}}) P_{c,\mathrm{N}}(\delta)), \tag{13.41}$$

where η denotes the small cell occupation probability and A_{L} is the likelihood that the user associates with a LOS macro BS, as derived in Eq. (13.21).

13.5.3 Numerical Evaluation

In this section, the performance of a typical user at the edge of a building, i.e., $r = R_{\mathrm{I}}$, is numerically evaluated. The results are provided in terms of normalized ergodic rate and are obtained by plugging $P_c(\delta|R)$ from Eq. (13.41) into Eq. (13.17).

Signal propagation along LOS- and NLOS links is modeled by

$$\ell_{\mathrm{L}}(R) = \ell_{\mathrm{O}}(R)\, L_{\mathrm{L}}, \tag{13.42}$$

$$\ell_{\mathrm{N}}(R) = e^{-\beta_{\mathrm{B}} R (1 - L_{\mathrm{B}})}\, \ell_{\mathrm{O}}(R)\, L_{\mathrm{N}}, \tag{13.43}$$

where L_{L} and L_{N} denote the wall penetration losses, $\exp(-\beta_{\mathrm{B}} R(1 - L_{\mathrm{B}}))$ accounts for the shadowing, as referred from Sect. 13.2.5, and $\ell_{\mathrm{O}}(\cdot)$ refers to the log-distance path loss, as defined in Sect. 13.2.5. Note that, in general, the intercept b_{O}, the constant c_{O} and the path loss exponent α_O will be different in the LOS- and NLOS case. These models are based on the following findings from measurement campaigns.

For analytical convenience, the characteristics of an urban environment are often condensed into different variances of a log-normally distributed RVs, which account for the shadowing [37]. However, the authors of [30] observed considerable deviations from this model in LOS scenarios, where signal characteristics are largely dominated by free space propagation as long as the first Fresnel zone is not obstructed

[30–34]. Breaking distances of 160–800 m have been reported from measurements in metropolitan areas [33, 38]. Using the concept of a *LOS ball* as defined in [28] yields equivalent circular LOS areas of radius 49.1 m for $p_I = 0.8$, 96.2 m for $p_I = 0.5$ and 266 m for $p_I = 0.2$, respectively. Hence, it is considered reasonably accurate to employ a single-slope free-space path loss law for LOS signal propagation, i.e., $\alpha_O = 2$ in Eq. (13.42).

In accordance with the 3GPP LTE-A standard [39] and numerous measurement campaigns [31–34, 40, 41], NLOS propagation alters the path loss exponent and adds an additional shadowing component. In this section, $\alpha_O = 4$ and $L_B = 10^{-1}$, respectively. Furthermore, measurement results in [30, 35, 36] indicate that signals from a NLOS BS experience a lower wall penetration loss. Intuitively, this is caused by the fact that multi path components approach the building more frontally after multiple reflections. The wall penetration loss for NLOS links is set $L_N = 10^{-1}$ whereas the loss for LOS links, L_L, is varied as specified below. Table 13.2 summarizes the parameters for numerical evaluation.

In order to verify the accuracy of the *virtual building approximation*, Monte Carlo simulations are carried out with the same system model. BS are distributed over a field of $15\,R_C \times 15\,R_C$ and their density is chosen such that the inscribing ball of the typical cell has a radius of 250 m. The results are assessed from averaging over 500 spatial- and 500 fading realizations.

Figure 13.7 depicts normalized ergodic rate $\tau(R_I)$ over indoor coverage ratio p_I. The curves correspond to two small cell occupation probabilities, $\eta = 0.2$ and $\eta = 0.8$, and two wall penetration loss values for the LOS signal, $L_L = 10^{-1}$ and $L_L = 10^{-3}$, respectively.

It is observed that

- Unlike results in Sect. 13.4, the normalized rate does not increase uniformly with the indoor coverage ratio but rather exhibits certain minima. This is explained by the facts that (i) a low indoor coverage ratio favors LOS connections and (ii) the likelihood to experience a LOS link rapidly decreases with higher indoor coverage ratio as the exponent of the LOS probability $v(R)$ linearly scales with the building density.

Table 13.2 Parameters for numerical evaluation

Parameter	Value
Macro-to-small cell power ratio	$P_S/P_M = 10^{-2}$
Macro BS density	$\mu_M = 4.61 \times 10^{-6}\,\text{m}^{-2}$
Radius of building area	$R_I = 25\,\text{m}$
Intercept of path loss law	$b_O = 1$
Constant of path loss law	$c_O = 1$
LOS path loss exponent	$\alpha_O = 2$
NLOS path loss exponent	$\alpha_O = 4$
Building penetration loss	$L_B = 10^{-1}$
NLOS wall penetration loss	$L_N = 10^{-1}$

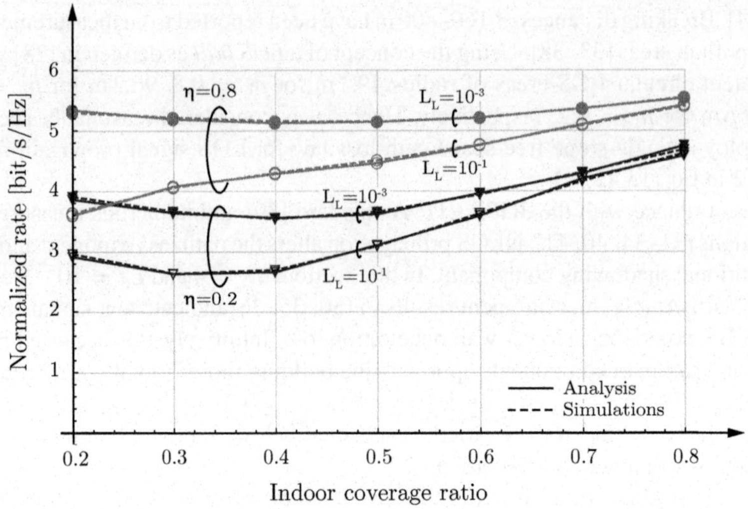

Fig. 13.7 Normalized rate in [bit/s/Hz] over indoor coverage ratio. *Solid*- and *dashed lines* correspond to results from analysis and simulations, respectively. *Curves* are depicted for varying small cell occupation probability η and wall penetration loss for LOS links, L_L, respectively

- In a scenario with low wall penetration loss for LOS links, small cell BSs are weakly isolated from the outdoor environment. Hence, for a high small cell occupation probability, they considerably contribute to the aggregate interference. As a result, the normalized rate for $L_L = 10^{-1}$ and $\eta = 0.8$ monotonically increases rather than exhibiting a minimum.
- The impact of the wall penetration loss becomes inferior with higher indoor area coverage. The result is based on the effects that (i) the likelihood of experiencing LOS interferers reduces with decreasing LOS probability and (ii) a higher building density establishes a safeguard against NLOS interferers. This corroborates previous findings that it is *the amount of blockages rather than their exact penetration loss which dominates the effect of indoor-outdoor partitioning*.
- The curves precisely fit with results from Monte Carlo simulations. In comparison to Sect. 13.4, the consideration of LOS- and NLOS macro BSs improves the accuracy of the model.

As shown in Chap. 14, these effects cannot be identified with the widely used log-normal shadowing model.

13.6 Conclusion

This chapter introduces a novel system model for two-tier heterogeneous cellular networks in urban environments. The focus lies on indoor users. Analytical expressions for the coverage probability in buildings with- and without small cell deployment are derived. The proposed *virtual building approximation* considerably improves the tractability of the analysis. Its accuracy is confirmed by simulation results. Numerically evaluations are carried out to investigate the performance of a *typical* indoor user in terms of normalized ergodic rate. The results reveal essential effects of an urban environment. Observations such as the blockage safeguard and the vanishing impact of LOS BSs and wall isolation with increasing building density are missed by overly simplistic models, as demonstrated in the next chapter.

References

1. V. Chandrasekhar, J. Andrews, Spectrum allocation in tiered cellular networks. IEEE Trans. Commun. **57**(10), 3059–3068 (2009)
2. V. Chandrasekhar, J. Andrews, Uplink capacity and interference avoidance for two-tier cellular networks, in *IEEE Global Telecommunications Conference*, Washington (2007), pp. 3322–3326. doi:10.1109/GLOCOM.2007.630
3. P. Xia, V. Chandrasekhar, J.G. Andrews, Open vs. closed access femtocells in the uplink. IEEE Trans. Wireless Commun. **9**(12), 3798–3809 (2010)
4. F. Baccelli, B. Blaszczyszyn, P. Muhlethaler, Stochastic analysis of spatial and opportunistic ALOHA. IEEE J. Sel. Areas Commun. **27**(7), 1105–1119 (2009). doi:10.1109/JSAC.2009.090908
5. H. Wang, M. Reed, Tractable model for heterogeneous cellular networks with directional antennas, in *Australian Communications Theory Workshop*, Wellington (2012), pp. 61–65. doi:10.1109/AusCTW.2012.6164907
6. S. Mukherjee, UE coverage in LTE macro network with mixed CSG and open access femto overlay, in *IEEE International Conference on Communications Workshops (ICC)*, Kyoto (2011). doi:10.1109/iccw.2011.5963549
7. V. Chandrasekhar, M. Kountouris, J. Andrews, Coverage in tiered cellular networks with spatial diversity, in *IEEE Global Telecommunications Conference*, Honolulu (2009). doi:10.1109/GLOCOM.2009.5425918
8. V. Chandrasekhar, J. Andrews, Z. Shen, T. Muharemovict, A. Gatherer, Distributed power control in femtocell-underlay cellular networks, in *IEEE Global Telecommunications Conference*, Honolulu (2009)
9. V. Chandrasekhar, M. Kountouris, J.G. Andrews, Coverage in multi-antenna two-tier networks. IEEE Trans. Wirel. Commun. **8**(10), 5314–5327 (2014)
10. J. Andrews, F. Baccelli, R. Ganti, A tractable approach to coverage and rate in cellular networks. IEEE Trans. Commun. **59**(11), 3122–3134 (2011). doi:10.1109/TCOMM.2011.100411.100541
11. H.-S. Jo, Y.J. Sang, P. Xia, J. Andrews, Heterogeneous cellular networks with flexible cell association: a comprehensive downlink SINR analysis. IEEE Trans. Wireless Commun. **11**(10), 3484–3495 (2012). doi:10.1109/TWC.2012.081612.111361
12. S. Mukherjee, Distribution of downlink SINR in heterogeneous cellular networks. IEEE J. Sel. Areas Commun. **30**(3), 575–585 (2012)
13. T. Bai, R. Vaze, R.W. Heath Jr., Analysis of blockage effects on urban cellular networks. IEEE Trans. Wirel. Commun. **13**(9), 5070–5083 (2014). doi:10.1109/TWC.2014.2331971
14. 3rd Generation Partnership Project (3GPP), Study on 3D channel model for LTE, 3rd Generation Partnership Project (3GPP), TR 36.873 (2014)
15. H. ElSawy, E. Hossain, Two-tier HetNets with cognitive femtocells: downlink performance modeling and analysis in a multichannel environment. IEEE Trans. Mob. Comput. **13**(3), 649–663 (2014). doi:10.1109/TMC.2013.36
16. H. ElSawy, E. Hossain, M. Haenggi, Stochastic geometry for modeling, analysis, and design of multi-tier and cognitive cellular wireless networks: a survey. IEEE Commun. Surv. Tutor. **15**(3), 996–1019 (2013)

17. M. Taranetz, M. Rupp, R.W. Heath Jr., T. Bai, Analysis of small cell partitioning in urban two-tier heterogeneous cellular networks, in *Proceedings of the International Symposium on Wireless Communications System (ISWCS'14)*, Barcelona (2014)
18. M. Taranetz, System level modeling and evaluation of heterogeneous cellular networks, Ph.D. dissertation, Institute of Telecommunications, TU Wien, 2015, http://theses.eurasip.org/theses/611/system-level-modeling-and-evaluation-of/
19. R. Cowan, Objects arranged randomly in space: an accessible theory. Adv. Appl. Probab. **21**(3), 543–569 (1989)
20. S. Chiu, D. Stoyan, W. Kendall, J. Mecke, *Stochastic Geometry and its Applications*, Wiley Series in Probability and Statistics (Wiley, Chichester, 2013)
21. F. Baccelli, B. Blaszczyszyn, *Stochastic Geometry and Wireless Networks: Volume I Theory*, Foundation and Trends in Networking (Now Publishers, Hanover, 2009). doi:10.1561/1300000006
22. B. Hanci, I. Cavdar, Mobile radio propagation measurements and tuning the path loss model in urban areas at GSM-900 band in Istanbul-Turkey. IEEE Veh. Technol. Conf. (VTC) **1**, 139–143 (2004). doi:10.1109/VETECF.2004.1399948
23. R.W. Heath Jr., M. Kountouris, T. Bai, Modeling heterogeneous network interference using Poisson point processes. IEEE Trans. Signal Process. **61**(16), 4114–4126 (2013). doi:10.1109/TSP.2013.2262679
24. G. Boudreau, J. Panicker, N. Guo, R. Chang, N. Wang, S. Vrzic, Interference coordination and cancellation for 4G networks. IEEE Commun. Mag. **47**(4), 74–81 (2009)
25. 3rd Generation Partnership Project (3GPP), Evolved Universal Terrestrial Radio Access (E-UTRA) physical channels and modulation, 3rd Generation Partnership Project (3GPP), TS 36.211 (2015)
26. M. Taranetz, J.C. Ikuno, M. Rupp, Sensitivity of OFDMA-based macrocellular LTE networks to femtocell deployment density and isolation, in *International Symposium on Wireless Communications Systems*, Ilmenau (2013)
27. M. Taranetz, M. Rupp, Performance of femtocell access point deployments in user hot-spot scenarios, in *Australasian Telecommunication Networks and Applications Conference*, Brisbane (2012)
28. T. Bai, R.W. Heath Jr., Coverage analysis for millimeter wave cellular networks with blockage effects, in *IEEE Global Conference on Signal and Information Processing (GlobalSIP)*, Austin (2013), pp. 727–730. doi:10.1109/GlobalSIP.2013.6736994
29. M. Di Renzo, Stochastic geometry modeling and analysis of multi-tier millimeter wave cellular networks, CoRR (2014). arXiv:1410.3577
30. R. Gahleitner, E. Bonek, Radio wave penetration into urban buildings in small cells and microcells. IEEE Veh. Technol. Conf. (VTC) **2**, 887–891 (1994). doi:10.1109/VETEC.1994.345218. Stockholm
31. Y. Nagata, Y. Furuya, E. Moriyama, M. Mizuno, I. Kamiya, S. Hattori, Measurement and modeling of 2 GHz-band out-of-sight radio propagation characteristics under microcellular environments, in *IEEE International Symposium on Personal, Indoor and Mobile Radio Communications (PIMRC)*, Sydney (1991), pp. 341–346. doi:10.1109/PIMRC.1991.571513
32. T. Schwengler, M. Glbert, Propagation models at 5.8 GHz - path loss and building penetration, in *IEEE Radio and Wireless Conference (RAWCON)*, Denver (2000), pp. 119–124. doi:10.1109/RAWCON.2000.881870
33. N. Papadakis, A. Kanatas, P. Constantinou, Microcellular propagation measurements and simulation at 1.8 GHz in urban radio environment. IEEE Trans. Veh. Technol. **47**(3), 1012–1026 (1998). doi:10.1109/25.704856
34. J. Andersen, T. Rappaport, S. Yoshida, Propagation measurements and models for wireless communications channels. IEEE Commun. Mag. **33**(1), 42–49 (1995). doi:10.1109/35.339880
35. F. Kakar, K. Sani, F. Elahi, Essential factors influencing building penetration loss, in *IEEE International Conference on Communication Technology (ICCT)*, Hangzhou (2008). doi:10.1109/ICCT.2008.4716083

36. J.-E. Berg, Building penetration loss at 1700 MHz along line of sight street microcells, in *IEEE International Symposium on Personal, Indoor and Mobile Radio Communications (PIMRC)*, Boston (1992), pp. 86–87. doi:10.1109/PIMRC.1992.279958
37. A. De Toledo, A.M. Turkmani, Propagation into and within buildings at 900, 1800 and 2300 MHz. IEEE Veh. Technol. Conf. (VTC) **2**, 633–636 (1992). doi:10.1109/VETEC.1992.245316. Denver
38. H. Masui, T. Kobayashi, M. Akaike, Microwave path-loss modeling in urban line-of-sight environments. IEEE J. Sel. Areas Commun. **20**(6), 1151–1155 (2002). doi:10.1109/JSAC.2002.801215
39. 3rd Generation Partnership Project (3GPP), Evolved Universal Terrestrial Radio Access (E-UTRA); Further advancements for E-UTRA physical layer aspects, 3rd Generation Partnership Project (3GPP), TR 36.814 (2010)
40. T. Rappaport, *Wireless Communications: Principles and Practice*, 2nd edn. (Prentice Hall PTR, Upper Saddle River, 2001)
41. X. Zhao, J. Kivinen, P. Vainikainen, K. Skog, Propagation characteristics for wideband outdoor mobile communications at 5.3 GHz. IEEE J. Sel. Areas Commun. **20**(3), 507–514 (2002). doi:10.1109/49.995509

Chapter 14
Simulation of Two-Tier Heterogeneous Cellular Networks

Markus Rupp, Stefan Schwarz and Martin Taranetz

In this section, two-tier heterogeneous cellular networks are investigated, encompassing eNodeBs on macro sites and femtocell Base Stations (BSs), respectively.

Although numerous system level simulation campaigns have been carried out, the utilized system models such as the dual-stripe- and the 5×5 approach from [1] or other customized setups such as [2–7] are mostly too specific to *systematically* investigate the impact of a *femtocell enhancement* on the existing macro cellular deployment. On the other is commonly evaluated in terms of capacity and cannot directly be transfered to achievable throughput due to highly idealistic setups.

This section introduces a system model, which enables to analyze the impact of User Equipment (UE) distribution, femtocell deployment density and femtocell isolation on the performance of a two-tier heterogeneous cellular network in a *systematic* manner. Moreover, it allows to discuss results from Chap. 13. In contrast to all previous considerations in Chaps. 11–13, the focus of this chapter is on *network-wide* performance metrics, i.e., a *global-* rather than a *user-centric* point of view.

14.1 System Model

The macro cellular setup comprises a central site and one tier of hexagonally arranged neighbors, as illustrated in Fig. 14.1. Each site employs three eNodeBs, which are equipped with one sector antenna each. The antennas are arranged at a spacing of

M. Rupp (✉) · S. Schwarz · M. Taranetz
Institute of Telecommunications, TU Wien, Vienna, Austria
e-mail: mrupp@nt.tuwien.ac.at

S. Schwarz
e-mail: sschwarz@nt.tuwien.ac.at

M. Taranetz
e-mail: martin.taranetz@tuwien.ac.at

M. Rupp et al., *The Vienna LTE-Advanced Simulators*,
Signals and Communication Technology, DOI 10.1007/978-981-10-0617-3_14

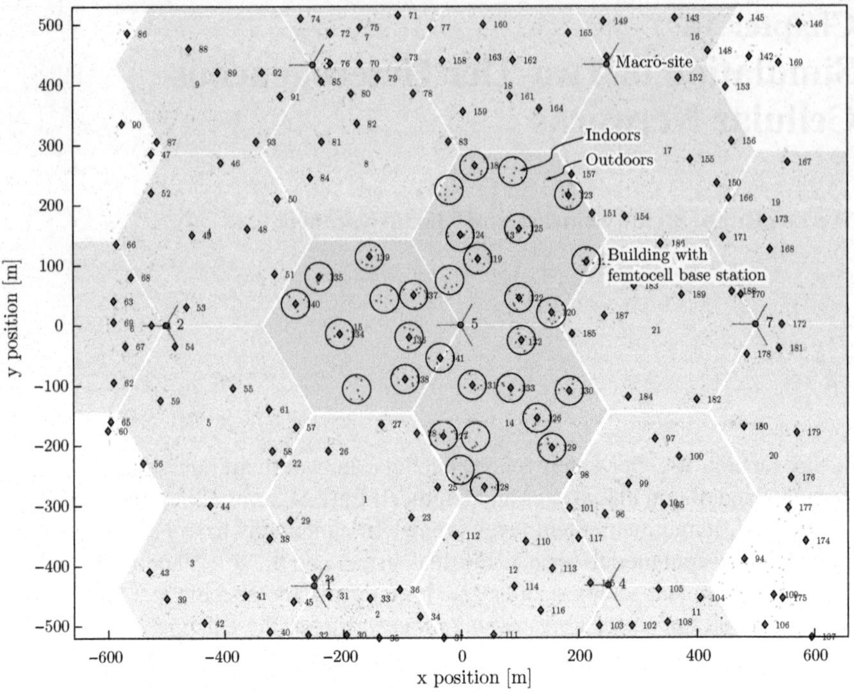

Fig. 14.1 Urban two-tier heterogeneous cellular network. UEs are located within annular regions around the centers of the randomly distributed buildings. Buildings are served by a femtocell BS with occupation ratio $\eta = 0.8$. Indoor- and outdoor environment are separated by wall penetration loss L_{W}

$2\pi/3$. Their radiation pattern is referred from [8] and is specified as

$$A(\theta) = -\min\left[12\left(\frac{\theta}{\theta_{3\mathrm{dB}}}\right)^2, A_m\right], \quad -\pi \leq \theta \leq \pi, \tag{14.1}$$

where $\theta_{3\mathrm{dB}} = \frac{65}{180}\pi$ and $A_m = 20\,\mathrm{dB}$. Applying maximum-Signal to Interference and Noise Ratio (SINR)-based UE association, such setup yields hexagonally shaped eNodeB coverage-regions, as illustrated in Fig. 14.1. Hereinafter, these regions are referred to as *macro sectors*.

After establishing the macro deployment, N_{C} circularly-shaped buildings of radius $R_{\mathrm{I}} = 25\,\mathrm{m}$ are uniformly distributed within each macro sector such that their footprints *do not overlap* each other. Along the lines of Sect. 13.2, a point is denoted as *indoors*, if it is covered by a building, and *outdoors* otherwise.

Each building hosts N_{U} UEs, which are uniformly distributed within an annular region of inner radius 5 m and outer radius 25 m around the center of the building. This procedure is equivalent to generating *UE hot-spots* according to an independent cluster process (see, e.g., [9] for nomenclature and further details on cluster processes). In this case, the parent process is constituted by the distribution of the building centers. A *UE-cluster* is formed by the UEs of a given building. In total, there are $N_S = N_C N_U$ UEs in each macro sector, hereafter denoted as *sector UEs*. By keeping N_S constant, the parameters N_C and N_U adjust the *degree of clustering*, also referred to as *extent of clustering* or *level of inhomogeneity* [10].

Femtocell BSs are deployed at the centers of buildings and equipped with omnidirectional antennas. Their occupation ratio η is tuned by N_F/N_C, with $0 \leq N_F \leq N_C$ (cf. Sect. 13.2.2). The parameter N_F denotes the number of randomly chosen buildings per macro sector, which are equipped with a femtocell BS. Figure 14.1 shows a snapshot with $N_F = 8$ and $N_C = 10$, respectively. Distributing the UEs within an annular region around the building centers guarantees a certain minimum distance to the femtocell BSs.

Both Open Subscriber Group (OSG)- and Closed Subscriber Group (CSG) mode are considered. In the OSG case, a femtocell BS serves all UEs within its coverage area (UE association regions are calculated according to a maximum SINR criterion), whereas in the CSG mode, only the UEs of the corresponding cluster are allowed to attach. UEs associated with an eNodeB from a macro site are denoted as *macro UEs*, UEs attached to a femtocell are referred to as *femto UEs*, respectively. Macro- and femtocell tiers are assumed to be uncoordinated and employ universal frequency reuse, i.e., reuse-1, thus representing a worst-case scenario in terms of interference.

Signal propagation in and out of a building is modeled by a constant wall penetration loss L_W. Depending on whether a signal originates from an eNodeB on a macro site or a femtocell BS, one of the following two path loss models is applied:

- *eNodeB on macro site*: The path loss model is referred from [8, subclause 4.5.2] and depicted as dashed line in Fig. 14.2.
- *Femtocell BS*: A dual-slope model is applied (solid line in Fig. 14.2). Within the building, the propagation model as specified in [11, subclause 5.2] is employed.

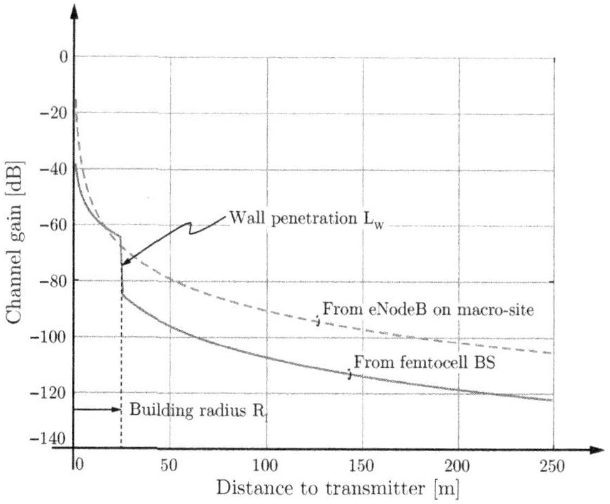

Fig. 14.2 Distance dependent channel gain from eNodeB on macro site and femtocell BS

Table 14.1 Specific simulation parameters for two-tier heterogeneous cellular network

Parameter	Value
eNodeB deployment	Three eNodeBs per macro site, $2\pi/3$ spacing
eNodeB antenna gain	$A(\theta) = -\min\left[12\left(\frac{\theta}{\theta_{3\,\mathrm{dB}}}\right)^2, A_m\right], \quad -\pi \leq \theta \leq \pi$
Femtocell BS Transmit power	$P_S = 26\,\mathrm{dBm}$
Femtocell backhaul	Unlimited, no delay
Femtocell access mode	{OSG, CSG}
Femtocell antenna gain	0 dB omni-directional

At distance R_I the signal is attenuated by the wall penetration loss L_W. For distances larger than R_I, again the propagation loss model from [8, subclause 4.5.2] is utilized.

Small scale fading and shadowing are modeled by time correlated Rayleigh- and spatially-correlated log-normal Random Variables (RVs), respectively. The simulation parameters are summarized in Tables 9.6 and 14.1, respectively.

The introduced system model serves as a basis for the subsequent investigations on altering UE distribution, building characteristics and urban environment. The presented simulation results are obtained by taking into account the three sectors of the central macro site and averaging over 100 scenario snapshots and 100 Transmission Time Intervals (TTIs) per snapshot, respectively. *Network-wide* performance is represented by *sector-wise* metrics.

14.2 Urban Two-Tier Heterogeneous Cellular Network

In this section, the performance of a typical indoor-UE in an urban two-tier heterogeneous cellular LTE network is evaluated. The target is to verify results from Chap. 13. Simulations are carried out with the system model from Sect. 14.1, which, however, differs from the model in Sect. 13.2 in two major aspects:

- Macro sites are arranged according to a hexagonal grid, as depicted in Fig. 14.1. This setup guarantees a certain macro sector size that is necessary to carry out systematic simulations, as detailed in Sect. 14.1.
- The characteristics of the urban topology, such as building density, are incorporated into the standard deviation of the log-normal RVs which represent the shadowing. Typical values in literature range from 6 to 10 dB [12–19]. The rationale for this model is to emphasize the difference to the exponential law in Eq. (13.5).

In accordance with Chap. 13, performance is evaluated for two wall penetration losses $L_W = \{-10, -30\}$ dB, and two femtocell occupation ratios $\eta = \{0.2, 0.8\}$, respectively. Figure 14.3 shows the average spectral efficiency of a *typical indoor UE* plotted over the shadow fading standard deviation. It is observed that

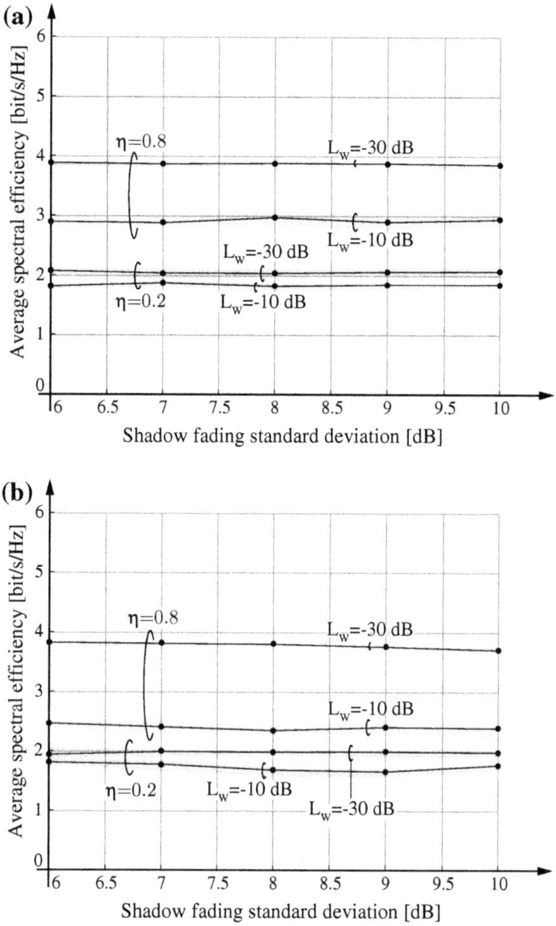

Fig. 14.3 Average spectral efficiency [bit/s/Hz] of typical indoor UE. Results are provided for femtocells operating in OSG- and CSG mode, respectively. Performance is evaluated for two wall penetration losses $L_W = \{-10, -30\}$ dB, and two femtocell occupation ratios $\eta = \{0.2, 0.8\}$, respectively. The *shaded regions* denote 95 % confidence intervals. **a** Femotcell OSG-mode. **b** Femotcell CSG-mode

- The standard deviation of the shadow fading has almost no effect on the performance. Thus, the impact of the urban environment topology is completely overlooked with the log-normal model. Moreover, the model neglects the differentiation between Line-Of-Sight (LOS)- and Non Line of Sight (NLOS) BSs, as investigated in Sect. 13.5. The importance of the latter is manifested through its inclusion into Rel. 12 of the Long Term Evolution (LTE) standard [20].
- The performance in OSG- and CSG-mode considerable deviate at high femtocell occupation ratio ($\eta = 0.8$) and low wall penetration ($L_W = -10$ dB). This is caused by the fact that, in the OSG case, UEs in a building without femtocell-BS can associate with a femtocell in a close-by building. In contrast, CSG operation forces these UEs to associate with the eNodeB on the macro site. Moreover, they will receive severe interference from the nearby femtocell-BS. This aspect is further investigated in Sect. 14.4 as it is not considered in the model of Chap. 14.

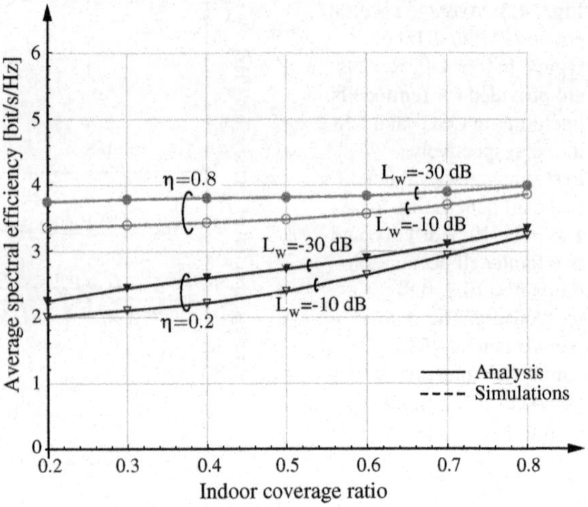

Fig. 14.4 Spectral efficiency [bit/s/Hz] of typical indoor user versus area ratio that is covered by buildings. Curves are obtained by numerical evaluating the theoretical model from Chap. 13 and applying the calibration parameters $\alpha_B = 0.85$ and $\alpha_{SIR} = 0.5$. *Curves* are depicted for varying small cell occupation probability η and wall penetration loss L_W

For comparison, a numerical evaluation of Eq. (13.17) with $P_c(\delta|r)$ from Eq. (13.16) is carried out, assuming that the users are uniformly distributed within an annulus of inner radius 5 m and outer radius $R_I = 25$ m, respectively. Since the simulator does not distinguish LOS- and NLOS BSs, it is refrained from employing the more elaborated model from Sect. 13.5. The curves are computed with the settings $P_S/P_M = 10^{-2}$ and $\mu_M = 4.61 \times 10^{-6}\,\text{m}^{-2}$ (according to the BS density in a hexagonal grid with an inter-site distance of 500 m), and the calibration parameters $\alpha_B = 0.85$ and $\alpha_{SIR} = 0.5$ from Sect. 12.5. According to Sect. 14.1, outdoor- and indoor path loss are specified as $\ell_O(R) = \min(10^{-7}, 1/10^{1.597}\,R^{-3.76})$ and $\ell_I(r) = \min(10^{-4.5}, 1/10^{3.846}\,r^{-2})$, respectively. Figure 14.4 shows the results.

It is observed that, in agreement with Sect. 12.5, the theoretical model tends to overestimate the performance due to the conservative calibration against the ideal system performance. The deviation is particularly pronounced at high indoor area coverage, where the building blockage provides a safeguard against interference. It is further remarkable that, in contrast to the simulation results, the analytical curves show a similar performance improvement when altering the wall penetration from $L_W = -10$ dB to $L_W = -30$ dB for *both* $\eta = 0.2$ and $\eta = 0.8$. This is caused by the fact that the model only takes into account interference from *neighboring* femtocells.

In conclusion, the model from Chap. 13 enables more subtle insights on the effects of building blockages than log-normally distributed shadowing, including the safeguard against interference as well as the decreasing impact of LOS BSs, which were not included in the simulations due to the novelty in the 3rd Generation Partnership Project (3GPP) standard [20]. On the other hand, it misses aspects of OSG and CSG-operation. The subsequent sections complete the picture by systematically evaluating *network-wide* performance. In the remainder of this chapter, shadowing is modeled by spatially correlated log-normal RVs with 8 dB standard deviation.

14.3 User Hot Spot Scenarios

In this section, the impact of *UE clustering* on the *global* performance of femtocell-enhanced macro cellular networks is investigated. Current research mainly focuses on the positioning of the transmitters while users are commonly considered uniformly distributed [21–28]. However, femtocell BSs are most effectively deployed at user hot-spots [29, 30]. Hence, appropriate models for the user distribution are essential to investigate the *performance limits* of a heterogeneous network. Based on [31], the contributions of this section are:

- A system model is presented, which enables to explicitly identify the effects of altering the *degree of UE clustering*.
- The importance of a fairness metric is stressed, as it is often disregarded in literature. By means of sum throughput, it is shown that a network-wide performance metric provides only limited view on the UE performance, since it conceals the distribution of the individual values.

In this section, the system model from Sect. 14.1 is exploited without separating indoor- and outdoor environment, i.e., $L_W = 0\,\text{dB}$. For a fair comparison of different UE distributions, the total amount of UEs per sector, N_S, is kept constant over all simulations. The degree of clustering is tuned by the parameters N_C and N_U, i.e., the number of UE clusters per sector and the number of UEs per cluster, such that $N_S = N_C N_U$.

The parameter settings for simulations are summarized in Tables 9.6 and 14.1, respectively. *Sector-wise sum throughput* is employed as a *global* performance metric. It is calculated by accumulating the throughput values of all N_S sector UEs, i.e., including macro- and femto UEs. Such metric is of particular interest for network providers when planning a femtocell roll-out.

The results are depicted in Fig. 14.5. Three scenarios are investigated, ranging from a low- to a high degree of UE clustering. The corresponding $\{N_C, N_U\}$-tuples

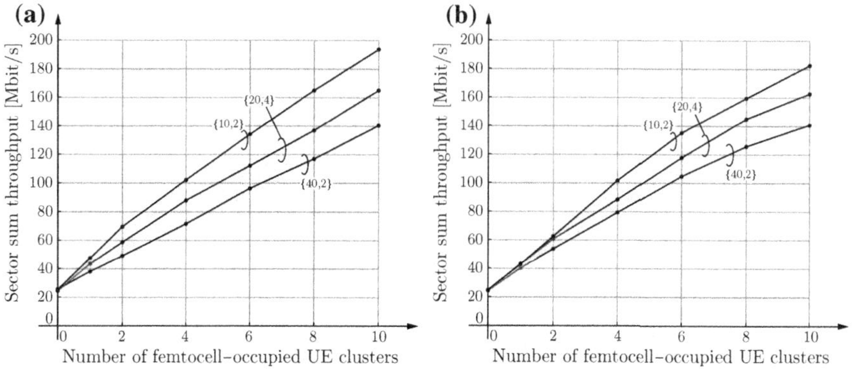

Fig. 14.5 Sector sum throughput [Mbit/s] over number of femtocell-occupied UE clusters. **a** Femotcell OSG-mode. **b** Femotcell CSG-mode

are specified as {40, 2}, {20, 4} and {10, 8}, respectively. It is observed that in all three cases the sum throughput strictly increases with additional femtocells, thus confirming the claim in [28]. The results however reveal that the efficiency of the femtocell operation considerably depends on the degree of clustering. The sum throughput increases steepest in a UE hot-spot scenario, and lowest in a close to uniform UE distribution. Remarkably, the curves show a slight saturation effect when increasing the number of femtocell BSs. This indicates that the femtocell deployment density does not perfectly compensate for the additional interference.

Sum throughput is suitable to measure the global performance of a femtocell-enhanced network. However, it conceals possible performance imbalances between the individual UEs. Therefore, measures for the distribution of the throughput values are imperative. In this section, Jain's fairness index is employed. For a given macro sector, it is expressed as

$$\mathrm{JFI}(\mathbf{t}) = \frac{\left(\sum_{i=1}^{N_\mathrm{S}} t_i\right)^2}{N_\mathrm{S} \sum_{i=1}^{N_\mathrm{S}} t_i^2}, \tag{14.2}$$

where t_i denotes the throughput as achieved by UE i.

Figure 14.6 depicts the fairness index plotted over the number of femtocell occupied UE clusters per macro sector. In a sparse femtocell deployment (i.e., low number of femtocell BSs), only few UEs achieve high throughput due to their vicinity to the femtocell BSs. The remaining UEs are attached to the eNodeB on the macro site, experiencing additional interference from the femtocells. Thus, low fairness is observed. The index increases with the number of employed femtocell BSs. At full femtocell occupation ($\eta = 1$), it lies below the reference case of a plain macro cellular network without femtocells ($\eta = 0$) for both OSG- and CSG mode. In accordance with the sum-throughput results, highest fairness is achieved at the largest extent of UE clustering.

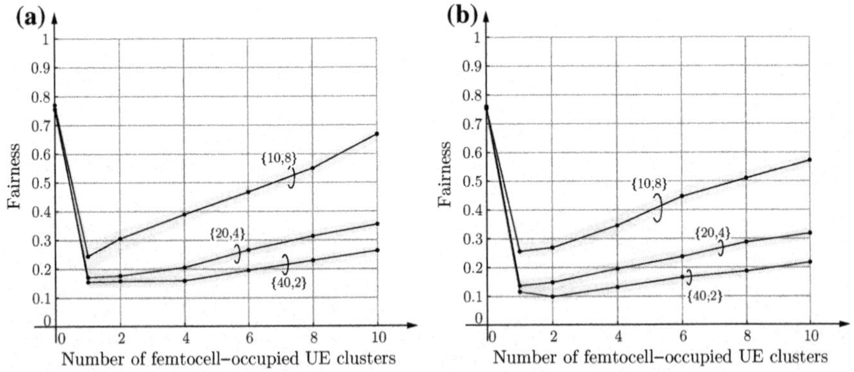

Fig. 14.6 Jain's fairness index over number of femtocell-occupied UE clusters. **a** Femotcell OSG-mode. **b** Femotcell CSG-mode

Hence, femtocells are most efficiently deployed in scenarios with a high degree of UE clustering. Motivated by the low fairness values, macro- and femto UE performance are evaluated separately in the next section. The particular focus is on the effect of isolating UEs in indoor areas from the outdoor environment by wall penetration loss.

14.4 Sensitivity on Femtocell Deployment Density and -Isolation

In this section, the impact of femtocell deployment density and -isolation on the downlink-performance of an LTE network is investigated. *Isolation* is defined as the separation between indoor- and outdoor environment by wall penetration loss.

Referring to my work in [32], the contributions of this section are:

- A system model is introduced, which allows to explicitly analyze the effects of varying femtocell density and -isolation.
- The individual UE performance, which would be concealed by network-wide performance metrics (cf. Sect. 14.3), is assessed by separately investigating macro- and femto UEs.

The system model which is largely based on the setup in Sect. 14.1. *High-* and *no-isolation* scenarios are investigated, corresponding to $L_W = -20$ dB and $L_W = 0$ dB, i.e., the worst case in terms of interference, respectively. According to results in Sect. 14.3, efficient balancing of UE throughput is only possible in scenarios with a high degree of UE clustering. Therefore, $N_C = 10$ and $N_U = 8$ in the remainder of this section.

The first metric of interest is the wideband SINR. It is defined as the ratio of the average receive power from the serving cell and the average aggregate interference from other cells plus noise [33]. Figure 14.7 depicts the corresponding Empirical Cumulative Distribution Function (ECDF) curves for OSG- and CSG mode, considering both high- and no isolation of the indoor areas. The curves are computed from the average wideband-SINR values of the N_S sector UEs. Various femtocell-occupation ratios $\eta = \{0, 0.2, 0.4, 0.8, 1\}$ are evaluated. Arrows indicate the directions of increasing η from 0 to 1. The case $\eta = 0$ serves as a baseline, representing a macro cellular network without femtocells.

The major observations are

- High- and no-isolation scenarios exhibit significantly different characteristics. On the one hand, for $L_W = -20$ dB the SINR almost consistently improves with increasing η. On the other hand, for $L_W = 0$ dB it hardly deviates from the baseline in the OSG case, while deteriorating in the CSG scenarios. These results indicate the system's high sensitivity to fluctuations of the femtocell isolation.
- The step-like behavior of the curves, which is particularly pronounced in OSG scenarios at high isolation, indicates a severe imbalance between UEs with

Fig. 14.7 SINR ECDFs for various femtocell-occupation ratios $\eta = \{0, 0.2, 0.4, 0.6, 0.8, 1\}$ and wall penetration losses $L_W = \{0, -20\}$ dB, respectively. *Arrows* denote the direction of increasing η. The case $\eta = 0$ serves as a baseline, corresponding to a macro cellular network without femtocells. **a** Femotcell OSG-mode. **b** Femotcell CSG-mode

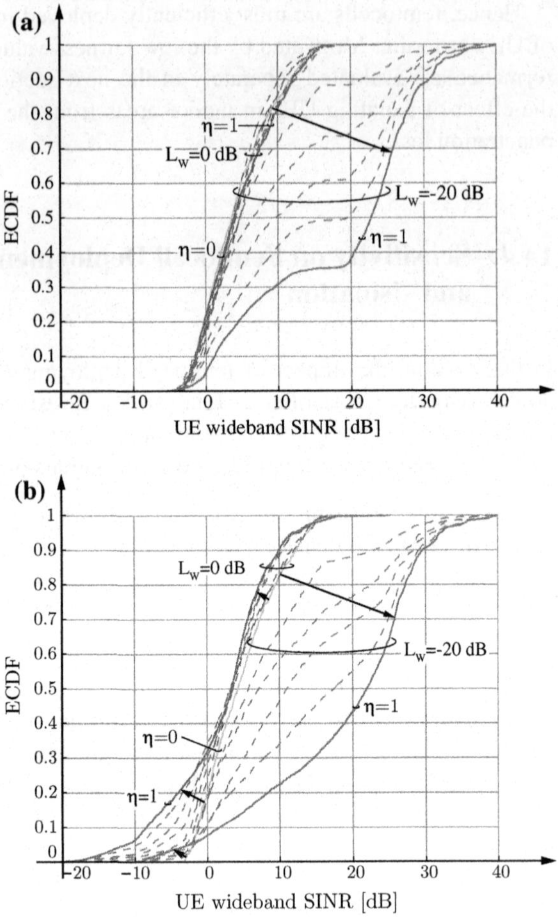

good- and UEs with bad channel conditions. Figure 14.8 depicts a snapshot of individual average-UE-throughput values over SINR for $L_W = -20$ dB, $\eta = 0.6$ and OSG mode. It allows to unambiguously identify the two groups as macro- and femto UEs, thus motivating their *separate* investigation.

Figure 14.9 shows throughput values as achieved by the *typical* macro-, femto- and sector UE, respectively. The results are obtained by averaging over the individual throughput values of the corresponding UE class. The axis of abscissas shows the femtocell occupation ratio η. It is found that

- The throughput of the typical sector UE monotonically increases with larger η and achieves its maximum at full femtocell occupation, i.e., $\eta = 1$.
- The performance of a typical femto UE is considerably higher for high isolation ($L_W = -20$ dB) than for no isolation ($L_W = 0$ dB). The latter might serve as a

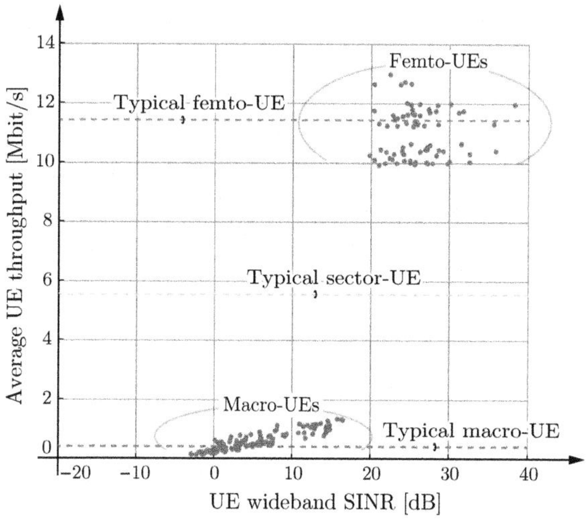

Fig. 14.8 Snapshot of individual average-UE-throughput values [Mbit/s] over wideband SINR [dB] for wall penetration loss $L_W = -20$ dB, femtocell occupation ratio $\eta = 0.6$ and OSG mode. The *dashed lines* denote throughput values of the *typical* macro-, femto- and sector UE as referred from Fig. 9a, respectively

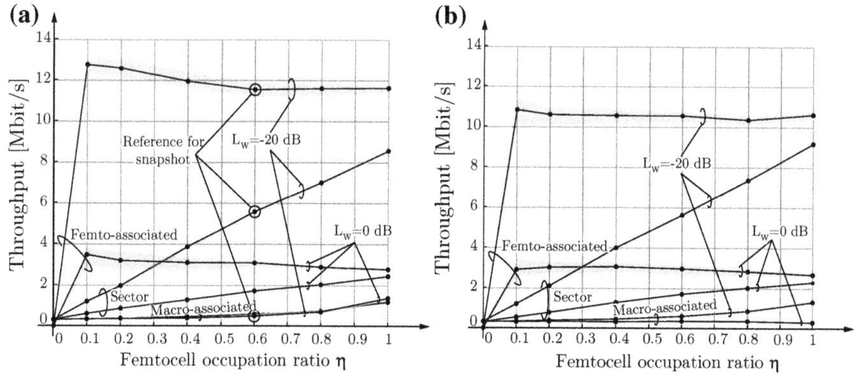

Fig. 14.9 Throughput of typical macro-, femto- and sector UE [Mbit/s] over femtocell occupation ratio η for $L_W = \{0, -20\}$ dB. *Shaded regions* denote 95 % confidence intervals. The *circled points* in (**a**) corresponds to the *straight dashed lines* in Fig. 14.8. **a** Femotcell OSG-mode. **b** Femtocell CSG-mode

warning scenario for network providers, considering open doors and windows throughout the whole scenario.

- In agreement with the results in Sect. 14.3 and the observations in the current section, the throughput of the typical sector UE considerably deviates from the typical macro- and femto-UE performance. This becomes particularly clear in OSG scenarios with $L_W = 0$ dB, where the throughput of the typical sector UE is enhanced while the performance of the typical macro UE deteriorates.

- Except for the aforementioned case, the throughput of the typical macro UE generally improves for increasing η. This is caused by the fact that handing off a growing amount of macro UEs to the femtocells compensates for the harm of additional interference.

Mapping the circled throughput values in Fig. 14.9a to Fig. 14.8 (straight dashed lines) substantiates the observation that the *individual* UE either performs much better or much worse than the *typical* sector UE. On the other hand, typical macro- and femto UE throughputs provide reasonably accurate indicators for the actual performances.

Hence, these two metrics bridge the gap between individual- and sector-wise performance. Figure 14.10 depicts typical femto- versus typical macro UE throughput. Each point represents a certain femtocell occupation ratio η, with lines connecting successive values. This depiction is conceptually similar to a *capacity*- or *throughput trade-off region* with the competing axes being typical macro- and typical femto UE throughput [34, 35], respectively. The following observations are equivalently obtained from Fig. 14.9, but become much clearer from Fig. 14.10:

- The typical femto UE throughput is almost constant for $\eta > 0$. High- and no-isolation scenario differ by a factor of about 3.8 in OSG- and about 3.5 in CSG-mode, respectively.
- In OSG mode the average macro UE throughput grows with increasing η at $L_W = 0\,\text{dB}$, even exceeding the performance at $L_W = -20\,\text{dB}$. In the CSG case, the throughput tuples exhibit a loop, yielding the lowest throughput of the typical macro UE at full femtocell occupation. Thus, operation in CSG mode is more resilient to fluctuations of L_W.

In conclusion, the results in this section confirm the assumption from Sect. 14.3 that the imbalance among individual UE performance values mainly results from

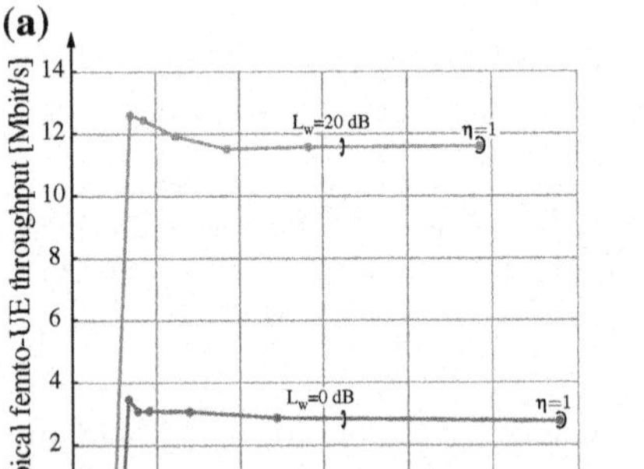

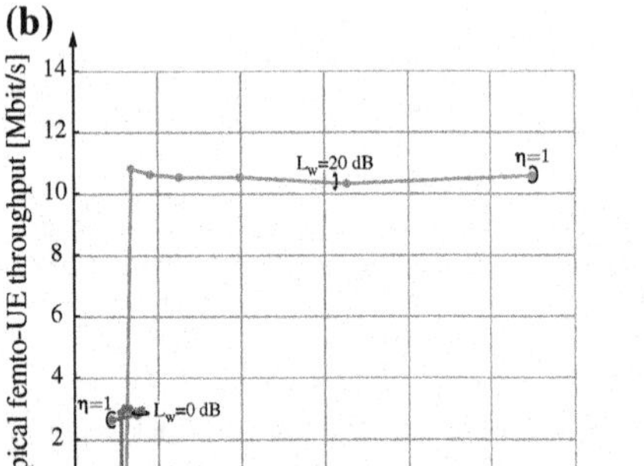

Fig. 14.10 Typical femto- versus typical macro UE throughput [Mbit/s]. Each point corresponds to a certain femtocell occupation ratio $\eta \in \{0, 0.1, 0.2, 0.4, 0.6, 0.8, 1\}$. *Lines* connect throughput tuples of successive η-values. **a** Femotcell OSG-mode. **b** Femotcell CSG-mode

the difference between macro- and femto UEs. On the other hand, performances within the corresponding UE class turned out to be relatively similar and motivated to introduce the notions of *typical macro-* and *typical femto UE*. Depicting their throughput values in a similar manner to a capacity region revealed that *femtocell isolation* mainly scales the throughput of the typical femto UEs while *femtocell density* and *femtocell access mode* predominantly affect the performance of the typical macro UE.

References

1. 3rd Generation Partnership Project (3GPP), Evolved Universal Terrestrial Radio Access (E-UTRA); Further advancements for E-UTRA physical layer aspects, in *3rd Generation Partnership Project* (3GPP), TR 36.814, March 2010
2. E. Pateromichelakis, M. Shariat, A. ul Quddus, R. Tafazolli, On the analysis of co-tier interference in femtocells, in *International Symposium Personal Indoor and Mobile Radio Communications(PIMRC)* pp. 122–126 Toronto (2011). doi:10.1109/PIMRC.2011.6139696
3. F. Capozzi, G. Piro, L. Grieco, G. Boggia, P. Camarda, On accurate simulations of LTE femtocells using an open source simulator. EURASIP J. Wirel. Commun. Netw. **1** (2012). doi:10.1186/1687-1499-2012-328
4. L.T.W. Ho, H. Claussen, Effects of user-deployed, co-channel femtocells on the call drop probability in a residential scenario, in *IEEE International Symposium Personal, Indoor and Mobile Radio Communications (PIMRC)*, September Athens (2007). doi:10.1109/PIMRC.2007.4394281
5. J. Ling, D. Chizhik, R. Valenzuela, On resource allocation in dense femto-deployments, in *IEEE International Conference Microwaves, Communications, Antennas and Electronics Systems(COMCAS)*, Tel Aviv, November (2009). doi:10.1109/COMCAS.2009.5385992
6. D. Calin, H. Claussen, H. Uzunalioglu, On femto deployment architectures and macrocell offloading benefits in joint macro-femto deployments. IEEE Commun. Mag. **48**(1), 26–32 (2010)
7. D. Lopez-Perez, A. Valcarce, G. de la Roche, E. Liu, J. Zhang, Access methods to WiMAX femtocells: a downlink system-level case study, in *IEEE International Conference Communication Systems(ICCS)*, pp. 1657–1662, Guangzhou, November (2008). doi:10.1109/ICCS.2008.4737463
8. 3rd Generation Partnership Project (3GPP), Evolved universal terrestrial radio access (E-UTRA); radio frequency (RF) system scenarios, in *3rd Generation Partnership Project (3GPP)*, TR 36.942, October 2014
9. A. Lawson, D. Denison, *Spatial Cluster Modelling* (Taylor & Francis, London, 2002)
10. U. Schilcher, M. Gyarmati, C. Bettstetter, Y.W. Chung, Y.H. Kim, Measuring inhomogeneity in spatial distributions, in *Vehicular Technology Conference(VTC Spring)*, pp. 2690–2694 Singapore, May (2008). doi:10.1109/VETECS.2008.589
11. 3rd Generation Partnership Project (3GPP), Evolved Universal Terrestrial Radio Access (E-UTRA); FDD Home eNode B (HeNB) Radio Frequency (RF) requirements analysis, in *3rd Generation Partnership Project (3GPP)*, TR 36.921, September 2014
12. S. Seidel, T. Rappaport, S. Jain, M. Lord, R. Singh, Path loss, scattering and multipath delay statistics in four european cities for digital cellular and microcellular radiotelephone. IEEE Trans. Veh. Technol. **40**(4), 721–730 (1991). doi:10.1109/25.108383
13. V. Erceg, L. Greenstein, S. Tjandra, S. Parkoff, A. Gupta, B. Kulic, A. Julius, R. Bianchi, An empirically based path loss model for wireless channels in suburban environments. IEEE J. Sel. Areas Commun. **17**(7), 1205–1211 (1999). doi:10.1109/49.778178

14. M.J. Feuerstein, K.L. Blackard, T.S. Rappaport, S.Y. Seidel, H. Xia, Path loss, delay spread, and outage models as functions of antenna height for microcellular system design. IEEE Trans. Veh. Technol. **43**(3), 487–498 (1994)
15. V. Abhayawardhana, I. Wassell, D. Crosby, M. Sellars, M. Brown, Comparison of empirical propagation path loss models for fixed wireless access systems. IEEE Veh. Technol. Conf. (VTC) **1**, 73–77 (2005). doi:10.1109/VETECS.2005.1543252
16. G. Durgin, T.S. Rappaport, H. Xu, Measurements and models for radio path loss and penetration loss in and around homes and trees at 5.85 GHz. IEEE Trans. Commun. **46**(11), 1484–1496 (1998)
17. J. Porter, I. Lisica, G. Buchwald, Wideband mobile propagation measurements at 3.7 GHz in an urban environment, in *IEEE Antennas and Propagation Society International Symposium*, vol 4 pp. 3645–3648 June (2004). doi:10.1109/APS.2004.1330136
18. T. Rautiainen, K. Kalliola, J. Juntunen, Wideband radio propagation characteristics at 5.3 GHz in suburban environments, in *IEEE International Symposium Personal, Indoor, and Mobile Radio Communications(PIMRC)*, September 2005
19. T. Schwengler, M. Glbert, Propagation models at 5.8 GHz - path loss and building penetration, in *IEEE Radio and Wireless Conference(RAWCON)*, pp. 119–124, Denver, September (2000). doi:10.1109/RAWCON.2000.881870
20. 3rd Generation Partnership Project (3GPP), Study on 3D channel model for LTE, in *3rd Generation Partnership Project (3GPP)*, TR 36.873, September 2014
21. S. Singh, H. Dhillon, J. Andrews, Downlink rate distribution in multi-RAT heterogeneous networks, in *IEEEInternational Conference Communications(ICC)*, pp. 5188–5193, Budapest, June (2013). doi:10.1109/ICC.2013.6655408
22. J. Andrews, F. Baccelli, R. Ganti, A tractable approach to coverage and rate in cellular networks. IEEE Trans. Commun. **59**(11), 3122–3134 (2011). doi:10.1109/TCOMM.2011.100411.100541
23. M. Haenggi, R.K. Ganti, *Interference in Large Wireless Networks, ser. Foundations and Trends in Networking*, vol. 3 (NoW Publishers, Hanover, 2009)
24. F. Baccelli, B. Blaszczyszyn, *Stochastic Geometry and Wireless Networks: Volume I Theory, ser. Foundation and Trends in Networking*, (Now Publishers, Hanover, 2009). doi:10.1561/1300000006
25. F. Baccelli, B. Blaszczyszyn, in *Stochastic Geometry and Wireless Networks, Volume II - Applications, ser. Foundations and Trends in Networking*, vol. 2, ed. by F. Baccelli, B. Blaszczyszyn, (NoW Publishers, Hanover, 2009), doi:10.1561/1300000026
26. H.-S. Jo, Y.J. Sang, P. Xia, J. Andrews, Heterogeneous cellular networks with flexible cell association: a comprehensive downlink SINR analysis. IEEE Trans. Wireless Commun. **11**(10), 3484–3495 (2012). doi:10.1109/TWC.2012.081612.111361
27. T. Zahir, K. Arshad, Y. Ko, K. Moessner, A downlink power control scheme for interference avoidance in femtocells, in *International Wireless Communications and Mobile Computing Conference(IWCMC)*, pp. 1222–1226, Istanbul, July 2011, doi:10.1109/IWCMC.2011.5982714
28. J. Andrews, Seven ways that HetNets are a cellular paradigm shift. IEEE Commun. Mag. **51**(3), 136–144 (2013). doi:10.1109/MCOM.2013.6476878
29. V. Chandrasekhar, J. Andrews, A. Gatherer, Femtocell networks: a survey. IEEE Commun. Mag. **46**(9), 59–67 (2008). doi:10.1109/MCOM.2008.4623708
30. H.-S. Jo, P. Xia, J.G. Andrews, Open, closed, and shared access femtocells in the downlink, in *EURASIP Journal on Wireless Communications and Networking*, vol. 1, December 2012, doi:10.1186/1687-1499-2012-363
31. M. Taranetz, M. Rupp, Performance of femtocell access point deployments in user hot-spot scenarios, in *Australasian Telecommunication Networks and Applications Conference*, Brisbane, 2012
32. M. Taranetz, J.C. Ikuno, M. Rupp, Sensitivity of OFDMA-based macrocellular LTE networks to femtocell deployment density and isolation, in *International Symposium Wireless Communication Systems*, Ilmenau, 2013

33. S. Ahmadi, *LTE-Advanced: A Practical Systems Approach to Understanding 3GPP LTE Releases 10 and 11 Radio Access Technologies, ser* (Elsevier Science, Philadelphia, 2013)
34. A. Adhikary, G. Caire, On the coexistence of macrocell spatial multiplexing and cognitive femtocells, in *IEEE Int. Conf. Commun.(ICC)*, pp. 6830–6834, June 2012
35. J. Andrews, R. Ganti, M. Haenggi, N. Jindal, S. Weber, A primer on spatial modeling and analysis in wireless networks. IEEE Commun. Mag. **48**(11), 156–163 (2010)

Chapter 15
Advanced System Level Applications

Martin Müller and Illia Safiulin

System level simulations have been proven to be a very important means when the behavior of many users need to be taken into account. Typical numbers can be 20, 100 or even 1000. In the following we demonstrate the feasibility of such studies on the example of communications in trains and in cars. In train communications we typically have a single train connected to a base-station but the train, e.g., German Inter-City Express (ICE), may carry 500 users at once. Similarly, along motorways there may be 500 cars with one user each in the neighborhood of a base-station.

15.1 Modeling Train Transmissions

Martin Müller

In the current market of train services, being able to provide mobile broadband access to costumers has become a main inducement for choosing this means of transportation. Due to the ubiquitous use of the Internet and the rapid adoption of novel devices such as smart phones and tablet computers, most passengers have become accustomed to experiencing high data rates and having the service following them no matter where they go. With the number of commuters being expected to increase, the high user mobility is also one of the most emphasized scenarios in the initiative for the fifth generation of wireless communications (5G). Long Term Evolution-Advanced (LTE-A), the contemporary standard for wireless communication, is not optimized for the challenges of High Speed Train (HST) scenarios.

M. Müller (✉) · I. Safiulin
Institute of Telecommunications, TU Wien, Vienna, Austria
e-mail: mmueller@nt.tuwien.ac.at

I. Safiulin
e-mail: illia.safiulin@tuwien.ac.at

M. Rupp et al., *The Vienna LTE-Advanced Simulators*,
Signals and Communication Technology, DOI 10.1007/978-981-10-0617-3_15

Hence, many train operators—mostly in collaboration with mobile operators—have increased efforts to satisfy the ever increasing requirements. There also exist international collaborations on a broad level to push for higher data rates and shorter latencies, for example the Shift2Rail initiative by the European Union [1].

Wireless communications in HST scenarios is confronted with unique conditions, which have a considerable impact on network planning. In particular, the scenarios are characterized by User Equipments (UEs) being densely packed inside of the train and moving at high speed, as well as the specific propagation effects in a diversity of different environments. More particularities are described in [2].

Employing dedicated hardware in the vicinity of the tracks allows to optimize the positioning of the stationary infrastructure. One promising approach for that matter is the utilization of Remote Radio Units (RRUs). Thus, also due to the smaller dimensions, the point of radiation can be brought closer to the tracks. Such distributed radiation points can be efficiently simulated with the Vienna LTE-A Dowlink System Level Simulator, as explained in Sect. 10.5. An additional advantage of RRUs is the possibility of collaboration among them, leading to a potential improvement of the overall system performance. The first part of this section deals with the description of the comparison of such collaboration schemes and is based on [3]. We introduce a theoretical model and compare it to the results obtained from simulations.

There exist two main approaches, when communication in HST scenarios is considered. Most publications on this topic assume a relay-based approach, assuming additional hardware installed on the train, i.e., a relay that communicates with the Evolved Base Station (eNodeB) as well as with UEs without them communicating directly [4, 5]. However, the direct communication between UEs and Base Stations (BSs) has its own advantages, but has not been given enough attention in literature. We discuss a comparison between the relay approach and the less studied direct link approach in the second part of this section. It is based on [6]. Next to a general comparison we also provide simulation results to compare the performance of both approaches.

15.1.1 Performance of Remote Unit Collaboration Schemes

15.1.1.1 System Model

First we introduce the system model. We consider a representative segment of an RRU deployment along a railroad track, which comprises four equidistantly spaced sites, each equipped with two RRUs, pointing in opposite directions, as shown in Fig. 15.1. The train is assumed to move between RRU_0 and RRU_1, and its center position is expressed relative to RRU_0 as $\delta\, d_B$, with $0< \delta <1$. The nodes RRU_{-1} and RRU_2 serve as dominant interferers. For simplicity, all other sources of interference are neglected. Due to antenna directivity, some RRUs are ignored in the analysis, as indicated by dashed arrows in Fig. 15.1.

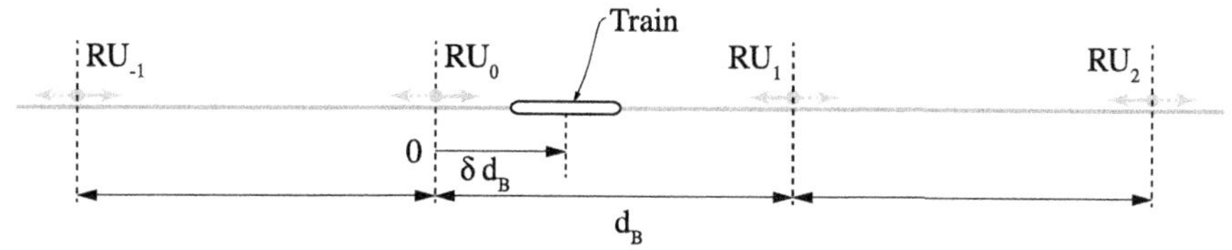

Fig. 15.1 Representative segment of a railroad track with four equidistantly spaced sites. Each site employs two RRUs pointing in opposite directions. RRUs depicted as *dashed lines* are omitted in the analysis. The UE is located at position $\delta\, d_B$ relative to RRU_0

There are several ways to associate eNodeBs with the RRUs. In this work, we investigate three schemes:

1. **Baseline**. All RRUs employ different cell IDs, i.e., are attached to different eNodeBs. Then,

$$\begin{aligned} \mathcal{S} &= \{\mathrm{RRU}_0\};\ \mathcal{I} = \{\mathrm{RRU}_{-1}, \mathrm{RRU}_1, \mathrm{RRU}_2\},\ 0 < \delta < 0.5 \\ \mathcal{S} &= \{\mathrm{RRU}_1\};\ \mathcal{I} = \{\mathrm{RRU}_{-1}, \mathrm{RRU}_0, \mathrm{RRU}_2\},\ 0.5 < \delta < 1. \end{aligned} \tag{15.1}$$

2. **Coordination**. RRU_0 and RRU_1 are attached to different eNodeBs and coordinate their transmission such that they do not interfere each other. Then,

$$\begin{aligned} \mathcal{S} &= \{\mathrm{RRU}_0\};\ \mathcal{I} = \{\mathrm{RRU}_{-1}, \mathrm{RRU}_2\},\ 0 < \delta < 0.5 \\ \mathcal{S} &= \{\mathrm{RRU}_1\};\ \mathcal{I} = \{\mathrm{RRU}_{-1}, \mathrm{RRU}_2\},\ 0.5 < \delta < 1. \end{aligned} \tag{15.2}$$

3. **Cooperation**. RRU_0 and RRU_1 are associated with the same eNodeB and coordinate their transmission such that both can be exploited as useful signal. Then,

$$\begin{aligned} \mathcal{S} &= \{\mathrm{RRU}_0, \mathrm{RRU}_1\};\ \mathcal{I} = \{\mathrm{RRU}_{-1}, \mathrm{RRU}_2\},\ 0 < \delta < 0.5 \\ \mathcal{S} &= \{\mathrm{RRU}_0, \mathrm{RRU}_1\};\ \mathcal{I} = \{\mathrm{RRU}_{-1}, \mathrm{RRU}_2\},\ 0.5 < \delta < 1. \end{aligned} \tag{15.3}$$

To make a fair comparison, it is assumed that the transmit power per eNodeB is limited to P_T, i.e.,

$$\sum_{\{i|\mathrm{RRU}_i \in \mathcal{S}\}} P_i^{\mathrm{Tx}} \leq P_{\mathrm{T}}. \tag{15.4}$$

In order to clarify the impact of the RRU deployment on the performance, no sophisticated power allocation techniques are applied. In the *cooperation* scheme, associated- and interfering RRUs transmit with half power in each direction.

15.1.1.2 Theoretical Model

For the analysis, the signal from RRU_i is assumed to experience a log-distance dependent path loss law $\ell_i(\delta)$ with exponent $\alpha = 2$ (i.e., free space propagation) and Gamma distributed small-scale fading, $G \sim \Gamma[N_{\mathrm{Tx}}, \theta]$, which corresponds to Maximum Ratio Transmission (MRT) with N_{Tx} transmit antennas and θ is the scale parameter of the fading. The relevance of the Gamma distribution is explained by the facts that it includes several important channel models such as Rayleigh and Nakagami-m as special cases and further allows to accurately approximate composite fading distributions such as Rayleigh-Lognormal [7].

The received power from RRU_i at a UE at relative position δ is expressed as

$$P_i^{\mathrm{Rx}}(\delta) = P_i^{\mathrm{Tx}}\, G_i\, \ell_i(\delta), \tag{15.5}$$

where P_i^{Tx} denotes the transmit power of RRU_i. Assuming that the penetration loss into the train carriage is constant, it can be omitted in the analysis since the scenario is considered interference limited.

Applying MRT corresponds to a best case scenario in terms of desired signal power and a worst case in terms of interference. The corresponding aggregate received powers are

$$S(\delta) = \sum_{\{i|\mathrm{RRU}_i \in \mathscr{S}\}} P_i^{\mathrm{Rx}}(\delta), \tag{15.6}$$

$$I(\delta) = \sum_{\{i|\mathrm{RRU}_i \in \mathscr{I}\}} P_i^{\mathrm{Rx}}(\delta), \tag{15.7}$$

where $\mathscr{S}$ and $\mathscr{I}$ refer to the set of desired and interfering RRUs as obtained from Eqs. (15.1)–(15.3), respectively. These expressions correspond to weighted sums of Gamma Random Variables (RVs). Sticking to integer-valued shape parameters, which correspond, e.g., to the number of antennas per RRU under the assumptions of MRT and Rayleigh fading, allows to employ the Generalized Integer Gamma (GIG) distribution [8]. It yields expressions of the form $\sum_k a_k x^{b_k} e^{c_k x}$ for both desired signal- and aggregate interference distributions, respectively. Then, the distribution of the Signal-to-Interference Ratio (SIR) $\gamma(\delta) = S(\delta)/I(\delta)$ generically formulates as (see Appendix of [3])

$$f_\gamma(\gamma; r) = \sum_s \sum_i a_s a_i \gamma^{b_s} (c_i + c_s \gamma)^{-i-b_s-b_i} \Gamma(i + b_s + b_i). \tag{15.8}$$

The spectral efficiency is defined by the well known expression $\tau(\gamma(\delta)) = \log_2(1 + \gamma(\delta))$. Its distribution is straightforwardly obtained by a non-linear, monotone transformation of RVs applied on Eq. (15.8) as

$$f_\tau(\tau; \delta) = \log(2) 2^\tau f_\gamma(2^\tau - 1; r). \tag{15.9}$$

Since the length of the train is typically in the same order of magnitude as the inter-RRU distance, it is not reasonable to represent all passengers by a single point. We apply the following procedure for performance evaluation. First, we determine the average spectral efficiency at each UE position between RRU_0 and RRU_1. Then, we assume the UEs to be uniformly distributed within the train. Intuitively, the train thus acts as a *sliding window* for the moving average over the single UEs' performances. The result is an average spectral efficiency value for each train position, also referred to as *train average spectral efficiency* in the remainder of this section.

For numerical evaluation, we consider a setup as illustrated in Fig. 15.1 with $d_B = 1000\,\text{m}$. The train has a length of 200.84 m and a width of 2.95 m, corresponding to a German "ICE 3" train. Each RRU employs two transmit antennas and each UE is equipped with one receive antenna. Results in terms of train average spectral efficiency for each RRU collaboration scheme are provided in Fig. 15.2, represented by the dashed curves and are discussed later in this section, together with the simulation results. The original results were obtained under the assumption that each RRU chooses its MRT beamformer independently. However, a LTE-A system only offers codebooks with a limited number of precoders, which typically distribute the available transmit power uniformly over all transmit antennas. As the simulator also applies LTE-A-compliant codebook based precoding, the numerical results of the theoretical model are scaled by a factor that accounts for the loss from utilizing codebook based precoding. This factor is constant over the train center position (with negligible variations).

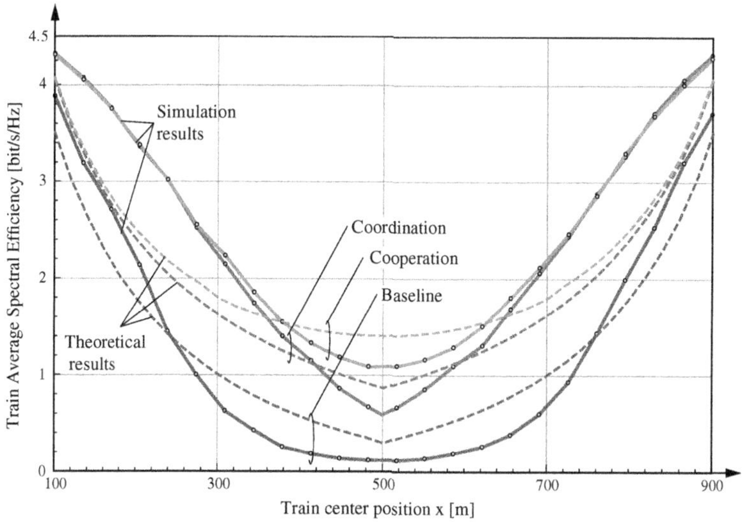

Fig. 15.2 Train average spectral efficiency in [bit/s/Hz] over train center position for baseline-, coordination- and cooperation scheme. *Dashed-* and *solid curves* refer to results of theoretical model and simulations, respectively (Confidence intervals are insignificantly small and therefore omitted in the figure.)

15.1.1.3 Simulation Setup

For simulations we consider the same ICE 3 train with the same dimensions as before and 460 seats in total. We assume 230 active UEs and employ a full-buffer traffic model. The UEs are distributed within the train according to a uniform random distribution.

RRUs dynamically assign to eNodeBs, according to the collaboration scheme. Each eNodeB has a total transmit power of $P_T = 40\,\mathrm{W}$. The signal between eNodeB and UE experiences path loss, fast fading and penetration loss. We employ the *rural* path loss model of the TR 36.942 specification [9], representing the typical propagation environment for a train traveling through the countryside. The small scale fading is modeled by an ITU 'Vehicular A' channel [10], which is designated for receivers inside moving metal objects. Although other literature on high speed train scenarios commonly employs a variation of the Wireless World Initiative New Radio (WINNER) Phase II model [11], it is not applicable in our case, since it was designed for moving objects with relay-units. Finally, we account for the penetration loss into the carriage by a constant of 30 dB (more on that in the following section). A summary of all simulation parameters is provided in Table 15.1.

15.1.1.4 Discussion of Results

Numerical results of the theoretical model and the simulation results are provided together in Fig. 15.2. The results of the theoretical model are displayed by solid lines, while the simulation results are represented by dashed lines, including the 95 %-confidence intervals. It is observed that the performance for each scheme respectively shows a close accordance of theoretical and simulation results. The major difference in the setup—MRT versus codebook based precoding—has been accounted for by the aforementioned correction factor. The remaining gap is caused by various practical design constraints of an LTE-A system such as Adaptive Modulation and Coding and scheduling (see [16]), which are not incorporated in the theoretical model.

Considering the three collaboration schemes individually, it is observed that the **cooperation** scheme universally yields the best performance. This finding has also been obtained in comparable studies, which assumed the trains to be equipped with relay nodes [11]. Note that this scheme completely avoids handovers in the region of interest. The **coordination** scheme shows a slightly lower performance than cooperation, particularly when approaching the middle point between two RRUs. In practical systems, coordination is typically far less complex than cooperation since only control data has to be exchanged between the eNodeBs. Even though the coordination scheme is less complex in its installation and only grants a slightly reduced performance, handovers occur more often, which might pose a problem for high speeds. Concretely, for a train, travelling at 200 km/h and an inter RRU distance of 1000 m, this would be the case every $1000\,\mathrm{m}/55.5\,\mathrm{m/s} = 18\,\mathrm{s}$. While theoretically supported

Table 15.1 Simulation parameters

Parameter	**Value**
System bandwidth	20 MHz
Carrier frequency	2.14 GHz
Inter-RRU distance	1000 m
eNodeB transmit power P_T	40 W
Antennas per RRU	2
MIMO mode	CLSM
Path loss model	TS 36.942 'rural' [9]
Channel model	ITU-R Vehicular A [10]
Train speed	200 km/h
Receiver type	Zero forcing
Noise power spectral density	−174 dBm/Hz
Penetration loss for direct communication	30 dB
Receiver noise figure	9 dB
Train length	200.84 m
Train width	2.95 m
Active UEs	230
UE distribution within train	Random uniform
Antennas per UE	1
Traffic model	Full buffer
Scheduler	Proportional fair
Channel knowledge	Perfect
Feedback	AMC: CQI, MIMO: PMI and RI
RRU backhaul connection	Radio over fiber, no delay

by the standard, such frequent handovers would cause considerable overhead on the air-interface and the backhaul. The kink in the coordination results at $\delta = 0.5$ stems from the RRU switching. The **baseline** scheme yields the worst performance even when the train is close to one of the RRUs, as RRU_0 and RRU_1 severely interfere each other. Again, the kink for the theoretical results at $\delta = 0.5$ stems from switching off RRU_0 and associating the user with RRU_1. The curve for the simulation results of the baseline scheme does not show a kink in the middle between two RRUs. This is caused by the fact that in the simulator, UEs are assigned to the eNodeB with the highest receive power. Consequently, for train center positions between 400 and 600 m, one fraction of the UEs associates to RRU_0, while the rest is assigned to RRU_1, respectively. If the UEs were forced to stay attached to one of the RRUs, the performance would even decrease in this region.

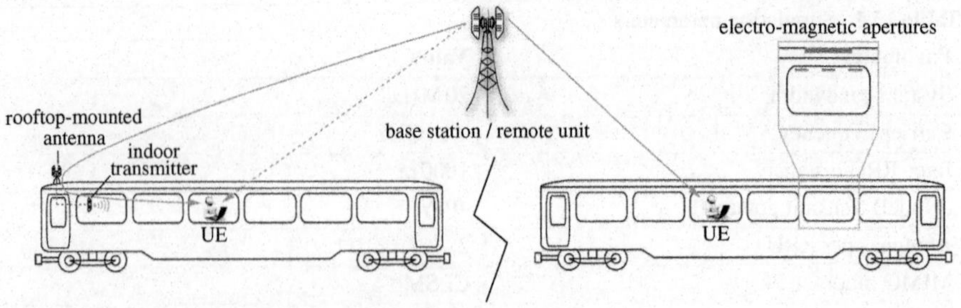

Fig. 15.3 Signal reception with relay- and direct-link approach. *Zoomed view* indicates possibilities for applying electro-magnetic apertures

15.1.2 Comparison of Relay-Based and Direct Communication

In this section we consider the downlink direction exclusively. Similar considerations are valid for the uplink direction. Generally speaking, there exist two opposing approaches to provide wireless communications to passengers of an HST. In the first case, the UE directly associates with the BSs along the tracks, while in the second case this link is established via a relay, as shown in Fig. 15.3. A comparison between these two approaches is drawn in this section and advantages and drawbacks on both sides are discussed. For evaluation purposes, we use the same system model as described in Sect. 15.1.1.

15.1.2.1 Relay Approach

In the relay scenario, one or several antennas are mounted on the outside of the train. These are connected to one or more relays which are then distributing the signal inside the train. This approach has the major advantage that the signal is not attenuated by the windows of the carriage. Moreover, this setup allows to configure the relay such that it appears as a single UE to the eNodeB, thus significantly reducing the number of handovers. Therefore, all traffic is aggregated by the relays and then distributed to the UEs.

While the quality of the experienced link might considerably be improved, the employment of relays does not come without cost. Firstly, relays have to be licensed for the specific band they are operating on. This is of minor consequence when no borders are crossed. In smaller countries such as, e.g., in most European, one train connection can easily span three or more countries. For each country, the relays have to be registered individually or have to be switched off, which leaves the system in a direct-link state. Another consequence is, that the employment of carriages becomes restricted to the countries/routes for which the relays on board are licensed and thus becomes less flexible.

A second issues appears with the choice of the frequencies for the RRU-to-relay and the relay-to-UE connection. If the same frequency is used on both link-sections, thus only bypassing the penetration loss of the carriage, the UE might still receive a considerable amount of the desired signal by a direct link through the window. The relative receive power of these two links depends on the individual position of the UE and the actual penetration loss. Due to the latency caused by the relay, the signal might be perceived via several multi-path components that cannot be equalized.

The aforementioned problem is completely avoided when two different frequencies are used on both sections of the link. For example, a dedicated frequency of a mobile operator is used from RRU to the relay and a second frequency, e.g., in the Industrial, Scientific and Medical (ISM) band, for the supply with WiFi inside. This setup is frequently referred to in literature [4]. Nonetheless, it only provides a data connection for the passengers. Since only the relay is visible as a single UE for the eNodeB, the passengers are not accessible for mobility management.

The performance of this setup may also considerably depend on the number of antennas and relays per train/carriage. It has to be scrutinized whether the relays should work individually (e.g., one per carriage) or if the received signals should be combined. This is also affected by the possibility to connect all relays to all carriages. Since a cable connection is feasible in practice, near field communication standards at higher frequencies than the traditional 6 GHz band (e.g., in the upper mm Wave band) may be considered for this task.

15.1.2.2 Direct-Link Approach

The direct-link approach assumes a direct connection between RRU and UE. In comparison to the above scenario, the signal does experience a severe penetration loss into the carriage in this case. As the chassis of the carriages is usually made of metal, the signal enters the train mainly through the windows. However, the penetration loss may greatly vary among window types, as they are mostly metal coated themselves. Attenuation values range from 20 to 40 dB for metal coated windows of a German ICE-train [12], but other train-types exhibit different values as observed, e.g., in [5, 13] with a combined range of 10–40 dB.

Note that these values reflect the situation for current carriages in use. Since the interest among train operators is increasing to provide best quality of experience to their customers, the design of future trains is likely to be adapted to the demands of the wireless link. Among various options is the possibility to introduce windows with small penetration loss (omitting the metal coating). Another option is to include apertures in the chassis or the window itself, by incorporating materials that are more permeable for electro-magnetic waves, for example carbon fiber materials. Examples for such apertures are indicated in Fig. 15.3.

15.1.2.3 Simulation Results

In order to compare the relay- and the direct link setup, system level simulations were carried out. Again, a fully occupied German ICE-train with 460 passengers is regarded, 10 % of them are having an active wireless connection. For the direct-link setup a penetration loss of 30 dB is applied referring to the mean value of the aforementioned penetration losses. *No* penetration loss is considered for the relay setup and the relay is modeled as *one* active UE. The simulation parameters are the same as in Table 15.1 with the difference of the train speed being set to 250 km/h and only 46 UEs. Simulations were again performed with the Vienna LTE-A Downlink System Level Simulator.

Simulation results are shown in Fig. 15.4 in terms of average UE throughput [Mbit/s] over center position of the train [m] (confidence intervals are insignificantly small, so that they were omitted). Since the throughput at the relay represents the cumulative throughput of all active UEs, the results for the relay-setup were normalized to 46 (UEs) for a fair comparison. Dashed lines represent the baseline scheme, solid lines the cooperation scheme. Comparing the performance of relay (circular markers) and direct-link setup (triangular markers), employing relays improves the performance roughly by a factor of two. This gain is surprisingly low, considering that the UEs in the direct setup experience a penetration loss of 30 dBs, and the relays are operating under ideal conditions (e.g., no processing delays, no overhead). Taking into account the already discussed advantages, the direct-link approach becomes a notable alternative. The throughput peaks in the vicinity of the base stations can

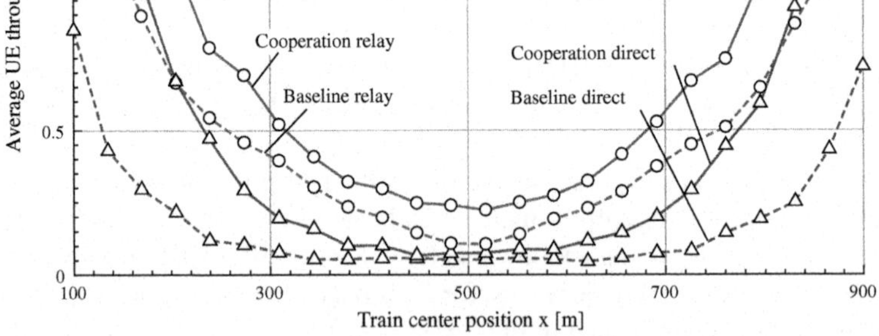

Fig. 15.4 Train average throughput [Mbit/s] versus train center position [m]. *Curves* refer to results for baseline- and cooperation scheme among RRUs for relay- and direct-link setup (Confidence intervals are insignificantly small and therefore omitted in the figure.)

be exploited by a sophisticated scheduler scheme. In regard of the traffic-type, UEs with delay sensitive data could be assigned resources such that their requirements are fulfilled. UEs with best-effort traffic models could mostly be served when the train is closer to an RRU and a higher total data rate is available.

15.2 Evolved Multimedia Broadcast Multicast Service

Illia Safiulin

3GPP introduced Multimedia Broadcast Multicast Service (MBMS) initially in Release 6 of Universal Mobile Telecommunications System (UMTS) [14]. MBMS enables to broadcast/multicast information in 3G networks, similar to classical radio/TV broadcast networks, targeting mobile TV as the main service offered. Commercially successful deployments, however, have not been reported to date. Nonetheless, development of MBMS has continued within 4G Long Term Evolution (LTE), evolving the technology to Evolved Multimedia Broadcast Multicast Service (eMBMS). eMBMS supports multicasting not only from single base stations, but even from multiple strictly time- and frequency-synchronized base stations, forming a so-called Multimedia Broadcast/Multicast Single Frequency Network (MBSFN) area (Fig. 15.5). In the current Release 12 of UMTS/LTE, MBMS/MBSFN is restricted to single antenna transmission; however, discussions are ongoing within 3GPP to enable multi-antenna support in future releases. MBSFN operation can provide substantial throughput gains at the cell edge for multicast users due to Signal to Interference and Noise Ratio (SINR) enhancement, since all signals received from base stations within the MBSFN area contribute useful information. Furthermore, MBMS/MBSFN transmission can enhance network efficiency, because common information can be delivered highly resource and cost effective to many users in parallel. Recent interest in the technology by network operators and equipment manufacturers exists especially in the context of venue casting [15], that is, multicasting/broadcasting at local events. This is confirmed by several technology trials conducted at sport events, such as, football games and car racing [16–18]. At such events, MBMS can be utilized to enhance the experience of spectators by providing additional information, e.g., live video feeds from inside race cars [17] or slow-motions of highlight scenes. Other applications envisioned include software/firmware/operating-system updates for mobile Internet of Things devices [19] as well as broadcasting of road safety related information to vehicles on highways and motorways [20]. Exchange of safety-relevant information for vehicular safety services imposes very stringent requirements on the employed communication technology in terms of dependability, that is, timeliness and reliability of data exchange. Mobile communications may not be suited as a substitute to dedicated vehicular communication technology such as ITS G5 [21], which is specifically designed for the potentially harsh vehicular communication environment, but should rather be considered as a supplement to enhance

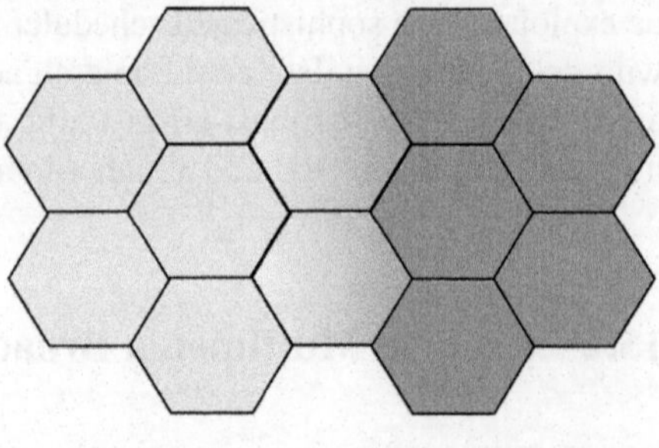

Fig. 15.5 Example of network with two MBSFN areas

the dependability of such dedicated systems. In [22], we have investigated the performance of MBSFN networks for exchange of Cooperative Awareness Messages (CAMs) [23] amongst high mobility cars in terms of packet-error-ratio, employing the Vienna LTE-A system level simulator.

We consider Single-Input Single-Output (SISO) transmission in the downlink of a cellular network. The input–output relationship of user i at subcarrier k in case of MBSFN transmission is

$$y_k[i] = \sum_{u \in \text{MBSFN}} h_k[u,i] \cdot x_{\text{MBSFN},k} + \sum_{l \notin \text{MBSFN}} h_k[l,i] \cdot x_k[l] + z_k[i], \tag{15.10}$$

where u denotes the base station index in the MBSFN area, $x_{\text{MBSFN},k}$ denotes MBMS data, which is the same for all multicast users, $z_k[i]$ is Additive White Gaussian Noise (AWGN) and $h_k[u,i]$ is complex channel coefficient, $h_k[u,i] = \gamma_u \cdot \widetilde{h}_k[u,i]$, where γ_u denotes macroscopic pathloss and shadow fading and $\widetilde{h}_k[u,i]$ represents microscopic fading.

Based on Eq. (15.10) we can expressed the SINR of MBMS user i as

$$\text{SINR}_{i,\text{multicast}} = \frac{|\sum_{u \in \text{MBSFN}} h[u,i]|^2}{\sigma_z^2 + \sum_{l \notin \text{MBSFN}} |h[l,i]|^2}. \tag{15.11}$$

In our work we assume delay- and error-free uplink transmission from vehicles to base stations and mainly focus on the downlink domain. According to Fig. 15.6 we assume that each car user generates MBMS data of size p_S bits at random starting time r and then produces packets every T ms. These data should be successfully distributed to all other vehicles in MBSFN area via multicasting or unicasting. The buffer size of car user i at time $\widetilde{n}$ can be calculated as

$$b_i[\widetilde{n}] = p_s - \sum_{m=1}^{n_p} p_t[\widetilde{n} - m], \tag{15.12}$$

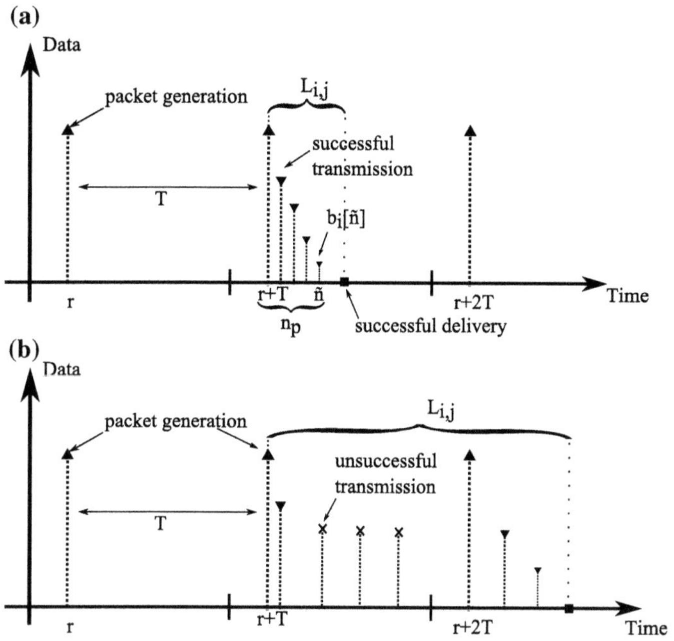

Fig. 15.6 Explanation of latency calculation

where $p_t[\widetilde{n}-m]$ is successfully transmitted packet of size p_t bits at time $[\widetilde{n}-m]$ and n_p is specified as

$$n_p = \begin{cases} n - \lfloor \frac{n}{T} \rfloor T - r, & r < \widetilde{n}, \quad \widetilde{n} = n - \lfloor \frac{n}{T} \rfloor T \\ n - (\lfloor \frac{n}{T} \rfloor - 1)T - r, & r > \widetilde{n}, \quad \widetilde{n} = n - \lfloor \frac{n}{T} \rfloor T, \end{cases} \tag{15.13}$$

which denotes the time difference between packet generation and time instance $\widetilde{n}$. We consider latency as time interval between data generation and successful delivery to all appropriate users within the MBSFN area. The latency value can be calculated for both cases of erroneous and error free transmission. According to the LTE standard, Hybrid Automatic Repeat reQuest (HARQ) is not specified in MBMS transmissions. It means that in case of the unsuccessful transmission of a packet, the packet will not be retransmitted but instead we accumulate the latency until we successfully receive the next packet from the same user. Additionally, if during waiting time new packets were generated, the old packets replace them. In Fig. 15.6 two cases of latency calculation are shown in more details. In Fig. 15.6a, the latency evaluation in case of success transmission is described, while in Fig. 15.6b the procedure of latency accumulation in case of unsuccessful transmission is explained. Assuming total number of MBMS users equal $N_{\text{m\_ue}}$ and each of them generate in total N_{packets} packets, we stack corresponding latency values into a large matrix L of size $N_{\text{packets}} \times N_{\text{m\_ue}}$, with elements

$$L_{s,i} = \begin{cases} t_i, & \text{error free transmission} \\ t_i + T \cdot j, & \text{erroneous transmission,} \end{cases} \tag{15.14}$$

where t_i is the time when $b_i[t_i] = 0$, i.e., complete transmission of packet s and j is the number of required retransmissions.

15.2.1 Latency Evaluation

We can define three different ways of latency performance indicators:

Combined latency CDF: We transform matrix $\mathbf{L}$ into a vector $\mathbf{l}$ of size $N_{\text{packets}} \cdot N_{\text{m\_ue}} \times 1$ and calculate the empirical cumulative distribution function (ECDF)

$$\text{CDF}_{\text{combined}} = \text{ECDF}(\mathbf{l}). \tag{15.15}$$

It should be noticed that the main contribution to this latency evaluation is added by the users that have high SINR and, as a consequence, represent significant amount of low latency receptions.

CDF of mean latency: We determine the mean latency for each user position (average over all latency values s obtained at a given user (i) and calculate the Cumulative Distribution Function (CDF) of these mean latencies.

$$\text{CDF}_{\text{mean}} = \text{ECDF}(\mathbf{l}), \tag{15.16}$$

where $[\mathbf{l}]_i = \frac{1}{N_{\text{packets}}} \sum_{s=1}^{N_{\text{packets}}} L_{s,i}$. Notice that this method does not represent the worst latency, which is, however, a critical indicator especially for safety-relevant applications (road-safety transmission).

Latency of individual users: We determine the latency Empirical Cumulative Distribution Function (ECDF) of each user position individually, i.e., we obtain $N_{\text{m\_ue}}$ CDFs corresponding to different car users within the network.

15.2.2 Network Utilization

Network utilization is considered as another important performance metric which gives us better understanding of the price to be paid in terms of throughput of ordinary unicast users for supporting MBSFN transmission. In our investigation we evaluated the network utilization as a percentage of resources to be used for sustaining the MBMS traffic. It can be calculated as

$$\text{Util} = \frac{p_s \cdot N_{\text{m\_ue}}}{N_{\text{RB}} \cdot N_{\text{RE}} \cdot \text{Efficiency}_{\text{CQI}[n]}} \cdot 100\,\% \tag{15.17}$$

where N_{RB} is a number of resource blocks, N_{RE} denotes number of recourse elements per resource block and $\text{Efficiency}_{\text{CQI}[n]}$ is the efficiency of the Channel Quality Indicator (CQI) (in bits per resource element) chosen for the transmission of MBMS data. With (15.17) we can calculate the number of subframes to be reserved for MBMS data transmissions.

15.2.3 CQI Adaptation

In our simulations, irrespective whether rate adaptation was applied, we reserve the same number of subframes for MBMS data, calculated with (15.17). However, for rate adaptation the real amount of used subframes could be reduced and unused MBMS subframes can instantaneously be reassigned for ordinary traffic, which may not be feasible in practice. The CQIs of all users in the MBSFN area are calculated according to proposed in [24] technique and stored in the vector CQI. From the vector CQI we choose the smallest CQI index for transmission, in order to support all users. During our research we found that using the smallest CQI index for transmission can cause traffic congestions and we should specify some lower bound ($\text{CQI}_{\text{bound}}$) to assure that the generated MBMS traffic can be sustained by the network. Therefore the CQI index at time n to be used for transmission can be calculated as

$$\text{CQI}_\text{n} = \max(\min_{\text{i}}(\text{CQI[i]}), \text{CQI}_{\text{bound}}). \tag{15.18}$$

15.2.4 Performance Evaluation

In this subsection we provide comparison of the performance of MBSFN transmission with standard unicast transmission in terms of Packet Loss Ratio (PLR), latency, network utilization and standard unicast user throughput. All results are based on [25, 26]. According to Fig. 15.7 we observe that transmission with low transmission rate (4 QAM with strong forward-error-correction coding) is necessary to achieve sufficient reliability and that the network overhead for distribution of Cooperative Awareness Messages (CAMs) is substantially reduced when employing MBSFN transmission instead of unicasting. Appropriate results on cell resource utilization are depicted in Table 15.2, simulation parameters in Table 15.3.

In Fig. 15.8 we compare the CDF of mean latencies for unicast and multicast transmissions of CAMs. We observe significant advantages of multicast transmissions comparing to the unicast case in terms of latency which is even more remarkable in terms of throughput of standard unicast users in Fig. 15.7. Such throughput reduction in case of unicasting is explained by resource consumption: transmission of CAMs via unicasting consumes 99.95 % of cell resources, while multicasting consumes 60 %. Hence we present mainly the behaviour of multicast transmissions in the MBSFN area (Table 15.3).

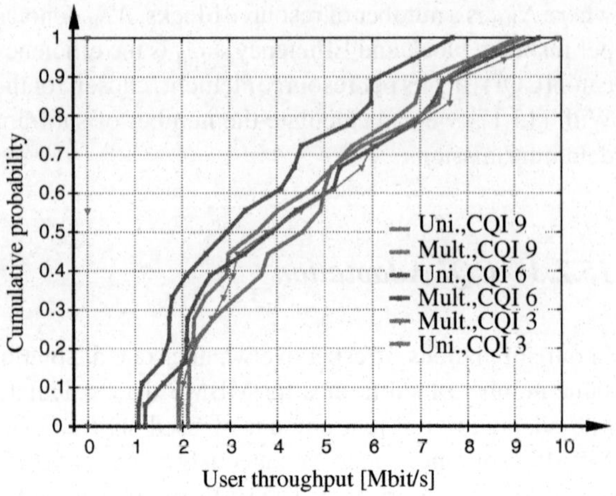

Fig. 15.7 Comparison of the throughput of ordinary users in case of unicast and multicast transmission

Table 15.2 Comparison of utilized resources

	Unicast (%)	Multicast (%)
CQI 3	99.95	60
CQI 6	52	19
CQI 9	26	8

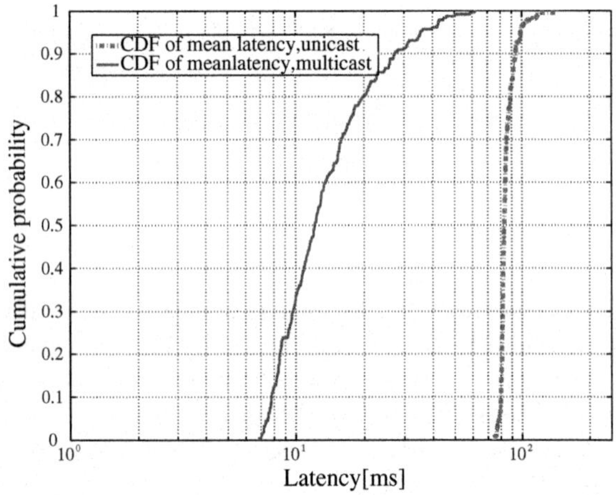

Fig. 15.8 Comparison of CDF of mean latency for unicasting and multicasting of CAMs at 5 MHz bandwidth and transmission with CQI3

15.2.5 Bandwidth Scaling

In Fig. 15.9a, b we compare the latency of multicast CAM transmission for both considered system bandwidths of 5 and 20 MHz and transmission with CQI3 (efficiency is 0.377). With 5 MHz bandwidth we have to reserve six subframes per radioframe for MBMS transmission. However, if not all of them are required, we reassign

Table 15.3 Simulation parameters

Parameter	Value
Center frequency	2.14 GHz
System bandwidth	5, 20 MHz
Channel	ITU-T VehA [10]
Number of base stations in MBSFN	7
Number of users per base station	6
Number of car users per base station	3
Speed of car users	100 km/h
Transmission	Unicast/multicast
Packet size to be transmitted	300 bytes
Packet generation rate (T)	10 Hz
Transmission rate	Rate adaptation/fixed rate for car users
	Rate adaptation for ordinary unicast users
Antenna configuration	1 × 1

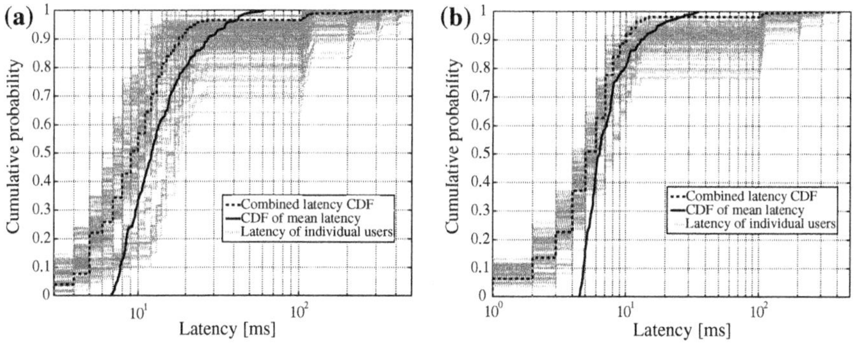

Fig. 15.9 Comparison of CDF of latency for multicast transmission with 5 and 20 MHz bandwidth. **a** Transmission with 5MHz bandwidth. **b** Transmission with 20MHz bandwidth

unused MBMS subframes for unicast transmission. With 20 MHz bandwidth, two subframes are sufficient. The corresponding mean values of main parameters are provided in Table 15.4. Given the multicast network utilization values from Table 15.4, the expected throughput improvement is:

$$\frac{R_{20}}{R_5} = \frac{(1 - 0.157) \cdot 100}{(1 - 0.52) \cdot 25} = 7.03$$

where R_{20} and R_5 are the number of resource blocks utilized for serving ordinary users at 20 and 5 MHz bandwidth. According the Table 15.4, the observed improvement equals 7.08. Hence, the throughput values scale very well with the bandwidth, provided the network utilization is considered. Considering transmissions with rate

Table 15.4 Summary of performance results with 5 and 20 MHz bandwidth and transmission with CQI3

Bandwidth (MHz)	Mean latency (TTI)	Mean throughput (Mbit/s)	Utilization (%)
5	14.7	0.72	52
20	8	5.1	15.7

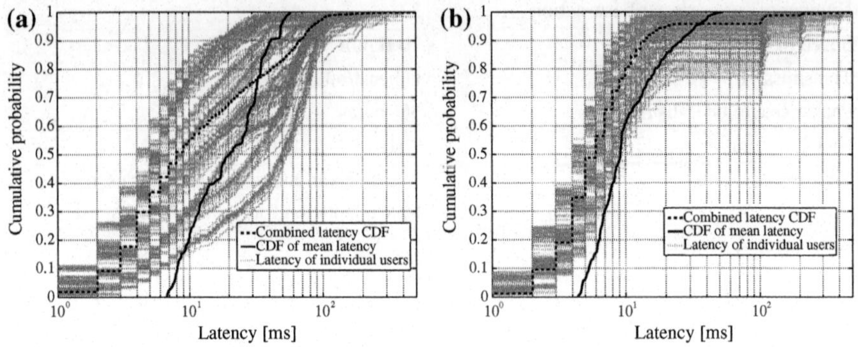

Fig. 15.10 Comparison of CDF of latency for multicast transmission with 5 MHz with and without CQI lower bound. **a** No CQI lower bound. **b** CQI lower bound of three

Table 15.5 Performance results for fixed CQI and adaptive CQI transmission with 5 MHz bandwidth

Transmission rate	Mean latency (TTI)	Mean throughput (Mbit/s)	Utilization (%)
CQI3	14.7	0.72	52
Adaptive	11.9	0.83	43

adaptation for multicast users, at first, we investigate how the performance changes if we do not apply a lower bound for the CQI, i.e., if we simply take the minimum CQI of all users even if we cannot sustain the traffic in this way. This is shown in Fig. 15.10a. Then we perform rate adaptation with the lower bound of CQI3. The corresponding results are illustrated in Fig. 15.10b and summarized in Table 15.4. We observe an improvement in the mean latency by a few Transmission Time Intervals (TTIs) with rate adaptation, which can be explained by the fact that we now require less Resource Blocks (RBs) for transmission in case the CQI of all users is high (exploiting channel diversity). This can also be seen in the average network utilization for multicast transmission, which reduces by almost 10 %. Correspondingly, the mean throughput of ordinary users improves from 0.72 to 0.83 Mbit/s (Table 15.5).

15.2.6 Conclusion

The introduction of MBSFN, however, also causes some new difficulties in the cellular network. As now many base stations broadcast the same signal in time and frequency, the overall signal delay can considerably increase and thus the cyclic prefix may not be of sufficient length any more. The straightforward solution proposed by 3GPP is to employ an extended cyclic prefix to avoid inter-symbol-interference, which is mandatory in LTE MBSFN transmission. Yet, this approach can be very inefficient especially at low Signal to Noise Ratio (SNR), since it sacrifices available resources. With the expected introduction of multi-antenna MBMS/MBSFN transmission in future releases of LTE, sophisticated beamforming/precoding schemes can be applied to optimize the efficiency of the network. Since the achievable transmission rate of multicasting is determined by the weakest user, well-known Coordinated Multi-Point (CoMP) transmission schemes for unicast situations do not perform well in multicast scenarios. We thus extended CoMP transmission in [27, 28] to multicast scenarios employing an interference-leakage control based distributed coordination algorithm. The method achieves substantial performance gains over other techniques; implementation complexity, however, can be an issue in practical realizations, since it involves iteratively solving complex optimization problems. Hence, we also provide some simple heuristic schemes in [28] that achieve a reasonable trade-off between performance and complexity.

References

1. Shift2rail: driving innovation on railways, European Commission (2014) [Online]. http://www.shift2rail.org
2. B. Ai, K. Guan, M. Rupp, T. Kürner, X. Cheng, X.-F. Yin, Q. Wang, G.-Y. Ma, Y. Li, L. Xiong, J.-W. Ding, Future railway services oriented mobile communications network. IEEE Commun. Mag. **53**, 78–85 (2015)
3. M.K. Müller, M. Taranetz, M. Rupp, Performance of remote unit collaboration schemes in high speed train scenarios, in *Proceedings of the 82nd Vehicular Technology Conference (VTC2015-Fall)*, Boston (2015)
4. Y. Zhou, Z. Pan, J. Hu, J. Shi, X. Mo, Broadband wireless communications on high speed trains, in *20th Annual Wireless and Optical Communications Conference (WOCC)* (2011). doi:10.1109/WOCC.2011.5872303
5. J.-Y. Zhang, Z.-H. Tan, Z. dui Zhong, Y. Kong, A multi-mode multi-band and multi-system-based access architecture for high-speed railways, *IEEE 72nd Vehicular Technology Conference Fall (VTC 2010-Fall), 2010* (2010), pp. 1–5. doi:10.1109/VETECF.2010.5594223
6. M.K. Müller, M. Taranetz, M. Rupp, Providing current and future cellular services to high speed trains. IEEE Commun. Mag. **53**, 96–101 (2015)
7. R.W. Heath, M. Kountouris, T. Bai, Modeling heterogeneous network interference using Poisson point processes. IEEE Trans. Signal Process. **61**(16), 4114–4126 (2013)
8. C.A. Coelho, The generalized integer gamma distribution - a basis for distributions in multivariate statistics. J. Multivar. Anal. **64**(1), 86–102 (1998). doi:10.1006/jmva.1997.1710
9. 3rd Generation Partnership Project (3GPP), Evolved universal terrestrial radio access (E-UTRA); radio frequency (RF) system scenarios, TR 36.942 (2014)

10. ITU, Recommendation ITU-R M.1225: Guidelines for Evaluation of Radio Transmission Technologies for IMT-2000, ITU, Technical report (1997)
11. Y. Qiu, H. Liu, P. Fan, Capacity enhancement using cooperative distributed antenna system under LOS MIMO channels in high speed train scenario, in *The Sixth International Workshop on Signal Design and its Applications in Communications* (2013), pp. 103–106. doi:10.1109/IWSDA.2013.6849073
12. M. Uhlirz, Adapting GSM for use in high-speed railway networks, Ph.D. dissertation, Institute für Nachrichtentechnik und Hochfrequenztechnik, Technische Universität Wien (1995)
13. C. Li, B. Li, An analysis on high speed train coverage solutions. Commun. World **33** (2008)
14. 3GPP, TSG RAN; introduction of the multimedia broadcast multicast service (mbms) in the radio access network (ran); stage 2 (release 6) (2006), http://www.3gpp.org/DynaReport/25346.htm
15. Ericsson, LTE broadcast: a revenue enabler in the mobile media era, Feb 2013, white paper
16. Valter D'Avino (Ericsson), Ericsson and Polkomtel test LTE broadcast in Poland, 1 (2014), press release
17. Deby Lewis (Verizon), Verizon delivers LTE multicast over commercial 4G LTE network in Indy, 21 May 2014, press release
18. Bill Smith (AT&T), AT&T LTE broadcast demo at college football national championship showcases exciting network future, 16 Jan 2015, press release
19. T. Lohmar, M. Slssingar, S. Puustinen, V. Kenehan, Delivering content with LTE broadcast. Ericsson Rev.: Commun. Technol. J. Since 1924 **2013**(1) (2013)
20. D. Valerio, F. Ricciato, P. Belanovic, T. Zemen, UMTS on the road: broadcasting intelligent road safety information via MBMS, in *67th IEEE Vehicular Technology Conference (VTC2008-Spring)* (2008), pp. 3026–3030
21. ETSI Standard ES 202 663 V1.1.0, Intelligent transport systems (ITS); European profile standard for the physical and medium access control layer of Intelligent Transport Systems operating in the 5 GHz frequency band (2010)
22. I. Safiulin, S. Schwarz, M. Rupp, System level simulation of LTE MBSFN networks with high mobility users, in *IWSSIP Conference*, London (2015)
23. ETSI TS 102 637–2 V1.2.1, Intelligent transport systems (ITS); vehicular communications; basic set of applications; part 2: specification of cooperative awareness basic service (2011)
24. S. Schwarz, M. Rupp, Throughput maximizing feedback for MIMO OFDM based wireless communication systems, in *IEEE International Workshop Signal Processing Advances in Wireless Communications (SPAWC)* (IEEE, San Francisco, 2011), pp. 316–320
25. I. Safiulin, S. Schwarz, M. Rupp, System level simulation of LTE MBSFN networks with high mobility users, in *2015 International Conference on Systems, Signals and Image Processing (IWSSIP)*, London (2015), pp. 21–24
26. I. Safiulin, S. Schwarz, T. Philosof, M. Rupp, Latency investigation of LTE MBSFN transmission for vehicular communications, in submitted (2015)
27. S. Schwarz, T. Philosof, M. Rupp, Leakage-based multicast transmit beamforming, in *Proceedings of the IEEE International Conference on Communications (ICC'15)*, London (2015)
28. S. Schwarz, M. Rupp, Transmit optimization for the MISO multicast interference channel. IEEE Trans. Commun. **63**, 4936–4949 (2015)

Index

M. Rupp et al., *The Vienna LTE-Advanced Simulators*,
Signals and Communication Technology, DOI 10.1007/978-981-10-0617-3